AF527832

Functional Dairy Foods :
Concepts and Applications

Functional Dairy Foods :

Concepts and Applications

Editors :

SUDHIR K. TOMAR
RAMESHWAR SINGH
ASHISH K. SINGH
SUMIT ARORA
R.R.B. SINGH

2011

SATISH SERIAL PUBLISHING HOUSE

403, Express Tower, Commercial Complex, Azadpur, Delhi-110033 (India)
Phone : 011-27672852, Fax : 91-11-27672046
E-mail :info@satishserial.com, hkjain1975@yahoo.com
Website : www.satishserial.com

Published by :

SATISH SERIAL PUBLISHING HOUSE
403, Express Tower, Commercial Complex, Azadpur, Delhi-110033 (INDIA)
Phone : 011-27672852 Fax : 91-11-27672046
E-mail : info@satishserial.com, hkjain1975@yahoo.com

ISBN 81-89304-90-9

Typeset at: Harminder *for* Typographiya # 9811482328
Printed at: Salasar Imaging Systems, Delhi-35

Foreword

Though in our country consumption of certain fermented milk products, known for their therapeutic qualities is an age old practice, yet in last few years there has been a paradigm shift in dairy and food processing sector towards "functional foods". In contrast to many other markets, the major format for functional ingredient has been probiotic as food supplements, rather than the most popular formats-spoonable and drinkable dairy foods. Probiotic supplements for example, account for 49 per cent of sales with foods and beverages at only four per cent. Globally, food and beverages account for 56 per cent of probiotic sales. Despite low levels of public awareness compared to western markets, functional foods and ingredients are finding growth in areas of India, as consumers switch on to the promise of healthier foods, beverages and supplements. There are many promising areas that require an integrated approach with the common objectives of improving the functionality, nutritional quality and health promoting effects of the food supply. India is developing guidelines that may mean probiotic strains will have to be backed by clinical trials conducted on Indian populations if they are to gain market approval. India's small but growing probiotic market is dominated by strains from non-Indian companies and so the mooted guidelines suggest probiotics should be tested on local populations before they receive safety approval because of possible differences in Indian gut flora. Besides, there is a supplemental need for India to develop specific, "indigenous probiotic strains for expressing optimal functionality". A draft is being finalized by Indian Council of Medical Research New Delhi with provision for assessment of efficacy, safety and health claims made by the probiotic foods being launched in India and defined parameters required for a strain/product to be classified as a probiotic. The research group has also proposed an adverse event surveillance system as part of the voluntary guidelines that have the potential to be mandatory at a later point of time.

The book entitled **"Functional Dairy Foods : Concepts and Applications"** encompasses a good number of chapters on various aspects of these specialized foods authored by scientists actively engaged in their respective area of specialization is poised to offer an excellent resource on the subject especially in Indian context.

A.K. Srivastava
Director & Vice Chancellor
National Dairy Research Institute Deemed University
Karnal

Preface

There is abundance of scientific evidence that diet has powerful impact on health and well being. This has revitalized the curiosity not only in consumers, but also interest among researchers and processors to develop formulated products, which is "natural," "functional" and "nutritional" as well. Estimates put the global probiotic foods and drinks market at about $10bn. Probiotics are major functional ingredients and India's probiotics market has been highlighted by researcher Frost & Sullivan as a *"major growth market of the future"*, with annual growth of 22.6 per cent and will be worth $1.3m by 2015. The presence of major functional dairy food suppliers and product makers such as Amul, Nestle, Danisco, Chr Hansen, Yakult and Danone is accelerating growth in this area which at present otherwise is in its infancy. Supporting this view, Mintel Global New Products Database (GNPD) statistics shows new product launches in India in weight management control category alone came in at 51 last year, compared to 22 in 2008 and 5 in 2005. Nevertheless, the public education program is required so that consumers are able to make the link between consuming probiotic products such as fermented milk products (*dahi,* yoghurt, cheese) drinks and supplements. Multinational companies have found the Indian market very challenging in the past owing to supply shortages, poor cold chain, storage facilities, poor infrastructure, and underdeveloped and complicated distribution channels. Performance in the food and beverage probiotic market is still dominated by issues in strain stability, low shelf life of products, and the development of probiotic application technology. Despite low levels of public awareness compared to western markets, functional foods and ingredients are finding growth in areas of India, as consumers switch on to the promise of healthier foods, beverages and supplements. The likes of probiotic-fortified yoghurt, buttermilk, omega-3-fortified health drinks and baby foods are winning over swathes of Indian consumers, especially those in the large conurbations such as Mumbai and New Delhi.

In the backdrop of these burgeoning issues, this book deals with various microbiological, biochemical, chemical and technological aspects of functional dairy foods especially in the national perspective.

Editors

Contents

□□□

Chapter 1

Functional Dairy Foods: From Concept to Product

A.K. SINGH AND A.K. SRIVASTAVA
National Dairy Research Institute, Karnal - 132 001

Introduction

Increasing awareness among consumers to know which specific molecules present in their food possess disease preventive or curative properties has led to the concept of "Functional Foods". Now the attention of scientific investigations has moved towards exploring the role of biologically active components on human health. Basic temptation in human being towards nature and the products that are natural, for every little disturbances related to health resulted in flourishing of market with products containing various therapeutic ingredients. Functional foods, pharma foods, designer foods and nutraceuticals are synonymous for foods that can prevent and treat diseases. Epidemiological studies and randomized clinical trials carried out in different parts of the world have been demonstrated or at least suggested numerous health effects related to functional food consumption, such as reduction of cancer risk, improvement of heart health, enhancement of immune functions, lowering of menopause symptoms, improvement of gastrointestinal health, anti-inflammatory effects, reduction of blood pressure, antibacterial & antiviral activities, reduction of osteoporosis etc. Nutritional significance of milk molecules is well documented and increasing cases of cancers, coronary heart diseases, osteoporosis and many other chronic diseases, have been attributed to our diet. But beyond these known nutrients i.e. vitamins, proteins, milk and milk constituents have clearly more to offer and scientists are scurrying to discover exactly which milk components might fend off specific diseases. An ever-expanding array of previously unknown such molecules with hard to pronounce names is being uncovered. But there exact metabolic role and how these can be utilized in designer food, need to be elucidated.

Defining Functional Foods

The term "functional food" was first used in Japan, in the 1980s, for food products fortified with special constituents that possess advantageous physiological effects. Functional foods may improve the general conditions of the body or decreases the risk of some diseases and could even be used for curing some illnesses. Although the term "functional food" has already been defined by many scientific and regulatory bodies, so far there is no universally accepted definition for this group of food. In most countries there is no legislative definition of the term and drawing a borderline between conventional and functional foods is challenging even for nutrition and food experts. To date, a number of national authorities, academic bodies and the industry have proposed definitions for functional foods.. Functional foods have been defined as "foods that, by virtue of the presence of physiologically active components, provide a health benefit beyond basic nutrition".

However, The Institute of Medicine's Food and Nutrition Board (IOM/FNB, 1994), defined functional foods as "any food or food ingredient that may provide a health benefit beyond the traditional nutrients it contains". To qualify as functional food, it should meet following three highlighted conditions:-

- It is a food (not capsule, tablet or powder) derived from naturally occurring ingredients.
- It can and should be consumed as part of daily diet.
- It has particular when ingested, serving to regulate a particular body process. Such as

- Improvement of biological defense mechanisms
- Prevention and recovery of specific disease.
- Control of mental and physical conditions
- Retarding the ageing process

Significance of Functional Foods

In recent years, there has been a vast and rapidly growing body of scientific data showing that diet plays an important part in diseases. Diet is thought to contribute to six of the 10 leading causes of death. Nutrients and nonnutritive food components have been associated with the prevention and/or treatment of chronic diseases such as cancer, coronary heart disease (CHD), diabetes, hypertension, and osteoporosis. According to an estimate about 70% of certain cancers are directly related to the type of food we eat. As the data supporting the role of diet in health promotion and disease prevention continue to mount, it is likely that the quantity of enhanced foods will expand substantially. There is an increasing demand by consumers for quality of life, which is fueling the nutraceutical revolution. Functional foods are viewed as one option available for seeking cost-effective health care and improved health status. Moreover, the large segment of the population is aging and considerable health care budget in most country is focused on treatment rather than prevention. Thus, the use of nutraceuticals in daily diets can be seen as means to reduce escalating health care costs that will contribute not only to a longer lifespan, but also more importantly, to a longer health span.

Trends in Functional Foods & Nutraceuticals

Companies involved in investigating and developing nutraceutical products mainly belong to the food (55%) and pharmaceutical (35%) industries. Strategic alliances are already in place among pharmaceutical and food companies. In India, nutraceuticals are marketed as Indian System of Medicine (ISM) drugs under the OTC category. No clinical validation of their safety or efficacy is required if therapeutically usefulness is mentioned in the text. Currently, functional foods market is estimated at $70 billion or 4% of processed food market and is growing at three times the rate. In developed markets, higher consumer awareness on health and wellness is being addressed through product innovations and marketing prowess of large players. While the ageing population needs more engineered foods the younger population is demanding more fortified foods to get extra energy. Health-related issues -obesity and coronary heart disease -are forcing food processors to launch campaigns to promote low carbohydrate diets or other such foods. Japan is the single largest market with per capita consumption of $140+ with the US and Europe following at $95 and $60 respectively. Despite static or reducing population, functional foods will continue to retain the market (or grow) due to growing health concerns. Regulations have ensured the segment's organized growth.

There are a lot of products sold in the name of nutraceuticals in the Indian market. Close to around 100 products are even listed on the Internet along with the global companies and around 20 Indian companies have a record of producing nutraceuticals and marketing them globally. India is relatively a new market. The size of the Indian nutraceutical market is estimated to be about Rs 1,600 crore in 2001. All major pharma players are in the process of entering this market. The level of exports from India is still small, estimated to be perhaps less than Rs 750 crore, if one excludes Psyllium. The major markets for India are the US, Europe and Japan. India can become leader in this field as we hold key expertise as well as we are rich with the biodiversity.

Milk Based Functional Foods and Nutraceuticals

Since time immemorial, dairy products have been an integral part of human diet. Milk is the only food, which has got the power to sustain life in all the stages of development, and is considered an important part of a balanced diet. Besides being a source of quality proteins and energy-rich fat, it contains important micronutrients like calcium, potassium, sodium, magnesium and vitamins, which are vital for overall development of the human body. Also, several health attributes are associated with milk or its constituents, such as the role of calcium in controlling hypertension and colonic anticarcinogenicity, protective roles of carotenenoids and conjugated Linoleic acid (CLA) against cancers. Butyric acid, the short chain fatty acid of milk fat has been shown to regulate cell growth and enhance the anti-tumor activities. Certain minor milk components either naturally occur or formed during processing have also been endowed with many unique health benefits. Examples include lactoferrin, lactulose, galacto-oligosaccharides (GOS), β-lactoglobulin, and bioactive peptides. Some of the important segments of functional dairy foods and nutraceuticals have been discussed hereunder.

a) Probiotic Dairy Foods

Human gastrointestinal tract (GIT) harbour more than 100 trillion microorganisms belong to 400 different bacterial species. The number cells are almost 10 times than the rest of the body cells. A delicate balance exists between beneficial and harmful bacteria [resent in GIT and any disturbance may lead to abnormalities. About 70% of the body's immune system is localized in GIT. Incorporation of beneficial bacteria into foods to counteract harmful organisms in the GIT has been the most visible component of this new area. Such microorganisms are termed as "Probiotics". There is growing scientific evidence to support the concept that beneficial gut microflora may provide protection against gastrointestinal disorders including gastrointestinal infections, inflammatory bowel diseases, and even cancer.

The basis for selection of probiotic micro-organisms include safety, functional aspects (survival, adherence, colonization, antimicrobial production, immune stimulation, antigenotoxic activity and prevention of pathogens) and technological details such as growth in milk and other food base, sensory properties, stability, phage resistance and viability. Newer avenues as carriers of probiotic organisms are being sought. Fermented milk pridcuts being a 'live' food, is potentially an excellent vehicle for these beneficial microbial cultures. Recognizing the beneficial properties of probiotic organisms and challenged with the possibility that these organisms may produce end products that may be different from those produced by the normal starters in these products; several attempts have been made to manufacture probiotic milk products like probiotic dahi, probiotic cheese, probiotic yoghurt and yoghurt drinks. Probiotic dahi developed at our Institute containing *Lactobacillus acidophilus* and *Lactobacillus casei* was found to delay the onset of glucose intolerance, hyperglycemia, dyslipidemia, and oxidative stress in high fructose induced diabetic rats. It indicates reduced risk of diabetes and its complications.

b) Fortified Milk Products

Milk in its natural form is almost unique as a balanced source of man's dietary need. The various steps in processing and storage have a measurable impact on some specific nutrients. Milk also provides a convenient and useful vehicle for addition of certain nutrients to our diet and has following benefits:

- Easier quality control measure implementation
- Wider consumption by all age groups
- Cost is affordable by target population.
- Higher stability and bioavailability of the added micronutrients
- Addition of fortificants usually caused minimum change in colour, taste and appearance.

Liquid milk fortification with vitamins A and/D is mandatory in several countries. â-carotene is added as a colour-enhancing agent to some milk products such as butter. Dried milk is often fortified with vitamins A and D, calcium, and iron. Milk based infant formula and weaning foods are fortified with a range of vitamins, minerals, and other nutrients such as polyunsaturated fatty acids. Powdered milk used for complementary feeding in Chile is fortified with vitamin C, iron, copper and zinc. However, the milk fortification usually impaired its sensory and processing quality characteristics. Moreover, bio-availability of fortified nutrients is another major concern. Investigations carried out at NDRI suggest possibilities of fortification of liquid milk with calcium and iron.

c) Whey Proteins and Peptides

Whey proteins are termed as "wonder proteins" and may find applications in nutritional, functional, dietetic, sports and infant foods. They possess excellent amino acids profile, high PER and biological value, easier digestibility and assimilability, bland taste and excellent functional properties. Dietary whey proteins have a number of putative, biological effects when ingested. The ability of whey proteins to increase the level of natural anti-oxidants within the body and possibly in stabilizing DNA during cell division is emerging as premier contribution to population health. The anticarcinogenic properties of whey proteins are related to compounds rich in sulphur containing amino acids, methionine and cysteine. They contain $\tilde{\gamma}$-glutamyl-cysteine residue, which makes cysteine readily available for synthesis of glutathion, (γ glutamyl- cysteinyl- glycine) a strong xenobiotic deactivating and anti-neoplastic agent. Methionine is utilized for glutathion synthesis in times of cysteine deficiency and it also acts as methyl donor. Hypomethylation of a DNA is an important risk factor for cancer at number of sites. Glutathion is, believed to act as an antioxidant, anticarcinogenic and in stabilization and repair of DNA.

Challenges in Development of Functional Dairy Foods

In India, we have traditional products touted as functional but have little scientific validation. Regulations will thus have to evolve to promote R&D, ensure validation and prevent exploitation of consumers. Companies will also have to be sincere and honest in their claims while marketing and communicating with consumers till appropriate regulations for scientific validation are evolved. Processors will need to provide an optimal merger between taste, convenience and health attributes. Companies will require expert knowledge in flavour masking fortification know-how and delivery systems.

a) Technological Challenges

Four different technological hurdles have to be overcome before a product containing bioactive substances is ready to consider marketing:

- Isolation of the desired components,
- Pre-establishment of the biological activity,
- Incorporation of the bioactive components into a formulated product,
- Verification of efficacy and safety of final product.

Such a sequence of experimental events is also required for the introduction of new food additives; additional is a thorough proof of the claimed benefit of the bioactive components. This applies especially when the bioactive component is a completely new substance never consumed before in significant amounts. Separation, purification and production at industrial level of such nutraceuticals must be thought in terms of integrated and high added value. Membrane technologies (MF, UF, NF) provide key opportunities to manufacture milk nutraceuticals in native state. The incorporation of bioactive components into processed foods, its delivery and bioavailability are other important issues that need reprisal.

b) Scientific validation of functional foods

The scientific evidence for functional foods and their physiologically active components can be categorized into four distinct areas: (a) clinical trials, (b) animal studies, (c) experimental in vitro laboratory studies, and (d) epidemiologic studies. Much of the current evidence for functional foods lacks well-designed clinical trials; however, the foundational evidence provided through the other types of scientific investigation is substantial for several of the functional foods and their health-promoting components.

c) Safety Issues

Although increasing the availability of healthful foods including functional foods in the diet is critical to ensuring a healthier population, safety is a critical issue. The optimal levels of the majority of the biologically active components currently under investigation have yet to be determined. The benefits and risks to individuals and populations as a whole must be weighed carefully when considering the widespread use of physiologically-active functional foods. Knowledge of toxicity of functional food components is crucial to decrease the risk: benefit ratio. The safety issues related to probiotic microorganisms that has to be considered include

- Intrinsic properties of the probiotic strains
- Pharmacokinetics of probiotic strains
- Interaction between probiotic strains and the host
- Knowledge of toxicity of functional food components is crucial to decrease the risk: benefit ratio.

Conclusion

Consumer interest in the relationship between diet and health has increased the demand for information on functional foods. Rapid advances in science and technology, increasing healthcare costs, changes in food laws affecting label and product claims, an aging population, and rising interest in attaining wellness through diet are among the factors fueling interest in functional foods. Credible scientific research indicates many potential health benefits from milk components. It should be stressed, however, that functional foods are not a magic bullet or universal panacea for poor health habits.

❑❑❑

Chapter 2

Folate Production by Lactic Acid Bacteria

S.K TOMAR AND RAMYA IYER
Dairy Microbiology Division, National Dairy Research Institute, Karnal-132001

Introduction

Folate, an important B-group vitamin, participates in many metabolic pathways such as DNA / RNA biosynthesis and amino acid inter-conversions. Human life can not exist without folate since this B-group vitamin is involved in essential functions of cell metabolism such as DNA replication, repair, methylation, synthesis of nucleotides, vitamins and some amino acids. Mammalian cells cannot synthesize folate; therefore, an exogenous supply of this vitamin is necessary to prevent nutritional deficiency. Folate deficiency has been implicated in a wide variety of disorders from Alzheimer's to coronary heart diseases, osteoporosis, neural tube defects, increased risk of breast and colorectal cancer. Many reviews have shown the health benefits associated with increased folate intakes and by consequence many countries now possess mandatory folate enrichment programs. Lately, a number of studies have shown that high intakes of folic acid, the chemically synthesized form of folate, but not natural folates, can cause adverse affects in some individuals such as the masking of the hematological manifestations of vitamin B12 deficiency, leukemia, arthritis, bowel cancer and ectopic pregnancies. For these reasons, many researchers have been looking for novel methods to increase concentrations of naturally occurring folate variants in foods. This chapter will focus on folate production by lactic acid bacteria (LAB) and levels of folate present in foods fermented by/or containing these valuable microorganisms. The proper selection and use of folate producing microorganisms is an interesting strategy to increase "natural" folate levels in foods. Also, folate production levels in LAB can be modulated by gaining insight in the genes, pathways and metabolites that are involved by metabolic engineering.

Folates

Chemical structure and bioavailability

The generic term folate will include the complete group of all folic acid derivatives, including the polylglutamates naturally present in foods and folic acid that is a synthetic folate form which is commonly used for food fortification and nutritional supplements. Folic acid or pteroyl glutamic acid (PGA) is comprised of *p*-aminobenzoic acid (PABA) linked at one end to a pteridine ring and at the other end to L-glutamic acid (**Fig. 1a**). The naturally occurring forms of folate differ in the extent of the reduction state of the pteroyl group, the nature of the substituents on the pteridine ring and the number of glutamyl residues attached to the pteroyl group (**Fig. 1b**). The naturally occurring folates include 5-methyltetrahydrofolate (5-MTHF), 5-formyltetrahydrofolate (5-formyl-THF), 10-formyl tetrahydrofolate(10-formyl-THF), etc. Most naturally occurring folates are pteroylpolyglutamates, containing two to seven glutamates joined in amide (peptide) linkages to the α-carboxyl of glutamate.

Bioavailability is defined as the proportion of a nutrient ingested that becomes available to the body for metabolic processes or storage. The bioavailability of dietary folate may be hampered by the polyglutamate chain to which most of the natural folate is attached. This polyglutamate chain must be removed, except for the proximal glutamate moiety, by the enzyme α-glutamyl hydrolase or human conjugase that is present in the brush border of the small intestine. Folate can be absorbed and transported as a monoglutamate into the portal vein. The available data suggest that the polyglutamate form is 60–80% bioavailable compared with the monoglutamate form.The relative bioavailability of dietary folates is estimated to be only 50% compared with synthetic folic acid.

Furthermore, folate-binding proteins from milk may act to increase efficiency of folate absorption by protecting dietary folates from uptake by bacteria in the gut, thus increasing absorption in the small intestine.

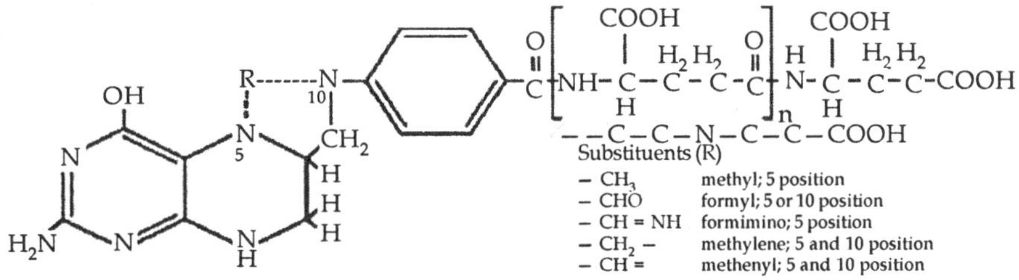

Fig. 1 Structure of a) folic acid (pteroyl-L-glutamic acid), and b) native food folates, e.g. reduced, one-carbonsubstituted forms of polyglutamates.

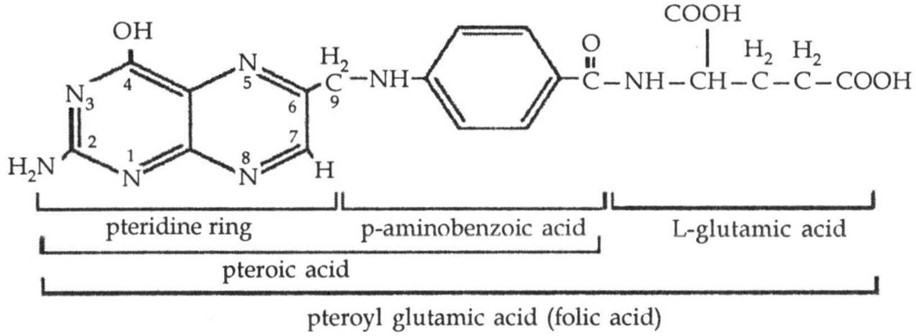

Measurement of Folate in Foods

During the last four decades, numerous reports have been published regarding food folate content and the various methods of folate determination including: microbiological, radiobinding or radiometric assay,electrochemical or spectrophotometric methods, some of which are carried out in combination with gel or high-pressure-liquid chromatography (HPLC). Microbiological assay has been considered to be one of the best and most versatile methods for determining food folates. *Lactobacillus* (*L.*) *rhamnosus* ATCC 7469 is the most commonly used and most accepted organism for folate analysis of natural products. It responds to natural folate forms present in foods, and does not respond to pteroic acid, a common folate degradation product. Microbiological assays serve as the traditional method of folate analysis and are based on the nutritional requirements of microorganisms. It is the only food folate method given official status by American Association of Analytical Chemists (AOAC).

Microbiological methods of is based on the nutritional requirement of a microorganism for a certain vitamin. This allows the formulation of a basal medium that provides all of the growth requirements for the organism except for the vitamin to be assayed. When aliquots of the sample containing the vitamin being quantified are added to the initially translucent medium, followed by inoculation with test organism, the organism reproduces in proportion to vitamin content, which can be measured photometrically (**Fig. 2**). Despite the advent of several alternative methods for determining folates in foods and biological samples, the microbiological assay using *L. rhamnosus* in casein based media remains the standard method for most applications. *L. rhamnosus* has greater capacity for response to the á-glutamyl folate polymers compared to the other assay organisms; however, its response is limited to no greater than three glutamates with much lower response to higher polymeric folates. Lack of response to the higher á- glutamyl folate polymers requires treatment with folate conjugase in order to hydrolyze folate polyglutamates to folates with shorter glutamyl residues such as mono- or diglutamates, which can be utilized by the assay microorganism. Treatment with conjugase alone is not much effective; thereby the use of additional enzymes, proteolytic or amylolytic, has shown to liberate folate from foods. Microbiological assay with an additional trienzyme extraction method in the order of protease, alpha amylase and finally conjugase along with the standard microbiological assay using *Lactoacillus rhamnosus* as an assay organism in casein based media for the estimation of folate was evaluated for its applicability in milks from different Indian milch species. Most current studies now determine food folate concentrations in response to growth of *L. rhamnosus* using 96-well microtiter plates.

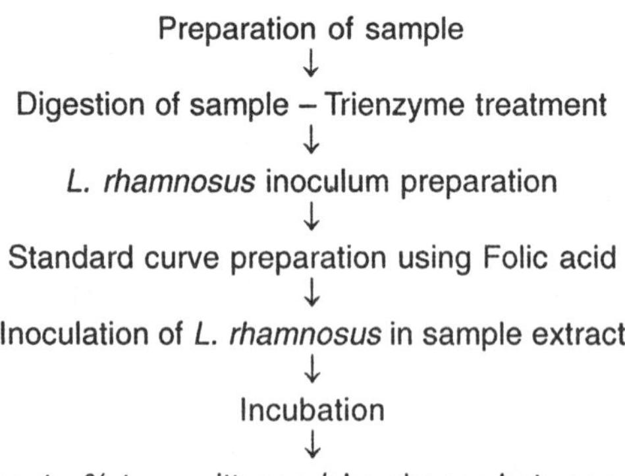

Fig.2. Sample preparation and extraction procedure of Microbiological assay

The limitation of *L. rhamnosus* and other assay organism in differentiating folate derivatives in the folate extract has prompted the use of chromatographic techniques. These techniques involve two distinct steps: separation and purification of deconjugated extract and detection and quantification of eluted monoglutamates. High Performance Liquid chromatography is a popular technique for separation and purification of individual folates. One of the major difficulties associated with the HPLC method is the rigorous sample clean-up procedure prior to final injection. The limitation of HPLC detectors to identify some of the folate derivatives and complex sample extraction and purification procedures resulting in the loss of sensitive folate compounds night have caused the lower folate result using the HPLC analysis.

Extraction by homogenization and Enzyme treatment (amylase and protease)

↓

Deconjugation by Conjugase Enzyme

↓

Purification by Filteration or solid phase extraction

↓

HPLC Detrmination by C18 column and using 10% acetonitrile plus 90% buffer aqueous phase with UV detector

Fig. 3. High performance Liquid chromatography determination of folate from sample

The most recent method for quantification of individual folate derivatives in foods and biological material is coupling HPLC with mass spectrometry (MS) along with modern LC systems. The primary advantage of LC analysis is the ability to quantify the different folate forms, a superior specificity not obtainable by other methods. Drawbacks for LC-MS methods used for food analysis are expensive equipment, absence of suitable commercially available internal standards, and so far, low sensitivities for some of the folate compounds.

Immunoassays are highly sensitive and specific as a result of the interaction of an antibody molecule with its target, a high affinity interaction that occurs even in complex matrices. The Enzyme Linked Immuno Sorbent Assay principle is based on recognition of folate a polyclonal antibody raised against folic acid. In ELISA, antibodies against the analyte are coupled to the well bottom of a microtiter plate. After addition of the analyte, a second portion of biotinylated antibodies is added, followed by a substrate addition yielding a colour shift measured spectrophotometrically. The concentration of folate in an unknown sample is calculated from a calibration curve prepared from a standard folic acid/folate. One of the major limitations of biospecific procedures is its failure to analyze total folate content in foods. Generally these kits have much lower response to folate derivatives other than folic acid, thereby underestimating the natural folate content of foods. It is best suited for analysis of the folate content of fortified foods where folic acid is the major folate derivative.

Folate Requirements

Humans cannot synthesize folate; in consequence, it is necessary to assimilate this vitamin exogenously. Although folate is omni-present in a normal human diet, folate deficiencies still occur frequently, even in well-developed countries. Folate deficiency in humans is associated with several health problems, such as cancer, cardiovascular diseases as well as neural tube defects in newborns. The daily recommended intake (DRI) is set at 200 and 400 μg/day for adults and women in the periconceptional period, respectively.

There are several ways to increase the folate levels of food products: i) fortification of food products, ii) selection of special plant cultivars, or fruits with increased folate pools, iii) fermentation fortification. Methods such as fortification have proven to be very useful in reducing health problems associated with folate mal-intake but high-level intake of chemically synthesized folate might have some adverse health effects as highlighted by the FDA includes masking of the early hematological manifestations of vitamin B12 deficiency , potential unknown risks for pregnant women etc. It was proposed that these problems could be circumvented by consumption of a mixed diet containing folate-rich foods. Natural folates (such as tetrahydrofolates produced by microorganisms) do not cause "masking" of pernicious anemia that occurs at high concentrations of folic acid and therefore, naturally produced folates seem to be more rationale for fortification purposes, and for increasing folates in foods containing microorganisms. Bacterial strain selection can be an important contributor to increased folate intake levels, e.g. in fermented food products. Hence, this is a way to increase folate content in food products is by fermentation fortification. Fermentation fortification is a new concept which can be improved in several ways: i) bacterial strain selection, ii) delivery engineering and iii) metabolic engineering.

There is a growing interest in using LAB for improvement of the nutritional value of fermented food products and/or adding health benefits to these products.

Folate Concentration in Fermented Dairy Products

Milk is a well known source of folates. The food composition tables and review papers based on microbiological assay report folate values in cow's milk in the range of 5–7 μg/100 g (**Table 1**). However, many dairy products are processed using microbial fermentations in which folate can be synthesized, significantly increasing folate concentrations in the final product (**Table 1**). Therefore, fermented milk products are reported to contain even higher amounts of folate as a result of the production of additional folates by bacteria. Also folate binding proteins of milk improve folate stability and its bioavailability in these fermented milk products. However, many bacteria synthesize this cofactor by themselves from simple precursors but some auxotrophic bacteria, including many lactic acid bacteria, have a strict growth requirement for folic acid. The production and consumption of folate by the microorganisms applied will be probably the most important factor determining the folate level in fermented milk products.

Table 1. Folate concentrations in dairy products and its contribution to the reference daily intake (RDI)

Product	Folate (μg/L)	Folate per serving (μg/240 mL))	%RDI (3 serving)
Milk	40 ± 10	10 ± 2	6-8
Buttermilk	90 ± 20	22 ± 5	13-20
Yogurt	80 ± 20	19 ± 5	11-18
Kefir	50 ± 10	12 ± 2	8-11
Ropy-milk	110 ± 20	26 ± 5	16-23
Sour Buttermilk	75 ± 15	18 ± 4	11-17
Acidophilus milk	50 ± 10	12 ± 2	8-11
Bifidus milk	75 ± 15	18 ± 4	11-17

United States Department of Agriculture (USDA) recommends the consumption of at least 3 servings of milk products as part of a healthy daily diet. Taking into account this recommendation, and considering that a normal serving consists of 240 ml, currently available fermented milk products could contribute significantly to the reference daily intake of folates (up to 23% of RDI based on **Table 1** values). Besides fermented dairy products, microorganisms are capable of increasing folate content in a wide variety of foods as wine, beer, rye, bread and fermented vegetables.

Folate Producing Microorganisms

The property of folate biosynthesis can be found in two of the three domains of the phylogenetic tree. The exception is the domain of the Archaea. Species in this domain use tetrahydromethanopterin (THMPT) as one carbon carrier instead of THF. However, exceptions of the ability to synthesize folate can also be found within the group of the Bacteria and Eukarya. The ability to produce THF is not only found in bacteria but also in green plants, fungi and certain protozoa. Natural folates are produced de novo by plants, yeasts and some species of bacteria through the folate biosynthesis pathway. Folates produced by microorganisms, contribute to the final content in various fermented foods, such as yoghurt, kefir and bread. The ability to produce folate can differ remarkably between different LAB (2-214 µg/L folate). Numerous researchers have reported that lactic acid bacteria, such as the industrial starter bacteria *Lactococcus* (*Lc.*) *lactis* and *Streptococcus* (*S.*) *thermophilus* have the ability to synthesize folate. For this reason, some fermented dairy products, including yogurt (200 µg/l), are reported) to contain even higher concentrations of folate than non-fermented milk (20 to 60 µg/l). This level depends on the strain of *S. thermophilus* and *Lactobacillus. bulgaricus* used, because the latter organism has been shown to use and to degrade folate during its growth. It is therefore of utmost importance to select the optimal combination of *S. thermophilus* and *L. bulgaricus* strains leading to an organoleptically acceptable yoghurt with a concomitantly increased folic acid concentration. In general, *Lactobacillus* strains do not produce folates with the exception of *Lactobacillus plantarum* . It is now known that not only the yogurt starter cultures and *Lc. lactis* have the ability to produce folate but also this important property exists in other LAB **(Table 2)** as *L. acidophilus, Leuconostoc lactis, Bifidobacterium longum* and some strains of *Propionibacteria*, well known vitamin B12 producer, can also produce large amounts of folate.

S .thermophilus has a strain specific ability of folate production and has been reported to produce higher quantity of folate in comparison to other LAB; majority of which is excreted into milk. By application of different *S. thermophilus* strains used in yoghurt productions, the folate contents ranging from 2 up to 15 μg/100 g were found. The culture NCDC177 which was observed in our laboratory (Tomar SK, Srivatsa N, Iyer R and Singh R, accepted for publication) to be the best folate producer among cultures of *S.thermophilus* available at National Collection of Dairy Cultures, National Dairy Research Institute, Karnal (India)was evaluated for the factors influencing its folate production. Also among the 120 lab isolates of *S. thermophilus* from milk and fermented milks of Indian origin, screened for folate production, about 10 % was found to produce folate in the range of 40-50 $\mu g L^{-1}$ (prolific producer) whereas about 40% produced in the range of 20-30 $\mu g L^{-1}$ (moderate producer) and rest were low producer (4 -16 $\mu g L^{-1}$). Also, *Saccharomyces cerevisiae*, has high folate content ranging from 1-4 mg/ 100 g yeast. The main forms are H4folate and 5-CH3-H4folate in exclusively polyglutamylated forms. Folate content in different strains of *S. cerevisiae* shows large differences among strains which clearly indicate the importance of choosing the proper starter culture when optimising folate content in yeast-fermented food. Apart from using yeast as a biofortificant in fermented food, high producing strains may be used as biocatalysts for biotechnological production of natural folates.

A recent study provided new perspectives on the specific uses of folate-producing probiotics such as bifidobacteria and propionibacteria, to prevent the localized folate deficiency that is associated with premalignant changes in the colonic epithelia. The oral administration of probiotic strains may confer a more efficient protection against inflammation and cancer, both by exerting the beneficial effects and by delivering folate to colonic-rectal cells. In humans, folate produced by the microbiota in the small intestine is absorbed across the large intestine and incorporated into the liver and kidneys. It is therefore possible that probiotic bacteria active in the intestinal tract may be able to contribute to the folate requirement of colonic epithelial cells.

Folate Biosynthesis Pathway and Folate Related Reactions

The folate biosynthetic pathway in LAB includes eight consecutive steps, in which the precursor GTP is converted via several steps into the biological active cofactor tetrahydrofolate (**Fig. 5**). Folate is a tripartite molecule, composed of a pterin residue, *p*ABA and a γ-linked glutamate moiety. These three structural elements reflect the three pathways (pterin, *p*ABA and glutamate pathway) that are involved in the synthesis of the building blocks of folate (**Fig. 3**). The pterin pathway reflects the conversion of GTP by GTP cyclohydrolases (I or II). The tetrahydrofolate that is produced is converted in the one-carbon pathway into one-carbon carriers involved in several reactions in the synthesis of purines, pyrimidines etc. Another remarkable feature of the folate biosynthesis pathway is that folate is involved in the synthesis of its own precursor, GTP.

Table 2. Folate produced by microorganisms grown in chemically defined folate-free medium

Microbial Species	Extracellular (μg)	Intracellular (μg)	Total (μg)
Lactococcus lactis subspp			
Lc. Lactis subsp cremoris	8-46	59-99	92-116
Lc. Lactis subsp lactis	5-26	47-269	57-291
Lc. Lactis subsp lactis biovar diacetylactis	11-21	65-81	79-100
Lactobacillus spp			
L.plantarum	27	18	45
L.helveticus	-1-3	-1-90	2-89
L.acidophilus	0	1	1
L.casei	-45	32	-13
L.casei subsp *rhammosus*	-98	34	-63
L.delbrueckii subsp. *Bulgaricus*	12	41	54
Propionibacterium spp.			
P. thoenii	28	8	36
P.acidipropionici	58	22	36
P.jensenii	51	-11	40
P. frendenreichii ssp. *Shermanii*	0-93	-20-11	17-78
P. sp.	13-31	-22-16	9-29
Bifidus species			
B. adolescents	1-65	10-40	70-110
B. animalis	26	-	-
B.bifidum	1	-	-
B.breve	1-3	-	-
B. catenularum	3	-	-
B. dentium	29	-	-
B. infantis	27	-	-
B. longum	2	-	-
B. pseudocatenulanum	12-82	5-35	75-90
Streptococcus species			
S. thermophilus	23-40	4-179	29-202
Other species			
Leuconostoc lactis	37	7	45
Leuconostoc paramesenteriodes	33	10	44

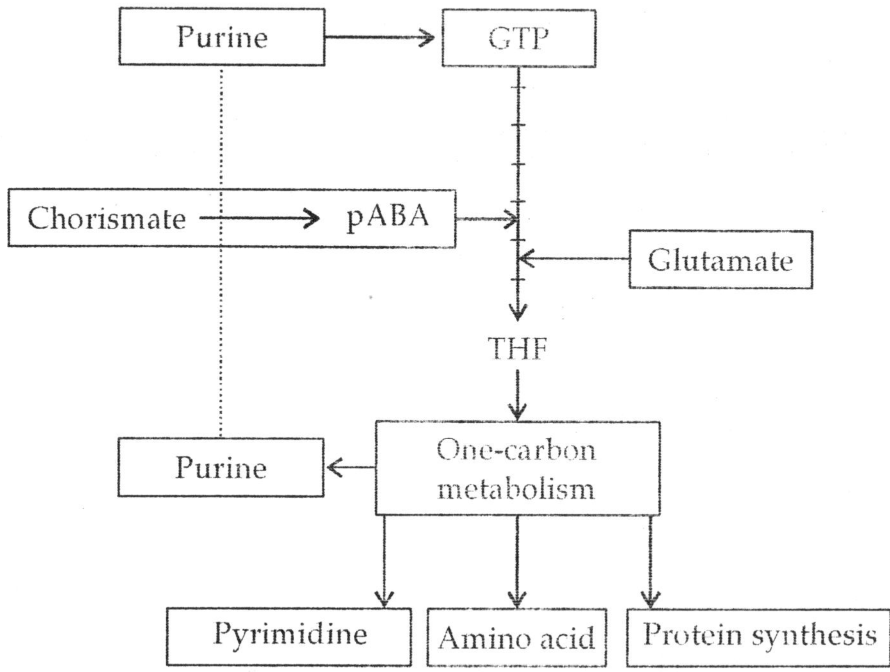

Fig 4. Schematic representation of the folate and one-carbon pathway. Whereby, GTP is guanosine triphosphate and *p*ABA is *para*-aminobenzoic acid.

Folate Biosynthesis Genes in Lactic Acid Bacteria

In order to modulate the folate content in fermented products it is crucial to have knowledge of the genes involved in folate biosynthesis. Up to now, *Lactococcus lactis* is the only species for which all folate biosynthesis genes are annotated. Folate gene cluster is a cluster of five genes of approximate size of 10kb, present on chromosome (Fig. 5). This folate gene cluster contains *folA,* encoding dihydrofolate reductase, *folB,* predicted to encode dihydroneopterin aldolase, *folK* and *folE,* encoding the biprotein 2-amino-4-hydroxy-6-hydroxymethyl-dihydropteridine pyrophospho-kinase and GTP cyclohydrolase I, *folP* predicted to encode dihydropteroate synthase, and *folC* encoding the bifunctional protein folate synthetase and polyglutamyl folate synthetase. The other genes present in the folate gene cluster *clpX, ysxc,* and *ylgG,* are not likely to be involved in folate biosynthesis. The gene *folE,* encoding GTP cyclohydrolase I, was always identified as an independent gene. The location of the folate genes on the chromosome can be a relevant factor for metabolic engineering purposes, since organization of the genes in a single gene cluster facilitates metabolic engineering strategies.

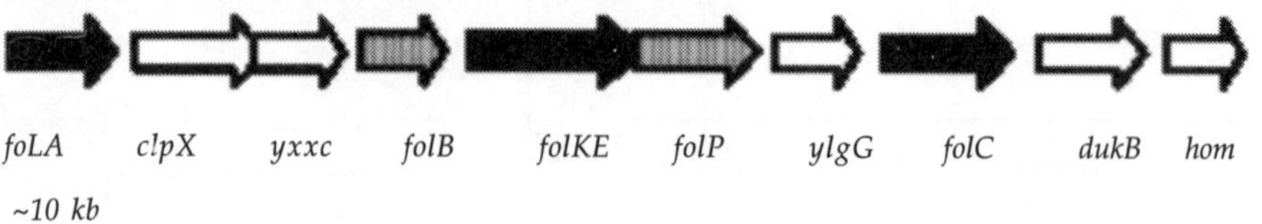

Fig .5. Schematic representation of the *Lactococcus lactis* folate gene cluster as identified in *L. lactis* MG1363

Metabolic Engineering of Production Strains

Metabolic engineering is used to alter metabolite or protein levels in the cell. This can be done by overexpression or disruption of metabolic or regulatory genes. In LAB several metabolic pathways have been rerouted or overexpressed; e.g. those for alanine, riboflavin, sorbitol, mannitol and folate production. Traditionally, a single pathway is engineered and many aspects of this pathway are investigated. However, the development of metabolic models and functional genomic tools can help to investigate impact of metabolic engineering on the overall performance of the host organism. Thus, folate production levels in lactic acid bacteria can be modulated by gaining insight in the genes, pathways and metabolites that are involved.

The genes for folate biosynthesis have been identified in some LAB such as *Lc. lactis* and *L. plantarum* , and recently it has been shown that *L. bulgaricus* possesses all the folate biosynthesis genes . However, some LAB, cannot synthesize folate, because at least some of the genes involved in folate biosynthesis are lacking in the genome, this is the case for *L. gasseri L. salivarius* , *L. acidophilus* and *L. johnsonii* .Recently, it has been shown that metabolic engineering can be used to increase folate levels in *Lc. lactis* , *L. gasseri* , and *L. reuteri* . The folate-consuming *L. gasseri* was transformed into a folate producer by the transfer of a broad-host-range plasmid containing the folate gene cluster from *Lc. lactis. Similarly, Lc. lactis* strains have been modified to produce intracellularly folates with a short glutamyl tail length (average polyglutamyl tail length of 3) or with a long polyglutamyl tail length (average polyglutamyl tail length of 8), which generates an increased retention of folate in the cells. These strains were evaluated in animal models in order to determine if the folate glutamyl tail length affects bioavailability. In contrast to monoglutamyl folate, polyglutamyl folates cannot be transported across the cell membrane. Hence, the release of intracellular polyglutamyl folate depends on the disruption of the cells during passage through the gastro-intestinal tract. The clear responses of lactococcal cells added to a folate free diet of deficient rodents on the folate levels in organs and blood indicate that these cells lyse after consumption and that the bacterial folate becomes available for absorption in the gastro-intestinal tract of the rat . This study provided the first animal trial with food containing living bacteria that were engineered in order to increase the intracellular accumulation of folate or to change the average polyglutamyl tail length compared to a wild-type lactococcal strain. Thus knowledge of the folate biosynthesis genes and the metabolic pathway in several folate-producing LAB enabled us to modulate the folate levels more than a million-fold.

Conclusions

Further research, exploration and development of folate producing LAB and yeast strains and combinations of them for food applications should be encouraged. By a careful testing of the folate production ability of microbial strains used in the production of fermented milk starters, formulations may be optimised and lead to natural enhancement of the folate content in food products. The increase in folate levels in yogurts and fermented milks are possible through judicious selection of the microbial species and cultivation conditions. The food industry should now take the next step to use this information for selecting folate-producing strains as part of their starter cultures in order to produce fermented products with elevated levels of this essential vitamin. Such products would provide economic benefits to food manufacturers since increased "natural" folate concentrations would be an important value-added effect without increasing production costs. Consumers would obviously benefit from such products since they could increase their folate intakes while consuming products that form part of their normal lifestyle. The proper selection of probiotic folate producing strains provides a strategy for the development of novel functional foods with increased nutritional value.

Suggested Reading

Crittenden, R. G., Martinez, N. R., and Playne, M.J. 2003. Synthesis and utilization of folate by yogurt starter cultures and probiotic bacteria. *International Journal of Food Microbiology* **80:** 217-222.

Forssen, M. 2000. Folates and dairy products: A critical update. *Journal of the American College of Nutrition* **19:** 100-110.

Iyer R, Tomar S.K., Singh, R., Sharma R. 2009. Estimation of folate in milk by microbiological assay using trienzyme extraction method. *Milchwissenschaft* **64(2):** 125-127.

Iyer, R., Tomar, S.K., Kapila, S., Mani, J & Singh, R., 2010. Probiotic properties of folate producing *Streptococcus thermophilus* strains. *Food Research International* **43:** 103–110.

Iyer, R., and Tomar S.K .2009. Folate: A Functional Food Constituent. *J. Food Sci.* **74(9):** R114-122.

Iyer, R., Tomar S.K., Maheswari, U., Singh, R. 2010. *Streptococcus thermophilus* strains: Multifunctional lactic acid bacteria. *International Dairy Journal* **20:** 133–141.

Le Blanc, J.G., Giori, G.S., Smid, E.J., Hugenholtz, J., Sesma, F. 2007. Folate production by lactic acid bacteria and other food-grade microorganisms. Communicating Current Research and Educational Topics and Trends in Applied Microbiology. 329-339.

Lin, M.Y., and Young, C.M., 2000. Biosynthesis of Folates by *Streptococcus thermophilus* and *Lactobacillus delbruekii* ssp. *bulgaricus*. *Journal of Food and Drug Analysis* **8:** 195-199.

Smid, E.J., Starrenburg, M.J.C., Mierau, I., Sybesma, W., Hugenholtz, J., 2001. Increase of folate levels in fermented foods. *Innovations in Food Technology* **10:** 13-15.

Sybesma, W., Starrenburg, M., Tijsseling, L., Hoefnagel, M.H.N., Hugenholtz, J., 2003. Effects of cultivation conditions on folate production by lactic acid bacteria. *Applied and Environmental Microbiology* **69:** 4542 – 4548.

Tamura T., 1998. Determination of food folate. *Nutritional Biochemistry* **9:** 285-293.

Tomar S.K., Srivatsa N., Iyer R., Singh R. Estimation of Folate Production by *S. thermophilus* using Modified Microbiological Assay. Milchwissenschaft, **64(3):** 260-263.

❑❑❑

Chapter 3

Bacteriocins of Lactic Acid Bacteria

R.K. MALIK , NARESH KUMAR AND ARUN BHARDWAJ
Dairy Microbiology Division, National Dairy Research Institute, Karnal (HR)

Introduction

Food safety is perhaps one of the greatest concerns of the consumers since antiquity and encompasses over all the issues surrounding food preparation, processing, packaging and storage. Food preservation is a vital parameter for ensuring the food safety. Globally research groups are conducting extensive research for extending shelf life and developing safety measures in this sector. Chemical preservatives and other additives have generally been used to combat specific microorganisms and in recent years their used in food items have increased exponentially to several thousands. Reports on possible health hazard from the consumption of these have shaken the faith of consumers, thus prompting them to question the safety and wholesomeness of food they eat. Moreover, the effect of actual consumption of these chemicals in multiple products over substantial length of time should be an important consideration in judging their safety and consumers do not prefer foods that contain any preservative(s) or additives which safety if questionable.

Biopreservation has been a common practice in the history of mankind for the extension of storage life and enhancing the safety of foods, using microorganisms and/or their anti-microbial products. It is viewed as 'natural' and recommended for improvement of food safety. Therefore, the basis of selection of antimicrobial biopreservatives to be used in foods should not only be their effectiveness but also their proven safety records and their acceptance by the health conscious consumers and health regulatory agencies. Among the compounds that have generated considerable interest as biopreservatives in the recent years are several antimicrobial metabolites of lactic acid bacteria used to produce, or associated with, fermented foods for millennia.

Antimicrobial Potential of Lactic Acid Bacteria (LAB)

LAB play a defining role in the preservation and microbial safety of fermented foods by the production of wide range of antimicrobial metabolites **(Fig. 1)**. Amongst these, lactic acid and acetic acid is obviously the most important. However, certain strains of LAB are further known to produce bioactive molecules such as ethanol, formic acid, fatty acids, hydrogen peroxide, diacetyl, reuterin, and bacteriocins. Among all the anti-microbial substances produced by LAB, bacteriocins offer the best incentive for the food preservation and food safety by undesirable microorganisms.

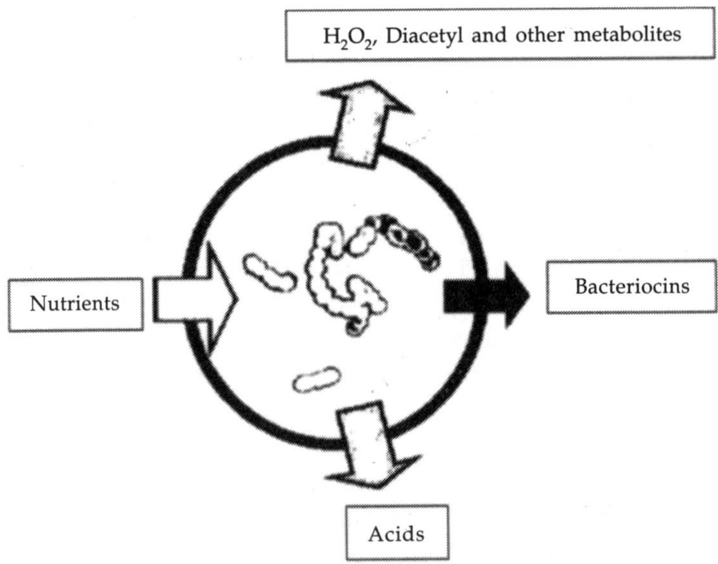

Fig. 1 Production of various antimicrobial metabolites by a LAB

Bacteriocin- a promising biopreservative

Bacteriocins are antimicrobial peptides which inhibit, by a bactericidal or bacteriostatic mode of action, micro-organisms that are usually closely related to the producer strain. A bacteriocin producer protects itself against its own antimicrobial compound by means of a system referred to as immunity, which is expressed concomitantly with the antimicrobial peptide. The bacteriocin family includes a diverse group of proteins in terms of size, microbial targets, modes of action, and immunity mechanisms. They differ from traditional antibiotics in one critical way; they have a relatively narrow killing spectrum and are only toxic to bacteria closely related to the producing strain **(Table 1).**

The bacteriocins produced by LAB offer several desirable properties that make them suitable for food preservation:

- They are generally recognised as safe (GRAS) substances,
- They are not active and nontoxic on eukaryotic cells,
- They are usually pH and heat-tolerant and become inactivated by digestive proteases, having little influence on the gut microbiota,
- They have a relatively broad antimicrobial spectrum, against many food-borne pathogenic and spoilage bacteria,
- They show a bactericidal mode of action, usually acting on the bacterial cytoplasmic membrane: no cross resistance with antibiotics,
- Their genetic determinants are usually plasmid-encoded, facilitating genetic manipulation.

Table 1: Characteristic aspects of bacteriocins and other conventional antibiotics

CHARACTERISTICS	BACTERIOCINS	OTHER ANTIBIOTICS
Application	Foods	Clinical
Synthesis	Ribosomally	Secondary Metabolism
Activity	Limited Spectrum	Wide Spectrum
Presence of immune cells in the host	Present	Absent
Action mode	Mostly through the channel formation in the cell cytoplasmic membrane	Specific Target
Toxicity/other effects in eukaryotic cells	Absent	Present

Classification of Bacteriocins of LAB

Most LAB bacteriocins are small (< 6 kDa), cationic, heat-stable, amphiphilic, membrane permeabilizing peptides that may be divided into three main groups: (i) **Class I-** lantibiotics or small, heat-stable, lanthionine- containing, single- and two-peptide bacteriocins, whose inactive prepeptides are subject to extensive post-translational modification; (ii) **Class II-** peptide bacteriocins or small, heat-stable, non-lanthionine-containing bacteriocins, including pediocin-like or *Listeria*- active bacteriocins **(class IIa)**, two-peptide bacteriocins **(class IIb)**, and circular bacteriocins **(class IIc)**, and, arguably, (iii) **Class III-** bacteriolysins or large, heat-labile, lytic proteins, often murein hydrolases. Extensive efforts have been made to resolve the relationship between structure and function for both class I and class II bacteriocins.

The majority of the class I and class II bacteriocins are active in the nanomolar range, causing membrane permeabilization, leading to the dissipation of membrane potential and the leakage of ions, ATP, and other vital molecules from the target bacteria. A fourth group **Class IV** with complex bacteriocins carrying lipid or carbohydrate moieties is often included in bacteriocins classifications. Recently a **Class V** of cyclic bacteriocins has been included in the classification scheme with lesser amount of modified amino acids. Within any class of bacteriocin, there may be amino acid sequence homologies not only within the mature peptide, but also in the N-terminal leader region and the associated proteins involved in bacteriocin secretion and processing.

Bacteriocins Production by LAB

Enterococcus

Among several species of *Enterococcus,* bacteriocins known as enterocins are mainly produced by *E. faecalis* and *E. faecium* that are linked to food bio-preservation. Enterocins usually belong to class II bacteriocins, i.e. they are small and heat-stable and membrane active non-lantibiotics, being stable in milk. They are insensitive to rennet, have stability over a wide range of pH values, and a general compatibility with other starter LAB species. The enterocins have been found to display inhibitory spectrum towards foodborne pathogens such as *Listeria spp.* and *Clostridium spp.* Activity towards Gram-negative bacteria such as *E.* coli and *Vibrio cholerae* has been shown as well. Many reports consider possible use of enterocins/ bacteriocinogenic enterococci as protective starters in different food models, where it can efficiently serve as a barrier in a hurdle technology. Well-characterised enterocins are A, P, CRL35, 1071A, and B; bacteriocin 31, 32, RC714, and T8; enterolysin A, AS-48, EJ97, RJ11, Q, L50A and B. Enterocins with broad inhibitory spectra against food spoilage and pathogenic organisms undoubtedly indicate their use as potential biopreservative agents in food industry.

Lactobacilli

The bacteriocinogenicity has been described for several lactobacilli viz. *Lb.acidophilus, Lb. helveticus, Lb. plantarum, Lb. sake* and *Lb. brevis.* The bacteriocins from the *L. helveticus* include helveticin J and helveticin V-1829. Lactacin B, a bacteriocin produced by *Lb. acidophilus* N2. A *Lb. plantarum* strain LTF154 produced a bacteriocin designated as plantacin 154. A bacteriocin producing *Lb. plantarum* strain LC74 from goat raw milk and named the bacteriocin as plantaricin LC74. Production of brevicin 286 has been recently reported in *Lb. brevis* VB286. The human isolate *Lactobacillus gasseri K7* produces bacteriocin with a wide range of inhibition.

Lactococci

Several workers have reported bacteriocinogenicity among the different strains of the three most economically important lactococcal species: *Lactococcus lactis* subsp. *lactis, Lc. lactis* subsp. *cremoris* and *Lc. lactis* subsp. *lactis* biovar. *diacetylactis* that are generally used as starter culture in dairy products. Nisin, the most extensively characterised bacteriocin from lactic acid bacteria, is produced by several strains of *Lc. lactis* subsp. *lactis.* The production of several lactococcins has been described in several other *Lc. lactis* subsp. *lactis* which include: Lactococcin by *Lc. lactis* subsp. *lactis* ADRI 85L030; lactococcin G by *Lc. lactis* subsp. *lactis* LMG2081; lactococcin 972 by *Lc. lactis* subsp. *lactis;* lactococcin 484 by *Lc. lactis* subsp. *lactis* 484.

Leuconostocs

First, Harding and Shaw in 1990 reported the production of bacteriocin by a strain of *Leu. gelidum* that was active against *Listeria monocytogenes. Leuconostoc paramesenteroides* OX was also found to produce a bacteriocin named as leuconocin S. Bacteriocins, carnosin 44A, carnocin LA54A and leucocin B-Ta11a, produced by *Leu. carnosum* LA44a isolated from sausage. Bacteriocinogenic strains of *Leuconostoc* spp. have also been isolated from milk and milk products. Malik et al.

1994 reported the detection and activity of a novel bacteriocin, leucocidin R1, produced by *Leu. paramesenteroides* NM14 isolated from an aged cream sample.

Streptococcus thermophilus

The bacteriocins of *S. thermophilus* strains are commonly known as thermophilins. In general, these thermophilins are stable and active over a wide range of pH and temperature, unlike nisin, which can not be used in neutral or alkaline foods. First report on production of bacteriocin by *Streptococcus thermophilus* 347 was given by Villani et al., (1995) and the bacteriocin was known as thermophilin 347. In the same year Thermophilin A, the bacteriocin like substance produced by *S. thermophilus* ST134 has been isolated from raw milk cheese. Aktypis et al. (1998) purified bacteriocin thermophilin T from *Streptococcus thermophilus* ACA-DC 0040.

Mode of Action

Biological action of bacteriocins occurs through the specific receptors located on the target microbial cell surface. After binding with these receptors, various mechanisms act, by isolated or concomitant way, causing the microbial cell killing. Microbial cell killing due to bacteriocins' action could occur as a consequence of unbalanced cytoplasmic membrane function (affecting energy synthesis and permeability), inhibition of nucleic acid synthesis, interference with the protein synthesis and changing cell translator mechanism. While many bacteriocins (e.g. Nisin) have been shown to induce pore formation in sensitive microorganisms that disrupt the proton motive force and the pH equilibrium causing leakage of ions and hydrolysis of ATP resulting in cell death **(Fig. 2)**.

Production and Purification of Bacteriocins

Although bacteriocins can be produced in the food matrix during food fermentation, but at low amounts as compare to production under *in vitro* conditions. The higher *in vitro* production is due to the absence of limiting factors, such as strong diffusion limitations, inactivation by proteases, and the adsorption to food particles. However, an influence of the environmental process conditions on the obtained bacteriocin activity can be seen. Temperature and pH as well as nutrient availability seem to play a crucial role in bacteriocin production, whereas the presence of elevated amounts of sodium chloride usually decreases production levels. In general, the cultivation conditions directly affect bacteriocin production as such and, indirectly, through biomass production. This is to be explained by the fact that bacteriocin production is a growth-dependent physiological trait and hence follows primary metabolite kinetics.

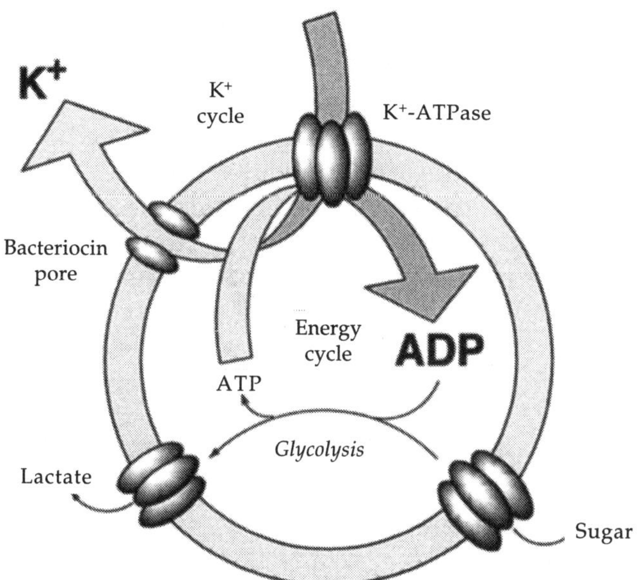

Fig 2: Proposed model of cell killing by pore-forming bacteriocins.

Three major methods for the purification of bacteriocins by LAB can be distinguished:

First, purification can be done by conventional subsequent steps of ammonium sulfate precipitation, ion exchange, hydrophobic interaction, gel filtration, and reversed-phase high-pressure liquid chromatography.

Second, a simple three-step protocol has been developed, including (i) ammonium sulfate precipitation, (ii) chloroform/methanol extraction and (iii) reversed-phase high-pressure liquid chromatography, the sole chromatographic step involved.

Third, bacteriocins can be isolated through a unique unit operation, i.e. expanded bed adsorption, using a hydrophobic interaction gel, after maximizing the bioavailable bacteriocin titer through pH adjustment of the crude fermentation medium. Latter two methods are more rapid and successful than the first conventional method.

Application of Bacteriocins in Foods

Two main applications concerning bacteriocins are possible i) addition of purified bacteriocins to foods ii) bacteriocins may be produced directly in the food as a result of starter culture or co-culture activity. Application of bacteriocin-producing starter cultures in sourdough (to increase competitiveness and hence establish a desired microbial population), in fermented sausage (anti-listerial effect to meet the zero-tolerance policy in ready-to eat foods), and in cheese (anti-listerial and anti-clostridial effects), have been studied during in vitro laboratory fermentations as well as on pilot-scale level. Results of these studies were highly promising and underline the important role that functional, bacteriocinogenic strains of LAB may play in the food industry as starter cultures, co-cultures, or bioprotective cultures, to improve food quality and safety. Bacteriocins often have synergies effects with other treatments, and can be used in hurdle technology that combines different preservation methods to improve the safety of food. For example, use of nisin along with pulsed electric field (PEF) nisin for the increase cell permeability. Bacteriocins in combination with chelators such as EDTA used for the effectiveness of bacteriocins against gram negative cells.

Future Prospects

A large number of bacteriocins from LAB have been characterized to date, for their potential use in food preservation. Since the efficacy of bacteriocins in foods is dictated by environmental factors, there is a need to determine more precisely the most effective conditions for application of each particular bacteriocin. Bacteriocinogenic cells may also act as living factories in foods. The antimicrobial effects of bacteriocins and bacteriocinogenic cultures in food ecosystems must be understood in terms of microbial interactions. Genomic studies can provide valuable information on the bacteriocinogenic potential of LAB development of adaptation or resistance. Genetic engineering of already identified strains may help to make them more efficacious. Continued study of the physical and chemical properties, mode of action and structure-function relationships of bacteriocins is necessary if their potential in food preservation is to be exploited. Further research into the synergistic reactions of these compounds and other natural preservatives, in combination with advanced technologies such as Pulsed Electric Field and Ultra High Pressure could result in replacement of chemical preservatives.

Suggested Reading

De Vuyst L., Leroy F. 2007. Bacteriocins from lactic acid bacteria: production, purification, and food applications. *Journal of Molecular Microbiology and Biotechnology*, **13:** 194–199.

Deegan L.H., Cotter P.D., Hill C., Ross P., 2006. Bacteriocins: biological tools for bio-preservation and shelf-life extension. *International Dairy Journal*, **16:** 1058–1071.

Gálvez A., Abriouel H., López . and Ben Omar N. 2007. Bacteriocin-based strategies for food biopreservation, *International Journal of Food Microbiology* (*In Press*).

Malik R.K., Chauhan P. 2008. Bacteriocins of Lactic Acid Bacteria in Food Biopreservation. In: *Biotechnology Concepts and Applications*, Narosa Publisher, Chennai, pp 116-160.

□□□

Chapter 4

Biothickners from Lactic Acid Bacteria

RAMESHWAR SINGH; ANBUKKARASI, K; PRADIP BEHARE; AND PRATHIMA, P.C
Dairy Microbiology Division, National Dairy Research Institute, Karnal

Introduction

Microbial exopolysaccharides (EPS) are extracellular which are either associated with the cell surface in the form of capsules or secreted into the extracellular environment in the form of slime. They are referred to as capsular or slime exopolysaccharide. EPS in their natural environment are thought to play a role in the protection of the microbial cell against desiccation, phagocytosis and phage attack, antibiotics or toxic compounds (e.g. toxic metal ions, sulfur dioxide, ethanol), predation by protozoans, osmotic stress, adhesion to solid surfaces and biofilm formation, and also in cellular recognition (e.g. via binding to a lectin). In pathogenic bacteria, such as *Streptococcus pneumoniae* and *Streptococcus agalactiae,* capsular EPS and O-antigen lipopolysaccharides are involved in the immune response.

Examples of industrially important microbial exopolysaccharides are dextrans, xanthan, gellan, pullulan, yeast glucans and bacterial alginates. Novel microbial biopolymers may fill the gap in market-available polymers or may replace the traditional product in terms of the improved rheological and stability characteristics. Now, microbial polysaccharides represent only a small fraction of the current biopolymer market. Factors limiting the use of microbial EPS are their production, which requires a thorough knowledge of their biosynthesis and an adapted bioprocess technology, and the high cost of their recovery. Xanthan is a microbial EPS approved in the food industry, mainly because of its unique rheological properties in foods and the possibility of low-cost production. It is produced in high amounts by *Xanthomonas campestris*, a phytopathogenic bacterium that is not generally recognized as safe (GRAS).

EPS produced by lactic acid bacteria (LAB) are receiving increasing attention because these microorganisms have a 'food-grade' status. The rheological properties of EPS produced by LAB have found a major application in the manufacture of fermented dairy products such as yogurt, cheese, or fermented milks, although they also play a role in the elaboration of fermented meats and vegetables. In addition, they have also been proven to have beneficial effects on human health such as cholesterol-lowering, antitumoral, inmunomodulating activities and prebiotic effects.

Classification

EPS from LAB can be subdivided into two major groups: (i) the homopolysaccharides, composed of one type of monosaccharide, and (ii) the heteropolysaccharides, composed of a repeating unit that contains two or more different monosaccharides. Based on their structures, four groups of homopolysaccharides can be distinguished: á-D-glucans, â-D-glucans, â-D-fructans, and others, such as polygalactan. Heteropolysaccharides are produced with more variety concerning the chemical composition, monomer ratio and molecular structure of the repeating unit, as well as the molecular mass of the polymer. The repeating units most often contain a combination of D-glucose, D-galactose and L-rhamnose, but in different ratios. LAB produce heteropolysaccharides are mesophilic (e.g.,*Lactococcus lactis, Lactobacillus sakei, Lb rhamnosus* and *Lb casei*) and thermophilic (e.g., *Lb acidophilus, Lb delbrueckii* subsp. *bulgaricus, Lb helveticus* and *Streptococcus thermophilus*). According to their monomer compositions, three major groups have been distinguished: (i) EPS containing only galactose (*Lb paracasei* 34-1) (ii) EPS containing galactose and glucose (*S thermophilus* Sfi39) and (iii) EPS containing galactose, glucose, and rhamnose (*S thermophilus* Sfi12).

The molecular masses of the EPS from LAB range from 10 kDa and over 200 kDa to >1,000 kDa. Importantly, both structure and molecular mass influence the rheological properties of a polysaccharide. Also, environmental factors (carbohydrate source, nitrogen source, and carbon/nitrogen ratio of the growth medium) can influence the production, monomer composition, and molecular mass of the EPS produced by a particular strain. Heteropolysaccharides receives renewed interest, since they play an important role in the rheology, texture and body, and mouth feel of fermented milk drinks. For instance, the creamy, smooth texture is one of the aspects of the quality of yogurt which seems to be improved by the ability of the yogurt bacteria to produce EPS, even though only small amounts of EPS are being produced. In addition, EPS from LAB have one of the largest technical potentials for development of novel and improved products such as low-milk-solid yogurts, low-fat yogurts, creamier yogurts, etc. Therefore EPS from LAB have potential for development and exploitation as functional food ingredients with both health and economic benefits.

Biosynthesis of Exopolysaccharides

In these gram-negative systems, assembly of the basic repeating unit occurs at the cytoplasmic membrane and involves sequential transfer of sugar nucleotide diphospho precursors to an isoprenoid lipid carrier, undecaprenyl phosphate. Once the basic repeating unit is assembled, the lipid-linked intermediates are usually translocated across the membrane and polymerized outside of the cell. Then, the EPS may be covalently linked to the cell surface to form a capsule, or released into the medium as slime. Cellular EPS biosynthesis is, an energy intensive process that involves enzymes for production of sugar nucleotide precursors, a glycosyl-1-phosphate transferase that transfers the first sugar group onto a phosphorylated carrier lipid, one or more glycosyltransferases that sequentially add new sugar groups to the growing repeat unit, and additional enzymes that participate in *eps* gene regulation, membrane translocation, and polymerization/chain length determination functions. These reactions commonly involve additional proteins with functions that are not unique to this process. Examples include housekeeping enzymes involved in the synthesis of molecules, such as lipid carrier and sugar nucleotide precursors, which are required for the assembly of other glycan-containing polymers, like peptidoglycan, teichoic acids and lipoteichoic acids.

Genetic Engineering of EPS

In order to improve properties of EPSs, genetic engineering could result in the production of designed polysaccharides' targeting predefined macroscopic properties. The biosynthesis of heteropolysaccharides involves several steps directly connected to the cell's carbon metabolism. Significant progresses have been made in the genetic analysis of EPS biosynthesis of various LAB. The genes required for the biosynthesis, have been identified in different clusters. Genetic analysis of EPS-producing strains has revealed a common operon structure, strongly suggesting that a common mechanism might be involved in their biosynthesis. The gene sequence in all the gram-positive cell-surface polysaccharide-biosynthesis clusters identified to date as follows: regulation, chain-length determination, biosythesis of repeating unit, polymerization and export. For instance, in *Lb bulgaricus* Lfi5, a genomic region containing 14 genes (epsABCDEFGHIJKLMN) involved in EPS biosynthesis was characterized. Putative functions for most of the *eps* gene products could be assigned by sequence similarities: regulation for EpsA, biosynthesis of the repeating unit for EpsE, EpsF, EpsG, EpsH, EpsI, EpsJ and EpsM, EPS polymerization and chain-length determination for EPsB, EpsC and EpsD and polymerization and export for EpsK and EpsN. Transcription of the *eps* gene cluster was shown to be driven by a single promoter immediately upstream of epsA. All genes are transcribed as a single polycistronic mRNA of approximately 15.5kb. Large transcribes encompassing biosynthetic genes for polysaccharide production were similarly reported for the *eps* cluster from *S thermophilus* and *L lactis* , as well as for the *ams* gene cluster from *Erwinia amylovora* and the *cps* gene cluster from *Klebisella pneumonia.* The sequential transfer of sugar residues from activated sugar donors onto a lipophilic carrier is directly followed by assembly of the oligosaccharide-repeating unit by specific glycosyltransferases. Subsequently, the completed repeating unit is exported, polymerized and released (in the case of secreted polysaccharides).

Based on the available genetic information, genetic modification of *eps* genes should lead to EPSs with a different repeating unit or with a different chain length. Based on the available genetic information, heterologous production of an EPS has been successful by transfer of the complete *eps* gene cluster alone of a LAB strain into a non EPS-producing heterologous host, provided that the heterologous host processed all necessary genetic information for precursor synthesis. Gene shuffling with glycosyltransferases is a very promising tool to produce different linkages or use different sugars. Heterologous expression of glycosyltransferase also allows to add other sugars at strategic positions to generate EPSs with new properties.

Applications in Dairy Industry

The intentional and controlled use of EPS from LAB as natural food additives or of functional starter cultures, i.e. strains producing interesting EPS, could result in a safe, natural end product, and may have an important impact on the development of novel food products (both fermented and non-fermented food products), especially food products with enhanced, rheological properties, improved texture and stability, and/or water retention capacity. For instance, low viscosity, gel fracture or high syneresis (whey separation) problems, which may occur during yogurt manufacture, can be solved in both ways. An EPS-producing *S thermophilus* strain was also responsible for an increased moisture level in low-fat mozzarella cheese. Since the use of LAB (GRAS microorganisms) is historically safe, production of in situ novel functional EPS means that toxicological testing will not be required and the products can be brought to the market more quickly.

a) Fermented milks

A very large variety of fermented milks are offered to the consumer. They vary with the type of milk used, i.e. cow, goat or buffalo milk, the bacterial starter culture used and the process. EPS producing LAB are not only important for yoghurt and yoghurt-like drinks but also for cheese production. For example several strains of *Lb helveticus* used for mozzarella production, produce EPS and contribute to water retention in the product. Lactic acid produced during the fermentation process lowers the pH in the milk and consequently the increased solubility of calcium and phosphate contained in micelles causes the micelles to destabilize. In addition the electric charge of casein micelles decreases to that of the isoelectric point. Hence, casein precipitates due to hydrophobic interactions and forms a continuous network with cavities filled with serum. When mechanical forces break such a gel, the casein network breaks up into gel particles dispersed in the serum. The viscosity of the dispersion of the gel particles is proportional to the viscosity of the serum phase. As a consequence, the viscosity of stirred yoghurt fermented with a non-EPS producing starter strain is low and the texture tends to be granular.

Nordic ropy milk is the generic name for fermented milks with mesophilic cocci which produce slime. They are mainly popular in Scandinavia, Iceland, Russia and Mongolia. While traditionally produced at home, these products are also industrially manufactured. The major LAB used are *L lactis* subsp. *cremoris, lactis* and *lactis* biovar *diacetylactis, Leuconostoc mesenteroides* subsp. *cremoris* and *dextranicum* for the mesophilic bacteria, *S thermophilus, Lb delbrueckii* subsp. *bulgaricus* and *Lb helveticus* for the thermophilic bacteria. In addition lactose fermenting yeasts are found in kefir and the mould *Geotrichum candidum* is associated with Finnish viili. Viili is a fermented milk consumed in Finland. The process consists of fermenting milk with *L. lactis* subsp. *lactis* biovar *diacetylactis, Ln mesenteroides* subsp. *cremoris* and the mould *G candidum.* The composition of the EPS produced by a strain of *L lactis subsp. cremoris* isolated from viili. The polysaccharide contained the sugar moieties rhamnose, glucose and galactose at molar ratio 1: 1.45: 1.75 and a phosphate group was attached to the d-galactopyranosyl residue.

Kefir is a traditional self-carbonated slightly alcoholic fermented milk from Eastern Europe. It contains about 1% lactic acid and 0.1–1% ethanol depending on the fermentation by yeasts. Milk is inoculated with kefir grains (1–10mm diameter) containing homofermentative and heterofermentative LAB, yeasts and acetic acid bacteria. Some typical species are *L lactis* subsp. *lactis* and *cremoris, Ln mesenteroides* subsp. *dextranicum* and *S thermophilus,* yeasts of the genera *Saccharomyces, Kluyveromyces, Candida, Mycotorula, Torulopsis, Cryptococcus, Torulaspora, Pichia,* and the acetic acid bacteria *Acetobacter aceti* and *Acetobacter racens.* The cells are embedded in kefiran, a slimy polysaccharide composed of glucose and galactose in equimolar proportions. Kefiran is the "glue" of the grains and confers a slimy texture to the product. A dissertation was carried out in our laboratory to study the environmental and cultural growth conditions affecting EPS production by *Lactococcus lactis subsp lactic* (NCDC 191), a strain of indigenous origin commonly used as dahi starter. The highest EPS production was achieved when the cultures were incubated at 25^{0}C in deproteinized whey medium maintained at 6.8 supplemented with 1.0% Casein hydrolyzate and 10 mg of glucose. Similarly, another PhD project was carried out in application of exopolysaccharides producing cultures in reduced fat fermented milks. Five promising EPS$^+$ cultures of *L.lactis* subsp.*lactis, S.thermophilus* and *Lactobacillus fermentum* has been exploited for preparation of fermented milks. All the EPS^{+} cultures have shown significant improvement in the rheological, sensory properties and control of syneresis in fermented milks prepared from different grades of milk as compared to EPS^{-} cultures. The use of EPS cultures improved the rheological and sensory properties of low-fat/fat-free products so as to reduce their total fat and energy content. The cultures may also be tested under field conditions for production of low-fat/fat-free fermented milks.Different formulation of these cultures can be prepared and studied for production of low-fat cheeses.

Yogurt consists of a network of casein micelles that associate to form a gel-like structure. This protein network has interstitial spaces that contain the liquid phase as well as larger void spaces containing starter bacteria. By trapping the whey, the casein matrix creates the typical viscoelastic properties of yogurt texture. Stirred yogurt is a product resulting from further processing of set yogurt. The retention of whey is vital to product integrity and not only the contribution of acid secretion and microbial proteinase is essential for texture development, but also exopolysaccharides secreted by the yogurt strains contribute to this. The use of ropy cultures results in yogurts with a more stable texture for shear. When subjected to a shear force changes occur in the microstructure of ropy yogurt. An increase of shear rate first disrupts the attachment of polymer to the bacterial surface, but the polysaccharide material remains incorporated with the casein where it continues to influence viscosity of yogurt. If interactions of EPS and protein require more energy to break than interactions among the protein that are responsible for yogurt gel texture, an initial decrease in viscosity, when examining shear stress against shear rate, could be due to the disrupted interactions between proteins, and a reduced rate of shear thinning observed later could be explained by the increased energy required to disrupt interactions among the bacteria, EPS and protein.

b) Mozzarella Cheese

While the application of EPS^{+} starter lactic acid bacteria has been largely confined to the production of yogurt and fermented milks, some companies have used ropy strains to improve the texture and functionality of cheese. Interest in EPS^{+} starters for cheese making has been restricted, however, by the fact that EPS often accumulates in the cheese whey, thereby increasing its viscosity. This outcome is undesirable because it retards the efficiency of membrane processing and slows whey protein concentration and drying processes. Consumer demand for reduced-fat foods has created interest in the development and manufacture of reduced- fat, low-fat, and nonfat Mozzarella cheese, but fat removal has several undesirable effects on cheese physical properties. Specifically, Mozzarella cheese becomes tough and rubbery, more heat is required for melting, and the cheese loses pliability rapidly during cooling. Research has revealed that these properties are heavily influenced by cheese moisture level. Examination of cheese microstructure shows that in full fat or part-skim Mozzarella cheese, much of the water is contained in channels formed in the protein matrix by entrapped fat globules. Because there are very few fat globules to break up the protein matrix in low-fat Mozzarella, these channels become much narrower with less space available for water retention. This results in cheese with a lower moisture level and, as a consequence, a tough, rubbery texture and poor melt and stretch properties. Thus EPS have excellent water-binding properties (moisture retention) which is vital to functionality in low fat cheese. Perry *et al.* (1997) investigated the influence of an EPS-producing starter pair on the moisture and melt properties of low fat (6%) Mozzarella cheese. Low-fat Mozzarella cheese manufactured with *S thermophilus* MR-1C and *Lb bulgaricus* MR-1R contained significantly ($P < 0.05$) more moisture and had better melting properties than cheese made with an EPS– commercial starter pair (*S. thermophilus* TAO61 and *Lb helveticus* LH100). Petersen *et al.* (2000) investigated the effect of capsular and ropy EPS$^+$ *S thermophilus* starters on part-skim Mozzarella cheese and whey. They showed that cheese moisture levels were significantly higher in Mozzarella made with EPS$^+$ versus EPS$^-$ streptococci, but whey from cheese made with the ropy starter was significantly more viscous ($P < 0.01$) than whey from cheeses made with the CPS$^+$ or EPS$^-$ strains. More importantly, no significant difference was detected in the viscosity of 5-fold concentrated whey from cheese made with CPS$^+$ or a commercial EPS$^-$ starter. These findings indicate that CPS$^+$, but not ropy, *S thermophilus* starter bacteria can be used to enhance Mozzarella cheese moisture level, yield, and melt properties without deleteriously affecting whey viscosity.

Suggested Reading

Aslim B, Beyatli Y and Yuksekdag Z.N. (2006). Productions and monomer compositions of exopolysaccharides by *Lactobacillus delbrueckii subsp. bulgaricus* and *Streptococcus thermophilus* strains isolated from traditional home-made yoghurts and raw milk. *International Journal of Food Science and Technology*, **41:** 973–79.

Behare, P.V. (2007).Application of Exopolysaccharide-producing cultures in reduced-fat fermented milks.Ph.D Thesis, NDRI Deemed University, Karnal (Haryana), India.

Broadbent JR, McMahon DJ, Welker DL, Oberg CJ and Moineau S. (2003). Biochemistry, Genetics, and Applications of Exopolysaccharide Production in *Streptococcus thermophilus*: A Review. *J. Dairy Sci.,* **86:** 407–23.

De Vuyst L and Degeest B. (1999). Heteropolysaccharides from lactic acid bacteria. *FEMS Microbiology Reviews,* **23:** 153-77.

Jolly L and Stingele F. (2001). Molecular organization and functionality of exopolysaccharide gene clusters in lactic acid bacteria. *International Dairy Journal,* **11:** 733–45.

Jolly L, Vincent SJF, Duboc P and Neeser JR. (2002). Exploiting exopolysaccharides from lactic acid bacteria. *Antonie von Leeuwenhoek,* **82:** 367-74.

Lamothe GT, Jolly L, Mollet B and Stingele F. (2002). Genetic and biochemical characterization of exopolysaccharide biosynthesis by *Lactobacillus delbrueckii* subsp. *bulgaricus. Arch Microbiol,* **178:** 218–28.

Meulen RV, Grosu-Tudor S, Mozzi F, Vaningelgem F, Zamfir M, de Valdez GF and De Vuyst L. (2007). Screening of lactic acid bacteria isolates from dairy and cereal products for exopolysaccharide production and genes involved. *International Journal of Food Microbiology,* **118:** 250–58.

Prathima, P.C (2006) Optimization of Exopolysaccharide production by *Lactococcus lactis* ssp *lactis.* M.Sc Thesis, NDRI Deemed University, Karnal (Haryana), India.

Purwandaria U, Shaha NP and Vasiljevica T. (2007). Effects of exopolysaccharide-producing strains of *Streptococcus thermophilus* on technological and rheological properties of set-type yoghurt. *International Dairy Journal,* **17:** 1344–52.

Savadogo A, Ouattara CAT, Savadogo PW, Barro N, Ouattara AS and Traoré AS. (2004). Identification of exopolysaccharides-producing lactic acid bacteria from Burkina Fas fermented milk samples *African Journal of Biotechnology,* **3:** 189-94.

□□□

Chapter 5

Conjugated Linoleic Acid : Metabolism, Role in Prevention of Pathologies and Mechanism of Action

VINOD K. KANSAL AND RENU SAGWAL
Animal Biochemistry Division, National Dairy Research Institute, Karnal (HR)

Introduction

Conjugated linoleic acid (CLA) is a mixture of positional and geometric isomers of linoleic acid, found naturally in the food. In contrast to linoleic acid, which has two cis double bonds at carbons 9 and 12, CLA contains cis and trans isomers at carbons 8 and 10, 9 and 11, 10 and 12 or 11 and 13. CLA is especially abundant in food lipids derived from ruminants such as beef, lamb and dairy products and, due to increased conversion of linoleic acid to CLA by rumen microorganisms. The cis-9, trans-11 isomer, the principal dietary form of CLA, is produced in the rumen of animals during the microbial biohydrogenation of dietary linoleic and linolenic acids by a linoleic acid isomerase from the rumen bacteria *Butyrivibrio fibrisolvens*. The predominant trans fatty acid in milk fat, i.e., vaccenic acid (trans 11-18:1) is converted into CLA in tissues via the D^9-desaturase. Dietary conditions causing an increase in trans-10, cis-12 CLA in milk are a high-grain, low-fiber diet. Human milk contains mostly cis-9, trans-11 CLA with only small amounts of the trans-10, cis-12 CLA isomer. The cis-9, trans-11 CLA is also the major isomer in beef products.

CLA has received more attention in the last decade because of CLA pleiotropic biological activities. There are multiple potential isomers, but the cis-9, trans-11 and trans-10, cis-12 isomers are thought to be active as anticarcinogenic (Ip *et al.*, 1991; 1994) and antiobesity (Park *et al.*, 1997). While it is chemically related to linoleic acid, CLA appears to have opposite effects in certain important areas. For instance, linoleic acid stimulates fat formation (lipogenesis) in adipose tissue, while CLA inhibits fat formation; linoleic acid tends to promote tumor growth, while CLA is an excellent inhibitor of tumor growth. Substantial interest has developed in the biochemical actions of CLA and its potential application to foods, feeds and pharmaceuticals.

CLA and Body Fat Metabolism

Animal studies

Dietary CLA has been shown to affect reduction in body fat and enhancement of fat-free mass in mice (Park *et al.*, 1997; Sagwal and Kansal, in press), rats and pigs (Cook *et al.*, 1998). Adiposity index is reduced, while body mass index increased (Sagwal and Kansal, in press) in CLA-fed mice. The effect of CLA in body fat reduction was enhanced by inclusion of non-fat milk in the diet, suggesting that CLA is more effective when presented in milk (Sagwal and Kansal, in press). The reduction of body fat stores without any reduction of energy intake was observed in mice (Sagwal and Kansal, in press), hence, the mobilization and loss of body fat stores certainly do not result from a reduction in energy consumption. Conjugated linoleic acid treatment significantly reduced adipose depot weights in the inguinal, epididymal and the retroperitoneal depots (West *et al.*,2000; Sagwal and Kansal, in press). In the case of lactating cows, trans-10, cis-12 isomer of CLA is responsible for reduced milk fat synthesis (Griinari and Bauman, 1999) i.e., milk fat depression.

Reports have shown that CLA-fed mice developed fatty liver (Sagwal and Kansal, in press). Sagwal and Kansal (in press) observed that CLA feeding increased liver, spleen and heart weight in mice. The increase in heart weight needs attention as it may represent the initial stage of cardiac hypertrophy. Further, the adverse effect of CLA on organ weight was eliminated by inclusion of non-fat milk in the diet. Therefore, a detailed investigation on synergistic effect of food components and CLA is essential.

Human Studies

Evidence on the effect of CLA on body fat in humans is contradictory. While Smedman and Vessby (2001) showed in the human subjects, given 50:50 (cis-9, trans-11 and trans-10, cis-12) CLA for a period of 12 weeks, a decrease in body fat by 3.8%, Zambell *et al.* (2000) found no effect in women fed 3 g CLA per day for 9 weeks. Blankson *et al.* (2000) observed a decrease in body fat mass in overweight or moderately obese men and women when consumed 3.4 or 6.8 g of a CLA preparation containing equal proportions of the cis-9, trans-11 and trans-10, cis-12 isomers for 12 weeks. These results are in contrast to those of Atkinson (1999), who found no difference between the CLA group (2.7g CLA/d) and the placebo group in a 6-mo placebo-controlled, randomized double-blind study in obese volunteers.

Mechanism of Action for Body Fat Reducing Effect of CLA

The ability of CLA to reduce adipose tissue could be explained by one or more of the following mechanisms: the induction of adipocyte apoptosis (Evans *et al.*, 2000), reduced accumulation of fatty acids in adipocytes due to an inhibition of lipoprotein lipase and increase in Carnitine palmitoyltransferase (Park and Pariza, 2001), the binding to peroxisome proliferator-activated receptor-??present in fat tissue and modification of the signaling cascade to down-regulate the expression of leptin (Kallen and Lazar, 1996) and the prevention of the triacylglycerol accumulation in adipocytes (Granlud *et al.*, 2003), or the modification of the energy expenditure, the metabolic rate, or both.

Several *in vitro* studies have reported an increase in lipolysis in response to CLA. CLA has been shown to inhibit lipoprotein lipase activity and stimulate lipolysis *in vitro* (Park *et al.*, 1997). Associated with their fat reduction properties, CLA potentially enhance the channeling of the fatty acids into the mitochondrial b-oxidation pathway in mice through an increase in Carnitine palmitoyltransferase (CPT) activity in epididymal fat pad and skeletal muscle of mice (Park *et al.*, 1997; Sagwal and Kansal, in press). Based on these findings it was proposed that the physiological mechanism of body fat reduction in mice by CLA involved inhibition of fat storage in adipocytes coupled with both elevated b-oxidation in skeletal muscle and epididymal fat pad.

CLA and Atherosclerosis

Effects of CLA supplementation on plasma lipids are highly inconsistent. A couple of animal studies suggested that CLA supplementation decreased plasma triglycerides and cholesterol and development of early atherosclerotic lesions in rabbit and hamsters (Lee *et al.*, 1994a; Nicolosi *et al.*, 1997). A subsequent studies reported increase or no change in plasma cholesterol and triacylglycerol in rabbits (Kritchevsky *et al.*, 2000) and mice (Sagwal and Kansal, in press).

Similarly, data on the role of individual isomer of CLA in altering lipid profile and atherosclerosis is highly inconsistent. In one study (McLeod *et al.*, 2004), both isomers of CLA, cis-9, trans-11 and trans-10, cis-12 (when fed alone) were effective in reducing atherosclerotic lesions in hamsters. However, the plasma triacylglycerol and HDL-cholesterol concentrations were higher in animals supplemented with trans-10, cis-12 isomer than in linoleic acid supplemented animals, and the differences were not significant between cis-9, trans-11 isomer and linoleic acid supplemented groups. In another study with hamsters fed with cis-9, trans-11 CLA or trans-10, cis-12 CLA or with an equal mixture of both. The mixed isomer preparation and the trans-10, cis-12 CLA preparation decreased LDL-cholesterol and HDL-cholesterol but increased VLDL-triacylglycerols. The cis-9, trans-11 CLA by itself had no significant effect on plasma lipids.

Human Studies

Blankson *et al.* (2000) found that the cholesterol level (total cholesterol, LDL and HDL) was significantly reduced in human subjects treated with the two-isomer preparation (cis-9, trans-11 and trans-10, cis-12) for 12 wk. Mougios *et al.* (2001) showed that 8 wk of CLA supplementation decreased plasma cholesterol, triacylglycerols, and HDL-cholesterol in normolipidemic volunteers. However, other studies found no significant effect on either HDL or LDL (Noone *et al.*, 2002). A significant decrease in plasma triglycerides concentrations with the 50:50 blend mixture of the two isomer preparation and a significant reduction in the VLDL-cholesterol with the 80:20 preparation enriched with the cis-9, trans-11 isomer have been observed (Noone *et al.*, 2002). Benito *et al.* (2001), however, showed no beneficial effect in terms of plasma lipoproteins and tissue fatty acid composition in humans. It is possible that the effect may not be visible in normolipidemic subjects and one may get better results, if studies are done on unhealthy subjects. Besides this, subject under treatment may already have a certain amount of CLA accumulated in their body and already have a cardioprotective lipid profile, and additional CLA may not exert the effect. If the effect of a dietary modification is to be seen in humans, the studies should be of longer periods.

CLA Acid and Cancer

Pariza *et al.* (1979) were first to report the anticarcinogenic property of CLA, when one fraction of the extract of heated ground beef containing a mixture of four isomers of CLA inhibited mutagenesis. Subsequently, it was tested in the benzo(a)pyrene-induced mouse forestomach neoplasia model, wherein it significantly reduced tumor incidence as well as tumor multiplicity in mice (Ha *et al.*, 1990).

Dietary CLA inhibited mammary tumors induced by 7,12-dimethylbenz(a)anthracene (DMBA) in rats in a dose dependent manner up to 1.0 % CLA in the diet (Ip *et al.*, 1991). A lower dose of DMBA (5 mg/animal), as little as 0.1 % CLA in diet was sufficient to reduce significantly the number of tumors (Ip *et al.*, 1994). When CLA was removed from the diet, CLA in neutral lipids and phospholipids returned to basal levels in about 4 and 8 weeks, respectively The rate of disappearance of neutral lipid CLA (rather than phospholipid CLA) subsequent to CLA withdrawal paralleled more closely the rate of occurrence of new tumors in the target tissue, thus suggesting that neutral lipid CLA might be more sensitive marker of tumor protection than phospholipid CLA (Ip *et al.*, 1999).

Inhibitory potential of CLA in human breast cancer is not well established. Postmenopausal breast cancer cases reported lower dietary intakes of CLA and had lower serum CLA concentrations than did the controls (Aro *et al.*, 2000). Other studies, however, failed to confirm this association (Chajes *et al.* 2002).

Mechanism of CLA Action in Inhibition of Carcinogenesis

Number of studies shows that the CLA suppresses the cancer by reducing the tumor cell proliferation. Dietary CLA reduced the bromodeoxy labeling index (proliferation index) in lobuloalveolar compartment of mammary gland (Ip *et al.*, 1994) and decreased the proliferation of terminal end bud and lobuloalveolar bud structures, the site at which tumor forms in both rat and human mammary cancer.

Number of studies show that CLA could reduce tumor cell proliferation by modifying cell cycle proteins that regulate this process (Belury,2002b). Feeding CLA up regulated the expression of p53, a tumor suppressor gene involved in monitoring the quality of DNA after G1 phase and, if DNA is damaged, it will block entry of the cell into S phase by altering the expression of genes involved in growth arrest and promotion. Ip *et al.* (2001) demonstrated that feeding CLA or c-9, t-11 CLA rich butter fat for 4 weeks reduced the expression of cyclins D and A in the terminal end buds and alveolar clusters of the mammary epithelium.

Apoptosis

Apoptosis or programmed cell death appears to be an important mechanism in deletion of tumor cells rather than increased cell proliferation. Feeding CLA has been reported to induce apoptosis in mammary (Ip *et al.*, 2000), colon (Park *et al.*, 2001) and adipose (Tsuboyama-Kasaoka *et al.*, 2000) tissues. Conjugated linoleic acid may induce apoptosis either by down regulating ErB3 signalling and the phosphoinositide 3-kinase (PI3K) and serine/threonine kinase (Akt) pathway (Cho *et al.*, 2003) or by decreasing expression of bcl-2, a gene involved in suppression of apoptosis (Ip *et al.*, 2000).

CLA Modulates via Prostaglandins Synthesis

The anticarcinogenic action of CLA may, in part be explained by its ability to inhibit arachidonic acid derived prostaglandin, since mammary tumors can be inhibited by agents that interfere with the archidonic acid cascade. Conjugated linoleic acid reduces the arachidonic-derived prostaglandin by one of the two mechanisms. First, the CLA displaces the arachidonate metabolites with (Thompson *et al.*, 1997) conjugated $C_{18:3}$ and $C_{20:3}$ fatty acids in mammary tissue. These unusual isomers of PUFA may inhibit the metabolism of linoleic acid and could also be precursors of unusual prostaglandins or inhibit the synthesis of normal prostaglandins derived from linoleic acid arachidonic acid cascade. These normal prostaglandins have been implicated in mammary tumor development.

A second explanation for reduction of arachidonate-derived prostaglandins by CLA may be through inhibition of cyclooxygenase (COX) enzyme. Cyclooxygenase catalyzes the rate-limiting step in the synthesis of prostaglandin from arachodonic acid. The isoform COX-2 is up regulated in response to growth factors, tumor promoters and cytokines (Herschaman, 1996). Prostaglandins produced by COX-2 (e.g., PGE2) induce inflammation and are potent mediators of a number of signal transduction pathways that are implicated in carcinogenesis (Vane *et al.*, 1998).

Cyclooxygenase-2 is related to cancer outlook through direct and indirect mechanisms. Prostaglandins may directly stimulate mitogenesis through a direct effect on fibroblasts, osteoblasts, and mammary cells. Cyclo-oxygenase-2 indirectly affects mutagenesis, angiogenesis, and increased cell migration and apoptosis as reviewed by Mazhar *et al.* (2006).

Peroxisome Proliferators Activated Receptor- Gamma (PPAR-g)

One potential mediator of the effects of specific fatty acids and fatty acid metabolites on mammary tumorigenesis is peroxisome proliferators activated receptor-gamma (PPAR-g) a ligand activated transcription factor expressed in normal and malignant mammary epithelial cells (Yee *et al.*, 2003). PPAR-g ligands have been reported to induce cell differentiation and apoptosis in several types of cancer, suggesting potential application as anticancer agents (Heaney *et al.*, 2003). Conjugated linoleic acid (CLA) is a high affinity ligand and an activator of PPAR-g (McCarty, 2000). When treated with CLA, PPAR-g expression is increased and adenomatous polyposis coli (APC) and c-myc (cellular- viral oncogene of the avian myelocytomatoris retroviruses) proteins are down-regulated in the human colon cancer cells, and finally proliferation of cancer cells is inhibited (Bozzo *et al.*, 2007).

Suggested Reading

Aro, A., Mannisto, S., Salminen, I., Ovaskainen, M.L., Kataja, V. and Uusitupa, M. 2000. Inverse association between dietary and serum conjugated linoleic acid and risk of breast cancer in postmenopausal women. *Nutr. Cancer*, **38**: 151-157.

Belury, M.A. 2002b. Inhibition of carcinogenesis by conjugated linoleic acid : Potential mechanisms of action. *J. Nutr.*, **132**: 2995-2998.

Benito, P., Nelson, G.J., Kelley, D.S., Bartolini, G., Schmidt, P.C. and Simon, V. 2001. The effect of conjugated linoleic acid on plasma lipoproteins and tissue fatty acid composition in humans. *Lipids*, **36**: 229-236.

Blankson, H., Stakkestad, J.A., Fagertun, H., Thom, E., Wadstein, J. and Gudmundsen, O. 2000. Conjugated linoleic acid reduces body fat mass in overweight and obese humans. *J. Nutr.*, **130**: 2943-2948.

Chajes, V., Lavillonniere, F., Ferrari, P., Jourdan, M.L., Pinault, M., Maillard, V., Sebedio, J.L. and Bougnoux, P. 2002. Conjugated linoleic acid content in breast adipose tissue is not associated with the relative risk of breast cancer in a population of French patients. *Cancer Epidemiol. Biomarkers Prev.*, **11**(7): 672-673.

Cho, H.J., Kim, W.K., Kim, E.J., Jung, K.C., Park, S., Lee, H.S., Tyner, A.L. and Park, J.H. 2003. Conjugated linoleic acid inhibits cell proliferation and ErbB3 signaling in the HT-29 human colon cell line. *Am. J. Physiol. Gastrointest. Liver Physiol.*, **284**(6): G996-G1005.

Cook, M.E., Jerome, D.L., Crenshaw, T.D., Buege, D.R., Pariza, M.W., Albright, K.J., Schmidt, S.P., Scimeca, J.A., Lofgren, P.A. and Hentges, E.J. 1998. Feeding conjugated linoleic acid improves feed efficiency and reduces whole body fat in pigs. *FASEB J.*, **12**: A836.

Evans, M., Geigerman, C., Cook, J., Curtis, L., Kuebler, B. and McIntosh, M. 2000. Conjugated linoleic acid suppresses triglyceride accumulation and induces apoptosis in 3T3-L1 preadipocytes. *Lipids*, **35**: 899-910.

Granlud, L., Juvet, L., Pedersen, J. and Nebb, H. 2003. Trans10, cis12-conjugated linoleic acid prevents triacylglycerol accumulation in adipocytes by acting as a PPAR gamma modulator. *J. Lipid Res.*, **44**: 1441-1452.

Griinari, J. M. and Bauman, D. E. 1999. Biosynthesis of conjugated linoleic acid and its incorporation into meat and milk in ruminants. In: Advances in Conjugated Linoleic Acid Research (M.P. Yuraweez, M.M. Mossoba, J.K.G. Kramer, M.W. Pariza and G. Nelson, eds.). Am. Oil Chem. Soc., Champaign, IL, pp.180-200.

Ha, Y.L., Storkson, J. and Pariza, M.W. 1990. Inhibition of benzo(a)-pyrene-induced mouse forestomach neoplasia by conjugated dienoic derivatives of linoleic acid. *Cancer Res.*, **50**: 1097-1101.

Heaney, A.P., Fernando, M. and Melmed, S. 2003 PPAR-gamma receptor ligands: novel therapy for pituitary adenomas. *J Clin Invest.*, **111**: 1381-1388.

Herschman, H.R. 1996. Prostaglandin synthase2. *Biochimica et Bophysica Acta*, **1299**: 125-140.

Ip, C., Banni, S., Angioni, E., Carta, G., McGinley, J., Thompson, H.J., Barbano, D. and Bauman, D. 1999. Conjugated linoleic acid – enriched butter fat alters mammary gland morphogenesis and reduces cancer risk in rats. *J. Nutr.*, **129**: 2135-2142.

Ip, C., Chin, S.F., Scimeca, J.A. and Pariza, M.W. 1991. Mammary cancer prevention by conjugated dienoic derivative of linoleic acid. *Cancer Res.*, **51**: 6118-6124.

Ip, C., Dong, Y., Thompson, H.J., Bauman, D.E. and Ip, M.M. 2001. Control of rat mammary epithelium proliferation by conjugated linoleic acid. *Nutr. Cancer.*, **39**: 233-238.

Ip, C, El-Bayoumy, K., Upadhyaya, P., Ganther, H., Vadhanavikit, S. and Thompson, H. 1994. Comparative effect of inorganic and organic selenocyanate derivative in mammary cancer chemoprevention, *Carcinogenesis.*, **15:** 187-192.

Ip, C., Ip, M.M., Loftus, T., Shoemaker, S. and Shea-Eaton, W. 2000. Induction of apoptosis by conjugated linoleic acid in cultured mammary tumor cells and premalignant lesions of the rat mammary gland. *Cancer Epidemiol. Biomarkers Prev.*, **9**: 689-696.

Kallen, C. and Lazar, M. 1996. Antidiabetic thiazolidinediones inhibit leptin ob gene expression in 3T3-L1 adipocytes. *Proc. Natl. Acad. Sci. USA*, **93**: 5793-5796.

Kritchevsky, D., Tepper, S.A., Wright, S., Tso, P. and Czarnecki, S.K. 2000. Influence of conjugated linoleic acid (CLA) on establishment and progression of atherosclerosis in rabbits. *J. Am. Coll. Nutr.*, **19**: 472S-477S.

Lee, K.N., Kritchevsky, D. and Pariza, M.W. 1994a. Conjugated linoleic acid and atherosclerosis in rabbits. *Atherosclerosis*, **108**: 19-25.

Mazhar, D., Ang, R. and Waxman, J. 2006. COX inhibitors and breast cancer. *British J. cancer*, **94:** 346-350.

McCarty, M.F. 2000. Activation of PPAR gamma may mediate a portion of the anticancer activity of conjugated linoleic acid. *Med. Hypotheses*, **55**: 187-188.

Mougios, V., Matsakas, A., Rings, S., Petridou, A., Sagredos, A., Melissopoulou, A. and Tsigilis, N. 2001. Effects of supplementation with conjugated linoleic acid on human body fat and serum lipids. *J. Nutr. Biochem.*, **12**: 585-594.

Nicolosi, R.J., Rogers, E.J., Kritchevsky, D., Scimeca, J.A. and Huth, P.J. 1997. Dietary conjugated linoleic acid reduces plasma lipoproteins and early aortic atherosclerosis in hypercholesterolemic hamsters. *Artery*, **22**: 266-277.

Noone, E.J., Roche, H.M., Nugent, A.P. and Gibney, M.J. 2002. The effect of dietary supplementation using isomeric blends of conjugated linoleic acid on lipid metabolism in healthy human subjects. *Br. J. Nutr.*, **88**: 243-251.

Pariza, M.W., Ashoor, S.H., Chu, F.S. and Lund, D.B. 1979. Effects of temperature and time on mutagen formation in panfried hamburger. *Cancer Lett.*, **7**: 63-69.

Park, Y. and Pariza, M.W. 2001. the effects of dietary conjugated nonadecadienoic acid on body composition in mice. *Biochem. Biophys. Acta*, **1533**: 171-174.

Park, Y., Albright, K.J., Liu, W., Storkson, J.M., Cook, M.E. and Pariza, M.W. 1997. Effect of conjugated linoleic acid on body composition in mice. *Lipids*, **32**: 853-858.

Sagwal, R and Kansal, V. K. Synergistic effect of synthetic conjugated linoleic acid and non-fat milk on fat ddeposition and lipid metabolism in mice, Ind. J. Med. Res. (in press)

Smedman, A. and Vessby, B. 2001. Conjugated linoleic acid supplementation in humans-Metabolic effects. *Lipids*, **36**: 773-781.

Thompson, H., Zhu, Z., Banni, S., Darcy, K., Loftus, T. and Ip, C. 1997. Morphological and biochemical status of the mammary gland as influenced by conjugated linoleic acid: Implication for a reduction in mammary cancer risk. *Cancer Res.*, **57**: 5067-5072.

Tsuboyama-Kasaoke, N., Takahashi, M., Tanemura, K., Kim, H.-J., Tange, T., Okuyama, H., Kasai, M., Ikemoto, S. and Ezaki, O. 2000. Conjugated linoleic acid supplementation reduces adipose tissue by apoptosis and develops lipodystrophy in mice. *Diabetes*, **49**: 1534-1542.

Vane J. R., Bakhle Y. S. and Botting R. M. 1998. Cyclooxygenases 1 and 2. *Annu Rev Toxicol* **38:** 97-120.

Yee, L. D., Guo, Y., Bradbury, J., Suster, S., Clinton, S. K. and Seewaldt, V. L. 2003. The antiproliferative effects of PPAR-g ligands in normal human mammary epithelial cells. *Breast Cancer Res. Treat.*, **78**: 179- 192.

West, D.B., Blohm, F.Y., Truett, A.A. and DeLany, J.P. 2000. Conjugated linoleic acid persistently increases total energy expenditure in AKR/J mice without increasing uncoupling protein gene expression. *J. Nutr.*, **130**: 2471-2477.

Zambell, K.L., Keim, N.L., Van Loan, M.D., Gale, B., Benito, P., Kelley, D.S. and Nelson, G.J. 2000. Conjugated linoleic acid supplementation in humans: Effects on body composition and energy expenditure. *Lipids*, **35**: 777-782.

□□□

Chapter 6

Functional Indian Traditional Dairy Products

DHARAM PAL AND P. NARENDER RAJU
Dairy Technology Division, National Dairy Research Institute, Karnal (HR)

Introduction

With the changing lifestyle and dietary patterns, the occurrence of non-communicable diseases like obesity, diabetes, cardiovascular diseases and cancer have increased worldwide. The major health consequences associated with overweight and obesity are type-2 diabetes, coronary heart diseases, hypertension, gall bladder, certain types of cancer, dyslipidaemia and insulin resistance (Ramachandran, 2004). The World Health Organization estimated that worldwide approximately 1.6 billion adults (age 15+), at least 400 million adults and 180 million people were overweight, obese and diabetes, respectively in the year 2005 (WHO, 2006 a & b). In India non-communicable diseases caused 5.10 million deaths in the year 2002, of which cardiovascular diseases were responsible for 2.78 million deaths (Beaglehole and Yach, 2003). The dietary factors such as high intake of fats, sugars, milk and its products and low intake of fruits and vegetables were ascribed for the role in the cardiovascular disease mortality (Gupta *et al.*, 2006).

Traditionally, food products have been developed for taste, appearance, value, and convenience of the consumer. The development of new products to confer a health benefit is a relatively new trend, and recognizes the growing acceptance of the role of diet in disease prevention and treatment. Among different terms that have been used to describe the many natural products currently being developed for health benefit, viz. nutraceuticals or functional foods are the commonly used terms. However, both the terms are often used interchangeably. In Japan, where interest in functional foods has been growing since the early 1980s, these foods are defined as "Foods for Specified Health Use" (Goldberg, 1994; Bawa and Khanum, 2003). In Finland, such products are registered as PARNUT (Foods for Particular Nutritional Purposes) products (Moreau, 2004). In India, we do not have definition for functional foods under the statutory requirements. The different food ingredients that are recognized as functional ingredients are dietary fiber; oligosaccharides; sugar alcohols; polyunsaturated fatty acids; bioactive proteins and peptides; glycosides, isoprenoides and vitamins; alcohols and phenols; cholines; lactic acid bacteria; minerals and others.

Indian traditional dairy products and sweets have great mass consumer base. These products are an integral part of Indian culture and have great social, religious, cultural, medicinal and economic importance. Each of these products has its unique flavor, texture, and appearance. In general, most of the Indian traditional dairy products contain high fat and high sugar. With the growing awareness of the link between diet and health among consumers, the demand for low-fat, low-sugar, fiber and mineral fortified health-based dairy products is increasing. With the advancement in the functional ingredients it has now become possible to manufacture traditional dairy products that have specific health benefit without compromising the taste and texture (Pal, 2008). Products like probiotic *dahi*, diabetic *rasogolla* and *burfi*, fruit *lassi* and *shrikhand* are already available in the market. Some of these functional traditional dairy products have been discussed in the present paper.

Dietetic *Burfi*

Burfi is the highly popular *khoa*-based confection prepared throughout India. Sugar and other ingredients are added in different proportions to *khoa* according to the demand of consumers. Different types of *burfi* for example plain, chocolate, *kaju*, *pista*, *kesar*, etc. are prepared and each has its typical organoleptic attributes depending on the type of the ingredient used. However, a product with high fat that impart creamy (rich) taste and often much sugar is preferred by consumers. That is why *burfi* prepared from buffalo milk has superior quality than that made from cow milk. Good quality *burfi* is characterized by moderately sweet taste, soft and slightly greasy body and smooth texture with very fine grains which is attained from buffalo milk *khoa*. Colour, unless it is chocolate *burfi*, should be white or slightly yellowish. Traditionally *burfi* is prepared by adding sugar to hot *khoa* and vigorous blending in a shallow kettle till a homogenous, smooth and fine grains mass is achieved. In hot condition it is spread in shallow trays for setting (Pal, 2000).

Burfi contains about 15% moisture, 20.5% fat, 30% sugar, 16.2% protein, 15.8% lactose and 2.4% ash (Palit, 1998). Since *burfi* contains high amounts of fat and sugar, the obese, diabetics and people on weight management diets cannot relish the taste of this delicacy. With a view to make *burfi* an acceptable delicacy to obese and diabetics, successful attempts were made by Prabha (2006) to develop a technology for the production of sugar-free *burfi* prepared from buffalo milk having 3.5% fat. The various ingredients used to prepare this dietetic *burfi* are: whey protein concentrate (WPC), sorbitol, maltodextrin and sucralose. Response surface methodology (RSM) was used for standardization of the levels of different ingredients. The product was found to be highly acceptable by the consumers. Shelf life studies revealed that vacuum packaged dietetic *burfi* can be stored without spoilage for 12 days at 30°C and 40 days at 5°C.

Dietetic *Rasogolla*

Rasogolla, a chhana-based delicacy, is also extremely popular throughout the country. Now-a-day many branded canned *rasogolla* are being exported to the countries where large sub-continent population has migrated. *Rasogolla* balls made by kneading soft *chhana*, are cooked as well as stored in sugar syrup. Snow-white in colour, *rasogolla* possesses a spongy and chewy body and smooth and juicy texture. Soft and freshly made cow milk *chhana* is most suitable for *rasogolla* preparation. *Rasogolla* contains about 32–55% sugar which is of major concern to the calorie conscious and diabetic people. Jayaprakash (2003) developed a technology for preparing low-calorie or diabetic *rasogolla* using a high intensity sweetener (aspartame) and a bulking agent (sorbitol). Attempts were also made to develop functional *rasogolla* by fortifying *chhana* with selective functional ingredients such as omega-3 fatty acids and bioactive peptides to impart health attributes (Natarajan, 2007). In that study flaxseed oil (2%) and WPC (4%) were used as sources of omega-3 fatty acids and bioactive peptides, respectively.

Functional *Dahi*

Dahi (curd) is a well known fermented milk product consumed throughout India, either as a part of the daily diet along with the meal or as a refreshing beverage or it may be converted into *raita* (seasoned with onion, spices, etc.). Being a widely consumed dairy product, *dahi* was chosen as a vehicle by some workers to incorporate different nutraceuticals and combat chronic and non-communicable diseases in India. Fortification with minerals viz. calcium (Singh *et al.*, 2005; Ranjan *et al.*, 2006), incorporation of dietary fiber in the form of fruits (Pandya, 2002) and incorporation of probiotic organisms (Yadav *et al.*, 2005) are some of the applications reported. Studies conducted on calcium fortification of cow and buffalo milks for *dahi* making revealed that among the three salts studied viz. calcium chloride, calcium lactate and calcium gluconate, the quality of *dahi* made from cow milk enriched with calcium gluconate (Singh *et al.*, 2005) and buffalo milk enriched with calcium gluconate and calcium lactate (Ranjan *et al.*, 2006) were comparable with the *dahi* made from the non-calcium fortified milk.

Artificially Sweetened *Misti Dahi*

Misti dahi, also called as *mishti doi* or *lal dahi* or *payodhi* is a sweetened variety of *dahi* popular in Eastern India. It is characterized by a creamish to light brown color, firm consistency, smooth texture and pleasant aroma. Various market survey reports on the quality of *misti dahi* sold in different parts of the country revealed wide variations in the fat (1-12%) and cane sugar (6-25%) contents. Besides several useful virtues as a fermented dairy product, *misti dahi* contains high milk fat and high sugar, which are causes of concern for health conscious consumers. Recently, a technology has been developed to manufacture reduced fat *misti dahi* from buffalo milk (Raju and Pal, 2008). It was reported that highly acceptable reduced fat *misti dahi* can be produced from buffalo milk standardized to 3.0% milk fat and 15.0% milk solids-not-fat (MSNF) and addition of 14% cane sugar and 0.1% caramel powder. Further studies led to the successful development of artificially sweetened *misti dahi* from reduced fat buffalo milk using high intensity sweeteners such as aspartame and acesulfame-K in binary combination along with a bulking agent (Raju and Pal, 2007).

Others

Kumar (2000) developed *lassi* for calorie-conscious and diabetic people using an artificial sweetener and reported that aspartame at a rate of 0.08% on curd basis was the most acceptable level to prepare low-calorie *lassi*. Recently, Khurana (2006) developed suitable technologies for the manufacture of mango, banana and pineapple *lassi* along with their low-calorie counterparts using artificial sweeteners.

Suggested Reading

Beaglehole, R. and Yach, D. (2003). Globalization and the prevention and control of non-communicable disease: the neglected chronic disease of adults. *Lancet*. **362**: 903-908.

Gupta, R., Misra, A., Pais, P., Rastogi, P. and Gupta, V. P. (2006). Correlation of regional cardiovascular disease mortality in India with lifestyle and nutritional factors. *International Journal of Cardiology*. **108**: 291-300.

Jayaprakash, K. T. (2003). Technological studies on the manufacture of *rasogolla* using sweeteners. M.Tech. Thesis submitted to National Dairy Research Institute (Deemed University), Karnal.

Natarajan, A. M. (2007). Development of low calorie rasogolla fortified with functional ingredients with therapeutic value. Souvenir of the 19th Indian Convention of Food Scientists and Technologists held at Indian Institute of Technology, Kharagpur from 31st Dec, 2007 to 2nd Jan, 2008. pp: 41-43.

Pal, D. (2000). Technological advances in the manufacture of heat desiccated traditional Indian milk products. *Indian Dairyman*. **52**(10): 27-35.

Pal, D. (2008). New innovations in the processing of traditional Indian dairy products. *Indian Dairyman*. **60**(3): 127-131.

Palit, C. (1998). Application of selected unit processes for commercial production of *burfi*. M.Tech. Thesis submitted to National Dairy Research Institute (Deemed University), Karnal.

Pandya, C. N. (2002). Development of technology for fruit *dahi*. M.Tech. Thesis submitted to National Dairy Research Institute (Deemed University), Karnal.

Prabha, S. (2006). Development of a technology for the manufacture of dietetic *burfi*. Ph.D. Thesis. National Dairy Research Institute (Deemed University), Karnal.

Raju, P. N. and Pal, D. (2007). Effect of bulking agents on the quality of artificially sweetened *misti dahi* prepared from reduced fat buffalo milk. Poster presented at the 19th Indian Convention of Food Scientists and Technologists held at Indian Institute of Technology, Kharagpur from 31st Dec, 2007 to 2nd Jan, 2008.

Raju, P. N. and Pal, D. (2008). The physico-chemical, sensory and textural properties of *misti dahi* prepared from reduced fat buffalo milk. *Food and Bioprocess Technoogy*. DOI 10.1007/s11947-008-0137-z. (http://www.springerlink.com)

Ramachandran, A. (2004). Diabetes and obesity – The Indian angle. *Indian Journal of Medical Research*. **120**: 437-439.

Ranjan, P., Arora, S., Sharma, G. S., Sindhu, J. S., and Singh, G. (2006). Sensory and textural profile of curd (*dahi*) from calcium enriched buffalo milk. *Journal of Food Science and Technology*. **43**(1): 38-40.

Singh, G., Arora, S., Sharma, G. S., Sindhu, J. S., Rajan, P. and Kansal, V. K. (2005). Sensory and textural profile of *dahi* from calcium enriched cow milk. *Indian Journal of Dairy Science*. **58**(1): 23-26.

Yadav, H., Jain, S. and Sinha, P. R. 2005. Preparation of low fat probiotic *dahi*. *Journal of Dairying, Foods and Home Science*. **24**(3/4): 172-177.

□□□

Chapter 7

Vitamin Binding Proteins in Milk: Natural Functional Ingredients

RAJAN SHARMA[1] AND Y.S. RAJPUT[2]

[1] Dairy Chemistry Division,[2] Animal Biochemistry Division, NDRI, Karnal (HR)

Introduction

Vitamins play a very important role in the human health and nutrition, and their deficiency in the diet is responsible for numerous diseases. Milk and milk products have played an important role in the recognition and early development of knowledge of the vitamins. By now, it is well established that milk contains all the known vitamins, although some are present in small quantities, but for others, milk is very rich source - indeed one of the best. Milk contains proteins which can specifically bind folic acid, vitamin B_{12}, retinol, vitamin D and riboflavin and accordingly these proteins are named. There are evidences that these proteins facilitate the absorption of the appropriate vitamin and presumably have a specific function.

Folic acid binding proteins (FBPs)

The existence of a 'milk folate binder' was originally suggested by Ghitis et al (1969). FBPs in bovine milk exist in concentration about 8 mg/L (Salter *et al.*, 1972) and binds approximately 1 mol of folate per mol protein at pH 7.2 (Salter *et al.*, 1981). In a recent study (Nygren *et al.*, 2004), the FBPs levels have been shown to vary between 0.3 to 11 mg/L in bovine milk and were observed to be high (up to 44 mg/L) in beginning of lactation but levels off after 30 days of lactation. The complete amino acid sequence of this minor whey protein was determined by Svendsen *et al* (1984) and the molecule consists of a single polypeptide chain of 222 amino acid residues and eight disulphide bridges. The protein contains about 10% carbohydrate, attached to Arg residues, probably 49 and 141, and had a molecular weight of ~ 30,000 Da, including carbohydrate. Bovine milk FBPs and FBPs purified from human milk exhibit a high degree of amino acid sequence homology (Svendesen *et al.*, 1982).

Folates in milk are completely bound by an excess of FBPs (Nygren *et al.*, 2003). The main function of FPBs is to sequester folate from blood to secure an adequate supply to the neonate. (Parodi, 1997). Several reports have shown that FBPs increases folate absorption in the gut (Ek & Manus, 1980; Colman *et al.*, 1981; Tani and Iwai, 1984; Jones *et al*, 2003). It has been found that folate bound to FBPs is less available to bacteria species, which colonise the gastrointestinal tract. This prevents folate from being used by bacteria, hence makes more folate available for absorption (IDF, 1991). Folate deficiency occurring during early pregnancy is known to cause neural tube defects (van der *et al.*, 2001). There is also increasing evidence of a connection between folate intake and occlusive vascular diseases (Selhub *et al.*, 1996). This has resulted in renewed interest in FBPs in recent years due to its positive effects on the bioavailability of folate and thus the ability to reduce risk of folate deficiency. Studies have indicated that nearly 50% of the FBPs exist in milk as apo-form and has been postulated that an excess of apo-FBPs from milk might be expected to bind ingested folates from other foods (and supplements) in the intestine, and thus increase the bioavailability (Nygren *et al.*, 2004). In one of the recent study, it has been observed that milk products are an appropriate matrix for fortification because of the high bioavailability of synthetic folic acid and 5-methyltetrahydrofolate added milk products. However, addition of excess of FBPs to milk decreases the bioavailability of folate (Verwei *et al.*, 2005).

Pasteurization of milk has been reported to result in heat-induced alterations of the FBPs molecule and thus its antibacterial activity is reduced (Ford, 1977; Gregory, 1982).

Vitamin B_{12} Binding Protein

Three different proteins are required for the uptake of vitamin B_{12} (cobalamin) in mammals. Gastric intrinsic factor (IF) binds the free vitamin released from food by digestion. The IF-cobalamin complex then enters the ileal mucosal cells by a receptor – mediated mechanism where the vitamin is transferred to another protein, transcobalin (TC). The TC-cobalamin complex and unsaturated TC are released into the portal plasma along with the third cobalamin binding protein, haptocorrin (HC), the function of which in plasma is not clear (Haggarty, 2003; Rosenberg and Selhub, 2006).

In 1954, Gregory showed that vitamin B12 in the milks of cow, goat, sheep, sow, human and rat was almost entirely bound to a whey protein which was partially purified from porcine milk and characterized by Gregory and Holdsworth (1955 a, b). The vitamin B_{12}-binding protein was isolated from human milk by Burger and Allen (1974) using affinity chromatography on B_{12}-Sepharose. The protein which was R-type B_{12} binding protein, had a molecular weight of 61,000 - 63,000 and contained ~ 35% carbohydrate. Later, Sandberg *et al* (1981) indicates that there are two vitamin B_{12} binding proteins in human milk, the principle one being R-type to which most or all of the vitamin B_{12} is bound, and the other being transcobalamin II, which is present at a low concentration and is devoid of cobalamin. In cow milk, Fedosov et al (1996) have purified and characterized cobalamin binding protein and they have classified this protein as TC rather than HC because N terminal sequence revealed homology between the isolated milk protein and TC from man and rabbit. The possible nutritional significance of this protein was investigated by Ford (1974). The protein probably plays two roles: aids absorption of vitamin B_{12} and acts as a bactericidal agent. Animal experiments indicate that protein-bound vitamin B_{12} is efficiently absorbed from the intestine, both before and for some time after cessation of transport of intact protein across the neonatal gut epithelium (Ford *et al.*, 1975; Trugo *et al.*, 1985). However, Sandberg et al (1981) suggest that cobalamin must be released from its binding protein by proteolysis and then combined with an intrinsic factor for absorption. Trugo and Turvey (1987) showed that vitamin B_{12} binding protein attaches to the brush border membrane where it stimulates absorption of vitamin B_{12}. Human milk has the capacity to bind around 43 ng cyanocobalmin/ ml but in milk most of the protein is unsaturated (Ford *et al.*, 1977). Human milk inhibits the uptake of cyanocobalamin by intestinal bacteria (Ford *et al.*, 1977). It appears likely that this protein protects vitamin B_{12} against bacterial uptake in the gut and may strongly influence the ecology of the intestinal microflora and the vitamin nutrition of the suckling infants (Ford, 1974). Recently Adkins and Lonnerdal, (2003) showed that human HC exposed to enzymes did not show a decrease in molecular weight, which indicated that HC can survive proteolytic degradation. It was also observed that HC exposed to digestive enzymes and undigested HC inhibited the growth of enteropathogenic *E.coli* and did so to a similar extent. Thus, HC in vitro not only retains its structure after exposure to proteases but also exhibits antimicrobial activity. It has been observed that HC exist in milk in an unsaturated form and thus have the potential role in protecting the neonate against pathogenic bacteria that require cobalamin for their survival.

The inhibitory effect of vitamin B_{12} binding protein against intestinal microorganisms is progressively lost on heating up to 75°C/ 15 min but binding capacity increases again after more severe heat treatment (Ford *et al.*, 1977). The authors suggest the minimum stability of vitamin B_{12} binding protein at 75°C may be due to the action of milk enzyme (presumably proteinase) which is itself denatured at higher temperature. They have also postulated that in refrigerated stored milk, there should be considerable loss of vitamin B_{12} binding protein due to the activity of native proteases. However, this fact was never verified.

Vitamin D-Binding Protein

Vitamin D-binding protein (DBP) is present in the plasma of most vertebrates. It is a glycoprotein with an apparent molecular weight of 52 000 Da. It has a strong structural homology with serum albumin and also, like albumin, has the ability to bind long-chain fatty acids. It has been detected in the mature milk of several species, but at much lower levels than in serum (~ 2%). The concentration of DBP is higher in bovine colsotrum and early milk than in mature milk; however, the protein is probably derived from serum as no synthesis of DBP has yet been observed in the mammary gland (Ena *et al.*, 1992; Haggarty, 2003).

Riboflavin Binding Protein

A riboflavin binding protein (RfBP), a phosphoglycoprotein has been reported in egg white yolk and blood of laying hens. This protein was found to be necessary for transport of riboflavin to the egg (Miller *et al.*, 1982, 1984). Later, similar protein has been reported in bovine milk (Kanno and Kanchara, 1985). The protein was purified by Toyosaki *et al* (1987) using DEAE cellulose, gel filtration and gel electrophoresis and the estimated molecular weight of protein was 38,000 Da. Since the RfBP of egg white and yolk are derived from blood plasma, it is likely that RfBP of milk also originates in the blood (Fox and Flynn, 1992). Riboflavin bound to RfBP has good antioxidant properties, similar to those of the protein isolated from avian egg where it has an important nutritional role (Toyosaki *et al* 1987; Toyosaki & Mineshita, 1988). The true physiological function of milk RfBP has not yet been determined.

Retinol Binding Protein

β-lactoglobulin in milk is capable of binding several hydrophobic molecules, including retinol, and it has been proposed that *in vivo* β- lactoglobulin functions to protect retinol against oxidation and to transport it through the stomach to the small intestine (Hambling *et al.*, 1992). The binding of retinol by bovine β-lactoglobulin has led to speculation about possible role of β-lactoglobulin in binding and protecting retinol in milk (Creamer and MacGibbon, 1996).The crystal structure of â-lactoglobulin shows a primary binding site in the central cavity of the molecule and other surface hydrophobic region capable of interacting with organic molecules and this binding property has also been exploited using this protein as a carrier to protect the stabililize organic compounds that are benificial to health and food quality (Jegasothy and Stockmann, 2003). It has also been postulated that β-lactoglobulin plays some role in absorption and transportation retinol in small intestine. The resistance of β- lactoglobulin to peptic hydrolysis is probably significant in this regard (Fox and Flynn, 1992).

Conclusion

The functions of many minor components present in milk are becoming known. Vitamin binding proteins, although present in milk in low amount, helps in improving bioavailability even of vitamins fortified to milk and milk products. In general, vitamin binding proteins in milk can bind over and above vitamins present in milk and therefore, milk can serve as better vehicle for vitamin fortification.

Suggested Reading

Adkins, Y., & Lonnerdal, B. (2003) Potential host-defense role of a human milk vitamin B_{12} binding protein, haptocorrin, in the gastrointestinal tract of breastfed infants, as assessed with porcine haptocorrin in vitro. *Am J. Clin. Nutr.*, **77:** 1234-40.

Creamer, L.K. & MacGibbon A.K.H. (1996). Some recent advances in the basic chemistry of milk proteins and lipids. *Int. Dairy J.* **6:** 539-568.

Ena, J.; Perez, M.D.; Aranda, P.;Sanchez, L. And Calvo, M. (1992). Presence and changes in the concentration of vitamin D-binding protein throughout early lactation in human and bovine colostrum and milk *The Journal of Nutritional Biochemistry*, **3:** 498-502

Fedosov, S.N.; Petersen, T.E. and Nexo, E. (1996). Transcobalamin from cow milk : isolation and physico-chemical properties. *Bioch. Biophys. Acta-Protein Structure and Molecular Enzymology*, **1292:** 113-119.

Ford, J. E. (1974) Some observations on the possible nutritional significance of vitamin B_{12} and folate-binding proteins in milk. *Br. J. Nutr.*, **31:** 243-57.

Ford, J. E., Law, B. A., Marshall, M. E., & Reiter, B. (1977) Influence of the heat treatment of human milk on some of its protective constituents. *J. Pediatrics*, **90:** 29-35.

Fox, P.F. and Flynn, A. (1992) Biological properties of milk proteins. In: Advanced Dairy Chemistry -1: Proteins, (Ed. P.F. Fox). Elsevier Science Publishers, London. Pp 255-284.

Fugate, R.D. & Song, P.S. (1980) Spectroscopic characterization of â-lactoglobulin-retinol complex. *Biochim. Biophys. Acta*, **625:** 28-42.

Ghitis J., Mandelbaum-Shavit, F. & Grossowicz, N. (1969) Binding of folic acid and derivatives by milk. *Am J Clin Nutr.*, **22:** 156-162.

Gregory, M. E. (1954) The microbiological assay of Vitamin B_{12} in the milk of different animal species. *Br. J. Nutr.*, **8:** 340-7.

Gregory, I. J. F. (1982) Denaturation of the folacin-binding protein in pasteurized milk products. *Journal of Nutrition*, **112:** 1329-1338.

Gregory, M. E.& Holdsworth, E. S. (1955a) The occurrence of a cyanocobalamin binding protein in milk and the isolation of a cyanocobalamin protein complex from sow's milk. *Biochem. J*, **59:** 329-34.

Gregory, M. E.& Holdsworth, E. S. (1955b) Some properties of the cyanocobalamin protein complex from sow's milk, and the mode of linkage of cyanocobalamin with protein. *Biochem. J*, **59:** 335-40.

Haggarty, N.W. (2003) Minor proteins, bovine serum albumin and vitamin binding proteins. In: Encyclopedia of Dairy Sciences (H.Roginski, J.W. Fuquay and P.F. Fox). Academic Press, London. pp 1942-1944

Hambling, S.G.; McAlpine, A .S. and Sawyer, L. (1992). â-lactoglobulin. In: Advanced Dairy Chemistry -1: Proteins, (Ed. P.F. Fox). Elsevier Science Publishers, London. Pp 141-90.

International Dairy Federation (1991) Significance of the indigenous antimicrobial agents of milk to the dairy industry. *Bulletin of the International Dairy Federation*, **264:** 2-19.

Jegasothy, H. and Stockmann, R. (2003) Binding of organic compound to β-lactoglobulin: use of the protein as a carrier. *Aust. J. Dairy Technol.* **58:** 191.

Jones, M.L., Treloar, T., & Nixon, P.F. (2003) Dietary interactions influence the effects of bovine folate-binding protein on the bioavailability of tetrahydrofolates in rats. *The Jorunal of nutrition*, **133**, 489-495.

Kanno, C. & Kanehara, N. (1985) Influence of riboflavin-binding protein in raw cow milk. Proc. Ann. Meeting Soc. Heric. Chem. Jpn, p.718 (Abstr.), cited from Toyosaki et al (1987).

Miller, M.S., Benore-Parsons, M. & White, H.B. (1982) Dephosphorylation of chicken riboflavin-binding protein and phosvitin decreases their uptake by oocytes. J. Biol. Chem., **257:** 6818-24.

Nygren, L., Sternesjo, A., & Bjorck L. (2003) Determination of folate-binding proteins from milk by optical biosensor analysis. *International Dairy Journal*, **13:** 283-290.

Nygren, L., Sternesjo, A., & Bjorck L. (2004) Factors influencing levels of folate-binding protein in bovine milk. *International Dairy Journal*, **14:** 761-765.

Parodi, P. W. (1997) Cow's milk folate binding protein: Its role in folate nutrition. *The AustralianJournal of dairy Technology*, **52:** 109-118.

Rosenberg, I.H. and Selhub, J (2006) Folate and vitamin B12 transport systems in the developing infant. *The Journal of Pediatrics*, **149:** S62-S6

Salter, D. N., Ford, J. E., Scott, K. J. & Andrews, P. (1972) Isolation of the folate-binding protein from cow's milk by use of affinity chromatography. *Febs Letter*, **20:** 302-306.

Salter, D. N., Scott, K. J., Slade, H. & Andrews, P. (1981) the preparation and properties of folate-binding protein from cow's milk. *Biochemical Journal*, **193:** 469-476.

Sandberg, D.P., Begley, J.A & Hall, C.A., (1981) The content, binding, and forms of vitamin B_{12} in milk. Am. J. Clin. Nutr., **34:** 1717-24.

Selhub, J., Jacques, P. F., Bostom, A. G., D' Angostino, R. B., Wilson, P. W. F., Belanger, A. J., O'Leary, D. H., Wolf, P. A., Rush, D., Schaefer, E. J., & Rosenberg, I. H. (1996) Relationship between plasma homosysteine, vitamin status and extracranial carotid-artery stenosis in the Franingham study population. *Journal of Nutrition*, **126:** 1258S-1265S.

Svendsen, I., Hansen, S. I., & Lyngbye, J., (1984) The complete amino acid sequence of the folate-binding protein from cow's milk. *Carlsberg Research communications*, **49:** 123-131.

Svendesen, I., Hansen, S. I., Holm, J., & Lyngbye, J.(1982) Amino acid sequence homology between human and bovine low molecular weight folate binding protein isolated from milk. *Carlsberg Research communications*, **47:** 371-376.

Tani, M., & Iwai, K. (1984) some nutritional effects of folate-binding protein in bovine milk on the bioavailability of folate to rats. *Journal of Nutrition*, **114:** 778-785.

Toyosaki, T., Yamamoto, A. & Mineshita, T. (1987) Partial purification of an antioxidizing component in raw cow milk. *J Food Sci.*, **52:** 88-90.

Toyosaki, T. & Mineshita, T. (1988) Antioxidant effects of protein-bound riboflavin and free riboflavin. *J Food Sci.*, **53:** 1851-3.

Trugo, N.M.F., Ford, J.E.,& Salter, D.N. (1985) Vitamin B_{12} absorption in the neonatal piglet. 3. Influence of Vitamin B_{12} binding protein from cows milk on uptake of Vitamin B_{12} by microvillus visicles prepared from small intestine of the piglet. *Br. J. Nutr.*, **54:** 269-83.

Trugo, N.M.F., & Turvey, A. (1987) Immunofluorescent localization of Vitamin B_{12} binding protein from sows milk in the ileal mucosa of piglets. Brazilian J. Med. Biol. Res., **20:** 285-8.

van der Put, N. M. J., van Straaten, H. W. M., Trijbels, F. J. M., & Blom, H. J. (2001) Floate, homocysteine and neural tube defects: An overview. *Experimental Biology and Medicine*, **226:** 243-270.

Verwei, V.; Arkbage, K.; Groten, J.P.; Witthoft, C.; vandenBerg, H. and Navenaar, R. (2005) The effect of folate-binding proteins on bioavailability of folate from milk products. *Trends in Food Science & Technology*, **16:** 307-310.

□□□

Chapter 8

Immunomodulatory Effects of Bioactive Peptides of Milk Origin

SUMAN KAPILA AND RAJEEV KAPILA
Animal Biochemistry Division, NDRI, Karnal-132001

Introduction

The immune system plays a central role in host protection against bacterial, viral, parasitic and fungal infections also cancers. Deficiencies in any aspect of the immune system can predispose an individual to a greater risk of infection and may enhance the severity of a disease. At birth, the immune system of the newborn is functionally immature and exhibits qualitative and quantitative deficiencies in several aspects of immune responsiveness. As a result, newborn infants exhibit enhanced susceptibility to infectious diseases during the first year of life. Milk contains a variety of components that provide immunological protection and facilitate the development of neonatal immune competence. This has been well corroborated from the fact that breast fed infants are less susceptible to infectious illness as compared to formula fed infants.

Two main categories of milk compounds are associated with immunological activity. The first category includes cytokines, which neonates do not produce efficiently. Cytokines present in milk are protected against intestinal proteolysis and could alleviate immunological deficits; aiding immune system maturation. The second category of milk compounds includes milk protein peptides. Milk peptides may affect mucosal immunity possibly by guiding local immunity until it develops its full functionality. Milk protein peptides are inactive within the native milk protein. They become active once they are released from the parent protein by enzymatic proteolysis e.g. during gastrointestinal digestion or fermentation of milk with proteolytic starter cultures or through the action of enzymes derived from proteolytic microorganisms during food processing.

Immunodefensing Peptides

The systems involved in the human body's defense against invaders are rather complex; diet is known to play an important role therein. The two main activities are the immunomodulatory (stimulation of immune system) and antimicrobial (inhibition of bacterial pathogens). Several casein and whey protein derived peptides display an immunomodulatory role in which case a totally separate cascade of host defense responses is initiated. Healthy humans have two immune mechanisms: acquired (specific) immunity, which responds to specific stimuli (antigens) and is enhanced by repeated exposure; and innate (nonspecific) immunity, which does not require stimulation and is not enhanced by repeated exposure. Innate immune mechanisms consist of physical barriers, such as mucous membranes, and the phagocytic and cytotoxic function of neutrophils, monocytes, macrophages, and lymphatic cells (NK cells). Acquired immunity can be classified into two types on the basis of the components of the immune system that mediate the response, ie, humoral immunity and cell-mediated immunity. Humoral immunity is mediated by immunoglobulins produced by bone marrow-derived lymphocytes (B lymphocytes) and is responsible for specific recognition and elimination of extracellular antigens. Cell-mediated immunity is mediated by cells of the immune system, particularly thymus-derived lymphocytes (T lymphocytes).

Casein Derived Immunomodulatory Peptides

Several casein- derived peptides play a significant role in the stimulation of the immune system. The structure of these immunopeptides has been characterized and they have been found to exert a protective effect against microbial infections and to enhance some functions of the immune system.

Bovine α_{s1}-casein enhanced the mitogen-stimulated proliferation of murine splenic T-lymphocytes. The peptide rich hyrdrolysates of

α_{s1}-casein modulate immune functions with different modulatory effects attributable to varying enzymatic digestion. Pancreatin and trypsin digests of bovine α_{s1}-casein and β- casein inhibited the proliferative responses of murine splenic lymphocytes and rabbit peyer's patch cells; whereas digests derived from pepsin and chymotrpsin treatment had no effect. It has been reported that pronase digests of bovine milk α_{s1}-casein inhibited the proliferative response of mouse spleen lymphocytes. Peptide residues derived from the pepsin/typsin hydrolysis of α_{s1}-casein and β- casein also suppress mitogen induced proliferation of human peripheral blood mononuclear cells *in vitro*. In contrast, trypsin treatment alone of bovine α_{s1}-casein results in release of peptide TTMPLW (194-199), which has been shown to promote antibody formation and accelerated phagocytosis *in vitro*, as well as reducing *Klebsiella pneumoniae* infection in mice *in vivo*. Chymosin digestion of bovine α_{s1}-casein results in the immunoenhancing product isracidin (N-terminal Sequence, 1-23) which is found protective to mice against infection by *Staphylococcus aureus* and cows and sheep against mastitis. C-terminal sequence of 193-209 of β- casein obtained from pepsin/chymosin digest of bovine casein has been shown to directly stimulate proliferation of rat lymphocytes *in vitro* in the absence of extraneous mitogens or antigens. Depending on the concentration, β-casomorphin-7 (60-66) and β-casokinin 10 (193-202) exhibit bipolar modulatory effects on human peripheral blood lymphocyte proliferatrion.

These immunoactive peptides are situated in a part of the α-casein molecule in which the β-casomorphins are also situated. Hence, a 'strategic zone' containing biologically active peptides has been proposed in the fragment 51 to 63 of human and 60-70 of bovine β-caseins.

Kappa-casein has been shown to have stimulatory effect on proliferation human-human hybridoma HB4C5 cells and immunoglobulin production. However the tryptic digests stimulated only proliferation of hydridoma cells. Digestion of κ-casein with chymosin resulted into fractionation of para-kappa-casein and glycomacropeptide, both fragments stimulated proliferation of the cells, but only para-kappa-casein stimulated IgM production. Tyr-Gly (38-39 of κ-casein) isolated from tryptic digests was shown to enhance cellular proliferation. Recently a peptide from trypsin digests of bovine κ-casein named κ-casecidin (17-21, Phe-Phe-Ser-Asp-Lys) was isolated. This was found to be cytotoxic towards mouse spleen cells and some mammalian cells including human leukemic cell lines.

Recent studies have focused on immunoenhancing properties of caseinophosphopeptides. Immunostimulatory action of phosphopeptides α_{s1}-CN *f*(59-79) 5P, α_{s2}-CN *f*(1-32)4P and β-CN *f*(1-25)4P which enhanced immunoglobulin IgG production in mouse spleen cell cultures was reported. Moreover, the level of serum and intestinal antigen specific IgA was higher in the mice fed the caseinophosphopeptides than those fed the control diet. Also, it has been suggested that immunomodulatory milk peptides may alleviate allergic reactions in atopic humans and enhance mucosal immunity in the gastrointestinal tract.

Another group of peptides which may be implicated in the stimulation of immunosystem are ACE inhibitors. Inhibition of ACE favors bradykinin formation and thus acts as immunomodulators. Bradykinin, known as mediator of the acute inflammation process, is able to stimulate macrophages to enhance lymphocyte migration and increase secretion of lymphokinines. The peptide fragments α_{s1}-casein f(194-199) and β-casein f(60-66) and f(193-202) have shown to have both immunostimulatory and ACE inhibitory activities.

Table 1: Immunomodulating peptides derived from milk proteins

Source	Peptides	Activity
Human β- Casein	VEPIPY (54-59)	Activates phagocytosis of sheep red blood cells by mice peritoneal macrophages, *in vivo* protection against *K.pneumonia.*
Human α-lactalbumin	GLF (51-53)	Stimulates *in vitro* phagocytosis
Bovine β- Casein	PGPIPN (63-68) LLY (191-193) C-terminal peptide (192-209)	Enhances proliferation of rat lymphocytes
Bovine α_{s1}-casein	TTMPLW (194-199)	Protection against infection.
Bovine α-lactalbumin	YGG (18-20)	Modulate proliferation of human peripheral blood lymphocytes.
Bovine κ-casein	YG (38-39) CMP (106-169)	Inhibition of the proliferation of B lymphocytes
Bovine ê-casein	FFSDK (17-21)	Cytotoxic towards mouse spleen cells and some mammalian cells including human leukemic cell lines

Whey Protein Derived Iopeptides Peptides

Several studies have focused on the *in vitro* and *in vivo* modulation of lymphocyte prolieration by individual whey proteins and mixtures of whey proteins. However, few studies have evaluated the effect of peptides naturally present in WPCs or those produced by ezymatic hydrolysis of whey proteins. It was reported that mitogenic activity of β-LG was stronger following a 3h digestion with pepsin, trypsin, chymotypsin or pancreatin, supporting the view that peptides might be responsible for this activity. In a study, it was showed that peptic hydrolysate of bovine lactoferrin promotes cell proliferation, especially splenocytes and peyer's patch cells. Two synthetic peptides corresponding to the sequences f (50-51) [Tyr-Gly] and f(18-20) [Tyr-Gly-Gly] of α-LA also enhance both the *in vitro* proliferation and protein synthesis of ConA-stimulated human peripheral blood lymphocytes. Another synthetic peptide Gly-Leu- Leu, corresponding the sequence f(51-53) from α-LA, significantly increases phagocytosis of SRBC by murine peritoneal macrophages and protects against *Klebsiella pneumoniae* infections.

Fermented Milk and Immunomodulation

Much evidence now exists that fermented milks are more efficient at stimulating the immune response than are non-fermented milks. A great number of studies demonstrate the ability of fermented milk peptides to induce immune function towards pathogen-derived antigens and suppress immune function in certain instances (tolerance to "harmless"antigens).

Induction of Immune Function

A number of recent studies have examined the effects of fermented milks on immune function. These effects can be attributed to the live bacteria present in the fermented milk but also to the presence of metabolites such as peptides or exopolysaccharides produced during fermentation. In certain cases, the effect can be attributed to the generation of specific peptides; in other cases, the effect is multifactorial. The main mechanism of protection against pathogen-derived antigens at the mucosal level is mediated through IgA(+) cells and secretory IgA, which is capable of neutralizing and, thus, preventing the entry of potentially harmful antigens in the host. The immunomodulatory activity of products derived from milk fermentation by kefir microflora in a murine model has been investigated. Feeding mice with products of milk fermentation resulted in an increase in interleukin-6 (IL-6) secretion, which is necessary for terminal differentiation of B cells to IgA(+)-secreting cells in the gut. In an earlier study, it was found that consumption of kefir-containing viable bacteria resulted in a Th-1 response in mice that was controlled by Th-2 cytokines. The consumption of pasteurized kefir, which did not contain viable bacteria, would induce both a Th-1 and a Th-2 response. Thus, the presence of viable bacteria and/or metabolites could modulate the immune response in the gut.

The effect of peptides released during fermentation of milk with the microorganism *L. helveticus* on humoral immunity was evaluated. Three fractions of peptides were generated by size-exclusion HPLC and were then fed to mice. All three peptides were capable of increasing the number of IgA(+) cells in the intestine and reduced the growth of fibrosacroma. Thus, bioactive peptides released in milk following fermentation possess immunoenhancing and antitumor properties. Feeding mice fermented milk increased the number of IgA(+) cells in the small intestine and the bronchial tissues. The increase in the cells both in the intestine and in the bronchial tissues indicated activation of the IgA cycle. The notion that peptides were indeed responsible for this effect was strengthened by the finding that a protease-deficient derivative of *L. helveticus* was ineffective.

Suppression of Immune Function

Prioult et al (2004) investigated whether the microorganism *Lactobacillus paracasei* was capable of suppressing immune function by generating peptides from the hydrolysis of β-lactoglobulin (β-LG). They reported that peptidases generated by *L. paracasei* were capable of further hydrolyzing peptides generated initially by hydrolysis of β-LG by trypsin-chymotrypsin. Furthermore, these peptides suppressed lymphocyte proliferation and increased IL-10 production, which acts as a major immunosuppressant. They concluded that *L. paracasei* was capable of inducing oral tolerance to β-lg by producing peptidases that can hydrolyze β-LG. Pecquet et al (2000) researched the effect of peptides generated by tryptic hydrolysis of β-LG various immune functions in mice. They reported that mice fed β-LG hydrolysates or fractions of the hydrolysate developed a tolerance to β-LG. Specific serum and intestinal IgE levels were reduced. Furthermore, delayed-type hypersensitivity and proliferative responses were inhibited. It was studied whether the microorganism *L. rhamnosus* GG was capable of suppressing immune function by generating peptides from the hydrolysis of casein. They reported that digests of casein by peptidases produced by *L. rhamnosus* inhibited protein kinase C translocation and downregulated IL-2 expression. Taken together, these results indicate suppression of T cell activation by casein digests.

Bioactivity of the Released Peptides

The structure-activity relationship and mechanisms by which milk-derived peptides exert their immunomodulatory effects is not yet defined. The proline-rich immunostimulating peptides such as the hepatapetide (PGPIPN) and its homologue from human casein are able to resist hydrolysis of gastric and pancreatic enzymes; their action is thus prolonged. The presence of peptide also enhances the hydrophobicity of the peptide and it was thought that by this mechanism the peptide could cross the biological barrier. It has been suggested that arginine in the N-or C-terminal region of peptide is important structural entity recognized by specific membrane bound receptors. The immunostimulatory activity of caseinophosphopeptides was attributable to phosphoseryl residues and the phosphorylation site appears to be an allergenic epitope in caseins. The results obtained with human lymphocytes suggest that opioid peptides may affect the immunoreactivity of lymphocytes via opiate receptor. Therefore there is a remarkable relationship between the immune system and opioid peptides, because opioid ì receptors for endorphins are present on lymphocytes. It has been shown that glutamine-containing peptides can substitute for the free amino acid glutamine which is required for lymphocyte proliferation and utilized at a high rate by immunocompetent cells, even in a resting state. Therefore such peptides exert a non-specific immunostimulation as a result of their trophic properties.

Immunopeptides have potential applications as supplements in the maintenance of immune health. For example, they can potentially provide some protection against infections involving bacteria, viruses and parasites. Alternatively, immunosuppressive peptides could be considered in some medical applications such as the prevention of graft or transplants rejection and in the regulation of inflammation process involved in various autoimmune disorders.

Milk Peptides and Immune Function

The Evidence from *in vitro* and *in vivo* Studies using Mice Model

A number of experiments have been performed in our laboratory at the National Dairy Research Institute. The caseins and their fractions were hydrolyzed with proteolytic enzymes (digestive and microbial sources) either alone or in combinations. Immunomodulatory activity as indexed by the effect on proliferative response of murine splenic lymphocytes cultured in the presence of different mitogens showed that rennet casein (unhydrolyzed) inhibited proliferative response of all mitogen stimulated splenocytes at 100 and 200ug/ml. In case of unhydrolyzed casein fractions, α- and β-caseins exhibit stimulatory effect whereas κ-casein exhibits suppressive effect on lymphocytes. When the casein fractions were hydrolyzed by different proteolytic enzymes, highest immunosuppressive effect was obtained in κ-casein hydrolysates from both cow and buffalo caseins followed by casein (Deepthi, 2004).

The immunomodulatory potential of oral administration of *Lactobacillus helveticus* - fermented milk and cell free supernatant after fermentation in mice was studied (Meena et al, 2007). Mice were fed for a period of 2, 5 or 8 consecutive days with *L. helveticus* fermented milk, cell free supernatant. The control group was fed only milk. The immune response was assessed by analyzing lysosomal enzymes viz. β-galactosidase and β-glucuronidase, phagocytic activity of peritoneal macrophages and lymphocyte proliferation index at the end of feeding periods. The mice fed with *L. helveticus* fermented milk and cell free supernatant showed significant increased lysosomal enzymes levels and phagocytic activity in peritoneal macrophages in comparison to control group. Similarly, lymphocyte proliferation index was also higher in fermented milk and cell free supernatant group. The results suggested that lactic acid bacteria as well as immunoactive component released in milk had positive impact on non specific (phagocytic activity) and specific immune response (lymphocyte proliferation).

Evaluation of immunoregulatory potential of *Lactobacillus helveticus* fermented milk and its cell free supernatant in mice has been carried out. Allergic conditions were induced in mice by injecting ovalbumin intra-peritoneally on day zero and 14. The mice were divided into four groups and fed milk, *L. helveticus* fermented milk and its cell free supernatant for a period of 21days and compared with an allergic control group. It was observed that on feeding fermented milk and cell free supernatant the production of total IgE and ovalbumin-specific IgE in serum declined significantly in comparison to allergic as well as milk controls. The splenocytes were cultured and cytokines were estimated by ELISA kits. It was observed that levels of Th1 (T helper Type1) cytokines increased whereas the levels of Th2 (T helper Type2) cytokines decreased. Thus decline in IgE production was associated with increase in production of Th1 (IFN- ã, IL-2) cytokines and decrease in Th2 (IL-4, IL-6) cytokines(Kapila et al,2008).

In another study whey and whey supplemented with WPC were fermented with *L. helveticus* and were analyzed for immuno-modulatory activity under *in-vitro* as well as *in-vivo* conditions. *In-vivo* study was carried out using allergic mice model. The parameters checked for allergic response were total IgE, ova-specific IgE levels in serum and cytokines namely Interlukin-2 (Th-1) and Interlukin-4 (Th-2) in lymphocytes after culturing for 48h. The results obtained showed that the lymphocyte proliferation index (PI) increased significantly in fermented whey hydrolysates than ultrafiltered fermented whey protein hydrolysate and whey. Feeding of fermented whey products resulted in significant decrease in total IgE and ova-specific IgE levels in allergy induced mice by shifting the production of more IL-2 and decline in IL-4 levels. Thus fermented whey hydrolysates were able to control the ovalbumin induced allergy in mice (Kapila et al, 2008).

Bioactive Peptides from *Lactobacillus rhamnosus* GG were evaluated for their protective effect on intestinal epithelial cells from apoptosis induced by Camptothecin. Intestinal epithelial cells were isolated from rats and were grown in DMEM medium. After 72 h, these cells were treated with bioactive peptides isolated from LGG (10-100 µmol/ L) for 2 h, and were incubated with camptothecin (20 µmol/ L) for 2 and 4 h, and cell viability was measured by MTT assay. Caspase activity assay was carried out to determine caspase 3 and caspase 9 activity. Moreover, cell lysates were also pretreated with caspase 3 inhibitor (200 µM). DNA fragmentation assay was carried out to determine the increase in DNA fragmentation. Bioactive peptides significantly prevented camptothecin induced apoptosis in rat intestinal epithelial cells. The treatment of intestinal epithelial cells with camptothecin for 4h resulted in five-fold increase in DNA fragmentation, and resulted in three-fold increase in caspase 3 activity, compared with controls. Pre-treatment with bioactive peptides (10 µmol/L) significantly attenuated the Camptothecin-induced DNA fragmentation by 29% ($P<0.001$), and significantly inhibited caspase 3 and caspase 9 activity by 35% and 39%, respectively, after 4 h of camptothecin exposure ($P<0.001$). Hence, it was proved that novel bioactive peptide from LGG were found to be potential anti-apoptotic agents and hence could be administered as therapeutic molecule for treatment of enteropathy resulting from chemo- and radiaton- therapy (Poddar et al, 2008).

Conclusion

The potential of milk derived bioactive peptides as immunomodulatory health products remains to be demonstrated since a number of critical factors still have to be studied. The fate of bioactive peptides during gastric transit, the impact of immunomodulating peptides on mucosal immunity in the gastrointestinal system and toxicity of such products are unknown, while clinical data on parameters such as the effect of dosage, mode of administration and systemic effects are lacking.

Nevertheless, milk peptides may have some potential application as supplements in the maintenance of immune health. For example, they can potentially provide some protection against infections involving bacteria, viruses and parasites. Alternatively, immunosuppressive peptides could be considered in some medical applications such as the prevention of graft or transplants rejection and in the regulation of inflammation processes involved in various autoimmune disorders.

Suggested Reading

Prioult, G., Pecquet, S., & Fliss, I. (2004). Stimulation of interleukin-10 production by acidic β-lactoglobulin-derived peptides hydrolyzed with Lactobacillus paracasei NCC2461peptidases. Clinical and Diagnostic Laboratory Immunology, **11:** 266–271.

Deepthi, K.(2004).Comparative studies on casein derived ACE inhibitory and immunomodulatory peptides from cow and buffalo milk. Ph.D Thesis NDRI (Deemed University), Karnal, India.

Meena, K., Kapila, S., Chand, R., Haque, E., Singh, S and Aggarwal, P. K.(2007).Immunomodulatory effect of oral administration of milk fermented with *Lactobacillus helveticus. Milchwissenschaft* **62(4):** 415-417.

Kapila, S., Jain, S.,Haque, E. Sabikhi, L and Dang A.K.(2008).Immunoregulatory response of *Lactobacillus helveticus* fermented milk in mice *Milchwissenschaft* **63(4):** 367-369.

Kapila, S., Jabadolia, L.N., Dang, A.K., Kapila, R. and Arora.,S.(2008). Augmentation of biofunctional properties of whey protein on fermentation with *Lactobacillus helveticus. Milchwissenschaft* (in press). 838.

Poddar, D., Kumar, M., Kapila, S., Raj, D. and Aggarwal, P.K. Bioactive Peptides From *Lactobacillus rhamnosus* GG Protects Intestinal Epithelial Cells From Apoptosis Induced By Camptothecin. 30th European Peptide Symposium (30EPS) 31 August – 5 September 2008, Finlandia Hall, Helsinki, Finland. Journal of peptide research **119:** P22021-145

❑❑❑

Chapter 9

Application of New Emerging Probiotic Yeast in the Development of Therapeutic Dairy Products

D.N. GANDHI AND KALPANA DIXIT
Dairy Microbiology Division, National Dairy Research Institute, Karnal (HR)

Introduction

Probiotic organisms are sold mainly in fermented foods and dairy products and play a predominant functional role as carriers of probiotic. These foods are well suited in prompting the positive health image of probiotics. Lactic acid bacteria had been explored as major probiotics worldwide. However, recent work recognizes *Saccharomyces* as non-pathogenic yeast which has been used as both a preventive and therapeutic agent for the treatment of variety of diseases. Among yeasts Saccharomyces genus is one of the yeast genus having biotherapeutic properties. Yeasts are predominantly unicellular fungi. Taxonomically, yeasts are rather heterogeneous and belong to the division Eumycota. *Saccharomyces* is the representative of ascosporogeneous yeasts and historically one of the most familiar microorganisms for human. Among yeasts, *Saccharomyces* is the only potential probiotic genus with known non-pathogenic nature, which can grow well at 37°C i.e. human body temperature (Mcfarland, 1996). It is a unique probiotic as it can survive under gastric acidity and is not adversely affected by antibiotic(s). Probiotic *Saccharomyces spp* are sensitive to non absorbable anti-fungals such as nystatin while they are safely administered with reabsorbable anti-fungal like flucanazole. *Saccharomyces* have resistance for antibacterial antibiotic because of absence of sites for antibiotic action, so if *Saccharomyces* is use as a probiotic, there is no problem of resistance genes and lateral gene transfer in pathogens, and further, it does not alter the normal flora of the host bowel.

In the genus *Saccharomyces, S. boulardii* is being used as a commercial probiotic by pharmaceutical companies for human beings. Pharmaceutical preparations containing *Saccharomyces boulardii* are present in Indian market like SB Kid (Paris Dakner Microspherules, Chennai), Derolac (Aristo Pharmaceuticals, Mumbai) as well as in USA, Florastor (BIocodex Pharmaceutical, San Francisco) used against different type of diarrhea. Controlled clinical trials have shown the efficacy of *Saccharomyces spp (S.cerevisiae/ S. boulardii)* for the prevention of and /or the treatment of various intestinal disorders including antibiotic-associated diarrhea (AAD), recurrent *Clostridium difficile* disease (Surawicz *et al.,* 2000), *Clostridium difficile-* associated enterocolopathies in infants , acute in adults, children, in critically ill tube-fed patients , traveler's diarrhea (Buts *et al.,* 2005), AIDS related, and relapses of Crohn's disease or ulcerative colitis (Guslandi *et al.,* 2003). *Saccharomyces spp* (*S. cerevisiae* and *S. boulardii)* also inhibit the pathogenic bacteria of gastrointestinal tract eg. *Vibrio cholerae, Salmonella enteritidis, Clostridium difficile* etc. without affecting the natural microflora (Czerucka *et al.,* 2002).

Acidophilus milk is a sour product that has been allowed to ferment under conditions that favor the growth and development of a large number of *Lactobacillus acidophilus* organisms. *L. acidophilus* is well known probiotic culture. This Lactic acid bacteria (LAB) ferment milk lactose through hydrolysis to glucose and galactose. The glucose moiety is fermented to lactic acid. *Saccharomyces spp* can be benefited by the metabolic products of milk sugar fermented by lactic acid bacteria and the presence of organic acids, which can easily be utilized. The beneficial aspect of this behavior was utilized in the preparation of acidophilus-yeast milk. Acidophilus-yeast milk differs from acidophilus milk in that in addition to *L.acidophilus,* yeast (lactose fermenting) or brewers' or bakers' yeasts (which ferment sucrose but not lactose) were used for fermenting the milk. Incorporation of probiotic *Saccharomyces spp* showed synergistic effect by the enhanced growth of lactic culture in fermented dairy products with an increased probiotic property. Products of mixed fermentation such as acidophilus yeast milk, being rich in alcohol and carbon dioxide, excite the respiratory and central nervous system. In the erstwhile Soviet Union, acidophilus yeast milk is widely used in treating tuberculosis. However no literature has been found regarding addition of probiotic *Saccharomyces spp* in the preparation of acidophilus-yeast milk. Incorporation of probiotic *Saccharomyces spp* in acidophilus milk enhances its demand as probiotic drink because of addition of beneficial properties like fast recovery from diarrhea as well as from tuberculosis.

Occurrence of *Saccharomyces Spp* in Milk and Milk Products

Saccharomyces spp such as. *S.burnetii, Skouteri, S.byanus, S.rosinii S.cerevisiae* and *S.boulardii* may be isolated from a variety of dairy products, including milk, yogurts, cream, dahi, cheeses and kefir. Yeasts rarely grow in milk stored at refrigeration temperature because they are out-grown by psychotropic bacteria. However, in sterilized milk in the absence of competition, *Saccharomyces spp* exhibits weak lipolytic and proteolytic activity and is capable of growing to reach populations of 10^8-10^9 c.f.u. /ml. *Saccharomyces spp* often present in soft mold ripened cheeses. It has been reported in Italian Stretching cheese at 3.1% of the yeast population present. They may be present in semi-hard cheese including cheddar cheese. Growth of *Saccharomyces spp* in cheeses is thought to be related to its ability to use lipid and protein products form other species and possibly its ability to utilize lactic acid present in the cheese (Robinson *et al* 2000).

Mechanism of Probiotic activity of Yeast against Pathogenic Bacteria

S. cerevisiae and *S.boulardii* share a common mechanism of action against pathogenic bacteria. These are as given below.

Pharmacokinetics

S. cerevisiae and S. boulardii resist to gastric acidity and proteolysis and are able to achieve quickly high concentrations in the gastrointestinal tract and to maintain constant levels in a viable form. They are not permanently colonizing the colon and not easily translocate out of the intestinal tract. Further it can be detected alive throughout the digestive system, if they are given daily in its freeze-dried form. In gnotobiotic mice, a single dose of *S. boulardii* leads to colonization of the intestinal tract, the yeast being detectable at a constant level (10^7 c.f.u. / ml) for 60 days. In healthy human volunteers, which received a single oral dose of 1 g *S.cerevisiae* or *S. boulardii,* the time to reach maximum stool concentration was found within 36 to 60 hours, and the time to decrease below detectable levels was 2 to 5 days later.

Pharmacodynamics

The pharmacodynamics of *S.cerevisiae and S.boulardii* involves 3 different aspects.

a) Direct antagonism

S.boulardii reduces the growth of *Clostridium albicans, E. coli, Salmonella typh, Shigella dysenteriae, Vibrio cholerae, Salmonella enteritidis and. Clostridium difficile S.cerevisiae* reduces the growth of *E.coli, Shigella flexnerii, Clostridium difficile, Vibrio cholerae.*

The mechanism of action of *S.cerevisiae* and *S.boulardii* in experimental infections has been extensively studied. Both animal's studies and *in vitro* assays have shown that *S.cerevisiae* and *S.boulardii* have a protective effect and specific activities against various enteric pathogens and members of family enterobacteriaceae (Czerucka and Rampal. 2002).They show antagonistic effect mainly because of the competition for adhesion site in gastrointestinal tract.

b) Antisecretory effect on the binding of toxins to intestinal receptors

Pathogenic strain of C. *difficile* produces two well-characterized toxins A and B that cause mucosa damage and inflammation of colon. *S.cerevisiae* and *S. boulardii* significantly reduces the liquid secretion and mannitol permeability caused by inhibition of Clostridium *difficile* toxin A in rat ileum, compared with control group. Chromatographic fraction of *S.boulardii* filtered supernatant led to the identification of an active fraction that decrease toxin A-induced rat ileal secretion by 46%, and intestinal permeability by 74%and prevent toxin A-mediated inflammation and villous damage. It was demonstrated that this fraction was enriched in a protease that acts both by proteolysis of the toxin. This protease was identified as a 54-kDa serine protease, which exerts a proteolytic activity against toxin B molecule. *S.cerevisiae* and *S.boulardii* protease diminishes the ability of toxin A and B mediated inhibition of protein synthesis and reduction of colonic resistance.

c) Trophic effect

It has been reported that rats treated with *Saccharomyces spp* showed significant increases in sucrase. isomaltase, lactase and maltase activities. In their study on human volunteers used an in situ technique to measure brush border enzyme activities in snap-frozen biopsies. After treatment with *S.cerevisiae* and *S. boulardii*, an increase in lactase, -glycosidase and alkaline phosphatase activity was detected both at the basal part and apical part of the villi, which increases ranging from 22 to 55 % compared to the basal activities measured before treatment. Thus in humans as in rats, *S.cerevisiae* and *S. boulardii* enhance the expression of disaccharidases and alkaline phosphatase that may improve the absorption of carbohydrates, usually in acute and chronic diarrhea disorders .*Saccharomyces cerevisiae* and *S. boulardii* contains polyamine (spermine and spremidine) which have the same trophic effect on the intestinal mucosa, with an increase in diasaccharidase activity, as an equivalent amount of spermine and spermidine given to test animals.

In view of potential probiotic activity of new emerging yeast, various products have been developed based by incorporation of probiotic yeasts in pharmaceutical formulations and therapeutic fermented milks as are enlisted in **Table 1** and **Table 2** respectively.

Table.1: Commercial Preparation of *S. boulardii*

Product Name	Pharmaceutical companies
Saccharomyces boulardii	Allergy Research Group, Alameda
Saccharomyces boulardii (Classic series)	Advanced Orthomolecular Research (AOR) Inc, Canada.
Diastase	Wren Laboratories Ltd, U.K.
Saccharomyces boulardii	Claries Lab, Singapore
Florastor	BIocodex Pharmaceutical, San Francisco
SB Kid	Paris Dakner Microspherules, Chennai
Derolac	Aristo Pharmaceuticals, Mumbai

Table 2: Types of Fermented milk products

Fermented Product	Microorganisms Responsible for Fermentation	Description
Taetee	*Streptococcus lactis var. hollandicus, Saccharomyces taetee*	Moderately ropy sour milk slight flowing consistency
Acidophilus yeast milk	*Lactobacillus acidophilus, Kluveromyces lactis rarely Saccharomyces spp*	Made from whole or skimmed milk. *Lactobacillus acidophilus* produces acid while ethanol and CO_2 produces by yeast.
Yogurt	*Streptococcus thermophilus, Lactobacillus bulgaricus*	Made from milk in which solids are concentrated by evaporation of some water and addition of skim milk solids. Product has consistency resembling custard.
Kefir	*Streptococcus lactis, Lactobacillus bulgaricus,* yeasts	A mixed lactic acid and alcoholic fermentation; bacteria produce acid, and yeasts produce alcohol.
Koumiss	*Lactobacillus bulgaricus ,* Lactose fermenting yeast,	Combined acid and alcoholic fermented milk, traditionally prepared from Mare's milk

Survival of *Saccharomyces spp* in fermented milk products and preparation of yeast based fermented milk products

The frequent occurrence of yeast in dairy related products indicate the ability of yeasts to survive and metabolized milk constituents. *Saccharomyces spp* can't ferment lactose so it develops in milk as secondary flora, after bacterial growth. Lactic acid bacteria (LAB) ferment about 35% of the lactose in milk through hydrolysis of glucose and galactose. Glucose moiety is changed in to lactic acid. Lactic acid cause low pH of yoghurt and the ability of yeast to utilize organic acid create a selective environment for yeast growth in fermented milk products. Milk products that include yeasts in their starter culture are acidophilus-yeast milk, Kefir, Koumiss and Leban. Method for the preparation of acidophilus yeast milk has been standardized milk at

National Dairy Research .Institute. For this preparation, skim milk with 8% sugar content is given a heat treatment to 90C /15 mts and cooled to 37C. A combination of *Lactobacillus acidophilus* and *Saccharomyces cerevisiae* starter cultures is added at the level of 1% each and fermentation is carried out in crown corked bottles at 39C / 20-22 hrs. It is then stored under refrigeration temperature and consumed as chilled beverage.(Kalpana and Gandhi, 2008)

The present study carried out at NDRI has revealed that probiotic potential of *Saccharomyces* species, especially *S. boulardii* and *S. cerevisiae,* and their immense possibilities as a biotherapeutic agents. In the present age of revolutionizing trends in food and dairy sector, it has been an obvious fact that in the modern days consumers are very much aware of quality food and looking forward for the food that have additional functional properties beyond basic nutrition. The presence of yeasts in various fermented foods and dairy products offer exciting opportunities to explore the hidden potential of such yeasts as probiotic. The careful selection of commercially important strains of *Saccharomyces boulardii* and *Saccharomyces cerevisiae* will open new avenues in the era of fermented foods and dairy beverages.

Conclusion

Although their are several reports on the probiotic attributes of lactic cultures in the recent years, but keeping in view potential biotherapeutic properties of new emerging probiotic yeast, there is a need to generate more data on isolation, product development. and therapeutic properties of *Saccharomyces sp in vitro* and *in vivo* alone or in co-fermentation with lactic acid bacteria in dairy products.

Suggested Reading

Aloysins, L., Souza, D., Rajkumar, C., Cooke, J., Bulpitt, C.J. 2005. Probiotic in prevention of antibiotic associated diarrhea. *British Med. J.,* **8**: 1-6.

Buts, J. P., Bernasconi, P. 2005. *Saccharomyces boulardii*: Basic Science and Clinical Applications in Gasteroenterology. *Gesterointerol.Clin.N.Am.* **34**: 515-532.

Costalos, C., Skouteri, V., Gounatis, A., Sevastidou, S., Triandafilidou, A., Ekonomidous, C., KontaxaKi, F., Petrochilou, V. 2003. Enteral feeding of premature infants with *Saccharomyces boulardii, Earli Human Dev.,* **74**: 89-96

Czerycka, D., Rampal, P., 2002.Experimental effect of *Saccharomyces boulardii* on diarrheal pathogens.*Microbes and infection* **4**:733-739.Czerucka, D., Piche,

T., Rampal, P. 2007. Yeast as probiotics-*Saccharomyces boulardii. Alimentary Pharmacol & Therapeutics.* **26**(16): 767-778.

Dixit, Kalpana (2008); Biotherapeutic properties of Probiotic Saccharomyces spp and it's Survival in Acidophilus milk .Ph.D Thesis, National Dairy Research Institute (Deemed University), Karnal

Gulsandi, M., Giollo, P., Testoni, P.A. 2003. A pilot trial of *Saccharomyces boulardii* in ulcerative colitis. *Eur J Gasteroenterol Hepatol.* **15**: 697-698.

Lourens, A., Viljoen, B.C. 2001. Growth and survival of a probiotic yeast in dairy products.*Food Research International* **34:** 791–795

McFarland, L. V. 1996. *Saccharomyces boulardii* is not *Saccharomyces cerevisiae. Clin. Infect. Disease,* **22**: 200-201.

Periti, P., Tonelli, F.,2001.Preclinical and clinical pharmacology. of biotherapeutic agents: Saccharomyces boulardii, *J. Chemo-. Ther.* **13:** 473–493.

Robinson, R.K., Bat, C.A, Patel, P.D. (2000). Encyclopedia of Food Microbiology, **3** Academic Press, A Harwart Science and Technology.Co.,London

Surawicz, C. M., McFarland L. V, Greenberg, R. 2000. The search for a better treatment of *Clostridium difficile* disease: Use of high-dose Vancomycin combined with *Saccharomyces boulardii. Clin Infect Dis.*

□□□

Chapter 10

Synbiotic Foods: The Concept and Development

MALVIKA MALIK, ANSHU ANAMIKA, NOOR AFSHA, SURAJIT MANDAL, A.K. PUNIYA AND KISHAN SINGH
Dairy Microbiology Division, National Dairy Research Institute, Karnal (HR)

Introduction

Diet is a major focus of public health strategy aimed at maintaining optimum health throughout life, preventing early onset of chronic diseases such as gastrointestinal disorders, cardiovascular disease, cancer, osteoporosis, as well as promoting healthier ageing. The growing demand for 'healthy' foods is stimulating the innovation and new product development internationally. Recent development in food industry has a central role in facilitating healthier eating practices through the provision and promotion of healthy foods. Continuously increasing consumers' health consciousness and expenditure are socio-economic factors responsible for the expanding worldwide interest in functional foods. The functional food concept has, in recent years, moved progressively towards the development of dietary supplement that may affect the intestinal microbial composition and activities. Thus, probiotic strains can be successfully manufactured and incorporated into highly acceptable food products where they can retain their viability. Currently, industrial demand for technology for probiotic provides promising prospects for improved culture performance and due to survival and colonization difficulties "synbiotic" (probiotic + prebiotic) approach offers an attractive alternative. The growth in consumption of food outside the home has brought several challenges to the food industry; especially for products which are closely associated with the household kitchen, and in particular, the refrigerator. Perishability and the general need for refrigeration have been the two principal constraints in development of carry away synbiotic products. Today, there is a growing need for development of long shelf-life synbiotic products so as to meet the consumer demand and also to provide health benefits to more and more number of people.

Probiotics

The concept of probiotics was evolved at the turn of the 20th Century from a hypothesis proposed by Russian Scientist, Elie Metchenikoff, that the long healthy life of Bulgarian peasants was resulted from the consumption of fermented milk products. Probiotic is a "live microbial (feed) supplement, which beneficially affects the host (animal) by improving its intestinal microbial balance". Probiotics are important in inhibiting pathogens, maintaining a beneficial enteric pH, synthesizing nutrients and producing enzymes, bioactive compounds, etc. that enhance digestion and exert health beneficial effects. Probiotics have been reported to have several health benefits such as balancing of intestinal microflora, stimulation of the immune system, prevention of diarrhoea, and anticarcinogenic activity. Probiotics were re-defined as "living organisms that upon ingestion in certain number exert health benefit beyond inherent basic nutrition". Good viability and activity of probiotics are considered prerequisites for optimal functionality. Production of probiotics should be based on the microbial criteria, and the ability to withstand stress during processing and storage of products is important. Thermophilic/ thermotolerant probiotics are of great interest as they can have all the desired characteristics. The basis for selection of probiotics includes safety, functional properties such as survival, adherence, colonization, antimicrobial substances production, immunostimulation, antigenotixic activity, prevention of pathogens, etc. and technological characteristics including growth in milk and other food base, sensory properties, phage resistance and viability. Safety aspects include specifications such as origin (healthy human), non-pathogenicity and antibiotic resistance characteristics. Preferably, strains used should be of human origin as a bacterial strain can function better in an environment similar to that it was isolated. Probiotics should possess GRAS (Generally Regarded as Safe) status. To survive and grow in the *in-vivo* conditions of the desired site of administration, the probiotic strain must be able to tolerate low pH of stomach and high concentration of both conjugated and de-conjugated bile acids of small intestine, etc. Currently used probiotics include Lactobacilli, such as *Lactobacillus acidophilus, L. casei, L. delbrueckii* ssp. *bulgaricus, L. reuteri, L. brevis, L. cellobiosus, L. carvatus, L. fermentum* and *L. plantarum*; Gram-positive cocci, such as *Lactococcus lactis* ssp. *cremoris, Streptococcus thermophilus, Enterococcus faecium, S. diacetylactis* and *S. intermedius* and Bifidobacteria, such as *B. fifidum, B. adolescentis, B. animalis, B. infantis, B. longum* and *B. thermophilum.* Nonpathogenic microorganisms that occupy important niches in the host gut or tissues, such as yeasts, enterococci and *Enterobacteriaceae,* are also used as human and animal probiotics.

Prebiotics

Prebiotics are "non-digestible dietary components that pass through to the colon and selectively stimulate the proliferation and / or activity of populations of desirable bacteria *in-situ*". The concept of prebiotics has become very popular since its introduction in 1995. Food ingredients classified as prebiotics must not be hydrolyzed or absorbed in the upper gastro-intestinal tract, need to be a selective substrate for one or a limited number of beneficial colonic bacteria, must alter the microbiota in the colon to a healthier composition and should induce luminal or systematic effects that are beneficial to host health. The prebiotics are mainly non-digestible carbohydrates (oligo- and poly-saccharides), some peptides and proteins, and certain lipids. The prebiotics identified thus far are non-digestible carbohydrates including lactose, inulin and range of oligosaccharides that supply a source of fermentable carbohydrate for beneficial bacteria in the colon described food-grade oligosaccharides in commercial production; these include lactulose, galactosaccharides, fructo-oligosaccharides, isomalto-oligosccharides, lacto-sucrose, gentio-oligosaccharides and xylooligosaccharides. The current potential for prebiotics rests largely with the investigation of those compounds that can be extracted form foods and their impact on residing intestinal microflora.

Synbiotics

A synbiotic is a combination of the concepts of probiotics and prebiotics. It is defined as " a mixture of a probiotic and prebiotic that beneficially affects the host by improving the survival and the implantation of live microbial dietary supplements in the gastro-intestinal tract, by selectively stimulating the growth and/ or by activating the metabolism of one or a limited number of health-promoting bacteria. Synbiotic foods come under functional foods which, in addition to their basic nutritive value and natural being, will contain the proper balance of ingredients, which will help as functional, therapeutic and nutraceutical agents enabling the human to prevent the illness and diseases. The concept of synbiotic is in its infancy, and is yet to be applied to the full extent in developing new foods. When combining both a probiotic and a prebiotic in a single food product, the expected benefits are an improved survival during the passage of probiotic bacteria through the upper intestinal tract. The advantages of synbiotic are that a well established probiotic with known benefits can be used and the prebiotic aids the establishment of the organism in the complex colonic environment. The end result is an improved survival of probiotic, which has readily available substrate for fermentation as well as the individual advantages that each may offer. This approach could have particular application in infants and the elderly as both harbor highly unstable intestinal microbial population. There are various studies reporting the development of synbiotic foods A synbiotic low-fat soft curd cheese was developed by incorporation of dietary fibers, one of which was prebiotic and it was found that combination of prebiotic and wheat fiber increased both the sensory and the nutritional quality of the product. Use of inulin and oligofructose in fruit yoghurts, milk-based drinks, milk, spreads, cheese and ice cream has also been reported. Number of synbiotic products containing Bifidobacteria and lactulose are already available in Japanese markets. Synbiotic yoghurt was prepared by incorporating *L. acidophilus* and inulin which resulted in better growth and acid production by the organisms. The better acceptability of synbiotic yogurt on sensory evaluation and the higher viability of probiotics in the product during refrigerated storage were also reported. Synbiotic product having Bifidobacterium and inulin was found to be better than either pro or prebiotic alone. Interaction between the probiotic and the prebiotic *in vivo* might be favoured by an adaptation of the probiotic to the prebiotic substrate prior to consumption. This might result in a competitive advantage for the probiotic if it is consumed concurrently with the prebiotic. One of the currently used prebiotic is resistant starch. The benefits of using resistant starch extend beyond traditional prebiotics, since resistant starch can be used to ensure the viability of probiotic populations from the food to the large intestine. Resistant starch offers an ideal surface for adherence of the probiotics to the starch granule during processing, storage and transit through the upper regions of the gastrointestinal tract, providing robustness and resilience to environmental stresses. Bacterial adhesion to starch may also provide advantages in new probiotic technologies to enhance delivery of viable and metabolically active probiotics to the intestinal tract.

Health Benefits Associated with Synbiotic Foods

It is generally recognized that an "optimum" balance in microbial population in our digestive tract is associated with good nutrition and health. The microorganisms primarily associated with this balance are lactobacilli and bifidobacteria. . Intake of synbiotics has been demonstrated to modify the composition of the gut microflora, restore the microbial balance and therefore have the potential to provide health benefits. Various clinical studies have provided a clear evidence of health promoting effects of synbiotics such as prevention of antibiotic-associated diarrhea, treatment of acute diarrhea, inflammatory bowel disease and eradication of *C. difficile* infection and enhancement of intestinal immunity. It has been reported that there is a significant effect on the composition of fecal flora on feeding volunteers a daily supplement of a synbiotic composed of *Lactobacillus* fermented milk contining oligofructose. A synbiotic preparation consisting of mixture of strains of *L. paracasei* and *L. gasseri*, oligosaccharide, glutamine, vitamin B6 and zinc was studied with human volunteers and it was reported that synbiotic preparation has favorable effects on the intestinal bifidobacteria and lactobacilli in the preparation. Studies have been carried out on the effect of a lyophilized synbiotic preparation containing *Bifidobacterium* species and inulin in mice. Besides, the reduction of fecal clostridial counts, coliforms and carcinogenic b-glucuronidase enzyme activity in synbiotic fed mice was observed. A milk chocolate was prepared by incorporating free as well as encapsulated *L. casei* and inulin and it was found to increase the faecal lactobacilli, decreased coliform and b-glucuronidase enzyme.

Extending the Shelf Life of Synbiotic Products

With the changing life-style and consumer demands the need of hour is development of long shelf- life, ready to use functional food products, especially the synbiotic products. The challenge in developing such foods is doubled as there is need to extend the shelf- life of the product as well as the viability of the probiotic culture is also needed to be maintained. Along with these hurdles producer has to keep in mind the consumer acceptability in terms of sensory attributes of the product. In this area the most studied and successful approach is microencapsulation. Microencapsulation is a process of providing microorganisms with a physical barrier against external environment; this approach is currently receiving considerable interest. For encapsulation of probiotics different wall materials are being studied. Probiotic strains are encapsulated within a food grade matrix by different techniques, i.e., freeze drying, extrusion, emulsion and spray drying. Microencapsulation of *L. casei* in alginate beads results in better survival at low pH, high bile salt concentration and heat treatment. Thermal processing for obtaining a storable form of the probiotics and the choice of proper feed matrix suitable for retaining the viability of the probiotic during the production should to be appropriate. Now a days thermophilic/ thermotolerant organisms are getting attention. Thermophilic/ thermotolerant organisms have an advantage in that they withstand higher temperature during processing and storage. They have a better chance of remaining viable during the drying process required for prolonged storage and they lead to a distinctly effective product. Attempts are thus being made to select or isolate thermophilic/ thermotolerant probiotics from the available sources and examine the survival of these strains during food processing like spray drying and to define the optimal processing parameters for obtaining the best products. Spray-dried powder harboring high numbers of viable probiotics is a convenient means of storage and transport of probiotic cultures and their subsequent application in functional foods. While spray drying is an economical process for the large-scale preparation of probiotic cultures, and is commonly used for the preparation of food ingredients, it suffers from the disadvantage of causing bacterial cell injury and death. One approach used by a number of workers to improve probiotic performance in food systems is the addition of protectants to the media prior to drying. In recent studies the incorporation of the soluble fibre, gum acacia in milk-based medium prior to spray drying the probiotic *Lactobacillus paracasei*, increased probiotic viability during powder storage, compared with milk powder alone. The use of prebiotic gum- acacia has been reported to provide dual effect, one is promoting the growth of the probiotic and secondly by providing thermo-tolerance to the probiotic cells during the spray drying process. Additionally, it has been reported that exopolysaccharide-producing strains of bifidobacteria may be naturally protected. Probiotic suppliers have developed variety of proprietary techniques to preserve and protect the integrity of these tiny living organisms.

Conclusion

It is possible to manipulate the composition of the gut microbiota in infants and adults through dietary supplementation of synbiotics *i.e.* through regular intake of probiotics and prebiotics. Species specific prebiotics may be used in formulation of synbiotics. For successful delivery of probiotics in products to the intestine, the beneficial microorganisms must survive food processing, storage during product maturation and shelf life, as well as the stress conditions in the host gastrointestinal tract system. The selection of probiotics with all these characteristics is a technological challenge and is evaluated according to the viability and stability of probiotics in the final probiotic product. The use of diet to fortify certain gut flora components is a popular current aspect of functional food sciences. In this context, probiotics, prebiotics and synbiotics have a significant role. The efficacy of probiotic foods can be enhanced by exploiting synergistic interaction between functional ingredients, as is potentially the case with synbiotics. Modification of the gut flora by dietary means offers one of the most effective opportunities for development of gut targeted probiotic functional foods. The area of gut flora modulation seems to hold much promise for the prophylactic management and/or treatment of gut disorder. Species specific prebiotics may be used in formulation of synbiotics. Future human trials that exploit the use of modern molecular-based detection methods for gut bacteria will help in determining the efficacy for improved health through functional foods.

Suggested Readings

Goldin, BR (1998) Health benefits of probiotics. *British Journal of Nutrition* **80:** S203–S207.

Guarner, F and Schaafsma, GJ (1998) Probiotics. *International Journal of Food Microbiology,* **39:** 237–238.

James, L (2004) Development of a lyophilized synbiotic preparation containing *Bifidobacterium* species and inulin. Ph.D. thesis, NDRI, Karnal, India.

Linda, MO (2002) Improving the gut feeling. *Food Technology,* **56:** 67-70.

MacFarlane, G.T. and Cummings, J.H. (1999) Probiotics and prebiotics: can regulating the activities of intestinal bacteria benefit health? *British Medical Journal,* **318:** 999-1003.

Mandal, S; Puniya AK and Singh, K (2006) Effect of alginate concentrations on survival of microencapsulated *Lactobacillus casei* NCDC-298. *International Dairy Journal,* **16:** 1190-1195.

Mishra, V; Praveen, K and Prasad, DN (2001) Prebiotics- a new approach in health management. *Indian Dairyman,* **53:** 11-14.

Nagpal R, Yadav H, Puniya AK, Singh K, Jain S and Marotta F 2007 Potential of probiotics and prebiotics for synbiotic functional dairy foods: an overview. *International Journal of Probiotics and Prebiotics* **2:** 75-84

Rastall, RA and Maitin, V (2002) Prebiotics and synbiotics: towards the next generation. *Current Opinion in Biotechnology,* **13:**490-6

Roberfroid, MB (1998) Prebiotics and synbiotics: concepts and nutritional properties. *British Journal of Nutrition* **80:** S197–S202

Scheppach, W; Leuhers, H and Menzel, T (2001) Beneficial health effects of low-digestible carbohydrate consumption. *British Journal of Nutrition,* **85:** S823-S830

Schrezenmeir, J and De Vrese, M (2001) Probiotics, prebiotics, and synbiotics-approaching a definition. *American Journal of Clinical Nutrition,* **73:** 361S-364S

Warrand, J (2006) Healthy Polysaccharides. *Food Technology and Biotechnology,* **44:** 355–370

□□□

Chapter 11

Delivery Systems for Functional Food Ingredients

SURAJIT MANDAL
Dairy Microbiology Division, National Dairy Research Institute, Karnal (HR)

Introduction

Significant strategy in the development of functional foods evolves increasing the levels of specific nutraceuticals that are known to have health benefits. This can be through enhancement of levels of the desired component that is inherent in the food or by fortification of food products with functional ingredients, such as, dietary fibre, antioxidants, natural isoflavone, plant sterols/stanols, other phytochemicals or phytonutrients, bioactive peptide, þ-3 PUFA and probiotics and/or prebiotics, minerals and vitamins, etc., which are having pivotal physiological and nutritional role in human (Table 1).

Functional food ingredients should present in foods in sufficient quantities/ numbers at the time of consumption and reach to the action site for functional activities. Thus, these ingredients should withstand the processing and preservation treatments of foods and stable during storage of foods and during gastrointestinal tract transient. These should not affected by food matrixes and environmental factors prevailed in foods as well as they should not react with food components. But most of the food ingredients react with food components and also affected by food matrixes and environments. These lead their poor stability/ survival of functional ingredients to give desirable functional attributes. Different methods of stabilization or improvement of survival of functional food ingredients are including selection of suitable and stable ingredients, and food combinations, addition of protective agents, segregation by physical barriers from outside environments in food as well as in GIT, etc. It is very difficult to select as suitable ingredients for food application as the selection procedure is highly probabilistic in nature. Alternatively, addition of compatible protective/ stabilizing agents and segregation by applying barriers are the very suitable and promising means for delivering of these functional ingredients to the action sites. Controlled release of food ingredients at the right place and the right time is also a key for the functionality of active ingredients. A timely and targeted release improves the effectiveness of food additives, broadens the application range of food ingredients and ensures optimal dosage, thereby improving cost-effectiveness for the food manufacturer. Reactive, sensitive or volatile additives (vitamins, cultures, flavors, etc.) can be turned into stabilize by suitable means for proper delivery. The good and wholesome image of milk and dairy products make them attractive as delivery vehicles for a range of nutraceuticals. Among the different techniques, microencapsulation offers opportunities to improve the nutrient content of dairy and food products without affecting their sensory qualities. Microencapsulation may be sued for stabilizing a desirable component, reducing the level of an undesirable component in a food or enabling the targeted delivery of nutraceutical to specific sites within the body. Thus, it is a useful tool for the development of new functional food products.

Hurdles Affecting the Functional Food Ingredients during Processing, Preservation and Storage and Gastrointestinal Tract Transient and their Proper Delivery Systems

Long chain poly-unsaturated fatty acids

Long chain omega-3 and omega-6 fatty acid have many health benefits and have been implicated in reducing the risk of heart disease, inflammatory and immune disorders and has a role in improving early development. The incorporation of fish oils or microencapsulated fish oil ingredients in manufactured dairy products post-farm is one robust method for value addition. Numerous challenges exist in the production, transportation, and storage of poly-unsaturated fatty acids (such as w-3, w-6 fatty acids) fortified functional food products because oils high in poy-unsaturated fatty acids are extremely susceptible to oxidative deterioration. It has been a challenge for oil refiners to inhibit lipid oxidation of oils high in w-3 fatty acids during processing, shipping, and storage. Additional challenges exist in preventing the oxidation of w-3 fatty acids when they are incorporated into processed foods **(Table 1)**.

Table 1. Categories of different functional ingredients and their functions

Food Modifications	**Functionality**
Incorporation of phytochemicals (as plant ingredients or extracts)	Antioxidant, lower risk of CHD, cancer, and lower blood pressure
Addition of probiotics	Improved gastrointestinal function, enhanced immune system, lower risk of colon cancer and of food allergy
Additional of prebiotics	Improved gastrointestinal function, lower risk of colon cancer, enhanced immune system
Addition of bioactive proteins or peptides	Enhanced immune function and bioavailability of minerals, hypertensive function
Addition of dietary fibers	Prevention of constipation, lower risk of colon cancer and lowering of blood cholesterol level
Addition of þ-3 PUFA	Lower risk of heart attack, lower risk of some cancers, enhanced immune system
Removal of allergens	Reduce or eliminate allergy to specific foods
Hydrolysis of lactose by adding β-galaoctosidase	Enable digestion of lactose by lactose-intolerant persons

Techniques used to control the oxidation of w-3 fatty acids in bulk oils are limited to temperature control, oxygen-free atmospheres, and lipid-soluble antioxidants. The most effective antioxidants in bulk oils are polar-free radical scavengers. These polar antioxidants are often not effective once the oils are dispersed into water-based foods because the antioxidant will migrate from the lipid into the aqueous phase where it is not as effective at protecting lipids. Oil-in-water emulsions may be better systems to increase the oxidative stability of lipids than bulk oils because additional protective measures such as water-soluble antioxidants, chelators, and emulsion droplet interfacial engineering can be used to inhibit lipid oxidation. Oil-in-water emulsions may be easier to disperse into water-based foods such as beverages, dairy products, salad dressings, and muscle foods than bulk oil that could physically separate from the aqueous phase during storage. Because the emulsion would maintain similar physical structures when incorporated into water-based foods, their antioxidant protection systems (for example, emulsion droplet interfacial engineering) could remain intact.

Proteins represent GRAS food additives that can form physically stable oil-in-water emulsions while altering the properties of the emulsion droplet interface in a manner that increases the oxidative stability of the lipid core. When pH values are below the isoelectric point (pI) of the protein, the emulsion droplet interfacial proteins will have a positive charge that repels transition metals away from the oil droplet, thus increasing oxidative stability. If these protein-stabilized emulsions are to be used by the food industry to inhibit lipid oxidation, they must be able to withstand the rigors of food-processing operations and remain stable during prolonged storage. Whey proteins can be used to produce a low-viscosity oil-in-water emulsion at oil concentrations ranging from 5% to 30%. These oil-in-water emulsions have excellent physical stability that is not adversely affected by thermal processing.

The use of microencapsulated long-chain polyunsaturated oils eliminates fishy orour and taste and enables the development of improved products enriched with these fatty acids. A supplement comprising a blend of omega-3 fatty acids, omega-6 fatty acids (gama-linoleinc acid C18: 3n-6 and arahidonic acid C20:4n-6) amd evening primrose oil encapsulated in gelatin may be provided for addition to infant formualae to achieve a milk composition approximating that of human milk. Infant formulae preparations fortified with miroencapsualted spray dried marine oil powders have been successful in the market place. Yoghurts, feremted milks and processed cheese with tuna oil encapsulated with processed milk-proetein-carbohydrate films (Driphorm 50) made using MicroMAX technology have higher sensory scores than those fortified with an equivalent amount of encapsulated oil.

The development of spray-dried microencapsulated fish oil was under taken as part of EU FAIR contact 9CT 95-0085). The objective of this study was to establish a delivery system for fish oil in powdered form so that there would be a degree of protection from oxidation during storage, containment of fishy odour as far as possible, and finally to try to achieve the highest oil content possible in dry matter. Deodorized sand-ell oil stabilized with natural antioxidants was used as a fish oil source that was formulated in an emulsion system incorporating oil: protein: lactose: water in 10:10:10:70 ration, respectively. Both processing variables (homogenization pressure and number of passes, and spray-drying effects) and packaging (vacuum vs nitrogen flushing) were studied. Based on physical indicator, it was concluded that homogenization pressure and protein source (sodium caseinate, calcium caseinate, and skim milk powder) influenced free fat and surface fat contents. Skim milk powder as a protein source, however, gave better sensory scores when used in the formulation of the microencapsulated fish oil emulsion. The resulting micro-encapsulated fish oil powder had very good sensory properties and was stable for upto 6 months provided it was stored under chilled conditions.

Vitamin and Mineral Fortification

Vitamin and mineral fortification has been used to improve nutrient content of many foods. The levels of vitamins decrease during processing and storage. It is necessary to take into account the impact of the interactions between the added minerals and vitamins with other components in foods. Many encapsulated preparations for addition to a range of beverage and food products has been developed to overcome undesirable interaction of vitamins with the environment and other components in food during processing and storage.

Vitamins: Many of the vitamins are relatively unstable and their activity in foods depends on pH and their stability to heat, light, oxygen, oxidizing agents and enzymes. The use of microencapsulated forms for the delivery of vitamins to dairy and food products reduces loss of vitamins during storage. The method and stage of addition of microencapsulated ingredients into diary and food products also affects the stability of vitamins during storage. Higher levels of the added vitamin D are entrapped into the cheese curd during cheddar cheese manufacture when the cheese milk is fortified with vitamin D preparation entrapped in liposomes compared to when vitamin D homogenized in cream or a water-soluble emulsion of vitamin D is added to cheese-milk. Although vitamin D delivered in liposomes were effective methods for increasing the level of vitamin D in cheese immediately after manufacture, the stability of the vitamin D in liposomes that have been incorporated into the cheese decreases at a faster rate after five months' storage **(Table 2).**

Table 2: Stability of various vitamin preparations

Product	Type of vitamin preparation	% Loss
Instant skim milk powder	Gelatin-encapsulated vitamin a (after simulated instantising treatment)	60% in 40 weeks of storage
	Gelatin-encapsualted vitamin A (Dry blends with non-fat dry millk)	Approx. 10% in 40 weeks storage
Cheddar cheese	Water-soluble emulsion of vitamin D	16% in 7 months
	Vitamin D homogenized in cream	11% in 7 months
	Vitamin D entrapped in liposomes	40% in 7 months

Minerals: There is a growing interest in fortification of foods with different minerals such as Ca, Fe, Mg, Mn, Zn, etc. The addition of free mineral salts is having undesirable interactions between mineral salts and components in milk and milk products can lead to precipitation, colour and flavour problems and affect their bioavailability. Encapsulated mineral salts lessen the tendencies fro undesirable interactions. The choice between fortification with microencapsulated minerals or direct addition of mineral salts is depend on their relative costs, the bioavailability of added components and their impact on the sensory properties of final food products. The fortification of milk with iron presents different challenges. The most common iron salts (e.g. ferrous sulphate). The addition of these salts can affect the sensory properties of the food due to the taste of the iron salt or the catalytic effect of iron the oxidaiotn of fats. Microencapsulated iron ingredients can prevent off-flavour development and maintain bioavailability of the iron. Jackson and Lee (1992) showed that stearine-coated iron salts because less oxidation of fat in Harvati cheese compared to those fortified with unprotected iron salt. A nutritional iron composition that contains emulsifiers (e.g., lecithin) that is capable of forming liposomes that may be used for fortification of beverages while minimizing off-flavours and interactions with other components in the food. The use of ferrous sulphate encapsulated in lecithin (SFE-171, Biofer) is claimed to allowed effective fortification of fluid milk and dairy products, while preventing undesirable interactions with milk components and providing highly bio-available iron. Iron absorption from milk with the use of encapsulated ferrous sulphate SFE-171 is higher than that with the direct addition of ferrous sulphate. Some of the major problems encountered by direct addition of simple calcium salts, such as loss of heat stability due to increase in calcium activity can be overcome by using calcium complexing agents or insoluble calcium salts. A possible strategy for fortification of fluid milk includes the use of micro-crystalline cellulose-based ingredient co-processed with calcium carbonate and carboxymethylcellulose, which results in milks with good flavour and stability. Alternatively, liposomes may be used to protect calcium salts from interactions with proteins at higher temperature, as their use to prevent precipitation of soy proteins in calcium-fortified soy milk.

Probiotics

Probiotics are added as live cultures in a range of food products to improve the microbial balance of the human gut. Their survival in foods during processing, preservation and storage as well as during GIT transient can be issue for development of probiotic functional foods. The viability of live probiotics in any product at the point of consumption is an important consideration for their efficacy, as they have to survive during the processing and shelf life of food and supplements, transit through gastrointestinal tract, where high acidic conditions of the stomach and enzymes and bile salts in the small intestine affect adversely the probiotics' survival. The consumption of probiotics at a level of 10^8-10^9 cfu/ day is a commonly quoted figure for adequate probiotic consumption, equating to 100 g of a food product with 10^6-10^7 cfu/g. However, probiotic strains exhibit poor survival in traditional fermented dairy products in different countries. Probiotic survival in products is affected by a range of factors including pH, post-acidification (during storage) in fermented products, hydrogen peroxide production, oxygen toxicity (oxygen permeation through packaging), storage temperatures, stability in dried or frozen form, poor growth in milk, lack of proteases to break down milk protein to simpler nitrogenous substances and compatibility with traditional starter culture during fermentation. Probiotic suppliers have developed a variety of proprietary technique to preserve and protect the integrity of these tiny living organisms. Viability of probiotics can be improved by appropriate selection of acid and bile resistant strains, use of oxygen impermeable containers, two-step fermentation, stress adaptation, incorporation of micronutrients such as peptides and amino acids, sonication of yogurt bacteria and microencapsulation. Microencapsulation is the most suitable alternative technology to offer the best protection to the probiotic cells resulting from the freeze-drying and milling and such microencapsulated probiotics can be used in numerous food systems **(Table 3 and 4)**.

Table 3: Survival of prebiotics in milk and milk products

Products	Storage conditions	Form of bacteria added	Counts (cfu/ml)	
			Initial	After storage
Fermented milk	21 days at 5°C	Free *B. longum*	$5.6x10^7$	$1.3x10^7$
		Free *B. bifidum*	$4.5x10^6$	$7.5x10^6$
Yogurt	30 days at 4.4°C	Free *B. longum* B6	1.5110^9	$3.54x10^8$
		B. longum B6 enapsulated in k-carrageenan	$1.51x10^9$	$1.02x10^9$
		Free *B. longum* ATCC 15708	$1.51x10^9$	$4.35x10^8$
		B. longum ATCC 15708 enapsulated in K-carragenna	$1.51x10^9$	$1.48x10^9$
Milk with 2% fat	12 days at 4°C	Free *B. longum* Bb-46	$1x10^7$	$1x10^4$
		B. longum Bb-46 encapsulated in Ca alginate	$4x10^7$	$3x10^5$
		Free *B. lactis* Bb-12	$1x10^7$	$1x10^7$
		B. lactis Bb-12 encapsulated in Ca alginate	$2x10^8$	$2x10^8$
Milk chocolate	60 days at 7°C	Free *L. casei* NCDC 298	8.40 log cuf/g	8.55 log cuf/g
		L. casei NCDC 298 encapsulated in Ca alginate	8.38 log cuf/g	8.48 log cuf/g
	30 days at room temperature	Free *L. casei* NCDC 298	8.44 log cuf/g	5.52 log cuf/g
		L. casei NCDC 298 encapsulated in Ca alginate	8.46 log cuf/g	6.88 log cuf/g

Table 4: Immobilization/Encapsulation of cells for food/biotechnological application

Culture	Technique/Mechanism	Product
B. bifidum, B. infantis	Calcium alginate	Mayonnaise
L. paracasei	Milk fat	Cheddar cheese
Enterococcus faecium	Milk fat	Cheddar cheese
B. bifidum, B. adolescentis	Cream	White brined cheese
B. bifidum, B. infantis, and B. longum	Calcium alginate gels	Crescenza cheese
L. lactis subspp. Lactis	k-Carrageenan and locust bean gum	Fresh cheese
L. casei	Liquid core alginate capsule	Lactic acid
Lactobacilli	Calcium alginate	Frozen dessert
Lactococci	Calcium alginate	Cream
L casei	k-Carrageenan and locust bean gum	Yoghurt

Sustained and Target Release of Functional Ingredients

Immunoglobulins have potential in functional food development as they afford protection against gastrointestinal infections. However, they are prone to inactivation in the gut. Encapsulation of immunoglobulins may be sued to preserve their activity in certain environments. Milk immunoglobulin G (IgG) that was encapsulated in double emulsions, solid agar or calcium alginate gels had improved stability in aid and alkali environments as well as making them less susceptible towards the action proteinases. Sustained release of amino acids after ingestion of protein supplements is desirable in sports nutrition and for improved exercise performance. A process for using liposomal, ion-exchange whey protein as a protein supplement that maintains plasma amino acids at higher levels compared to when conventional proteins supplements are ingested. Liposome preparations containing cholesterol-lowering plant sterols and stanols have been developed for use in milk and dairy products. The use of these preparations is an alternative to the use of the free stanols and sterols that may have limited solubility in some food products.

Microencapsulation an Efficient Delivery System

Microencapsulation is a process where droplets of liquids, solids, or gases (core) are coated by thin films (coatings), which protect the core until it is needed. In the foods alone, a large number of substances have been microencapsulated, such as acidulants, amino acids, antimicrobials, bases, colorants, edible oils, flavor, enzymes, microorganisms, flavor enhancers, leavening agents, minerals, sugars, salts, and vitamins. The core can be released at different times as and when occasion demands by any desired mechanisms, such as disruption, dissociation, dissolution or diffusion and with any desired rates, such as instantaneous, delayed, controlled or sustained release depending on the properties of the coatings that are applied. The coating on a core is semi-permeable and protects the core from severe conditions and controls substances flowing into the core and the release of metabolites from the core.

Encapsulation involves the incorporation of food ingredients, enzymes, cells or other materials in small capsules. Applications for this technique have increased in the food industry since the encapsulated materials can be protected from moisture, heat or other extreme conditions, thus enhancing their stability and maintaining viability. Encapsulation in foods is also utilized to mask odours or tastes. Various techniques are employed to form the capsules, including spray drying, spray chilling or spray cooling, extrusion coating, fluidized bed coating, liposome entrapment, coacervation, inclusion complexation, centrifugal extrusion and rotational suspension separation **(Table 5)**. A wide variety of foods is encapsulated-flavouring agents, acids, bases, artificial sweeteners, colourants, preservatives, leavening agents, antioxidants, agents with undesirable flavours, odours and nutrients, among others. The use of encapsulation for sweeteners such as aspartame and flavours in chewing gum is well known. Fats, starches, dextrins, alginates, protein and lipid materials can be employed as encapsulating materials. Various methods exist to release the ingredients from the capsules. Release can be site-specific, stage-specific or signaled by changes in pH, temperature, irradiation or osmotic shock. In the food industry, the most common method is by solvent-activated release. The addition of water to dry beverages or cake mixes is an example. Liposomes have been applied in cheese-making, and its use in the preparation of food emulsions such as spreads, margarine and mayonnaise is a developing area. Most recent developments include the encapsulation of foods in the areas of controlled release, carrier materials, preparation methods and sweetener immobilization. New markets are being developed and current research is underway to reduce the high production costs and lack of food-grade materials.

Table 5. Various encapsulation techniques and the processes involved in each technique

No	Microencapsulation technique	Major steps in encapsulation
1	Spray-drying	a. Preparation of the dispersion b. Homogenization of the dispersion c. Atomization of the infeed dispersion d. Dehydration of the atomized particles
2	Spray-cooling	a. Preparation of the dispersion b. Homogenization of the dispersion c. Atomization of the infeed dispersion
3	Spray-chilling	a. Preparation of the dispersion b. Homogenization of the dispersion c. Atomization of the infeed dispersion
4	Fluidized-bed coating	a. Preparation of coating solution b. Fluidization of core particles c. Coating of core particles
5	Extrusion	a. Preparation of molten coating solution b. Dispersion of core into molten polymer c. Cooling or passing of core-coat mixture through dehydrating liquid
6	Centrifugal extrusion	a. Preparation of core solution b. Preparation of coating material solution c. Co-extrusion of core and coat solution through nozzles
7	Lyophilization	a. Mixing of core in a coating solution b. Freeze-drying of the mixture
8	Coacervation	a. Formation of a three-immiscible chemical phases b. Deposition of the coating c. Solidification of the coating
9	Centrifugal suspension separation	a. Mixing of core in a coating material b. Pour the mixture over a rotating disc to obtain encapsulated tiny particles c. Drying

Contd...

10 Co-crystallization	a. Preparation of supersaturated sucrose solution b. Adding of core into supersaturated solution c. Emission of substantial heat after solution reaches the sucrose crystallization temperature
11 Liposome entrapment	a. Microfluidization b. Ultrasonication c. Reverse-phase evaporation
12 Inclusion complexation	Preparation of complexes by mixing or grinding or spray-drying

Consideration of Materials for Micro-Encapsulation

The structure formed by microencapsulating agent around the core material is called the wall material which protects the core against deterioration, limits the evaporation of volatile core materials. The encapsulating agents should have certain ideal characteristics, depending on the objectives and requirements, process of encapsulation, chemical characteristics of the core material, the intended use of the core material, the conditions under which the product will be stored, and the processing conditions to which it will be exposed. Some typical encapsulation agents are dextrans, gums, starches or proteins **(Table 6)**. Many coating materials have been used for encapsulation of microorganisms. These include a mixture of k-carrageenan and locust bean gum, cellulose acetate phthalate, alginate, alginate-starch mixture, k-carrageenan.

Table 6: Coating materials used to produce microcapsules

Class of coating materials	Specific types of coatings
Gums	Gum arabic, agar, sodium alginate, carrageenan
Carbohydrates	Starch, dextran, sucrose, corn syrup
Celluloses	CMC, methylcellulose, ethylcellulose, nitrocellulose, acetylcellulose, cellulose acetate-phthalate, cellulose acetate-butylate-phthalate
Lipids	Wax, paraffin, tristearin, stearic acid, monoglycerides, diglycerides, beeswax, oils, fats, hardened oils
Inorganic materials	Calcium sulfate, silicates, clays
Proteins	Gluten, casein, gelatin, albumin

Additional Treatments of Microcapsules

Entrapment in hydrocolloid gels, such as alginate, k-carrageenan, etc. have some limitations due to less stability of microcapsules in the presence of chelating agents such as phosphate, lactate, citrate etc., which share the affinity for ions such as Ca^{+2}, K^{+}, etc. and destabilize the gel. The problems are encountered during lactic acid fermentation and cause cell release from the beads. In other matrix material, such as chitosan, the entrapped cells can be released form the beads during fermentation and cause low initial loading for the next fermentation. Therefore, additional treatments, such as coating the beads, are applied to improve the properties of beads. Coated beads not only prevent cell release but also increase mechanical and chemical stability. Cross-linking with cationic polymers, coating with other polymers, mixing with starch and incorporating additives improve stability of beads.

Conclusion

The uptake of functional food and drinks, there remain growth opportunities in the global health market as scientific studies uncover the benefits of emerging and existing ingredients. The discovery of the benefits of new natural ingredients is an important development in the market as consumers become increasingly conscious of what they are putting in and on their bodies. With carefully fine-tuned controlled release properties and stabilization of functional ingredients in food and during GIT transient is the key of development of functional foods. Among the different techniques, microencapsulation is no longer just an added value technique, but the source of totally new ingredients with matchless properties and can be applied in the development of new and novel functional foods. It is only one of a suite of technologies that may be applied to enhance the quality of healthy dairy foods and its suitability depends on the food product to be fortified, the need for protection of food components and timed release of nutraceuticals. It can provide novel solutions to problems encountered in the development of healthy properties of foods but it adds costs to the production of these foods.

Suggested Reading

Augustin, M. A. 2003. The role of microencapsulation in the development of functional dairy foods. *The Australian Journal of Dairy Technology*, **58**(2): 156-160.

Chen, L., Remondetto, G.E. and Subirade, M. 2006. Food protein-based materials as nutraceutical delivery systems. *Trends in Food Science and Technology,* **17**(5): 272-283.

Desai, K.G.H and Park, H.J. 2005. Recent Developments in Microencapsulation of Food Ingredients. *Drying Technology,* **23**(7): 1361 – 1394

Frankel EN, Satue-Gracia T, Meyer AS, German BJ. 2002. Oxidative stability of fish and algal oils containing long-chain polyunsaturated fatty acids in bulk and in oil-in-water emulsions. *Journal of Agricultural Food Chemistry,* **50**:2094–9.

Gouin, S. (2004). Microencapsulation industrial appraisal of existing technologies and trends. *Trends in Food Science and Technology,* **15**(7-8): 330-347.

Hu M, McClements DJ, Decker EA. 2003. Impact of whey protein emulsifiers on the oxidative stability of salmon oil-in-water emulsions. *Journal of Agricultural Food Chemistry,* **51**(5):1435–9.

James, L., Puniya, A.K. and Singh, K. 2002. Health promoting functional synbiotics - a mini review. In: *Proceedings of Summer Institute, 'Advances in applied dairy and food microbiology'*, Dairy Microbiology Division, National Dairy Research Institute, Karnal, June 12 - July 11, pp. 209-214.

Makhal, S., Mandal, S., Kanawjia, S. K. and Singh, S. (2005). Value addition to foods for health. In: *Indian Dairy Industry* published by Chawla Dairy Information Centre Pvt. Ltd., New Delhi-110092, pp: 157-162.

Makhal, S., Mandal, S., Kanawjia, S. K. and Singh, S. (2005). Value addition to dairy products. In: *Indian Dairy Industry: Annual dairy resource book.* (Vol. I), 1st edn. Eds. Chakraborty, G.C., Dr. Chawla Dairy Information centre Pvt. Ltd., New Delhi - 110092, pp: 120-128.

□□□

Chapter 12

Development of Bioactive Cheese: A Holistic Biotechnological Approach

S.K. KANAWJIA, A. CHATTERJEE AND K. KUNAL
Dairy Technology Division, National Dairy Research Institute, Karnal (HR)

Introduction

Diet-health link is now an integral part of healthy life style. The role of diet and specific foods for the prevention and treatment of disease and improvement of body functions is being investigated around the world. Since nowadays, consumers are more diverted towards the health driven foods, hence recent advances in research on bioactive peptides show much promise in new product development using these biomolecules to derive multifarious health benefits. Bioactive foods enriched with health giving bioactive peptide holds a very promising future because researchers are realizing the importance of the importance of the poly functional roles of bioactive peptides. The *in situ* production of bioactive peptides in fermented dairy products has now been conceptualized as a novel approach to improve their health value of the products. While cheese has long been associated with a high quality nutritional image, more recently research efforts devoted to the development of bioactive cheese with demonstrated health promoting properties of bioactive peptides have been made to push cheese into the "functional foods" category.

Bioactive Peptides: A Functional Biomolecule

Proteins possess biological and physico-chemical properties. For example, milk is known to contain a wide range of proteins, which either provide protection against enteropathogens or are essential for the manufacture and characteristics of certain dairy products. Research carried out during the last 10 years has shown that both major milk protein groups, caseins and whey proteins, can also be an important source of biologically active peptides. Functional or bioactive peptides are described as defined sequences of amino acids which are inactive within the native protein, but which display specific properties once they have been released by enzymatic activity. These peptides are in an inactive state inside the protein molecule and can be released during enzymatic digestion *in vitro* or *in vivo*. Bioactive peptides usually contain 3 to 20 amino acid residues per molecule, except for glycomacropeptides (GMP) which consists of 64 amino acid residues. The sequence of the amino acids in these small molecules is a crucial factor in their activity. The amino acid located in the C-terminal or N-terminal position is often important (Nayak *et al.*, 1999).

The formation and properties of milk protein-derived peptides have been reviewed in many recent articles. Peptides with biological activity can be produced from milk proteins in three ways:

- Enzymatic hydrolysis with digestive enzymes
- Fermentation of milk with proteolytic starter cultures
- Through the action of enzymes derived from proteolytic microorganisms

The bioactivities of peptides encrypted in major milk proteins are latent until released and activated by enzymatic proteolysis, e.g. during gastrointestinal digestion or food processing like during fermentation. The proteolytic system of lactic acid bacteria (LAB) can contribute to the liberation of bioactive peptides. *In vitro,* the purified cell wall proteinase of *Lactococcus lactis* was shown to liberate oligopeptides from β- and α- caseins, which contain amino acid sequences present in casomorphins, casokinines and immunopeptides. The further degradation of these peptides by endopeptidases and exopeptidases of LAB could lead to the liberation of bioactive peptides in fermented milk products. Opiod peptides are opiod receptor ligands, which can modulate absorption processes in the intestinal tract, angiotensin-I-converting enzyme (ACE) inhibitory peptides are hemodynamic regulators and exert an antihypertensive effect, immunomodulating casein peptides stimulate the activities of cells of the immune system, antimicrobial peptides kill sensitive microorganisms, antithrombotic peptides inhibit aggregation of platelets and caseinophosphopeptides (CPP) may function as carriers for different minerals, especially calcium. Bioactive peptides can interact with target sites at the luminal side of the intestinal tract (Meisel and Bocklmann, 1999).

It is now evident that during the fermentation of milk with certain dairy starters, peptides with various bioactivities can be formed and are detected in an active form even in the final products, such as fermented milks and cheese. Furthermore, they can be absorbed and then reach peripheral organs. Food-derived bioactive peptides are claimed to be health enhancing components, which can be used for functional food and pharmaceutical preparations. These peptides are likely to be produced industrially in the future in the form of hydrolysates or peptide mixtures, which can be used as ingredients for various dietary or pharmaceutical products or can be produced by conditioned fermentation in several fermented milk products.

Development of Bioactive Cheese

Dietary patterns of today's consumers have been changed and they are going for foods having polyfunctional health benefits. Hence, the important allure of the modern health conscious consumer is to consume functional foods to derive health-giving and curative functions, naturally imparted to foods. In an endeavour to expand the range of bioactive dairy products, a small number of researchers and companies have attempted to manufacture cheeses, which contain a considerable amount of these wonder molecules viz bioactive peptides. Cheese will have a number of advantages over the non-fermented products as a delivery system for bioactive peptides. Given the increasingly competitive scenery of the European market in food products, the cheese industries in Ireland and throughout Europe are trying to derive benefit from a marketing advantage, such as added-value probiotics and bioactive peptides containing cheese, which would afford a competitive edge over existing products. The development of bioactive cheese could thus lead to a major economic advantage.

Growing public awareness of diet-related health issues has also fuelled the claim for probiotic foods, which are currently restricted predominantly to fermented milk drinks and yoghurt harbouring beneficial probiotic cultures, such as bifidobacteria and lactobacilli associated with a plethora of health benefits. While cheese has long been associated with a high quality nutritional image, more recently research efforts devoted to the development of bioactive cheese with demonstrated health benefits of bioactive peptides have been made to make cheese more promising as functional foods. Generally, incorporation of probiotic bacteria into milk-based food systems, including cheese still remains at the Biblical age in terms of maintaining viability and probiotic functionality during manufacture and shelf life. Therefore, there is a good potential to incorporate some health attributes in cheese by *in situ* synthesis of bioactive peptides to derive some health benefits beyond its basic nutrients. This approach of enriching cheese with bioactive peptides seems to be more attractive in terms of its stability and maintaining functionality.

Diverse Health Attributes of Bioactive Peptides

The multifunctionality of physiologically active peptides derived from milk and their wide distribution among mammals could confer on them a role of messenger molecules. Their contribution to the health of the newborn would be three-fold: an easily assimilated source of organic nitrogen, a good source of essential amino acids and a potential source of bioactive molecules. Several casein-derived peptides may play a significant role in the stimulation of the immune system. They have been found to exert a protective effect against microbial infections and to enhance some functions of the immune system (Migliore-Sammour *et al.*, 1989).

Milk-derived peptides have been shown to exert several beneficial physiological effects. They serve to modulate metabolic processes like digestion, circulation, immunological responsiveness, cell growth and repair, and nutrient intake. These functional peptides display partial resistance to hydrolysis and can exert their effects either locally in the digestive tract or elsewhere in the body. The milk derived peptides with various biological activities, such as opioids, mineral-carriers, antihypertensives, immuno-stimulants, antithrombotics and antigastrics. These bioactive peptides are generated *in vivo, in vitro* and during food processing. Bioactive peptides are largely found in milk, fermented milks and cheeses. Proteolytic enzymes naturally occurring in milk and enzymes from lactic acid bacteria or from exogenous sources contribute to the generation of bioactive peptides. Dairy processing conditions, such as cheese ripening are also relevant (Smacchi and Gobetti, 2000).

In contrast to endogenous bioactive peptides, many milk-derived peptides have multifunctional properties. Regions in the primary structure of caseins contain overlapping peptide sequences, which exert different biological effects. These regions, considered to be strategic zones, are partially protected from proteolytic breakdown. For example, most â-casomorphins and casokinins arc both ΛCE-inhibitory and immunostimulatory and α- and β-lactorphin contain sequences with both opioid and ACE-inhibitory activities (Meisel 1998). It has been reported that physiologically active peptides also have the ability to promote the growth of bifidobacteria in the GIT, thus increasing the number of these beneficial organisms. This is attributed particularly to k-casein, which can affect bacterial adhesion as a receptor analogue (Lonnerdal, 1998).

Feeding mice with milk fermented by *B. longum, Lactobacillus casei* or *Lactobacillus helveticus* has shown a significant stimulation of phagocytosis by pulmonary macrophages Perdigon *et al.* (1998) suggest that the increased number of cells secreting IgA (but not IgG) in the large intestine of mice given yoghurt could contribute to limit the inflammatory immune response. Since IgA is considered to be an immune; barrier in colonic neoplasia. The modulation of mucosal inflammation by IgA is important to prevent the tissue damaging consequences of a permanent inflammatory response, which occurs during the development of tumors and neoplasia. This suggests that the mechanisms by which yoghurt inhibits tumour development could be through the decrease of the inflammatory response. Since the increase of the IgA response might be related to the proteolytic activity of *L. helveticus,* we are currently investigating the role of individual peptides on the immunostimulation and the regression of mucosal tumours.

They could in the future have interesting dietary and pharmacological applications. Biologically active peptides have been found to have specific activities, such as antihypertensive, antimicrobial, immunomodulatory, opioid or mineral-binding. Many milk-derived peptides reveal multi-functional properties, that is, specific peptide sequences may exert two or more different biological activities. Due to their physiological and physico-chemical versatility, milk-borne bioactive peptides are regarded as highly prominent ingredients for health-promoting functional foods or pharmaceutical preparations. **(Table 1)** exhibits different bioactive milk derived peptides and their biological activities:

Table 1: Bioactive Peptides Derived From Milk Proteins

Bioactive Peptides	Protein Precursor	Bioactivity
α-Casomorphins	α-Casein	Opioid agonist
β-Casomorphins	β-Casein	Opioid agonist
α-Lactorphin	α-Lactalbumin	Opioid agonist
α-Lactorphin	β-Lactoglobulin	Opioid agonist
Lactoferroxins	Lactoferrin	Opioid antagonists
Casoxins	X-Casein	Opioid antagonists
Casokinins	α-, β-Casein	ACE-inhibitory
Casoplatelins	X-Casein, Transferrin	Antithrombotic
Lactoferriein	Lactoferrin	Antimicrobial
Immunopeptides	α-, β-Casein	Immunomodulatory
Phosphopeptides	α-, β-Casein	Mineral carriers

Source: Meisel and Boekelmann (1999)

Cheese: Recognized Bioactive Peptides

Bioactive peptides can also be produced from milk proteins through fermentation of milk by starters employed in the manufacture of fermented milks and cheese. In particular, antihypertensive and mineral binding peptides have been identified in fermented milk, whey and ripened cheese. A few of these peptides have proved effective also in humans after ingestion of fermented milk products. LAB used in cheese manufacture might be indirectly involved in the modulation of the immune response of the host by contributing, during the fermentation process, to the release of peptides bearing hormone-like activities. During the fermentation of milk by LAB, casein undergoes a slight proteolysis capable of generating potentially bioactive peptides.

Bioactive peptides may be liberated during the manufacture of cultured milk products, such as yoghurt, *Dahi*, cheese, etc. Bacterial starter cultures contain several proteolytic enzymes that are responsible for the breakdown of protein into peptides and amino acids during cheese maturing. During fermentation of milk, various long oligopeptides are liberated by degradation of caseins, which would be precursors of peptides with biological activity when cleaved by other enzymes. In fermented milk products and cheese, intracellular peptidases of LAB will most likely contribute to further degradation after cell lysis. The specificities of known peptidases suggest that all peptide bonds in caseins may be cleaved partially. Various studies have reported on the casomorphins, ACE-inhibitory peptides and phosphopeptides found in several fermented milk products including cheese. Cheese contains phosphopeptides as natural constituents and extensive proteolysis during cheese ripening leads to the formation of other bioactive peptides, such as ACE-inhibitory peptides and precursors of b-casomorphins having several healths attributes.

The starter culture applied in the manufacture of "Festivo" cheese, a novel bioactive cheese, is a mixture of commercial starter cultures containing 12 different strains of the following genera or species" *Lactococcus* sp. and *Leuconostoc* sp. (BD-type culture), *Propionibacterium* sp., *Lactobacillus* sp. as well as *Lactobacillus acidophilus* and *Bifidobacterium* sp. It has been reported that the ACE-inhibitory activity increases during "Festivo" cheese ripening and decreases when proteolysis exceeded a certain level during the storage period. These results would suggest that ACE-inhibitory peptides and probably other biologically active peptides as well, are naturally formed in cheese. This called for the development of a totally new type of cheese with the health attributes of milk derived bioactive peptides. The new 'Festivo' cheese based on an innovative concept in the production of healthy foods has been commercially developed (Ryhanen *et al.*, 2001). This is the first time that a probiotic cheese along with the health benefits of bioactive peptides is being produced on an industrial scale. 'Festivo' is now manufactured commercially in Finland and has attracted growing interest among health-aware consumers. The new type of cheese, 'Festivo', developed for production on an industrial scale, has showed good results in organoleptic analysis. On the basis of this study, it is suggested that the new cheese may possessed multifunctional health effects, since it contain several beneficial components such as probiotics, bioactive peptides, conjugated linolein acid, and calcium.

Several bioactive peptides have been found in cheeses and may be due to the intense proteolysis during cheese ripening. Ripening conditions and the type of starters used affect bioactive peptides synthesis in cheese. LAB, used as starter microorganisms in dairy products or endogenous to milk microflora, have a proteolytic system, which include cell bound proteinases (CEP) and intracellular peptidases. CEP release many different oligopeptides into shorter fragments and amino acids, which contribute directly or as precursors to flavour. Some of these peptides are bioactive. The Festivo cheese was produced by using cheese starter cultures in combination with *Lactobacillus acidophilus* and bifidobacteria. The ripened cheese was found to contain bioactive peptides with potential antihypertensive effects (Ryhanen *et al.*, 2001).

Meisel *et al.* (1997) reported the presence of low-molecular weight ACE inhibitory peptides in several ripened cheeses. Proteolysis increases ACE inhibitory activity but only to a certain level after which the ACE inhibition index decreases. The ACE inhibitory activity in medium aged Gouda was about double that of the long ripened Gouda. A ás$_1$-casein derived antihypertensive peptide, isolated from 6 month-ripened Parmesan cheese (Addeo *et al.*, 1992), was not detectable after 15 months of ripening. Products having a low level of proteolysis (Quarg) have a low ACE-inhibition index. These results indicate that the bioactive peptides liberated by proteolytic enzymes from LAB during cheese ripening are degraded further to inactive fragments as a result of further proteolysis (Meisel *et al.*, 1997). The following table **(Table 2)** presents different bioactive peptides detected and identified in a number of cheeses by different workers.

Table 2: Bioactive Peptides Identified in Several Cheese

Product	Bioactivity observed
Quarg	ACE inhibitor
Parmesan Reggiano	β-casomprphin precursors
Comte'	Phosphopeptides
Cheddar	Phosphopeptides
Edam, Emmental, Gouda, Requefort, Tilsit	ACE inhibitory
Mozzarella, Italico, Crescenza, Gorgonzola	ACE inhibitory
(Italian varieties)	ACE inhibitory
Edam, Emmental, Turunmaa, Cheddar,	
Festivo (Finnish varieties)	
Cheddar, Edam, Swiss, Feta, Camembert,	Immunomodulatory, Antihypertensive
Blue vein (Australian varieties)	Antiamnesic, Opioid agonist
Gouda, Havarti, Emmental	ACE inhibitory

Source: IDF Bulletin 363

ACE inhibiting peptides have been isolated from several Italian cheeses characterized by short (Crescenza and Italico) and medium (Gorgonzola) ripening period (Smacchi and Gobbetti 1998). Another bioactive peptide containing the â-casomorphin-7 sequence, which has shown an antihypertensive activity, has been isolated from Crescenza and Cheddar cheeses. Proteolytic enzymes of LAB produce caseinophosphopeptides during ripening of cooked curd cheeses like Comte' or Grana Padano. Peptides derived from a cheese slurry prepared using *Lactococcus lactis* subsp. *lactis* as starter culture have anticarcinogenic effect (Kim *et al.*, 1995).

Bioactive Cheese: The Practical Revelation

Microbial starters, such as LAB in dairy products contribute to the synthesis of bioactive peptides. An increasing number of *in vitro* and *in vivo* studies reveal that biologically active peptides are released from bovine milk proteins upon microbial fermentation and hydrolysis by digestive enzymes. Peptides with different bioactivities can be found in fermented milks and cheese varieties, but the specificity and amount of peptides formed is regulated by starter cultures used for cheese ripening and the rate of proteolysis, which needs further extensive studies on the controlled maturation of cheese. Highly proteolytic starter cultures, for example, *Lactobacillus helveticus,* would seem to produce short peptides with various bioactivities in several cheeses. Food researchers are presently considering bio-active peptides as health-enhancing nutraceuticals for use in functional foods. Besides supplementation of food, production of desirable bioactive peptides during food processing, e.g. by use of specific enzymes or genetically transformed microorganisms is also a target area that seems to hold the interest of these workers. Presently, the development of bioactive cheese with the *in situ* production of bioactive peptides gets immense scientific interest. May in the forthcoming days, cheese with the enriched level of bioactive peptides would occupy place in market shelves. It is in this context that cheeses, a fermented milk product which is natural sources of peptides with their physiologically significance have a potential for future food research.

Bioactive peptides or related sequences proteolytic enzymes from LAB produce bioactive peptides in several dairy products, but, once liberated, bioactive peptides also influence the biochemical activity of the microbial communities in cheese. This role is probably been underestimated in dairy processing, which seems to be a challenge to the development of bioactive cheese. Processing conditions should be carefully controlled in order to synthesize specific bioactive peptide during cheese production by selected or genetically-modified micro-enzymatic processes. Given the advantages that enzymatic synthesis has over, chemical and recombinant DNA approaches to synthesize bioactive peptides, enzymes responsible for their generation in food processing may have a potential application. Trypsin, thermolisin, a-chymotrypsin, papain and subtilism are generally used for the preparative synthesis of biologically active peptides, such as angiotensin, caerulein, enkephalin, oxytocin and dynorphin. However, the restricted availability of suitable biocatalysts is still problem. With the modified stability, activity, and specificity via site-directed mutagenesis of well-known proteinases, proteolytic enzymes of LAB, due to their high regiospecificity and proven efficiency, may be assayed for *in situ* enzymatic synthesis of biopeptides in cheese.

In the selection of microbial starters, the sensitivity to peptides should be carefully considered because, after liberation in cheese ripening, these peptides may accumulate which could retard or block ripening due to enzyme inhibition. The use of specific peptide inhibitors may be helpful in order to understand and the role of different proteolytic enzymes in generating the desired flavour, removing bitterness in cheese and selecting LAB strains for starters or proteolytic enzymes for accelerating cheese ripening (Samacchi and Gobbetti, 1998).

Conclusion

Fermented food products have been consumed through many centuries by millions of human beings long before the existence of microorganism was scientifically demonstrated. Their nutritive value and therapeutic properties have been and still are widely recognized in all parts of the world and allow them to be classified as nutraceuticals. Bioactive peptides, isolated from fermented dairy products mainly from cheese, have strengthened the interest to develop bioactive cheese with controlled fermentation. One might foresee very promising future for the bioactive cheese with the health promoting effects of bioactive peptides in the field of functional foods and nutraceuticals particularly with regard to the prevention or attenuation of several symptoms related to physiological or infectious diseases. The combinations of special strains of bacteria, specific substrates and optimal fermentation/growth conditions are likely to yield synergies that would not be observed in conventional cheese ripening process. Molecular biology, biochemistry, microbiology, chemistry and many other fields of knowledge will certainly contribute to a better understanding of what can be considered as a new level of interaction between foods, microorganisms and the digestive system of most animal species including man. Development of bioactive cheese might have been successful in prevention of several physiological disorders of human beings. To decipher such a new reservoir of biological messages that is well encrypted in our food is quite a challenge. Fortunately, we can count on probiotics to help us better understand this new language and continue to improve our quality of life.

Suggested Reading

Berner, L.A. and O' Donnell, J.A. (1998). Functional foods and health claims Legislation: Application to dairy foods. *Int. Dairy J.*,**8**: 855-362.

Kanawjia, S. K., Makhal, S. and Harpreet, K. K. (2003) Novel bio-active cheese-Fads and Fantasy to Facts and Figures, CAS Compendium on Application of Biotechnology in Dairy and Food Processing, D. T. division, NDRI, Karnal. Pp 168-178.

Lonnerdal, B. (1998). Milk proteins and gut microflora of infants. In: Abstracts and Poster Presentations. 25th International Dairy Congress. Aarthus, Denmark. Sep. 21-24.

Makhal, S. and Kanawjia, S. K. (2005). Developments in cheese technology: A mini assessment. *Food and Pack*, **5**: 28-31.

Makhal, S., mandal, S. and Kanawjia, S. K. (2005). Probiotic in management of atherotic cardiovascular disease-A review, Indian J. dairy Sci., 58:147-161.

Meisel, H. and Bocklmann, W. (1999). Bioactive peptides encrypted in milk proteins: proteolytic activation and thropho-functional properties. *Antonie van Leeuwenhoek*, **76**: 207-215.

Meisel, H., Goepfert, A., and Gunther S. (1997). Occurrence of ACE inhibitory peptides in milk products. *Milchwissenschaft*, **52**: 307-311.

Nakamura, Y., Yamamoto, N., Sakai, K., Okubo, A. Yamazaki, S. and Takano, T. (1995) Purification and characterization of angiotensin I-converting enzyme inhibitors from a sour milk. *J. Dairy Sci.*, **78**: 777-83.

Nayak, S.K., Pattnaik, P., Arora, S. and Sindhu, J.S. (1999). Functional peptides from milk proteins. *Indian Dairyman*, **51** (8): 29-34.

Perdigon, G.,Valdez, J.C. and Rachid, M. (1998). Antitumor activity of yogurt study on possible immune mechanisms. *J. Dairy Res* **65**: 129-138.

Ryhanen, E., Leppala, A.P. and Pahkala, E. (2001). A new type of ripened, low fat cheese with bioactive properties. *Int. Dairy J.*, **11**: 441-447.

Smacchi, E. and Gobbetti, M. (2000). Bioactive peptides in dairy products: synthesis and interaction with proteolytic enzymes. *Food Microbiol.,* **17** (2):129-141.

□□□

Chapter 13

Functional Attributes of *Lactococcus lactis*: A New Insight

RAMESHWAR SINGH & POORNIMA
Dairy Microbiology Division, National Dairy Research Institute, Karnal (HR)

Introduction

Lactococcus lactis species is one of the most important group of lactic acid bacteria that are used in the dairy industry. It has two subspecies and a bv: *L. lactis* subsp. *lactis*; *L. lactis* subsp. *cremoris*; *L. lactis* subsp. *lactis* biovar *diacetylactis*. *L.. lactis* subsp. *lactis* and to lesser extent *L. lactis* subsp. *cremoris* have long been extensively used in starter cultures for dairy fermentation (*i.e* in cheeses, sour cream and butter), composed of single or multiple strains with or without other lactic acid bacteria. The pivotal role of *L. lactis* in dairy fermentation is acidification, mainly by producing L-lactic. They also contribute to the development of texture by producing exopolysaccharides, or to flavour by producing aromatic compounds (alcohols, ketones, aldehydes) or by citrate, amino acid or fat metabolism. Now-a-days, besides these technological properties, this bacterium is acknowledged through their functional attributes such as ability to produce organic acids and bacteriocins, their use as probiotics and for the vitamin and folate production. Research on the genetic and physiological properties of these bacteria has expanded rapidly in the last decade. Therefore, for the lucid understanding of functional attributes of *Lactococcus lactis* species, this review contained some of the up-to-date information on theirs crucial functional aspects so that lactococci with scientific evidence should be accepted in consideration to the development of a wide variety of new dairy products with health claims.

Probiotic Activity

The probiotic is commonly recognized as viable microorganisms that exhibit a beneficial effect on the health of the host when they are ingested. For example, the ability to reduce serum cholesterol levels, productivity of antimicrobial substrates, immunomodulatory activity, etc. is considered as effective properties of probiotics. The most widely used probiotic bacteria are lactobacilli and bifidobacteria, which have been isolated from the human gastrointestinal tract but the development of new probiotic strains, which are more feasible and beneficial organisms, is awaited in the dairy industry. In the selection of new probiotic microorganisms, the safety of probiotic microorganisms has been of prime importance. In this consent, lactococci would be promising species because of its GRAS status and their extensive usage in manufacturing dairy products such as cheese and fermented milk. Moreover, they can grow well in milk while some of the probiotic cultures, such as *Lactobacillus acidophilus* strains, do not grow during production of fermented milk products. So, in manufacturing fermented milk with *Lb. acidophilus* some additives, such as compounds which are resolved by protease, are used or dairy lactic acid bacteria are added as an accelerator. However, there have been few studies on the probiotic activity of lactococci since they are traditionally not considered to be natural inhabitants of the human gastrointestinal tract. But recently, several works showed the possibility of the presence of lactococci in the flora of the human or animal gastrointestinal tract.

a) Bile and pH tolerance

It has been reported that *L. lactis* subsp. *lactis* is capable of displaying adaptive response to the stress by media containing 0.3% bile salts, while in *Lc. lactis* subsp. *cremoris* the response is much weaker and also showed that *Lactcoccus lactis* ssp. *lactis* NIAI527 showed highest adhesion to cacao-2-cell lines and also tolerated low pH of 2·5 for 30 min and oxgall at 0·5–0·9%. Kimoto 2007 reported that *L. lactis* subsp. *lactis* bv. *diacetylactis* N7 had the best tolerance to 0.9% bile.

When the initial cell count of strain N7 was 9.2 log cfu/mL, it had a moderate survival rate (5.4 log cfu/mL) for 2 h at pH 2.0 or a better rate (7.8 log cfu/mL) for 2 h at pH 3.016. According to the result, strain N7 was tolerant to low pH and bile, indicating that strain N7 may reach the intestine alive after ingestion.

b) Anti-proliferative activity

It has been suggested that the peptidoglycans and cytoplasm fractions, as well as heat-killed whole cells of LAB, had significant anti-proliferative activities against several cancer cell lines. In particular, the cytoplasm fractions exhibited marked direct antiproliferative activities against colon and gastric cancer cell lines, whereas the peptidoglycans retarded growth of colon and bladder cancer cell lines. It has been also demonstrated that the prostate carcinoma cell line, PC3, was highly affected by only heat-killed whole cells of (*Lactococcus lactis* ssp. *lactis*), showed the decreased cell proliferation by 88%. The cytoplasm fractions of *Lactococcus lactis* ssp. *lactis* inhibited proliferation of two cancer cell lines by 50% at 23 μg ml-1 for SNUC2A (a human colon adenocarcinoma cell line) and 11 μg ml-1 for SNU-1 (a human gastric cancer cell line), respectively.

Delivery Vector and Immunomodulation

L. lactis is used as a live bacterial vector to deliver biological agents. Some advantages of using lactococci as a DNA delivery vehicle can be considered. (i) Oral administration of such a DNA vaccine vector could induce specific protection against food allergies by inducing a local IgA immune response. (ii) In contrast to immunization with naked plasmid DNA, no further plasmid amplification and purification steps are needed, considerably reducing cost and labor. Finally, (iii) an improved safety profile can be attained due to the fact that lactococci are food-grade bacteria that deliver genes encoding only the desired antigens, in contrast to similar delivery systems with pathogens. In a concise manner, some pivotal information related to the usage of *L. lactis* to deliver biological agents is as follows:

a) The production of IL-2 or IL-6 by *L. lactis* has been observed to stimulate early proliferation of mucosal T cells or augment B-cell growth and/or differentiation and immunoglobulin production.

b) It has been reported that Lactococci which transit with the diet are quite resistant to gastric acidity (90 to 98% survival) when mixed with food suggested that the food was providing protection to the bacteria. In contrast, only 10 to 30% of bacteria survive in the duodenum when ingested population was moderate. This property suggests that lactococci could be used as a vector to deliver specifically into the duodenum the proteins produced in the cytoplasm and showed appropriate alternative for targeting digestive enzymes such as lipase to treat pancreatic deficiencies.

c) Genetically modified *L. lactis* expressed heterologously recombinant vaccine antigens such as fragment C of the tetanus toxin and the E7 antigen of human papilloma virus (HPV) in consideration to elicit specific immune responses against the antigens in vaccinated mice.

d) It has been revealed that *L. lactis* subsp. *lactis* G50 induced the highest cytokine production (IL-12, IL-6 and TNF-á) in cultured macrophage. These results indicated that strain G50 showed an immune modulation by enhancing the Th1- type immune response. It also has been investigated that the total IgE antibody level suppressed in serum of immunized mice by oral administration of strain G50.

e) It has been demonstrated on the mucosal vaccination with CRR delivered in *L. lactis* produced immune responses, prevented pharyngeal infection with *S. pyogenes,* and promoted survival.

f) The phase I clinical trial using *L. lactis* expressing human IL-10 as a therapeutic treatment for inflammatory bowel disease demonstrated that mucosal drug delivery using *L. lactis* was effective and clinically well-tolerated. So, the use of *L. lactis* as a live bacterial delivery system is highly promising.

g) The activity of recombinant muIFN-g produced by genetically engineered *L. lactis* was confirmed in an antiviral assay using MoV cells infected with Vesicular Stomatitis Virus.. This recombinant strain could lead to the development of a new, well-tolerated vector to deliver active mu IFN-g at the mucosal level.

h) The genetically engineered *L. lactis* secreting a single-chain form of IL-12 was successfully used as an adjuvant in the context of mucosal vaccination against human papillomavirus type 16 (HPV-16), to modulate Th1/Th2 balance in an ovalbumin (OVA)-induced asthma model and to prevent an allergic reaction against the cow's milk allergen lactoglobulin (BLG).

Cholesterol Removal

Elevated cholesterol level causes a risk of hyper-cholesterolemic effect. Therefore, decreasing serum cholesterol levels is important to prevent the disease. It has been reported that cholesterol removal by some lactobacilli was only due to a disruption of the cholesterol micelles caused by the deconjugation and precipitation of cholesterol with the free bile salts as the pH of the media dropped by acid production during growth. It has been reported that *L. Lactis* subsp. *lactis* biovar *diacetylactis* N7 could remove as much cholesterol as *L. acidophilus* ATCC 43121 and achieved the highest cholesterol removal (97%, 68.1 microgram /ml). The manner of cholesterol removal by strain N7 corresponded to the manner of its growth. It has been found out that lactococci isolated from plant materials showed growth in broth containing 0.3% bile, and removed cholesterol from media during growth. It has been also reported on the two possible mechanisms to remove cholesterol from media. He suggested that the amount of cholesterol removal was strain specific. One is adhesion of the cholesterol to the cell surface. Another possible mechanism is an assimilation of cholesterol by the cells. It was reported that some strains of *Lb. acidophilus* incorporated some of the cholesterol into the cellular membrane. Similar mechanisms is involved in the cholesterol removal by *L. lactis* subsp. *lactis* bv. *diacetylactis* N7. It is evident that some cholesterol bound to the strain N7 cells because even the heat-killed cells of strain N7, which cannot take up cholesterol, could remove it from media. Moreover the amount of cholesterol removal during growth (i.e., growing cells) was significantly ($P < 0.05$) higher than that removed by heat-killed and resting cells. The cholesterol removed between viable cells and heat-killed cells suspension should reflect the incorporation of cholesterol into the cells during the growth. Thus, it seems that strain N7 can remove cholesterol from media both by binding of cholesterol to the cells and by the uptake of cholesterol into the cells during growth. It has been suggested that cholesterol removal by the lactococci was not related to the precipitation of cholesterol with free bile salts. All strains of lactococci could remove cholesterol from media without degrading cholesterol.

GABA Production

Aminobutyric acid (GABA) is synthesized by glutamate decarboxylase (GAD), a pyridoxal 5_-phosphate-dependent enzyme that catalyzes the irreversible decarboxylation of L-glutamate to GABA.

It is a non-protein amino acid, possesses well-known physiological functions such as neurotransmission, induction of hypotension, and diuretic and tranquilizer effects. Treatments for sleeplessness, depression, and autonomic disorders; treatment for chronic alcohol-related symptoms; and stimulation of immune cells have also been related to the administration of GABA. It has been proposed a general role for the inducible bacterial amino acid decarboxylases, including GAD, in the maintenance of the physiological pH and confers resistance to bacterial cells under acidic conditions. Production of GABA by *L. lactis* certainly occurred because the medium was acidified by bacterial growth. It has been reported on the GABA producing *L. lactis* designated as strain 0I-7 and showed the maximum activity of GAD in *L. lactis* at an acidic pH (4.7). Recently, Ith has been reported that *L. lactis* PU1 showed highest synthesis of GABA is 36 mg/kg and considered GABA as a strong secretagogue of insulin from the pancreas that may prevent diabetic conditions.

Antagonism

Certain strains of the species *L. lactis* produce a multitude of different antagonistic compounds, including antimicrobial proteins or bacteriocins.

a) Bacteriocins

Bacteriocins are proteins or protein complexes containing bacteriocidal properties. Lactococcal bacteriocins are small thermostable proteins that destroy closely related bacteria. The exception is the nisin molecule which activity is directed towards the wide range of gram-positive bacteria, including *Listeria monocytogenes* and food spoilage organisms. It is the best-known bacteriocin and belongs to the group of lantibiotics, and has a practical use in food preservation because of its non-toxicity. In april 1989, it was the first bacteriocin approved by FDA for use in the USA, the UK, and the European Union and is currently recognized as a generally-regarded-as-safe (GRAS) food additive and preservative mainly for processed foods. Besides, *L. lactis* subsp. *lactis*, the certain strains of *L. lactis* subsp. *cremoris* and *L. lactis* subsp. *lactis* biovar *diacetylactis* possess an ability to produce different bacteriocins with a range of inhibition. The bacteriocins are effective against gram-negative bacteria only if used in combination with chelating agents. Two natural variants of nisin such as nisin A and nisin Z are well known, between them the structural gene of nisin Z has been characterized and sequenced. It has been shown that nisin A permeabilizes the cytoplasmic membrane, thereby dissipating the membrane potential, inhibiting transport of amino acids, and causing release of accumulated amino acids from cells and membrane vesicles from bacteria such as *Staphylococcus aureus* and *Bacillus subtilis* but not gram negative bacteria It has also been reported on the nisin Z (3.5 kDa) produced by *Lactococcus lactis* CM1, an isolate from homemade "Dahi," was closely similar to that of nisin and inhibited a number of food spoilage and pathogenic bacteria. In contrast to nisin, the activity of diplococcin (bacteriocin produced by L. *cremoris* 346) is restricted to dairy lactococci. It has been demonstrated on the lactococcin R, bacteriocin produced by *L. lactis* subsp. *cremoris* R isolated from radish. It is active against *Bacillus, Clostridium, Listeria, Micrococcus, Staphylococcus Enterococcus, Lactobacillus, Leuconostoc, Streptococcus* and *Leuconostoc* species but not against gram-negative bacteria. It differs from nisin in having a low molecular weight, 2.5 KDa and sensitive to pepsin and Ü-chemotrypsin. It alos has been provided first description of lantibiotic nisin Z produced by *L. diacetylactis* UL719 strain and showed inhibitory activity against gram-psitive bacteria such as *listeria, clostridium* and *staphylococcus aureus* strains.

It has been demonstrated that Nip1 lactococci, however possess poor performance as cheese starters because produce acid at slower rates and have lower heat resistance and reduced proteolytic activity compared with those of commercial starters. In addition, Nip1 lactococci are more sensitive to bacteriophage attack than are strains which do not produce the antibiotic. For this reason, Nip1 lactococci have been combined with nisin-resistant starters to ensure adequate performance of the cheese starter. This system is effective only if the resistance mechanism does not result in nisin inactivation. For example, *Lactococcus lactis* DPC3147, a strain isolated from an Irish kefir grain, produced a bacteriocin designated as lacticin 3147 with a broad spectrum of inhibition and can acidify milk at sufficient rates to allow commercial manufacture of cheddar cheese. It also has been reported on the frequency of production varied from about 1% in *Streptococcus lactis* subsp. *diacetylactis* to 9 and 7.5% in S. *lactis* and *Streptococcus cremoris*, respectively. The bacteriocinogenic strains were also characterized on the basis of plasmid content showed the possession of plasmids between one and nine ranging from 1 to 50 megadaltons.

b) Hydrogen peroxide

Currently, hydrogen peroxide is widely used in the food industry for the aseptic packaging of fluid foods. In the USA, the use of H_2O_2 as a GRAS substance in milk (0.05%), whey (0.04%), starch (0.15%), and corn syrup (0.15%) is permitted. *L. lactis* subsp. *lactis* strains can accumulate H_2O_2 at higher amounts than lactobacilli. As paradigm, It has been reported that *L. lactis* subsp. *lactis* AI 62, which produced approximately 350 ppm H_2O_2, inhibited *Enterococcus faecalis, Ent. faecium,* enterotoxigenic *Escherichia coli, Listeria ivanovii, Staphylococcus aureus, Yersinia enterocolitica,* and *Aeromonas hydrophila.* After 1 h incubation at 30°C in the cell-free filtrate, the initial viable cell counts of the target bacteria (5.53- 6.00 log cfu/mL) were reduced by 0.12-5.00 log units, except in the case of enterococci. Among the species examined, *Aer. hydrophila* was the most sensitive to H_2O_2.

Vitamin production

a) Riboflavin

Riboflavin (vitamin B_2) is a water-soluble vitamin belonging to the B-complex group that is important for optimal body growth and red blood cell production and helps in releasing energy from carbohydrates and fatty acids. *Lactococcus lactis* is a commonly used starter strain that can be converted from a vitamin B_2 consumer into a vitamin B_2 'factory' by over-expressing its riboflavin biosynthesis genes. It has been reported on the two approaches were taken to develop riboflavin-overproducing strains of *L. lactis*: targeted metabolic engineering and isolation of spontaneous mutants to a toxic riboflavin analogue. It has been revealed that the animals supplemented with fermented milk B2-?? showed significantly lower growth rates (2.78 (SD?0.17)g/D) compared with those fed fermented milk B2+ or 0.5 mg of B2/kg (3.30(SD 0.20) and (3.19 (SD)?0.16)g/D, respectively), suggesting that the riboflavin-enriched product can exert a positive biological function. These results suggested that the regular consumption of the fermented product manufactured using the riboflavin-producing strain is able to decrease the relative liver weight and other abnormalities occurred due to vitamin B2 deficiency. The use of riboflavin-producing strains in the production of dairy products such as fermented milks, yogurts, and cheeses is feasible and economically attractive because it would decrease the costs involved during conventional vitamin fortification and satisfy consumer demands for healthier foods.

b) Folate production

Folate is an essential nutrient in the human diet. Folate deficiency leads to numerous physiological disorders, most notably anemia and neural tube defects in newborns and mental disorders such as psychiatric syndromes among the elderly and decreased cognitive performance. In addition, folate is assumed to have protective properties against cardiovascular diseases and several types of cancer. Due to the ability of lactic acid bacteria to produce folate, folate levels in fermented dairy products are higher than those in the corresponding non-fermented dairy product. The natural diversity among dairy starter cultures with respect to their capacity to produce folate can be exploited to design new complex starter cultures which yield fermented dairy products with elevated folate levels. So, *Lactococcus lactis* is by far the most extensively studied lactic acid bacterium because it is able to synthesize folate and accumulates large amounts of folate, predominantly in the polyglutamyl form. Only small amounts of the produced folate are released in the extracellular medium. Over the last decades a number of elegant and efficient genetic tools have been developed for this starter bacterium which showed critical importance in metabolic engineering strategies that aim at inactivation of undesired genes and/or (controlled) overexpression of existing or novel ones. The five genes involved in folate biosynthesis in a folate gene cluster in *L. lactis* MG1363 were characterized are *folA*, *folB*, *folKE*, *folP*, and *folC* and showed that the overexpression of *folKE* in *L. lactis* was found to increase the extracellular folate production almost 10-fold, while the total folate production increased almost 3-fold. The controlled combined overexpression of *folKE* and *folC*, encoding polyglutamyl folate synthetase, increased the retention of folate in the cell. The cloning and overexpression of *folA*, encoding dihydrofolate reductase, decreased the folate production two fold, suggesting a feedback inhibition of reduced folates on folate biosynthesis.

Conclusion

Lactococci are ubiquitous in the environment and in food. They are widely used as starters, with the long history of use for the sake of their small-scale manufacture to industrial-scale processes. It can be generally considered safe, in view of their extensive, daily consumption by humans and their low incidence in human infections. The metabolic properties of the strains within the *L. lactis* species have a direct or indirect influence on the organoleptic, nutritional and hygienic quality of fermented dairy products, which makes knowledge of their characteristics extremely important from an economic aspect. Apart from this, the findings which are based on the functional attributes of *L. Lactis* considered this bacterium is more than just lactic acid producer. Moreover, these outcomes will excite the interest of scientists and dairy experts, as well as presenting a challenge in numerous genetic, physiological and technological research projects and can contribute to an evolution of this societal position on the restricted use of genetically modified organisms for human health in the next years.

Suggested Reading

Bermúdez-Humarán. L. G., Cortes-Perez. N. G., Ah-Leung. S., Lefèvre. F., Yang. G., Pang. Q., Wu. C., Zeng. Y., Adel Patient. K.and Langella. P. 2008. Current Prophylactic and Therapeutic Uses of a Recombinant Lactococcus lactis Strain Secreting Biologically Active Interleukin-12. *J. Mol Microbiol Biotechnol;* **14:** 80–89.

Bermudez-Humaran. L. G., Cortes-Perez. N. G., Haridon. R. L. and Langella. P. 2008. Production of biological active murine IFN-c by recombinant *Lactococcus lactis. FEMS Microbiol Lett,* **280:** 144–149.

Cao. L. T., Wu. J. Q., Xie. F., Hu. S. H. and Mo. Y. 2007. efficacy of nisin in treatment of clinical mastitis in lactating dairy cows. *J. Dairy Sci.* **90:** 3980–3985.

Kim. J., Woo. H., Kim. Y. and Lee. H. 2002. Screening for antiproliferative effects of cellular components from lactic acid bacteria against human cancer cell lines. *Biotechnology Letters,* **24:** 1431–1436.

Kimoto-Nira. H., Mizumachi. K., Nomura. M., Kobayashi. M., the late Fujita. Y., T.Okamoto., the late Suzuki. I., Tsuji. N. M., Kurisaki. J. and Ohmomo. S. 2007. Lactococcus sp. as potential probiotic lactic acid bacteria. *JARQ* **41(3):** 181-189.

LeBlanc. J. G., Burgess. C., Sesma. F., Savoy de Giori. G. and Van sinderen. 2005. D. Ingestion of milk fermented by genetically modified *Lactococcus lactis* improves the riboflavin status of deficient rats. *J. Dairy Sci.* **88:** 3435–3442.

Mitra. S., Chakrabartty. P.K. and Biswas. S. R. 2007. Production of Nisin Z by *Lactococcus lactis* isolated from dahi. *Appl Biochem Biotechnol.,* **143**: 41–53.

Siragusa. S., De Angelis. M., Di Cagno. R., Rizzello. C. G., Coda. R. and Gobbetti. M. 2007. Synthesis of ã-aminobutyric acid by lactic acid bacteria isolated from a variety of Italian Cheeses. *Applied and Environmental Microbiology. Nov.* **73 (22):** No. 22. 7283–7290.

Sybesma. W., Starrenburg. M., Kleerebezem. M., Mierau. I., De Vos. W. M. and Hugenholtz. J. 2003. Increased production of folate by metabolic engineering of *Lactococcus lactis*. *Applied and Environmental Microbiology*, June **(6):** 3069–3076.

Yama. K., Pouliot. P., N'diayea M. M., Fourniera. S., Olivier. M . and Cousineau. B. 2008 Innate inflammatory responses to the Gram- positive bacterium Lactococcus lactis. *Vaccine*. **26(22):** 2689-2699.

□□□

Chapter 14

Direct-Fed Microbials: Enhancing Functional Attributes of Milk

ANIL KUMAR PUNIYA[1], SUMIT SINGH DAGAR[1], SANJAY KUMAR[1], JATINDER PAUL SEHGAL [2] AND KISHAN SINGH[1]

[1]Dairy Microbiology Division, [2]Dairy Cattle Nutrition Division, National Dairy Research Institute, Karnal (HR)

Introduction

Dairy products have long traditions in nutrition for promoting and maintaining human health. These products are rich in essential nutrients, such as oleic acid, conjugated linoleic acid, omega-3 fatty acids, vitamins, minerals and bioactive compounds that have positive health effects. In dairy products, concentration of many of these nutrients can be manipulated through feeding regimes and using "Direct Fed Microbials" (DFM). DFM have been used for the past few years to improve animal performance and in increasing many metabolites such as CLA, protein and vitamins milk and meat exert beneficial effect on the well being of the consumers. The bacterial and fungal feed additives and supplements, commonly categorized under animal probiotics, such as lactobacillus, yeasts or fungi or their extracts, have been shown to affect the rumen fermentation patterns significantly and thus animal performance and milk production.

Direct Fed Microbials

According to Fuller, a DFM/ probiotics is "a live microbial food supplement that beneficially affects the host animal by improving the intestinal microbial balance. In recent years, DFM have been given a more precise definition as "mono or mixed cultures of live microorganisms which, when applied to animal or man, beneficially affect the host by improving the properties of the indigenous microflora" As per FDA, DFM is "a source of live (viable) naturally-occurring microorganisms", and this includes bacteria, fungi and yeast. The claims that have been framed in favor of DFM are enlisted below:

- Better utilization of poorly digestible fibrous feeds.
- Enhanced milk production.
- Stimulation of immune system to improve animal's health.
- Improved resistance to infectious diseases.
- Increased growth rate of herd.

To be an effective feed additive, a DFM should meet the following criteria:

- Easy and economic to culture
- Exert a beneficial effect to host
- Genetically stable; be non-pathogenic and non-toxic
- Viable at the time of administration to get a positive effect
- Be capable of surviving and metabolizing in the gut
- Maintain viability under storage and field conditions for longer durations
- Should have "GRAS"(Generally recommended as safe) status

However, in support these claims, results are variable and inconsistent; and there is need to carry out more research in this area before reaching to final conclusion regarding beneficial effects of DFM.

Mechanism of Action of DFM

The Live microbial supplement not only improves animal health but also prevent colonization of pathogens in lower gut. The different approaches by which DFM benefits the ruminants are listed below:

- By secreting antimicrobial compounds (lactic/acetic/formic acids, bacteriocins, antibiotics, hydrogen peroxide)

- Elimination of undesirable microorganisms by competing for space and/or nutrients
- Production of growth promoting substances (e.g. amino acids, vitamins)
- Production of bioactive compounds such as CLA
- Production of fibre degrading/ biohydrogenating enzymes (i.e. cellulases, hemicellulases, xylanases, proteases, isomerase, reductase etc.)
- Detoxification of undesirable compounds (i.e. amines, endotoxins etc.)
- By stimulating host immune response

DFM for Ruminants

Bacteria: Efforts have screened certain microbial strains that facilitates in enhancing meat or milk production. Several bacteria-based DFM has been developed with their potential application in ruminant diets. DFM based on *Lactobacillus, Enterococcus, Streptococcus,* and *Bifidobacterium,* have been studied in young calves and their importance in maintaining ruminal microflora is well established. Infusion of *Lactobacillus* and *Streptococcus* suppress diarrhea and improve the growth rate of young ruminants. Most DFM are probably beneficial because they have effects in the lower gut rather than in the rumen. For example, *L. acidophilus* produces lactic acid that lowers the pH in small intestine so as to inhibit the growth of pathogens. Increased milk production has been reported from cows fed with *L. acidophilus* @ $1x10^9$ cfu animal^{-1} day^{-1} and cows fed *Lactobacilli* during transition period produced more milk, and had lower non-esterified fatty acids and higher glucose in blood than controls. Propionibacteria, have the ability to convert lactic acid to acetic and propionic acid, improves weight gain in calves and reduces nitrate toxicity.

Fungi: *Aerobic Fungi* - In contrast to most bacteria, fungi based DFM makes favorable changes in ruminal fermentation, hence, appear to be more beneficial. The number of rumen anaerobes and cellulolytic bacteria has been found to be increased with fungal extracts, hence, indirectly assists in increasing fibre digestibility. The other possible benefit is fungal DFM may prevent accumulation of excess lactic acid when ruminants fed high concentrate diets. Extracts of *A. oryza* and *S. cerevisiae* stimulated the uptake of lactic acid by rumen lactateutilizer *Selenomonas ruminantium* and *Megasphaera elsdenii* by providing a source of malic acid. Yeasts also stimulate acetogenic bacteria in presence of methanogens that might result in a more efficient fermentation. Another hypothesis suggests that *Aspergillus* extracts may improve fiber digestion as these contain esterases. It was also reported that young calves fed an *Aspergillus* extract weaned faster than controls and supplementation also increased numbers of rumen bacteria and volatile fatty acids. It has been well documented that feeding of *S. cerevisiae* has increased dry matter intake and milk production.

Anaerobic Fungi : Various researches in animal nutrition have suggested a positive relationship between ruminal anaerobic fungi and dry matter intake of low digestible diets in ruminants. Few studies have also concluded that the intake of forage by early weaned calves was 35% higher with *Neocallimastix* sp. and dosing fungus-free sheep with *Neocallimastix* sp. resulted in a 40% increased intake of a low quality straw-based diet. These fungi also supply the proteins to the host through activity of proteolytic enzymes and a proportion of the synthesized microbial protein passes to abomasum and intestines for absorption. The role of rumen fungi in the plant fibre degradation of has been investigated extensively. The rhizoids or bulbous holdfasts of vegetative thalli penetrates plant tissue far better than other ruminal microflora, so they gain access to plant material that is not available to bacteria and protozoa. Several reports suggested that such penetration leads to a faster and complete degradation of forage. Degradation of lignified walls of plant-cells is an important characteristic feature of rumen fungi. Zoospores of *Piromyces, Neocallimastix, Orpinomyces* and *Anaeromyces* species appear to colonize the lignified tissues preferentially on sclerenchyma and xylem cells. The greater ability of rumen fungi, compared to rumen bacteria, to weaken forage fibre may be important in enhancing forage utilization by the host animal.

Studies have also suggested increased *in vitro* dry matter digestibility, and decreased cell wall contents of straws with the use of different anaerobic fungi viz, *Orpinomyces, Piromyces* and *Anaeromyces* in comparison to control. There is 7-12% increase in voluntary intake of straw based diet, when the sheep were dosed with monocentric fungi isolated from other herbivores.

In a recent study, it has been concluded that *Piromyces* sp. WNG-12 could be a promising anaerobic fungi that can be exploited as direct-fed microbial in lactating buffaloes so as to enhance the digestibility of wheat straw based diets ultimately leading to higher milk production (8.48 kg/day) with better quality. Digestibility of dry matter, crude protein, neutral detergent fibre, acid detergent fibre, cellulose and digestible energy also increased significantly. The pH and NH_3-N were lowered, whereas total volatile fatty acids, total N, TCA-N and number of zoospores/ ml of rumen liquor were significantly (P=0.05) higher.

Genetically Tailored DFM

To increase the degradation of fibre components, recombinant organisms expressing specific polysaccharidases at a high level could be one of the possible strategies. The manipulation of predominant rumen bacteria such as *Prevotella ruminicola, Butyrivibrio fibrisolvens* or *Streptococcus bovis,* aims at the expression of heterologous polysaccharidase genes to increase fibrolytic activities in animals fed with fibre-rich diets. Such diets decrease rumen pH, consequently inhibiting the native cellulolytic bacteria that are very sensitive to low pH. Expression of gene from *Ruminococcus flavefaciens* in *S. bovis,* expression of an xylanase gene from *P. ruminicola* in another strain of the same species, and expression in *B. fibrisolvens* of a xylanase gene originating from the rumen fungus *Neocallimastix patriciarum* or from *Eubacterium ruminantium* have been attempted in recent past.

Role of DFM in improving Milk Quality

Milk and meat quality can be enhanced either by adding/ by increasing concentration of various bioactive compounds. The fatty acids in bovine milk are derived from two sources. The first source is fatty acids supplied to the udder by the blood, composed of fatty acids absorbed from the intestine produced as a result of lipolysis by microbial lipases and other enzymes. The second source is derived from circulating blood acetate and butyrate produced during fermentation in the rumen, again by the activity of microorganisms. So, rumen microorganisms play a key role in changing milk fat content. These rumen microbes secrete various enzymes for biohydrogenation of dietary unsaturated fatty acids which is a defence mechanism against the toxicity of these polyunsaturated fatty acids. Ingested lipids are first hydrolysed by microbial lipases. Lipases break dietary fats into smaller free fatty acids. These free fatty acids are then biohydrogenated by microbial isomerase and reductase into saturated fatty acids. First step of biohydrogenation i.e. isomerization of linoleic acid (Unsaturated

Fatty Acid) into CLA (*cis*-9, *trans*-11) is catalysed by isomerase, followed by the activity of reductase enzyme which produces *trans*-vaccenic acid; which is again reduced by reductase, finally to stearic acid (*cis*-9, *trans*-11). Here, some of CLA and *trans*-vaccenic acid accumulates and are directly absorbed in the intestine, ultimately incorporating in different tissues where tissue synthesis of CLA takes place. This endogenous synthesis of CLA occurs in adipose tissue and mammary gland of the lactating animals by the activity of ?9-desaturase enzyme, converting *trans*-vaccenic acid into CLA. CLA is associated with its diverse health benefits *viz.*, anti-carcinogenic properties, reduced atherosclerosis, enhanced immune system, antidiabetic effects, reduced body fat; increased body protein and enhanced bone formation etc

Among rumen microorganisms, bacteria play most important role in this conversion and hence in CLA production, followed by fungi. A number of CLA producing bacteria have been isolated from the rumen, human intestine and dairy products which also include novel strains of *Butyrivibrio fibrisolvens* and *Megasphaera elsdenii.* Most of the lactic acid bacteria, namely, *Lactobacillus acidophilus, Lactobacillus delbrueckii* ssp. *bulgaricus, L. delbrueckii* ssp. *lactis, Lactococcus lactis* ssp. *cremoris, L. lactis* ssp. *Lactis, Streptococcus thermophilus* and *Propionibacterium freudenreichii* have the ability to convert linoleic acid into CLA. *Bifidobacterium breve* of human and *L. brevis* of rumen origin have also been found to produce significant amount of CLA. Recently fungi have also been reported to produce *cis*-9, *trans*-11-CLA as the chief product of linoleic acid metabolism and Orpinomyces was found to be most bioactive. *N. frontalis* were also found to produce more cis-9, trans-11-CLA from 0.5% linoleic acid as compared to P. communis.

All vitamins of the B-complex are synthesized by the rumen microbes, normally in sufficient amounts to cover the animal's needs. So, the vitamin- B content in milk can also be enhanced through DFM administration.

Practical Applications of DFM

There are varieties of DFMs such as powder, paste, gel, and capsules available commercially. These different forms may be mixed in feed, top-dressed, given as a paste, or mixed into the drinking water or milk replacer. However, their use must be managed effectively as viability of organism can be largely affected on interactions with chlorine, water, temperature, minerals, flow rate, and antibiotics.

Bacterial DFM pastes are formulated with vegetable oil and inert gelling ingredients. Non-hydroscopic whey is generally used as a carrier for bacteria based DFM. Fungal DFM products are formulated with grain by-products as carriers. Some DFM are developed for one-time dosing while others are developed for feeding on a daily basis. Most DFMs contain live bacteria; however, some contain only bacterial or fungal extracts or fermentation by-products. The best response can be observed during the following situations: (a) when a newborn animal acquire beneficial bacteria from environment, (b) during weaning or dietary changes, (c) periods of stress i.e. shipping, vaccination, and other situations, and (d) antibiotic therapy.

The stability of DFMs is crucial because the microbes must be delivered live to the animal to be effective. For this, most DFMs require storage in a cool and dry area, away from heat, direct sunlight, and high levels of humidity. They must not only survive during processing and storage but also in the gut environment. The metabolites present in culture extracts have been suggested to be the "active" ingredients. It has been reported that pulse dosing alone or pulse dosing with daily feeding of *Streptococcus faecium* resulted in improved performance of incoming feedlot cattle.

Conclusion

The initial concept of feeding DFM to livestock was based on the potential for beneficial effects, including establishment of desirable gut microflora, prevention of pathogenic organisms and reducing ruminal acidosis. DFM may be genetically modified to secrete essential amino acids or secrete high levels of growth factors. In near future, microbes capable of detoxification of compounds like alkaloids, oxalates, tannins, or mycotoxins and enriching the dairy products with bioactive compounds may prove to be a better approach. However, commercial development of these organisms will not occur if they cannot be economically grown and stabilized, especially if daily feeding is not required. But, many questions with respect to type of enzymes, doses, interactions with maturity and moisture need to be answered. Research is still needed to fully understand the mechanisms and specific requirements of predominant ruminal microorganisms under different feeding conditions before DFM could be developed to take advantage of improve feed digestion.

Suggested Reading

Dayananda TL, Nagpal R, Puniya AK, Sehgal JP and Singh K 2007 Biodegradation of urea-NH_3 treated wheat straw using anaerobic rumen fungi. *J Anim Feed Sci.* **16:** 484- 489.

Dey A., Sehgal J.P., Puniya A.K., Singh K., 2004. Influence of ananaerobic fungal cultures (*Orpinomyces* spp.) administration on growth rate, ruminal fermentation and nutrient digestion in calves. *Asian-Austr. J. Anim. Sci.* **17:** 820-824.

Jenkins TC, Wallace RJ, Moate PJ and Mosley EE 2008 Recent advances in biohydrogenation of unsaturated fatty acids within the rumen microbial ecosystem. *J Anim Sci* **86:** 397–412.

Lee SS, Choi CK, Ahn BH, Moon YH, Kim CH and Ha JK (2004). *In vitro* stimulation of rumen microbial fermentation by a rumen anaerobic fungal culture. *Anim Feed Sci Tech* **115:** 215-226

Nagpal R, Yadav H, Puniya AK, Singh K, Jain S and Marotta F 2007 Conjugated linoleic acid: sources, synthesis and potential health benefits - an overview. *Current Topics in Nutraceutical Research* **5(2/3):** 55-66.

Nam IS, Garnsworthy PC 2007 Biohydrogenation of linoleic acid by rumen fungi compared with rumen bacteria. *J Appl Microbiol* **103:** 551–556.

Paul SS, Kamra DN, Sastry VRB and Agarwal N (2004). Effect of administration of an anaerobic gut fungus isolated from wild blue bull to buffaloes on *in vivo* ruminal fermentation and digestion of nutrients. *Anim Feed Sci Tech* **115:** 143-157.

Puniya AK, Chaitanya S, Tyagi AK, De S and Singh K 2008 Conjugated linoleic acid producing potential of lactobacilli isolated from the rumen of cattle. *J Ind Microbiol Biotechnol* (In press)

Sehgal JP, Jit D, Puniya AK and Singh K 2008 Influence of anaerobic fungal administration on growth, rumen fermentation and nutrient digestion in buffalo calves. *J Anim Feed Sci* (In press)

Thareja A, Puniya AK, Goel G, Nagpal R, Sehgal JP, Singh PK and Singh K 2006 *In vitro* degradation of wheat straw by anaerobic fungi from small ruminants. *Arch Anim Nutr* **60:** 412-417.

Tripathi VK, Sehgal JP, Puniya AK and Singh K 2007 Effect of administration of anaerobic fungi isolated from cattle and wild blue bull (Boselaphus tragocamelus) on growth rate and fibre utilization in buffalo calves. *Arch of Anim Nutr* **61:** 1-8.

□□□

Chapter 15

Low Carb Sweeteners from Lactic Acid Bacteria

S.K TOMAR AND FALGUNI PATRA
Dairy Microbiology Division, National Dairy Research Institute, Karnal-132001

Introduction

Lactic acid bacteria (LAB) are industrially important microbes that are used all over the world in a wide variety of industrial food fermentations. They are excellent ambassadors for an often maligned microbial world.Beyond the horizon of their conventional role in acid, flavor and texture development, they are being looked up on as burgeoning "cell factories" for production of host of functional biomolecules and food ingredients such as biothickeners, bacteriocins, vitamins, bioactive peptides and amino acids. Besides, endowed with their relatively simple metabolism, limited biosynthetic capacity, and apparent lack of gene multiplicity, these bioreactors are ideal tool for metabolic engineering to facilitate overproduction of desirable goodies compared to other microorganisms like yeast or fungi. To this list of metabolites, low calorie sugars have rather been recent addition and have attracted a great deal of the interest of researchers, manufactures and consumers for varied reasons.

Broadly speaking, low calorie sweeteners can be categorized in to two groups. One group of sweeteners consists of substances with a very intense sweet taste that are used in small amounts to replace the sweetness of a much larger amount of sugar e.g., acesulfame-K, aspartame, neotame, saccharin, and sucralose, alitame, cyclamate, and stevia/steviol glycosides. These are produced by chemical methods and are only can be used in reconstituted foods. These sweeteners lack in bulking properties, are not globally approved and also have differing degree of after taste.

Second category of sweeteners that can substitute for both the physical bulk and sweetness of sugar include the sugar alcohols (also called "polyols") sorbitol, mannitol, xylitol, isomalt, erythritol, lactitol, maltitol, hydrogenated starch hydrolysates, and hydrogenated glucose syrups and often termed as "sugar replacers" or "bulk sweeteners,". Other two sweeteners viz. trehalose and tagatose, although are actually sugars rather than sugar alcohols yet are similar in function to the polyols. Some of these substances, such as sorbitol, xylitol, and tagatose, also occur naturally in certain fruits or other food while some are produced by yeast, fungi and bacteria. Several health promoting effects have also been attributed to the addition of these ingredients and hence addition/fortification to foods can lead to products with functional benefits. These sweeteners can either be directly added to foods, or use of these sweeteners-producing microorganisms can lead to natural foods containing these sweeteners.

Polyols and other bulk sweeteners have a couple of potential advantages over sugar as food ingredients:

- Do not promote the development of dental caries (tooth decay)
- Produce a lower glycemic response than most sugars and starches do. Thus, their use may be advantageous for people with diabetes.
- Have low calorific values due to their poor digestion and poor absorption.

In this article an attempt has been made to discuss a number of such sugars and sugar alcohols produced by LAB in terms of their industrial application and health attributes.

Mannitol

Mannitol, a naturally occurring polyol, is found in animals and plants. It is present in small quantities in most fruits and vegetables.

Marine algae, especially brown algae, are also rich in mannitol. Furthermore, mannitol is commonly found in the mycelium of various fungi and is present in fresh mushrooms (about 1%). Also some fungi and bacteria produce mannitol. Moreover, small quantities of mannitol are found in wine. It is widely used in the food, pharmaceutical, medicine, and chemical industries. It is only half as sweet as sucrose. Mannitol also exhibits reduced caloric values compared to the respective value of most sugars e.g. the calorific values of mannitol and sucrose are 1.6 kcal/g and 4 kcal/g, respectively. Mannitol is considered safe for use in foods and it has a food additive status. Mannitol is presently on the U.S. FDA GRAS- /INTERIM (*Generally recognized as safe*) list.

Technological Application

At present, the main application for mannitol in the food industry is as a sweetener in sugar-free chewing gums and for dusting chewing gum sticks. In addition, mannitol is used as a bodying and texturizing agent, anticaking agent, and humectant. Mannitol has been shown to exhibit osmoprotecting effect during drying of LAB. It has also been shown to have antioxidant effect by scavenging off free hydroxyl radicals. Mannitol has the potential to extend the shelf life of various foodstuffs. As mannitol is only partially metabolized by humans and does not induce hyperglycemia, which makes it useful for diabetics and for so-called light foods. Besides, mannitol is commercially useful in making artificial resins and plasticizers.

Functional Application

Mannitol belongs to a group of drugs referred to as osmotic diuretics. In medicine, mannitol ("Osmitrol") is used to increase the formation of urine in order to prevent and treat acute renal failure, and also in removal of toxic substances from the body. Mannitol is also used to reduce both cerebral edema and intraocular pressure. Furthermore, it is used to alter the osmolarity of the glomerular filtrate in treating kidney failures. In the pharmaceutical industry mannitol is commonly used as a constituent in chewable tablets and granulated powders. Its sweet cool taste is used to mask the unpleasant taste of many drugs.

Microbial Production

Yeast, fungi, and especially LAB, are also known to produce mannitol. When compared fungal and bacterial processes for the production of mannitol, the latter one proved to be better. Some homofermentative LAB were found to produce small amounts of mannitol intracellularly e.g. *Streptococcus mutans*

- *Lb. leichmanii*
- Lactate dehydrogenase-negative mutant of *Lb. plantarum* and

Lactate dehydrogenase deficient mutant of *Lc. lactis*. Heterofermentative LAB belonging to the genera *Lactobacillus* and *Leuconnostoc* are the potent producer of mannitol. These bacteria contain the enzyme mannitol dehydrogenase to convert fructose into mannitol. The LAB reported to produce mannitol are *Ln. pseudomesenteroides; Lactobacillus spp* B001; *Lactobacillus spp* KY-107; *Lactobacillus spp.* Y-107 and *Leuconostoc spp.* Y-002; *Ln. pseudomesenteroides* ATCC 12291; *Lb. sanfranciscensis;* and most recently mannitol production by *Lb. intermedius* NRRL B-3693 using inulin as a substrate was reported. Another group studied the production of mannitol by heterofermentative LAB in a resting state. They developed a very efficient mannitol production process by combining membrane cell - recycle technology and high cell density batch cultures. Using the same initial biomass, a stable high-level production of mannitol was maintained for 14 successive bioconversion batches. Research has been done to constructed random mutants of *Ln. pseudomesenteroides* by chemical mutagenesis for improving mannitol production of the mutant lacking the fructokinase activity. An increased fraction of fructose was reduced to mannitol by mannitol dehydrogenase (increased yield of mannitol from 74 to 86 mol %).

Tagatose

Tagatose is an isomer of fructose that occurs naturally in some dairy products. In comparison to sugar it is metabolized differently, providing fewer calories and producing a smaller glycemic response. It is involved in browning reactions during heat treatment and decomposes more readily than sucrose at high temperatures. Tagatose is derived from lactose, the sugar found in milk. During the production of tagatose, after enzyme hydrolysis of lactose the two monomers, galactose and glucose, are chromatographically separated. The galactose fraction is then converted to tagatose under alkaline conditions. Tagatose is a functional sweetener and is very similar in texture to sucrose (table sugar) and is 92% as sweet, but with only 38% of the calories.

D-tagatose is a malabsorbing sugar, as it is poorly absorbed in the small intestine. D-tagatose has a sucrose-like taste with no cooling effect or aftertaste. It is similar to the polyols in having a low caloric value and tooth-friendly property. However, it has no laxative effect unlike polyols. D-tagatose has been found not to contribute to net energy as revealed by a number of growth studies on rats and human clinical trials shows that subjects gradually and consistently lose weight at medically desirable rates. As a result of these properties, D-tagatose is considered to be a potential reduced-energy sweetener.

Technological Application

D-tagatose can be used as a low-calorie sweetener (1.5 Kcal), as an intermediate for synthesis of other optically active compounds, and as an additive in detergent, cosmetic, and pharmaceutical formulation. Besides effecting enhancement of flavor, it provides the natural taste and texture of sugar. Using tagatose to replace sucrose in foods would reduce the amount of metabolizable sugars in the diet as well as provide the desirable prebiotic effect.

Functional Application

A number of health benefits have been attributed to tagatose such as promotion of weight loss, no glycemic effect, anti-plaque, non-cariogenic, antioxidant, anti-halitosis, prebiotic, and anti-biofilm properties, organ transplants, improvement of pregnancy and fetal development, treatment of obesity, reduction in symptoms associated with type 2 diabetes, hyperglycemia, anemia, and hemophilia.

Microbial Production

Biological manufactures of D-tagatose using several biocatalyst sources have been studied intensively in recent years. Among the biocatalysts, L-arabinose isomerase (EC 5.3.1.4) catalyzes the conversion of D-galactose to D-tagatose as well as the conversion of L-arabinose to L-ribulose, economically feasible tagatose manufacturing process. L-arabinose isomerase has been of interest for its potential application in galactose isomerization into tagatose and among other microorganisms; LABs were recognized as source of this enzyme. The LAB reported to contain L-arabinose isomerase are *Lb. gayonii, Lb. plantarum, Bifidobacterium longum.* Cheetham and Wootton (1993) invented a process to convert D-galactose to D-tagatose by using LAB. Further, Ibrahim and Spradlin (2000) have patented an enzymatic isomerization process using arabinose isomerase originating from a lactic acid bacterium.

Sorbitol

The most commonly used polyol in the United States is sorbitol, which is the standard sweetener in several sugar-free chewing gums and over-the-counter medicines. Sorbitol, also referred to as D-glucitol, is naturally found in many fruits e.g., berries, cherries, and apples. Sorbitol is sweet tasting, forms a viscous solution, stabilizes moisture, possesses bacterio-static property and is generally chemically inert. These features and properties make sorbitol an ideal and preferred ingredient in many products. Sorbitol is often used in modern cosmetics as a humectant and thickener. Sorbitol is used as a cryo-protectant additive. Sorbitol, together with other polyhydric alcohols such as glycerol, is one of the ingredients in alkyl resins and rigid polyurethane foams manufacturing. In tobacco industries, sorbitol may give mild effect in sniff, good humectant agent, and is also to avoid acrolein formation which formed in burned glycerine. In textile industries, sorbitol is used as softener and colour stabilizer, and as softener in leather industries.

Technological Application

This polyol has a relative sweetness of around 60% vis-à-vis sucrose and displays a 20-fold higher solubility in water than mannitol. Owing these properties, sorbitol is widely used in a range of food products such as confectionery, chewing gum, candy, desserts, ice cream, and diabetic foods. In these products, it imparts sweetness and play technological role such as a humectant, a texturizer, and a softener. Besides, sorbitol is the starting material for the production of pharmaceutical compounds such as sorbose and ascorbic acid.

Functional Application

In medication it is used as a laxative to treat occasional episodes of constipation. Vitamin C (ascorbic acid) is mainly semi-synthesized from sorbitol by fermentation process by *Baccillus suboxydant*. In oral or topical preparation, it is used as humectant, sweetener, bodying and viscosity agent, vehicle, anti-caplocking and texture improvement. In other case, sorbitol is useful to promote the absorption of certain minerals such as Cs, Sr, F and vitamins B12. In high concentration, sorbitol is a stabilizer for unstable vitamins and antibiotics. It is used as an excipient and intravenous osmotic diuretic in pharmaceutical fields. Sorbitol is also having antioxidant properties.

Microbial Production

Sorbitol is claimed to have important health-promoting effects. A recombinant strain of *Lb. casei* was constructed, cells of which when pre-grown on lactose, were able to synthesize sorbitol from glucose. Inactivation of the L-lactate dehydrogenase gene led to an increase in sorbitol production. *Lb. casei* is a lactic acid bacterium relevant as probiotic and used as a cheese starter culture. A sorbitol-producing *Lb. casei* strain might therefore be of considerable interest in the food industry. Another study reported the capacity of *Lb. plantarum*, a lactic acid bacterium found in many fermented food products and in the gastrointestinal tract of mammals, to produce sorbitol from fructose-6-phosphate by reverting the sorbitol catabolic pathway by over expressing sorbitol 6-P-dehydrogenase and the mutant strains deficient for both L- and D-lactate dehydrogenase activities.

Trehalose

Trehalose, also known as mycose, is a natural alpha-linked disaccharide formed by an á, á-1, 1-glucoside bond between two á-glucose units. Trehalose is found naturally in insects, plants, fungi, and bacteria; the major natural dietary source is mushrooms. It is implicated in anhydrobiosis-the ability of plants and animals to withstand prolonged periods of desiccation. Trehalose has the added advantage of being an antioxidant. Trehalose is a naturally occurring reducer of cell stress, protecting these organisms from extremes in heat shock and osmotic stress. It acts by altering or replacing the water shell that surrounds lipid and protein macromolecules. Trehalolipids (trehalose linked at C-6 and C-6′ to mycolic acid) is produced from *Rhodococcus erythropolis* and *Arthrobacter spp.* acts as biosurfactant. Trehalose has been accepted as a novel food ingredient under the GRAS terms in the U.S. and the European Union.

Technological Application

It is anti-cariogenic, synthesized by several microorganisms, only partially digested by humans, and therefore considered a dietetic sugar. It protects proteins against denaturation. It also exhibits antioxidant activities. Trehalose is used in wide range of products due to the multi-faceted effects of trehalose's, such as its inherently mild, sweet flavor;

its preservative properties, which maintain the quality of the three main nutrients (carbohydrates, proteins, fats); its powerful water-retention properties, which preserve the texture of foods by protecting them from drying out or freezing; and its ability to suppress bitterness, stringency, harsh flavors, and the stench of raw foods, meats, and packaged foods.

Functional Application

Several recent studies have revealed health promoting attributes of trehalose. In rats it was shown to almost completely suppress dental caries, and in humans it reduced significantly acidification in plaques. Furthermore, it has been demonstrated that trehalose has important stabilizing effects on human proteins, preventing protein aggregation as well as formation of pathological con formational forms. It stabilizes monoclonal antibodies and enzymes for medical use. It stabilizes proteins for inhaled use. It has been successfully used to preserve embryos, and cellular blood products. Cyopreservation of transplant cells and tissue in presence of trehalose has been shown to increase viability and decrease host immune response. As it inhibits lipid and protein misfolding, trehalose has become an attractive molecule for study in neurodegenerative disease characterized by protein misfolding and aggregate pathology. Such diseases include Alzheimer's and Parkinson's disease, and the less common triplet repeat diseases. Recent scientific publications describe trehalose benefit in model systems that recapitulate aggregate pathology that characterize Alzheimer's (AD), Huntington's (HD), and occulopharyngeal muscular dystrophy (OPMD).

Microbial Production

Trehalose is widespread within the genus *Propionibacterium*. Trehalose accumulation in *Propionibacterium* e.g. *P. acidipropionici* and *P. freudenreichii* subsp. *shermanii* has also been observed to occur in response to stress conditions. In particular, *P. freudenreichii* subsp. *shermanii* strain NIZO B365 accumulates trehalose to remarkable levels, and the trehalose content increases considerably in response to osmotic, oxidative and acid stress [up to 40% (w/w) of the cell protein]. In this organism, trehalose results from the conversion of glucose 6-P and ADP glucose via trehalose 6-P synthase to trehalose 6-P and its subsequent dephosphorylation by trehalose 6-P phosphatase. Alternatively, trehalose can be formed from maltose through the action of trehalose synthase.

Alanine

Alanine is an á-amino acid with the chemical formula $HO_2CCH(NH_2)CH_3$. The L-isomer is one of the 20 proteinogenic amino acids, i.e. the building blocks of proteins. It is classified as a non-polar amino acid. L-alanine is second only to leucine, accounting for 7.8% of the primary structure in a sample of 1,150 proteins. D-alanine occurs in bacterial cell walls and in some peptide antibiotics. L-alanine is a non-essential amino acid. Alanine is found primarily in poultry, beef, pork and fish. Almost all animal based protein sources are rich in alanine. Alanine plays a major role in transferring nitrogen from tissue sites in the body, to the liver. Alanine is also used by the body to draw upon blood sugar as an energy source.

Technological and Functional Application

Alanine is an industrially relevant food and pharmaceutical ingredient. Because alanine is used by the body to draw upon blood sugar as an energy source, it may prove useful for bodybuilders and other athletes who are attempting to reduce body fat levels. Alanine is also known to support prostate health. For this reason, elderly males may benefit from its use. L-alanine is major component of amino acid rich medical supplement and substrate for many biochemical synthesis reactions in industry. It is included, together with other amino acids, in preparations for parenteral administration as clinical preoperative and postoperative foods, as well as an animal feed supplement. Furthermore, alanine is used as a food additive on account of its sweet taste. It has comparable sweetening power as glucose and fructose as a supplement and can serve as low calorie sweetener in food as a result of its much lower caloric value.

Microbial Production

L-alanine is produced industrially by the decarboxylation of L-aspartate by means of immobilized cells or cell suspensions of *Pseudomonas dacunhae*. But this is an expensive substrate. D/L-alanine can also be prepared by chemical synthesis or by the direct fermentation of sugars by *Corynebacterium gelatinosium, Arthrobacter oxydans, Brevibacterium lactofermentum, Clostridium* sp. Currently *Corynibacterium glutamicum* has been used for commercial production of this amino acid. Metabolic engineering has been developed that uses carbohydrate and ammonium sources to produce high concentration of alanine. Using metaboloic engineering on the level of pyruvete metabolism high production of alanine was achieved in bothe *L lactis* and *Lactobacillus plantarum* (US patent 6627420). According to this invention in such a lactic acid bacterium can be applied as main-starter. They also used the alanine-producing *L. lactis* strains to produce buttermilk and also investigated in a Gouda cheese model system which showed promising results.

Conclusions

Production of low calorie sugars from food grade microorganisms is becoming more attractive because of the inherent problems of the chemical processes. The examples presented here clearly shows that LAB are ideal candidates for production of these low calorie sweeteners and for metabolic engineering. A diverse number of approaches have been employed successfully to reroute carbon metabolism towards the products of interest, resulting in high level production of both natural and novel compounds, or removal of undesirable metabolites. Some of these constructed strains will undoubtedly find their way into novel fermented products with increased nutritional value. Taking into consideration health benefits and industrial applications, the development of novel dairy products naturally enriched in these sweeteners during fermentation processes offers interesting perspectives. The use of these bulk sweeteners are also safe, but consumers need to be aware of their presence in food products so that they can limit their intake sufficiently to avoid gastrointestinal discomfort. These safe sweeteners can be exploited as vital food ingredient for development of low calorie foods with added functional values especially for children, diabetic patients and weight watchers.

Suggested Reading

Cardoso FS, Gaspar P, Hughenholtz J, Ramos A, Santos H. 2004. Enhancement of trehalose production in dairy propionibacteria through manipulation of environmental conditions. *Int J Food Microbiol* **91:** 195204.

Deok-Kun OH. 2007. Tagatose: properties, applications, and biotechnological processes. *Applied Microbiology and Biotechnology* **76:** 1-8.

Hols P, Kleerebezem M, Kuipers OP, Ferain T, Delcour JMAMG, De Vos WM. 2003. Process for the production of alanine by recombinant microorganisms. US Patent 6627420.

Hugenholtz J. 2008. The lactic acid bacterium as a cell factory for food ingredient production. *International Dairy Journal* **18:** 466-47.

Kiviharju K, Nyyssola A. 2008. Contributions of Biotechnology to the Production of Mannitol. *Recent Patents on Biotechnology* **2:** 73-78

Ladero V, Ramos A, Wiersma A, Goffin P, Schanck A, Kleerebezem M, Hugenholtz J, Smid EJ, Hols P. 2007. High-Level Production of the Low-Calorie Sugar Sorbitol by actobacillus plantarum through Metabolic Engineering. *Applied and Environmental Microbiology* **73:** 1864-72.

Lu Y, Levin GV, Donner TW. 2008. Tagatose, a new antidiabetic and obesity control drug. *Diabetes, Obesity and Metabolism* **10:** 10934.

Saha CB. 2006. Production of mannitol from inulin by simultaneous enzymatic saccharification and fermentation with Lactobacillus intermedius NRRL B-3693. *Enzyme and Microbial technology* **39:** 991-95.

von Weymarn N, Hujanen M, Leisola M. 2002. Production of d-mannitol by heterofermentative lactic acid bacteria. *Process Biochemistry* **37:** 120713.

□□□

Chapter 16

Biosurfactants of Lactic Acid Bacteria : Novel Functional Biomolecules

RAMESHWAR SINGH & UMA MAHESWARI T
Dairy Microbiology Division, National Dairy Research Institute, Karnal-132001

Introduction

Surfactants are amphipathic molecules with both hydrophilic and hydrophobic (generally hydrocarbon) moieties that partition preferentially at the interface between fluid phases with different degrees of polarity and hydrogen bonding such as oil/water or air/water interfaces. These properties render surfactants capable of reducing surface and interfacial tension and forming micro emulsion where hydrocarbons can solubilize in water or where water can solubilize in hydrocarbons. Such characteristics confer excellent detergency, emulsifying, foaming, and dispersing traits, which makes surfactants some of the most versatile process chemicals. Almost all surfactants currently in use are chemically derived from petroleum; however, interest in microbial surfactants has been steadily increasing in recent years due to their diversity, environmentally friendly nature i.e. lower toxicity and higher biodegradability, the possibility of their production through fermentation, higher foaming, high selectivity, specific activity at extreme temperatures, pH and salinity and the ability to be synthesized from renewable feed stocks.

Biosurfactants are a structurally diverse group of surface active molecules synthesized by large variety of microorganisms, which vary in their chemical properties and molecular size. They are produced on living surfaces, mostly microbial cell surfaces, or excreted extracellularly and contain hydrophobic and hydrophilic moieties that reduce surface tension and interfacial tensions between individual molecules at the surface and interface, respectively. They are complex molecules covering a wide range of chemical types including peptides, fatty acids, phospholipids, glycolipids, antibiotics, lipopeptides, etc. Microorganisms also produce surfactants that are in some cases combination of many chemical types referred to as the polymeric microbial surfactants (PMS). Many microbial surfactants have been purified and their structures are elucidated. The low molecular weight surfactants are often glycolipids, the high molecular weight surfactants are generally either polyanionic heteropolysaccharides containing covalently-linked hydrophobic side chains or complexes containing both polysaccharides and proteins. The major classes of biosurfactants include mycolic acid, glycolipids, lipopeptides and lipoproteins, phospholipids and fatty acids, polymeric surfactants, and or the microbial cell surface itself. Various micro organisms are known to produce specific kind of biosurfactants. This depends on mainly the molecular composition of the type of biosurfactants produced. The different types of biosurfactants are trehalose lipids, sophorolipids, rhamnolipids, glycolipids, cellobiose lipids, polyol lipids, phospholipids, sulfonylipids, surfactin, viscosin, diglycosyl diglycerides.

Classification of Biosurfactants

Different classes of biosurfactants are distinguished on the basis of their chemical structures. The biosurfactants that are most often isolated and most extensively investigated are the glycolipids, including the rhamnolipids produced by *Pseudomonas aeruginosa* and *Pseudomonas fluorescens*. Other groups of biosurfactants are the lipopeptides and protein like substances, phospholipids, substituted fatty acids, and lipopolysaccharides. Little is known about biosurfactants of LAB, although much information concerning bacteriocins from lactobacilli and other lactic acid bacteria is available. The main physiological role of biosurfactants is to facilitate the uptake of water-immiscible substrates by lowering the surface tension at the phase boundary, emulsification, and enabling the microbial cells to adhere to the organic compounds.

Biosurfactants of LAB

Lactobacilli are believed to interfere with pathogens by different mechanisms. Firstly competitive exclusion effects, then Lactobacilli have also been shown to produce antibacterial compounds, such as lactic acid, hydrogen peroxide, bacteriocin-like substances and possibly biosurfactants. Dairy *Streptococcus thermophilus* strains was reported to produce biosurfactants which cause their own desorption and oral *Streptococcus mitis* strains produce biosurfactants that inhibit adhesion of *Streptococcus mutans*. *Lactobacillus* and *Streptococcus* species have been shown to be able to displace adhering uropathogenic *Enterococcus faecalis* strains from hydrophobic and hydrophilic substrata in a parallel-plate flow chamber through biosurfactant production. Biosurfactants from *L. acidophilus* RC14 and *L. fermentum* B54 were richest in protein, while those from *L. casei* subsp. *rhamnosus* and ATCC 7469 had relatively high polysaccharide and phosphate content. The bacteriocin lactocin 27 was isolated from culture supernatants of *Lactobacillus helveticus* LP27 as a protein-lipopolysaccharide complex. The main physiological role of biosurfactants is to facilitate the uptake of water-immiscible substrates by lowering the surface tension at the phase boundary, emulsification, and enabling the microbial cells to adhere to the organic compounds. Biosurfactants can also exhibit antibiotic activity against various microbes.

Cell growth and the accumulation of metabolic products are strongly influenced by medium composition. The response surface methodology was used for the optimization of medium components for *Lactococcus lactis* and *Streptococcus thermophilus* growth, in order to optimize the production of a growth associated biosurfactant. For each lactic acid bacteria strain, six growth factors were analysed based on MRS medium for *Lactococcus lactis* (lactose, peptone, meat extract, yeast extract, ammonium citrate, potassium phosphate) or M17 medium for *Streptococcus thermophilus* (lactose, peptone, meat extract, yeast extract, soya peptone, natrium glycerophosphate). Response surface methodology proved to be a powerful tool in optimizing medium compositions for biosurfactant production by acid lactic bacteria. The optimal medium composition allowed biosurfactant production to be 1.3 times faster compared to traditional medium.

Biosurfactant production by probiotic strains, *Lactococcus lactis* 53 and *Streptococcus thermophilus* A, using conventional synthetic media and its applications was reported. By response surface optimization, the medium components for the production of biosurfactants by probiotic bacteria was optimized and concluded that it was possible to determine optimal operating conditions to obtain a higher cellular growth, thus a higher biosurfactant production yield. Moreover, both bacterial strains studied shown higher amounts of biosurfactant produced with the optimized medium; it would be possible to develop strategies for biosurfactant production from whey. In another study suitable kinetic models were established for several *Lactobacillus* strains biosurfactant producers using whey as an alternative medium. A great variety of alternative raw materials is currently available as nutrients for industrial fermentations, namely various agricultural and industrial by-products and waste materials. A good substrate for biosurfactant production is whey, as it is composed of high levels of lactose, protein, organic acids and vitamins. The success of biosurfactant production depends on the development of cheaper processes and the use of low-cost raw materials, which account for 10–30% of the overall cost. A great variety of agro-industrial wastes have been studied as potential substrates for biosurfactant production. A by-product of the sugar cane industry, molasses, represents an alternative for the biosurfactant production process as it is a relatively inexpensive raw material compared to other substrate sources, and it possesses other valuable compounds for the fermentation process. This alternative medium was also studied in the present work for biosurfactant production by probiotic bacteria. The biosurfactant production yields achieved with supplemented molasses medium were higher than the obtained whether with conventional or supplemented cheese whey medium.

The lactic acid bacteria *L. casei* CECT-5275, *L. rhamnosus* CECT-288, *L. pentosus* CECT-4023 and *L. coryniformis* subsp. *torquens* CECT-25600 were found to be biosurfactant-producing strains. Depending upon the nature of the biosurfactant and the producing microorganisms, several patterns of biosurfactant production by fermentation are possible. The pH reduction results of simultaneous production of lactic acid that changes drastically the media conditions and can be responsible for the biosurfactant production inhibition. The lowest values of surface tension were obtained at the end of fermentation, therefore, biosurfactant release by the selected lactobacilli strains is maximal for cells in the stationary cells. Four *Lactobacillus* strains were screened for biosurfactant production by surface tension determination, and the biosurfactant production as well as the time courses of glucose consumption and biomass growth modeled and found that *Lactobacillus acidophilus* RC14 and *Lactobacillus fermentum* B54 were strongly biosurfactant-producing strains. Other biomedical applications of the biosurfactants are the use of biosurfactants obtained from *L. lactis* 53 and from *S. thermophilus* A to prevent the microbial colonization of silicone rubber voice prostheses.

Biosurfactants by LAB are used more often for medical purposes, as ingredients of therapeutic agents playing a key role in the prevention and control of infections caused by pathogens representing various groups of microorganisms. Due to their significant role in the adhesion of microorganism cells to different surfaces, as well as biofilm formation, they are used as impregnants applied to the surface of medical devices. The antimicrobiological activity of post-culture fluids of selected lactic acid bacteria capable of synthesizing biosurfactants was estimated against strains of indicatory Gram-negative (*Escherichia coli, Klebsiella pneumoniae, Pseudomonas aeruginosa*) and Gram-positive (*Bacillus subtilis, Micrococcus roseus, Staphylococcus epidermidis*) bacteria. Their antimicrobial activity was determined by the agar well diffusion method. The lactic acid bacteria able to synthesize biosurfactants were tested for the capability to survive in the presence of bile salts and pH below 4. Selected lactic acid bacteria synthesizing biosurfactants can survive in the presence of bile acid salts and at pH below 4. The results suggested the ability of *Lactobacilli* to synthesize biosurfactants may be considered an indicator of their probiotic properties.

Potential Food Applications

Biosurfactants can be explored for several food-processing applications as follows:.

a) Food formulation ingredients

Apart from their obvious role as agents that decrease surface and interfacial tension, thus promoting the formation and stabilization of emulsions, surfactants can have several other functions in food. For example, to control the agglomeration of fat globules, stabilize aerated systems, improve texture and shelf-life of starch-containing products, modify rheological properties of wheat dough and improve consistency and texture of fat-based products. In bakery and icecream formulations biosurfactants act by controlling consistency, retarding staling and solubilizing flavour oils; they are also utilized as fat stabilizers and antispattering agents during cooking of oil and fats. Improvement in dough stability, texture, volume and conservation of bakery products and improvement in the properties of butter cream, croissants and frozen confectionery products is obtained by the addition of rhamnolipid surfactants. L-Rhamnose has considerable potential as a precursor for flavouring. It is already used industrially as a precursor of high-quality flavour components like furaneol.

b) Antiadhesive agents

A biofilm is described as a group of bacteria that have colonized a surface. The biofilm not only includes bacteria, but it also describes all the extracellular material produced at the surface and any material trapped within the resulting matrix. Bacterial biofilms present in the food industry surfaces are potential sources of contamination, which may lead to food spoilage and disease transmission. Thus controlling the adherence of microorganisms to food-contact surfaces is an essential step in providing safe and quality products to consumers. The involvement of biosurfactants in microbial adhesion and detachment from surfaces has been investigated. A surfactant released by *Streptococcus thermophilus* has been used for fouling control of heat-exchanger plates in pasteurizers, as it retards the colonization of other thermophilic strains of *Streptococcus* responsible for fouling. The preconditioning of stainless steel surfaces with a biosurfactant obtained from *Pseudomonas fluorescens* inhibits the adhesion of *L. monocytogenes* L028 strain. The bioconditioning of surfaces through the use of microbial surfactants has been suggested as a new strategy to reduce adhesion.

c) Immuno modulatory action

Sophorolipids are promising modulators of the immune response. It has been previously demonstrated that sophorolipids, decreased sepsis related mortality at 36 h in vivo in a rat model of septic peritonitis by modulation of nitric oxide, adhesion molecules and cytokine production and decreased IgE production in vitro in U266 cells possibly by affecting plasma cell activity. The results show that sophorolipids decrease IgE production in U266 cells by down regulating important genes involved in IgE pathobiology in a synergistic manner. These data continue to support the utility of sophorolipids as an anti-inflammatory agent and a novel potential therapy in diseases of altered IgE regulation.

Future Trends

Successful commercialization of every product depends largely on its bioprocess economics. At present, the prices of microbial surfactants are not competitive with those of the chemical surfactants due to their high production costs and low yields. Hence, they have not been commercialized extensively. For the production of commercially viable biosurfactants, process optimization at the biological and engineering level needs to be improved. Improvement in the production technology of biosurfactants has already enabled a 10–20-fold increase in productivity, although further significant improvements are required. However, the use of cheaper substrates and optimal growth and production conditions coupled with novel and efficient multi-step downstream processing methods and the use of recombinant and mutant hyperproducing microbial strains can make biosurfactants production economically feasible. Novel recombinant varieties of these microorganisms, which can grow on a wide range of cheap substrates and produce biosurfactants at high yields, can potentially bring the required breakthrough in the biosurfactant production process. Although a large number of biosurfactant producers have been reported in the literature, biosurfactant research, particularly related to production enhancement and economics, has been confined mostly to a few genera of microorganisms such as *Bacillus, Pseudomonas* and *Candida*. Biosurfactants are not only useful as antibacterial, antifungal and antiviral agents; they also have the potential for use as major immunomodulatory molecules, adhesive agents and even in vaccines and gene therapy. A judicial and effective combination of these strategies might, in the future, lead the way towards large-scale profitable production of biosurfactants. This will make biosurfactants highly sought after biomolecules for present and future applications as fine specialty chemicals, biological control agents, and new generation molecules for pharmaceutical, cosmetic and health care industries.

Suggested Reading

Desai JD & Banat I. M (1997) Microbial Production of Surfactants and Their Commercial Potential. *Microbiology and Molecular biology Reviews,* **61(1):** 47–64.

Dubey K & Juwarkar A (2001) Distillery and curd whey wastes as viable alternative sources for biosurfactant production. *World J. Microbiol. Biotechnol.,* **17:** 61–69.

Gan B. S., Kim J, Reid G, Cadieux P & Howard, J. C. (2002) *Lactobacillus fermentum* RC-14 inhibits *Staphylococcus aureus* infection of surgical implants in rats. *J. Infect. Dis.*, **185:** 1369- 1372.

Mulligan C. N. (2005) Environmental applications for biosurfactants. *Environ. Pollut.*, **133:** 183–198.

Muthusamy K, Gopalakrishnan S, Ravi KT & Sivachidambaram P (2008) Biosurfactants: Properties, commercial production and application. *Current Science*, **94(6):** 736-747.

Nitschke M & Coast S. G. (2007) Biosurfactants in food industry. *Trends Food Sci. Technol.*, **18:** 252–259.

Rahman PKSM & Gakpe E (2008) Production, Characterization and Applications of Biosurfactants - Review. *Biotechnology*, **7(2):** 360-370.

Rodrigues L., Moldes A, Teixeira J & Oliveira R (2006) Kinetic study of fermentative biosurfactant production by *Lactobacillus* strains. *Biochemical Engineering Journal*, **28:** 109–116.

Rodrigues L., Teixeira J & Oliveira R (2006) Low-cost fermentative medium for biosurfactant production by probiotic bacteria. *Biochemical Engineering Journal*, **32:** 135–142.

Rodrigues L., van der Mei H, Teixeira J & Oliveira R (2004) Biosurfactant from *Lactococcus lactis* 53 inhibits microbial adhesion on silicone rubber. *Appl. Microbiol. Biotechnol.* **66:** 306–311.

Rodrigues L., van der Mei H, Teixeira J & Oliveira R (2004) Influence of biosurfactants from probiotic bacteria on formation of biofilms on voice prosthesis. *Appl. Environ. Microbiol.* **70:** 4408–4410.

□□□

Chapter 17

Immunodiffusion - A Technique for Antigen Characterization

Y.S. RAJPUT[1], RAJAN SHARMA[2] AND REWA KUMARI[1]
[1]Animal Biochemistry Division,[2] Dairy Chemistry Division, NDRI, Karnal

Introduction and Principle

Antigen and antibody molecules, when placed in wells prepared in agarose gel plate, diffuse through the gel. During diffusion, the concentration of antibody and antigen changes with distance. At certain point in diffusion path, the concentration of antigen and antibody reaches at a level where large antigen-antibody complexes are formed. The sufficiently large size complexes (> 200 kD) can not move through the pores of agarose gel and sharp precipitation lines are formed. If an antiserum contains antibodies against more than one antigen, more than one-precipitation lines may be observed

Agar is an algal polysaccharide consisting of repeating units of galactose and 3,6-anhydrogalactose. Anionic groups such as sulphate and pyruvate are affixed in matrix. Agar is fractionated to obtain polysaccharide with lower contents of charged groups and fractionated agar is referred as agarose. Thus, agarose is purified linear galactan hydrocolloid isolated from agar or agar-bearing algae. Since, sulphate is major ionic group present in agarose, sulphate content is indicator of purity of agarose. Agarose melts at temperature, which is different from temperature where agarose forms gel. Agarose with < 0.20 % sulphate, gelling at 36 ± 1.5 °C and melting at 88 ± 1.5°C is suitable for immunodiffusion. This specification of agarose is also suitable for immuno-electrophoresis

Equipments and Materials

Glass plates (8 X 8 cm or 10 X 10 cm), agarose, humidified chamber, gel cutter

Phosphate Buffer Saline (PBS)

Na_2HPO_4,	M.W. 142	(10.1 mM)	1.44 g
KH_2PO_4	M.W. 136	(1.7 mM)	0.24 g
NaCl	M.W. 58.5	(137 mM)	8.0 g
KCl	M.W. 74.5	(2.7 mM)	0.2 g
NaN_3	M.W. 65	(3 mM)	0.2 g

Dissolve above ingredients in about 900 ml distilled water and adjust pH to 7.2. Make volume to 1 litre with distilled water.

Agarose Gel for Pre-Coating

Two g agarose is added to 100ml distilled water in 250 ml conical flask. The agarose suspension boiled for 5 to 6 min. and kept at 56°C in water bath till used for pre-coating of glass plates.

Agarose Gel for Coating

One g agarose is added to 100ml phosphate buffer saline in 250 ml conical flask. The agarose suspension is boiled for 5 to 6 min. and kept at 56°C in water bath till used for coating of glass plates.

Preparation of Plates for Pre-coating

Glass plates are cleaned with detergent, washed with tap water and then with distilled water. If glass plates contained greasy material, plates should be dipped overnight in chromic acid before these are cleaned with detergent. After drying, plates are placed in oven maintained at 50 to 60°C for about 30 min. Glass pipettes and Pasteur pipettes to be used for pre-coating and coating are also placed in oven (50 to 60°C) before use so that agarose does not get cooled while transferred from conical flask to glass plates.

Pre-coating

Place few drops of warm agarose (2 % in water) over glass plate towards one arm of square plate.With the help of another glass plate, the agarose is quickly spread over the entire plate (see fig. a and b).Uniform force should be applied during pre-coating. This way, a thin uniform layer of agarose is prepared and is referred as pre-coating. Let the gel dry-up at room temperature. It will be dried in few minutes. Observe the pre-coated plate against light. Correctly pre-coated plate will look transparent.

Coating

In order to have uniform and defined thickness (say 2 mm, 3 mm, 4 mm) of agarose gel, adhesive tape is pasted at all the four edges of glass plate such that a portion of tape sticks to glass plate and remaining portion is perpendicular to the surface of glass plate. The height of tape perpendicular to plate decides the thickness of agarose gel. For example, if you require 2 mm gel thickness, the tape should be up to a height of 2 mm from surface of glass plate.

Place glass plate on horizontal surface and pour warms (56°C) agarose gel (1 % in PBS). Let the gel gets solidified at room temperature. Then, plate is kept in refrigerator for 10 - 15 min. Now, gel plate is ready for making circular basins.

Preparation of Wells

Circles (4 mm diameter) at a distance of 10 mm (measured from centre of one circle to centre of adjacent well) in appropriate design are drawn on a sheet of paper. The gel plate is then placed over paper drawing. With the help of gel puncher (cork- borer or empty Jotter refill), gel is punched just above circular drawings. The gel from circular cut is carefully removed. Now, plate is ready for immunodiffusion.

Antigen and Antibody Reaction

Antigen (1 mg / ml PBS) and antibody solutions (1 mg / ml PBS) are prepared fresh. Anti-serum can be used as source of antibody. Serial dilutions of these solutions may be required for determining optimum concentrations. The individual well (basin) is filled with either antigen or antibody solution. These solutions should not over flow. A well-thought decision for using a particular well for antigen or antibody solution is essential in order to obtain maximum information from single plate. The plate is placed in humid chamber (plastic dessicator with sufficient water at the bottom can be conveniently used as humid chamber) for overnight at room temperature. Precipitation lines between wells can be observed when gel plate is placed against light. It has been observed that, in some cases, additional incubation for 24 h is required to see precipitation lines.

Staining of Precipitation Lines

Staining of precipitation lines is essential for better visualization or photography of precipitation lines. Besides antigen-antibody complex in the gel, soluble antigen and antibody molecules are present in the gel. Irrespective of any staining protocol adopted for staining of precipitation lines in the gel, removal of soluble antigen and antibody molecules is essential in order to minimize background staining. The soluble proteins are removed as described below.

Fill wells with distilled water and wet the surface of agarose gel with water.

Whatman No.1 filter paper is soaked in water and is placed over the gel such that air bubbles are not trapped.

Place several layers of ordinary filter. A glass plate of 10 mm thickness is placed over filter paper for uniform pressing. Wait for few minutes.

3. Replace ordinary filter papers if these have become wet (do not remove Whatman No.1 filter paper)
4. Continue to replace ordinary filter paper if these have become wet. When no more water is removed from the gel, remove ordinary and Whatman No.1 filter papers. At this stage gel will be thin and would stick to glass plate.
5. The gel plate is then dipped in 0.85 % NaCl for 15 min. at room temperature with mild shaking.

Repeat steps 2 to 5.

Repeat steps 2 to 4.

The gel plate is dipped in distilled water for 15 min. at room temperature with mild shaking. Repeat steps 1 to 4.

The precipitation lines are stained by coomassie brilliant blue R (CBB-R) as described below.

- Gel plate is kept in CBB-R solution for 30 min. at room temperature. The dye solution is prepared by dissolving 0.25 g CBB-R in 100 ml solution containing 25 ml methanol, 10 ml acetic acid and 65 ml distilled water.
- Gel is then destained in destaining solution, which is prepared by mixing 5 ml methanol, 7.5 ml acetic acid and 87.5 ml distilled water.
- Precipitation lines will appear blue in colour.

□□□

Chapter 18

Western Blotting - A Tool for Protein Characterization

Y.S. RAJPUT[1], RAJAN SHARMA[2] AND POONAM[1]
[1]Animal Biochemistry Division, [2]Dairy Chemistry Division, NDRI, Karnal

Introduction

For analysis based on antibody reactivity or nucleic acid hybridization, the separated molecules are made free of electrohoretic matrix. This can be simply achieved by slicing the gel followed by elution in buffer. But, this process is slow and resolution is also poor. An alternative efficient method is 'blotting' technique in which molecules are separated on a slab gel and separated molecules are eluted through the broad face of the gel onto a membrane that binds the molecules as they emerge. Proteins and nucleic acids stay on the surface and can be detected.

The membrane materials frequently employed in blotting are nitrocellulose, nylon and polyvinylidene difluoride (PVDF). The choice of membrane depends on the type of analysis and characteristics of detection system. Nitrocellolose is the most widely used since it works well with both protein and nucleic acids. Some nylons do not bind protein reliably. PVDF, is often used when bound proteins are analysed for sequencing.

The transfer of the proteins or nucleic acids from the gel to the membrane can be achieved by capillary flow of buffer or by transverse' electrophoresis. The use of capillary flow to transfer DNA from agarose gels to nitrocellulose membrane was first described by Southern (1975) and thus referred as Southern blotting. Using the same method for transfer of RNA is referred as Northern blotting. Western blot refers to transfer of protein from gel to membrane and this technique was described in 1979-80 by many workers but the method described by Towbin (1979) is most cited.

Western blotting essentially comprises of three techniques which are applied in sequence. The first one is referred as SDS-PAGE through which proteins are separated based on the molecular size of molecules in acrylamide gel. Sodium dodecyl sulphate (SDS) is an anionic detergent that denatures proteins by wrapping around the polypeptide backbone. This results in net negative charge to polypeptide in proportion to its length. Laemmli system (Laemmli, 1970) employing discontinuous buffer is most widely used electrophoretic system. The resolution in Laemmli's method is excellent because treated peptides are concentrated in stacking gel before entering the separating gel. The technique, which follows SDS-PAGE, is transfer of protein/from gel to membrane. There are two types of equipments for electrophoretic transfer of proteins: the semi-dry blotting apparatus and 'tank' buffer apparatus. The third technique used in sequence is for identification of protein (antigen) by performing" antigen-antibody (first antibody) reactions on the membrane itself. Second antibody enzyme conjugates were then allowed to interact with immobilized first antibody and , then using appropriate substrate, protein bands are detected. Although, antigen-antibody interactions are widely employed in Western blot, other kind of interactions such as glycoprotein-Iactin and biotin-avidin have allowed research w6rkers to employ this technique for other applications including carbohydrate staining of glycoprotein, protein sequencing etc.

Semi-dry Electrophoretic Transfer

In semi-dry electrophoretic transfer, a stack of wetted filter papers surrounding the gel and the blotting membrane is used as a buffer reservoir, instead of tank as in conventional electrophoretic transfer. The electrodes, consist of conductive plates made of graphite or stainless steel or a conducting polymer. The size of the plates is at least the same size as that of gel to provide homogeneous electric field. The main advantages with semidry transfer are the ease of handling, the short time (30 min, to 1 h) required for the transfer and low buffer consumption. Another important feature is that different buffers can be used at the anodic and cathodic sides to improve the transfer. The short electrode distance gives a high voltage gradient despite low power. Cooling is not normally required since heat production is negligible. Transfer can be performed from several gels at p time, either by placing them beside each other if the electrodes are large enough or by placing several transfer units on top of each other. Because of the short electrode distance, voltage applied is most often 10-20V. On the other hand, because of large cross-sectional area, the current passing through the transfer sandwich is fairly high, in the range 0.1-1A.

Tank-Buffer Electrophoretic Transfer

In tank-buffer electrophoretic transfer, transfer cassette is submerged in a 'tank' of. buffer. Gel, membrane, filter paper, porous foam sheet are arranged in cassette as per instructions of manufacturer. The details have been provided in write-up for practicals.

Blot Membranes

Numerous types of papers and membranes have been utilized for protein blotting. Nitrocellulose paper (film of nitric acid esterified cellulose) has been the most frequently used membrane. The binding of proteins to nitrocellulose is probably hydrophobic. For electrophoretic transfer of small proteins, membranes with 0.1 or 0.2μm pore size are selected. If membranes stick to low concentration gels' after transfer, membranes with pore size of 0.45 μm are selected. A drawback with nitrocellulose membrane is, however, that they are very brittle when dry.

The other membranes, which is in use, is polyvinyllidene fluoride (PVDF). PVDF membrane is a teflon-type polymer composed of the basic repeating unit (- $^{8+}$CH2-$^{8-}$CF2-)$_n$ and has good mechanical strength. Proteins interact with the polymer non-covalently through dipolar and hydrophobic interactions. PVDF is chemically compatible with the aqueous buffer systems. Since PVDF is resistant to most organic solvents and it can withstand harsh chemical conditions in which nitrocellulose membranes dissolve or decompose. These membranes are expensive. Some PVDF membranes have additional components. Westren consists of PVDF cast on a polyester web. The web does not interfere with electroblotting or alter the characteristics of the PVDF. Immobilon-

CD is PVDF membrane in which surface is chemically modified to have a cationic charge. Although hydrophobic and dipolar interactions with the Immobilon - CD may contribute to protein binding, the primary binding interaction is ionic. The membranes with high internal surface area (>2000cm^2 per cm^2 of frontal area) bind substantially more protein (400 μg BSA *I* cm2) as compared to membrane with low internal surface area (~400 cm2 per cm2 of frontal area) that binds to around 130 μg BSA *I* cm^2. Low internal surface area membranes usually function better in immunodetection. They are comparatively easy to block and antibodies are better able to penetrate the more open pore structure. Membranes with high internal surface structure are more difficult to block effectively and less open pore structure often limits antibody accessibility. Besides immunodetection, PVDF membranes are used for amino acid sequencing, amino acid analysis and peptide mapping. For these applications blocking is not required and there is no steric hindrance encountered by antibodies. Higher internal surface membranes and Immobilon CD are suitable for amino acid sequencing and amino acid analysis. Peptide mapping is more effective on low internal surface area membranes. PVDF membrane is compatible with protein staining and immuno chemical protocols.

Positively charged nylon membranes arc mechanically strong and have a high binding capacity. A disadvantage is their high non-specific binding which results in a high background after immunodetection. Most general protein stains are anionic dyes and can not be used with nylon membranes since they bind to these membranes.

Transfer Buffer

A major concern in transferring proteins onto nitrocellulose membrane is the composition of the transfer buffer. !he original protocol of Towbin *et al.* (1979) uses a transfer buffer containing methanol, which was added to ,counteract swelling *of* the gel. Methanol also decreases gel pore size, removes SDS from proteins. Methanol may precipitate the proteins within the gel, however it increases the capacity and the affinity of nitrocellulose membrane for proteins. PVOF membrane is activated by placing it in 100% methanol for 1-2 sec. This allows the hydrophobic surface of PVOF to wet with aqueous solvent. Addition of 20% methanol to transfer buffer is recommended for low molecular weight proteins. Methanol is not required for transfer to charged nylon membranes. Methanol facilitates the dissociation of SOS-Protein complexes and increases the hydrophobic interaction between protein and membrane. On the other hand, for high molecular weight proteins, methanol can decrease the elution efficiency by denaturing the proteins or retarding the elution from the gel. In contrast to low molecular weight proteins, high molecular weight proteins do not require methanol for adequate binding to the membrane.

The presence of SDS in transfer buffer increases the mobility of protein from gel to membrane. This is especially useful for transfer of protein after isoelectric focussing, when proteins have no net charge. However, 80S decreases the binding of the protein to both nitrocellulose and PVOF membrane. It is sometimes necessary to add SOS (0.01-0.02%) to aid transfer of high molecular weight proteins. Transfer buffer generally used is 25 mM Tris, 192 mM glycine, ph 8.3 and 20% methanol. If membrane is to be used for protein sequencing or amino acid', analysis, CAPS buffer (10 mM 3-(cyclohexylamino)-1 propanesul(onic acid, 10% methanol, pH 11.0 is recommended. Application of Protein Blotting Characterization of Antigens Characterization of antigen will require antigen specific antiserum. By simultaneously running molecular weight markers and proteins (extracted from biological materials) in SDS-PAGE and subsequent detection after electrophoretic transfer provides information about molecular weight of antigen. Antibodies should be specific otherwise cross-reaction is observed and interpretation is more difficult. Affinity purified antibodies or monoclonal antibodies provides good result. Through these reactions, one can detect presence or absence of such antigens in related and unrelated biological materials. Now tools are available for ascertaining carbohydrate moiety in proteins on membrane. These proteins can be oxidized by period ate resulting in generation of free aldehyde groups. The generated groups are reacted with biocytin hydrazide leading to biotinylation of glycoproteins. Using appropriate probe such as avidin-peroxidase and substrate, glycoproteins are detected. Alternately lactins specific for carbohydrate residues can be employed. In this approach antibody (against lactins) enzyme conjugates or lactin - enzyme conjugates can be used for staining glycoproteins. Asn-linked oligosaccharides can be cleared from protein onto membrane by hydrazinolysis. The released oligosaccharides can be characterized by using biochemical techniques. Proteins onto membrane can be hydrolyzed for determining amino acid composition. Peptide mapping and protein sequencing are other useful applications where proteins on membrane are the starting material for subsequent steps.

Suggested Readings

De Maio, A. (1994). Protein blotting and immunoblotting using nitrocellulose membranes *in* Protein Blotting - A Pratical Approach (Ed. B.S. Dunbar), IRL Press, Oxford pp. 11-32

Dunbar, B.S. (1994). Introduction and basic set-up for protein transfer *in* Protein Blotting - A Pratical Approach (Ed. B.S. Dunbar), IRL Press, Oxford, pp. 5-9

Jacobson, G. (1994). Protein blotting using semi-dry electrophoretic transfer equipment *in* Protein Blotting - A Practical Approach (Ed. B..S. Dunbar, IRL Press, Oxford, pp. 53-70

Laemmli, U.K. (1970). Cleavage of structural proteins during the assembly of the head of bacteriophage T4. *Nature* **227:** 680-685.

Mansfield, M.A (1994). Protein blotting using polyvinylidene fluoride membranes *in* Protein Blotting - A Practical Approach (Ed. B.S. Dunbar), IRL Press, Oxford, pp. 33-52.

Towbin, H., Staehelin, T. and Gordon, J. (1979). Electrophoretic transfer of proteins from polyacrylamide gels to nitrocellulose sheets: Procedure and some applications. *Pro Natl Aca. Sci* USA. **76:** 4350-4354.

□□□

Chapter 19

Vitamin Fortification as Dietary Supplements in Functional Dairy Products

SUMIT ARORA
Dairy Chemistry Division, National Dairy Research Institute, Karnal

Introduction

Malnutrition is a serious problem among children in developing countries. Caused by lack of food, improper diet and unsafe drinking water, malnutrition leads to deficiencies in micronutrients such as iron, iodine and vitamins. Of these, vitamin A deficiency is the most lethal. The risks become less in older children, but vitamin A deficiency reduces overall immunity and makes all children susceptible to diseases like measles and diarrhoea. UNICEF estimates that vitamin A deficiency is a public health concern in 72 countries in Asia and Africa.

According to World Health Organization (WHO) guidelines, vitamin A deficiency is a public health concern if the mortality rate of children below five years is greater than 70/1,000. This means that the death of more than 70 of every 1,000 children indicates vitamin A deficiency. In 1970, the mortality rate of children under five years in India was 130/1,000, thereby making vitamin A deficiency a public health concern (Hindu Health, December 8, 2001). Fortification of food with vitamin A is an approach that nutritionists suggest to overcome this deficiency. But this is possible only where food is sold centrally. In rural India, most people grow and consume their own food. Or food is bought and sold locally, making fortification difficult. Fortifying commodities like sugar and milk will probably work. But ensuring that children consume enough milk to get their supply of vitamins is not easy.

According to Sehgal (2008) every second child in India under the age of 5 suffers from malnutrition, and Vitamin A available in milk is a key component in fighting it. At present, both Vitamins A and D are being removed along with the fat content in milk. Mandatory milk fortification will ensure that Vitamin A is added to all toned and semitoned milk. Mandatory milk fortification will require an amendment to the Prevention of Food Adulteration (PFA) Act. Our policy makers are hopeful that the government will pass a Bill regarding fortification. An interaction of 15 MPs across party lines agreed on milk fortification. Health Minister Anbumani Ramadoss has referred the idea of making milk fortification mandatory to the Committee for Food Standards and Safety for its comments.

Soluble Vitamin A is manufactured locally in India and the incremental cost of using it is less than 1 paisa per litre. For those dairies that don't have homogenisers, the cost will work out to 2 paise per litre." The scheme was tried earlier in 1989-90 when the government reimbursed dairies for the cost of using a Vitamin A premix. But although some dairies are continuing with the fortification, most are no longer doing it. In Delhi, Mother Dairy, which sells 7.5 lakh litres a day, fortifies its toned milk but not milk being sold in packets. Although incremental costs are low, with India being the largest supplier of milk in the world the cost of fortification for milk suppliers will run into crores of rupees. "Amul alone sells 4 million litres of milk per annum; fortification will cost between Rs 2-3 crore. Feedback from the industry is that while they do not oppose it they will not volunteer to do it on their own," (www.infochangeindia.org, 2008). A recent survey by the National Nutrition Monitoring Board in Hyderabad has shown that 90% of children between 1-5 years of age have an intake of less than 50% of their required daily allowance of vitamins. Nutritionists maintain that the best method of combating micronutrient malnutrition is by supplying the right diet. But dietary diversification is difficult to achieve on a mass scale. Fortification of essential foods such as oil, wheat flour and milk will make an enormous difference to the health of both child and mother. Nutritionists point out that Parliament has passed only two bills regarding food fortification in the last 60 years. These are the fortification of vanaspati with Vitamin A, since 1953, and the policy of universal iodisation of salt, in 1995, though it became law only in 2005. The immediate benefit of this was to lower rates of goitre and mental retardation.

Milk and Milk Products as a Suitable Vehicle for Fortification

Milk in its natural form is almost unique as a balanced source of man's dietary need (Table 1). The various steps in processing and storage have a measurable impact on some specific nutrients (Table 2). Milk also provides a convenient and useful vehicle for addition of certain nutrients to man's diet and has following benefits:

- Since milk is centrally processed so that the quality control can be effectively implemented.
- Milk and milk products are widely consumed regularly in predictable amounts by people of all age groups.
- Cost is affordable by target population.
- The stability and bioavailability of the added micronutrients to the milk remains high.
- Since milk is nearly a complete food and all nutrients exist in almost fully available form, the bioavailability of added nutrients remains high.
- Addition of fortificants usually caused minimum change in colour, taste and appearance.

Nutrients Generally Added to Milk

Liquid milk fortification with vitamins A and/D is mandated in several countries. â-carotene is added as a colour-enhancing agent to some milk products such as butter. Dried milk is often fortified with vitamins A and D, calcium, and iron. Milk based infant formula and weaning foods are fortified with a range of vitamins, minerals, and other nutrients such as polyunsaturated fatty acids. Powdered milk used for complementary feeding in Chile is fortified with vitamin C, iron, copper and zinc.

Table 1: Micronutrient Content of Cow Milk Vs Recommended Dietary Allowances (RDA)

Micronutrient	Quantity/ Litre	RDA	
		Men	Women
Vitamin A (IU)	1300	1000	800
Vitamin D (IU)	42	5-10	5-10
Vitamin E (IU)	1.5	10	8
Vitamin K (ìg)	41	45-80	45-65
Vitamin B1(mg)	0.4	1.5	1.1
Vitamin B2 (mg)	1.7	1.7	1.3
Vitamin B6 (mg)	0.4	2	1.6
Folic acid (μg)	62	200	180
Niacin (mg)	1	19	15
Vitamin B12 (mg)	3	2	2
Vitamin C (mg)	15	60	60

Table 2: Micronutrient content of cow milk Vs Recommended Overages.

Micronutrient	Quantity/ Litre	Recommended Overages (%)			
		Pasteurised	UHT	Dry Milk	Milk Desserts
Vitamin A (IU)	1300	20	30	40	20
Vitamin D (IU)	42	20	30	40	20
Vitamin E (IU)	1.5	10	30	20	10
Vitamin K (ìg)	41	-	-	-	-
Vitamin B1(mg)	0.4	25	50	20	25
Vitamin B2 (mg)	1.7	15	40	20	15
Vitamin B6 (mg)	0.4	30	30	20-30	30
Folic acid (ìg)	62	20	40	40	20
Niacin (mg)	1	15	20	20	15
Vitamin B12 (mg)	3	15	30	40	20
Vitamin C (mg)	15	30	100	50	30

(OMNI, 2001)

Fortification of Milk & Milk Products with Vitamins

Under ambient conditions the water soluble vitamin C and vitamins of the B-complex group such as thiamin, riboflavin, vitamin B_6, niacin, pantothenic acid, folic acid, biotin and vitamin B_{12} are powdered and thus relatively easy to work with when producing most dairy products. The fat soluble vitamins which include vitamin A, D, E and K, however, exist either as oil or as crystals, which may cause processing difficulties during the production of certain types of dairy products (Mortensen and Gotfredson, 1996). One of the problem encountered with the vitamins, is their limited stability in presence of heat, humidity and oxygen. Among the water soluble vitamins, vitamin C, folic acid, vitamin B_6 and vitamin B_{12} are the less stable. While in the case of fat soluble vitamins vitamin A, D and E are least stable.

In order to improve the stability of these vitamins, a number of different coating technologies have been developed. One of the most important methods to protect the fat soluble vitamins is microencapsulation, which results in a highly sophisticated powder, where the vitamin is kept protected from degradation by the coating material used for the encapsulation. During microencapsulation, the fat soluble vitamins are brought from the form of oil or a crystal - which in some processes would be difficult to handle - to the form of a free flowing powder much easier to handle and mix with other dry ingredients (Mortensen and Gotfredson, 1996). When two or more vitamins are added to a food product at the same manufacturing stage, this is commonly done in the form of premix or as blend. Premix is a homogenous mixture of desired vitamins in a dry powder from, whereas a blend is the same for the fat soluble vitamins, but in an oily form. A premix can consist of both water soluble and fat soluble vitamins and carotenoids, in which case the fat soluble vitamins have to be microencapsulated.

Vitamin A

Foods which have been successfully fortified with vitamin A include margarine, fats and oils, milk, sugar, cereals, and instant noodles with spice mix. Moisture contents in excess of about 7-8% in a food are known to adversely affect the stability of vitamin A. Beyond the critical moisture content there is a rapid increase in water activity which permits various deteriorative reactions to occur. Repeated heating, as may be experienced with vegetable oils used for frying, is known to significantly degrade vitamin A. The hygroscopic nature of salt has prevented its use as a vehicle for vitamin A fortification in countries of high humidity. In developed countries vitamin A fortification is limited to milk and dairy products, margarine and fat spreads and breakfast cereals. Current levels of fortification of Vitamin

A are generally considered as safe. Carotenoids can be used as a source of vitamin A. However the conversion and bioavailability of the carotenoids is known to be affected by the vitamin A status of the individual and by dietary composition. A conversion factor of 6:1 (six milligrams of beta carotene equivalent to one milligram of retinol) is currently being applied in calculating intake. The cost factor in using carotenoids as the source of vitamin A activity in fortification is generally considered prohibitive.

Vitamin D

There are a limited number of food vehicles suitable for vitamin D fortification. These include margarine, vegetable oils and dairy products. Vitamin D metabolism is linked to calcium absorption and parathyroid hormone. Susceptible populations for marginal vitamin D status include children under three years of age, the elderly and any population where cultural practices or climatic conditions prevent exposure to sunlight.

Vitamin E

Vitamin E, as tocopherol acetate, is added to fats and oils including margarine and fat spreads and breakfast cereals. Vitamin E intake is known to be related to the total dietary intake of fat and it has been shown to enhance absorption and bio-conversion of dietary carotenoids to vitamin A. Emerging evidence indicates that daily intake of vitamin E greater than 30 IU may afford protection against the development of degenerative diseases.

Vitamin C

There are technologies available for vitamin C fortification of fruit juices, fruit juice drinks, other related beverages, dairy products and some breakfast cereals. Vitamin C, as used in foods, is known to improve iron bioavailability. High moisture content (greater than 7%) in the presence of oxygen is known to adversely affect the stability of vitamin C in cereals.

B-Complex Vitamins

There are no problems related to the technology of the addition of B vitamins to cereals and grains. The development of off-flavours due to the thermal instability of thiamine is easily circumvented by adding this vitamin following all heat treatments. Riboflavin is thought to be associated with anaemia prevention and its role in this association needs to be clarified. The status of B complex vitamins is known to be marginal in populations with low consumption of animal foods (milk, meat, liver) and/or those which use low extraction (highly refined) products of cereals and grains such as flour as well as in populations in transition from rural to urban settlements. Food fortification with B complex vitamins may be appropriate for such populations. A confirmed association between dietary folic acid intake and neural tube defects has been reported. Populations with marginal or low folic acid intake may therefore benefit from fortification of staples with folic acid. Furthermore, folic acid together with vitamin B6 and vitamin B12 have been reported to decrease plasma homocysteine levels; the latter is reported to be inversely associated with the development of ischaemic heart disease.

Considerations During Fortification of Milk & Milk Products

- *Bioavailability of commercial preparations*: Bioavailability of different compounds facilitates the selection of the optimal compound.
- *Nutrient–nutrient reaction*: Interaction among the nutrients and other food components is a key factor in nutrient addition. For example, Vitamin C will improve the absorption of iron (Kiran *et al*, 1977). On the other hand, the iron will accelerate vitamin degradation. Fortification of calcium in milk may interfere with absorption of iron or zinc (Weaver, 1998).
- *Nutrient-matrix reaction*: The added nutrient must not react with any component of the milk. For example, iron is a pro-oxidant and can accelerate the development of fat rancidity, destroy some of the vitamins and form coloured products.
- *Shelf-life & packaging*: Many of the fortified milk and milk products may have limited shelf life and thus may need different types of packaging which can be either oxygen impermeable or opaque to light. This is particularly true for the fortification of liquid milk with vitamin A as vitamin A fortified milk develops off flavour within 6 h when exposed to light, compared to 12 h for control (Fellman *et al*, 1991). All the fortified products require proper labelling on the pack.
- *Process considerations*: The stability of all the vitamins is well known during various processing conditions and the same

knowledge can be applied while processing the vitamin fortified milk.

- Cost factor: Cost may not be a crucial factor in the manufacture and marketing of fortified milk and milk products.
- Safety factor: There should be sufficient insurance against excessive intake of the fortificant. Unlike water soluble vitamins, fat soluble vitamins exhibited toxicity at higher concentrations.

Conclusion

Milk and milk products provide a convenient and useful vehicle for fortification with micronutrients. Fortification ensures a safest method by which manufacturers can deliver health promoting, nutritionally dense food products. It is considered as an emerging technology as it considers the issues of the role of foods in quality of life and the role of foods in reduction of the risk of chronic diseases. The risks associated with fortification are minimal except if good manufacturing practices are not followed and only isolated incidents of this type have ever been reported. Improved understanding of interactions between food ingredients and health and ingenuity of food technologists in food formulation and fabrication will contribute to the advances in food fortification.

Suggested Reading

FAO (1995) Micronutrient fortification of food: Technology and quality control. FAO TechnicalConsultation of Food Fortification: Technology and Quality Control. Rome, Italy, 20-23 November.

Fellman, R.L., Dimick, P,S and Hollender, R. (1991) Photooxidative stability of vitamin A fortified 2%low fat milk and skim milk. *J. Food Protection* **54:** 113-116.

Kiran, R., Amma. M. K. P. and Sareen, K. N. (1977) Milk fortification with a system containing both iron and ascorbic acid. *The Indian J. Nutr. Dietet.* **14:** 260-266.

Mortensen, B. L. and Gotfredsen, P. (1996) Fortification - nutritional improvement of dairy products. Danish Dairy Food Industry worldwide 64-65.

OMNI/Roche/USAID (2001) Fortification basics: Milk. http://www. roche.com

Weaver, C.M. (1998) Calcium in food fortification strategies. *Int. Dairy J.* **8:** 443-449.

Hindu Health, December 8, 2001

Sehgal Rashme (2008) Mandatory milk fortification mooted to help remove malnutrition InfoChange News & Features, March 2008.

www.infochangeindia.org (2008) Food security, not vitamin supplements, is vital for public health. Accessed on 1.11.08.

□□□

Chapter 20

Adaptation at Molecular Level in Hyperthermophiles

JAI K KAUSHIK
Animal Biotechnology Centre, National Dairy Research Institute, Karnal

Introduction

Several key genomes have been sequenced completely while for others a rough draft sequence has been already finished. However, the sequence analysis work may take years to relate a simple sequence to the function. There has been a conception that sequencing of DNA is the goldmine in the hands of the scientists who can provide a panacea for all questions posed by our incessant inquisitiveness about life. There is a big misconception that the genome sequence would enlighten us about all the questions posed. The genome sequence has nothing to do with the biological functions that are essential to our life. The genome sequence is composed of DNA containing only four bases. The expression of the genetic information stored in the DNA sequence involves the translation of a linear sequence of nucleotides into a co-linear sequence of amino acids in proteins according to a universal genetic code (codons). An amino acid sequence translated from the DNA sequence directs the unique conformation of the protein. The tertiary structure exerts its proper function, i.e. each function is determined by its amino acid sequence. Identifying all proteins produced by a genome is a gigantic task. The analysis of the variations in proteins and the linked changes in functions constitutes another important problem after the sequencing of the genomes. The new science, which is developing, may encompass more broader and challenging areas is the proteomics science, which is the study of protein expressions by a genome temporally as well as spatially. Also, it constitutes the comparison of similar proteins on the basis of biological functions and their divergence on the evolutionary tree.

If the principle of how the tertiary structures of proteins are defined depending on their amino acid sequence is revealed, we would be able to easily predict the tertiary structure from the amino acid sequence the gene encoded. Moreover, a new protein with a novel function or with improved functions can be constructed if using these principles. Also, knowing the exact structure of proteins should help drug designer devise chemicals to fit the slots on the proteins that either activate them or prevent them from interacting. However, these novel approaches are far from realization, because a protein structure is affected not only by a part of the amino acid sequence but also by extensive interactions. One approach to understand the effect of sequence alterations on the structure-function relationship of proteins is by replacement of the particular amino acids by site directed mutagenesis and then analyzing the changes in structure, stability and function, while another approach is the comparative study of proteins, with similar functions from different species or strains. Proteins with similar functions from related organisms usually differ at only few amino acid residues. However, comparative studies suffer from the fact that structures of proteins from different organisms show not only modifications necessary to increase the thermal stability of the investigated protein, but also species-specific variations. Nonetheless, it provides an attractive source of establishing the structure-stability-function relationship. The activity of an enzyme is related to environmental factors; most important among them is the temperature. Proteins differing only slightly can have very different conditions of optimum activity. These conditions are determined by how strong is a protein or by its stability profile, which again is determined by the structure of the protein. The stability of the native conformation of a protein under physiological conditions is only about 5-15 kcal/mol, which is quite marginal resulting from the comparatively large changes in both enthalpy and entropy. This means that the free energy of stability of proteins can be remarkably affected by subtle structural changes. In fact, the stability of a mutant protein, in which only a single amino acid is replaced, can be considerably different from that of the wild-type protein.

Recently, attention has been focused on proteins from archaea, a phylogenetically distinct evolutionary kingdom that includes many extremophiles and whose study has expanded our horizons on the limits of biodiversity. This genus includes hyperthermophiles that thrive at temperatures above boiling point of water and studies on proteins from such organisms, which are necessarily more thermostable, are expected to provide clues to understanding the molecular basis of thermal stability. The crystallographic analysis of a number of proteins from hyperthermophiles to mesophiles has permitted direct structural comparisons, homology based modeling studies and site directed mutagenesis, combining sequence data on extremophilic enzymes with structural data from their mesophilic counterparts. Studies show that similar proteins from mesophilic and thermophilic or hyperthermophilic organisms differ only slightly in interactions. Small adaptive changes are sufficient to alter the stability of a protein remarkably. In the following section we will analyze the interactions that impart stability to proteins, and how their interplay makes a protein stronger and rugged in comparison to their counterparts. The understanding of these factors is expected to help in making designers molecules with novel and enhanced functionalities. This is possible if the rules of the protein folding energetics and structure architecture are known precisely to model a new molecule.

Factors that Contributes to Protein Stability and Structure

To understand the architecture of proteins and how a particular fold of proteins help them achieve a unique biological function, it is important to understand the basic forces which contributes towards stabilization and folding of a protein into a three dimensional structure. The Gibbs energy changes between the native state and the denatured state result from some combination of terms, called the "stabilization factors". Several kinds of stabilization factors, such as the hydrophobic effect, hydrogen bonding, backbone and side-chain configurational entropy, and electrostatic interaction, have been proposed to contribute to the protein stability. Mutagenesis studies have been employed thoroughly to elucidate the contribution of individual interaction. However, these studies have shown that the stabilizing factors contribute towards the protein stability to varying extents depending upon the environment of the substituted residue. Replacement of a single amino acid may lead to changes not only at the place of substitution but at sites far away from the site of replacement. Also, the substitution of an amino acid residue with another differing in size may lead to steric hindrances or changes in packing density or creation of cavity in the protein, which may entrap water molecules. Such alterations lead to a very complicated situation, e.g. an entrapped water molecule may be stabilizing due to strong hydrogen bonding with the protein or it could be destabilizing due to loss of configurational entropy of the water molecule due to entrapment. Therefore, in spite of lot of experimental data it has not been possible to exactly quantitate these forces applicable to all situations. Nonetheless, extensive experimentation has produced enough data to correlate the experimental free energy with the calculated one, and it has been possible to parameterize the free energy of stabilization.

Hyperthermophiles Versus Mesophiles

Systematic study of similar proteins from different sources can help us to better understand the molecular basis of evolution and the strategies adopted by organisms to survive under diverse and adverse environments through adaptive mechanisms. It has been observed that a protein from a thermophilic or hyperthermophilic protein is much stronger than their counterparts from mesophiles not because of extensive changes but rather due to small adaptive changes, which leads to optimization of the stabilizing factors. Systematic comparative studies have shown that proteins from thermophiles have higher content of amino acids with higher helical and hydrophobicity propensities, complex charge-charge interactions, extensive hydrogen bonding, increased polarity of the exposed surface, low internal cavity volumes, decreased loop lengths, decreased solvent exposed surface area, and a propensity to form oligomeric structures. Structure of Pyrrolidone carboxyl peptidases (PCP) from *Pyrococcus furiosus*, a hyperthermophile, and the mesophilic counterpart, *Bacillus amyloliquefaciens*, show the role of complex electrostatic interactions in the stabilization of hyperthermophilic protein. In hyperthermophilic PCP, the number of salt-bridges is much higher and more complex than in PCP from mesophilic protein (Figure 1).

Hydrophobic effect is understood to be the major force stabilizing the tertiary structure of a protein, whereas the salt-bridges have been thought to be contributing negligibly or even negatively. This is based on the observations that the Gibbs free energy of stabilization is directly proportional to the buried surface area of a protein, which is mainly constituted by hydrophobic residues, and the desolvation energy of a salt bridge is higher than the coulombic interaction. On the other hand, salt bridges are present on the exposed surface area. It has been postulated that even if the charges on the surface are not able to make salt bridge with another charges on the protein surface, it can always interact with solvent to form a salt bridge, therefore contributing negligibly to the net free energy of stabilization of the native state. However, systematic studies show that the number of ion-pairs in a hyperthermophilic protein is higher in comparison to that in a mesophilic protein. Moreover, these ion-pairs usually form complex network of interactions so as to strengthen the charge-charge interactions by forming charge domains. The protein-protein ionic interactions are also considered to be stronger than the protein-solvent interactions, and charge-charge interactions are strengthened as the temperature increases; therefore even if they are not much effective at ordinary temperatures, they become considerably effective in thermophilic or hyperthermophilic proteins, which can still be native in the 80-110 °C zone of temperature. The height of the energy barrier, which increases with temperature, for breaking a salt bridge is significantly higher than a barrier observed for hydrophobic interactions. This phenomenon also points to the significance of salt bridges in stabilizing proteins at higher temperatures.

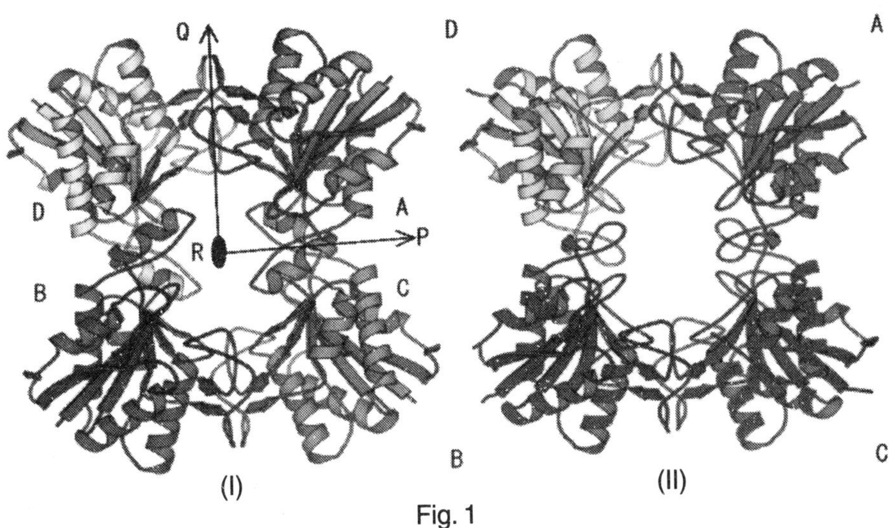

Fig. 1

contd....

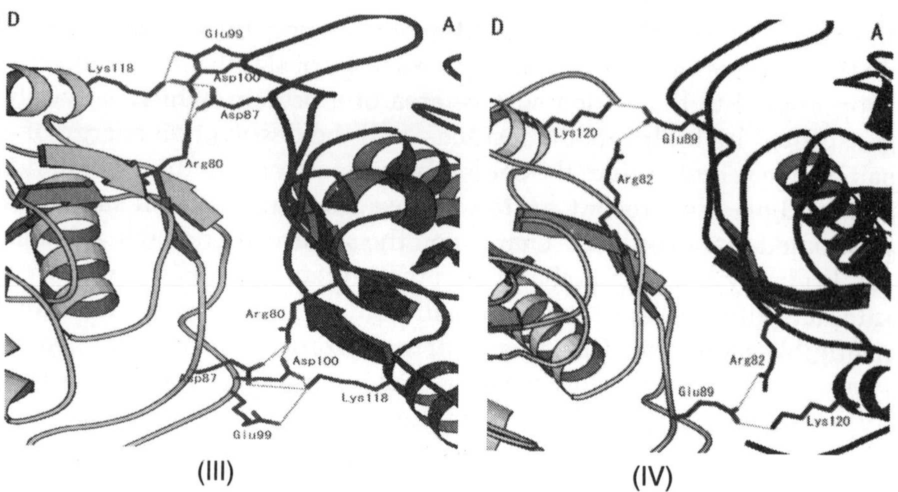

Fig. 1. Structure of Pyrrolidone carboxyl peptidase (PCP) from *Pyrococcus furiosus* (I), *Bacillus amyloliquifaciens* (II), and the salt bridges between subunit A and D of *Pf*PCP (III) and *Ba*PCP (IV). A, B C and D are the four subunit of PCP. Note the more complicated structures at the interfaces between subunit A and C or B and D in the case of *Pf*PCP.

Some studies indicate that only ion pairs in protein have been found to show statistically significant correlation with growth temperature of the organism, and other properties like hydrogen bonding, hydrophobic effect etc show much smaller statistical significance. However, on the other hand, site directed mutagenesis experiments indicate that hydrophobic interactions as well as hydrogen bonding contribute significantly to protein stability. Cavities and buried polar groups usually observed to destabilize the protein, whereas hydrogen bonds and secondary structural elements generally have a stabilizing contribution. The contribution depends upon the site of replacement and on the overall structure of the protein. It seems that there are several strategies adopted by hyperthermophiles to survive at higher temperatures through various molecular adaptations. It seems that a given protein family uses an individual 'strategy' to achieve a high thermal stability. That is, one family of proteins can achieve higher stability by increasing the number of ions pairs, while another family may do so by decreasing the internal cavities, or increasing the packing density or say by increasing the hydrogen bonding. For example, pyrophosphatase from *Thermus thermophilus* is stabilized by ion pairs but also somewhat destabilized by larger cavities in comparison to pyrophosphate from *E. coli*. However, for malate dehydrogenase from

Thermus flavus, it is just the reverse case: it has fewer ion pairs than its mesophilic counterpart from pig, but its cavities are smaller.

Apart from these structural changes, hyperthermophilic and thermophilic proteins show a bias in selecting amino acids which are less amenable to chemical degradation or modifications at higher temperatures. In thermophilic proteins, a significant decrease in the contents of methionine, asparagines, serine and glycine; whereas increase in the contents of proline, tryotophan, glutamic acid has been observed in comparison to proteins from mesophiles. Methionine, asparagines and glutamine are known to degrade at higher temperature, while glycine breaks the structure by introducing flexibility. The balance depends upon the growth temperature of the organism.

Moderately and extremely thermostable proteins appear to adopt different mechanisms to achieve greater stability, which could be dictated by various forces and their dependence on temperature. The growth temperature of the organism seems to be very important in selecting the strategy used by the proteins. This is also pertinent with the evolution which took place beginning from the bacterial and archaeal organisms, which are at the root of the phylogenetic tree. Most of the hyperthermophiles belong to the domain archaea and precede the mesophile counterparts. However, recent analyses suggest that extant hyperthermophiles evolved from mesophiles and via adaptation to high temperatures. It seems that protein utilized those interactions that can form relatively fast in the process of molecular evolution. Therefore, interactions forming on the outer surface of a protein are preferred. The knowledge gained through study of evolution of proteins structures can be used to engineer proteins with enhanced stability and also to design new molecules of industrial and therapeutic importance, which is precisely one of the reasons to sequence the genomes of various organisms.

□□□

Chapter 21

Genomics Based Biomarkers for Probiotic Functionality

SUNITA GROVER, RAJ KUMAR DUARY AND V. K. BATISH
Dairy Microbiology Division, National Dairy Research Institute, Karnal

Health benefits attributed to consumption of certain lactic acid bacteria have been known since times immemorial. However, it is only in recent years that the scientific basis of these claims have been investigated and have led to the concept of probiotics. Probiotic bacteria are of GRAS status and have positive effect on the health of the host. Lactobacillus and Bifidobacteria are the two key members of this group. Diet, stress, and modern medical practices have been implicated as factors capable of exerting an influence on human health and nutrition. As a result of these and other factors, the incidence of illnesses which may be induced by deficient or compromised microflora (e.g., gastrointestinal tract infections; constipation; irritable bowel syndrome (IBS); inflammatory bowel disease (IBD) - Crohn's disease and ulcerative colitis; food allergies; antibiotic-induced diarrhoea cardiovascular disease; and certain cancers) is increasing. Research has demonstrated that nutrition plays a crucial role in the prevention of chronic diseases, as most of them can be related to diet. The intestinal tract harbors a huge diversity of metabolically active aerobic and anaerobic bacteria that interact, forming a complex ecosystem. There are about 500 different species of nonpathogenic bacteria numbering up to 10^{11} cells per gram found in human intestine. Among them, Lactic acid bacteria constitute the key member because of its role in manufacturing fermented milk products; health promotion and other therapeutic properties. The positive effect of consumption of probiotics are: improvement of bowel habit; reduction of diarrhoea and constipation; modulation of lipid metabolism by normalizing cholesterol values; reduction of osteoporosis by improving mineral absorption; reduction of allergy risk through immune system modulation; reduction in colon cancer; *etc.* The beneficial role of the probiotic microflora in disease has led to an interest in the use of probiotics as a therapeutic modality as an alternate to drug therapy. One of the fundamental properties of a probiotic bacterium is its ability to survive in a sustained transient state in the gastrointestinal tract. A relationship between the normal flora and the host immune system exists with a mutual dependency between the two.

There are various potential probiotic and immuno-modulating markers for selecting a probiotic culture against various type of intestinal diseases. Among probiotic markers - the microorganisms must survive the acidic and high bile concentration environment in the stomach and intestine which is important for *in vivo* functionality as well as stable fermentation characteristics. Advances in genomics and proteomics have led to the identification of genes involved in acid as well as bile stress in *Lactobacillus* such as F_1F_0-ATPase (an important element in the response and tolerance to low pH) and Bile salt hydrolase (BSH, a detergent shock protein that protects the bacteria from the toxicity of bile acids by deconjugation, dehydroxylation, dehydrogenation, and deglucuronidation in the gastrointestinal tract) enzymes. Other probiotic markers include antibiotic resistance, which provides organism to survive in the gut in the presence of co-administered antibiotics. Various antibiotic resistance genes such as *tet*(W) for Tetracycline, *erm* for erythromycin, *lnu*(A) for lincosamide, *etc.* are responsible for various antibiotics resistance in Lactobacilli. Probiotic organisms also possess free radical scavenger properties by producing a heme-dependent catalase, which can degrade H_2O_2 at a very high rate, blocking the formation of peroxyl radicals. Another important enzyme â - Galactosidase provides biosynthesis of galactooligosaccharides and preparation of lactose-free milk which is important from health point of view (Lactose tolerance). Apart from these, probiotic organisms also produce bacteriocins and various antimicrobial components which prevent the pathogenic organisms to survive in the gut. One major criterion considered for selection of probiotic bacteria has been their capacity to adhere to the human intestinal epithelial cells. Maximum probiotic effect is achieved if the organisms adhere to intestinal mucosal cells. Probiotic bacteria display various surface determinants (extracellular and transmembrane proteins) that are involved in their interaction with intestinal epithelial cells. Extracellular protein like Sep (a cell surface associate) and surface proteins like *Mub* (Mucus binding protein), *Map*A (Mucus associate protein), adhesion-like factor EF-Tu, *Fbp* (Febronectin binding protein) and *CnBP* (Collagen binding protein) plays important role in binding of probiotic bacteria to the mucus layer of GI tract. The intestinal epithelium overlayed by mucus gel layer, prevents gastrointestinal pathologies. Mucus is mainly composed of complex glycoproteins called mucins which are encoded by various mucin (*MUC*) genes. A total of 16 *MUC* genes encoding for the peptide backbone till had been identified. The membrane-associated mucins (*MUC* 1, 3, 4, 11-13 and 15-17) have been proposed to have a role in interacting with intracellular proteins by way of the C-terminal cytoplasmic tail. The gel-forming mucins (*MUC* 2, 5AC, 5B and 6-9) are able to form polymers and are responsible for the viscosity of mucus. A Probiotic bacterium increases mucin expression preventing enteropathogenic micro-organism attachment in the host.

Crohn's disease and ulcerative colitis are collectively called IBD. Crohn's disease is an ongoing disorder that causes inflammation of the digestive tract, also referred to as the gastrointestinal (GI) tract. Crohn's disease can affect any area of the GI tract, from the mouth to the anus, but it most commonly affects the lower part of the small intestine, called the ileum. The swelling extends deep into the lining of the affected organ. Ulcerative colitis causes inflammation and ulcers in the top layer of the lining of the large intestine. Cytokines are key regulators of inflammation in IBD, and several pro-inflammatory and immune regulatory cytokines are dysregulated in the mucosa of IBD patients. Antibody-neutralization studies have implicated tumor necrosis factor (TNF) and IL-12 p40 in the pathogenesis of Crohn's disease. Activation of NFkB stimulates expression of numerous molecules relevant to the pathogenesis of IBD. These include molecules involved in the inflammatory response, such as IL-1, TNFá, IL-6, IL-8 and other chemokines. Expression of each of these proinflammatory molecules is increased in active IBD. Thus, the NFkB pathway was thought to have predominantly proinflammatory activities. Numerous bacterial adjuvants, most notably lipopolysaccharide, peptidoglycan, flagellin and nonmethylated DNA (CpG motif), can bind selectively to various TLRs (Toll like receptors) on innate immune cells, intestinal epithelial cells and mesenchymal cells. Ligation of these TLRs activates NFB and the mitogen-activated protein kinases, which stimulate the transcription of a host of proinflammatory and regulatory genes.

Probiotic-mediated immunomodulation represents an interesting option in the management of IBD and it was shown that both the systemic and mucosal immune systems can be modulated by orally delivered bacteria. Probiotics can prevent the relapse of chronic, relapsing pouchitis and ulcerative colitis, but are not useful in treatment of active disease. Probiotic down regulates pro-inflammatory ((IL-8, IL-6, IL-12, Tumor necrosis factor alpha (TNFá) and Interferon gamma (IFNã)) cytokines and up regulates anti-inflammatory cytokines (IL-4 and IL-10). Hence, these potential probiotic biomarkers can play a very important role for selection of promising probiotic strains with novel functional properties before their application in clinical studies. The role of these potential biomarkers in understanding the functionality of these probiotics will be the main focus of this presentation.

References

Ann, M. O'Hara., O'Regan, P., Fanning, A., O'Mahony, C., MacSharry, J., Lyons, A., Bienenstock, J., O'Mahony, L. and Shanahan, F. 2006. Functional modulation of human intestinal epithelial cell responses by *Bifidobacterium infantis* and *Lactobacillus salivarius*. *Immunology*, **118**: 202–215.

Donglai, Ma., Forsythe,P. and Bienenstock, John. 2004. Live *Lactobacillus reuteri* is essential for the inhibitory effect on Tumor Necrosis Factor Alpha-induced Interleukin-8 expression. *Infection and Immunity*, **72(9)**: 5308–5314.

Foligne, B., Nutten, S., Grangette, C., Dennin, V., Goudercourt, D., Poiret, S.,Dewulf, J., Brassart, D., Mercenier, A. and Pot, B. 2007. Correlation between *in vitro* and *in vivo* immunomodulatory properties of lactic acid bacteria. *World Journal of Gastroenterology*, **13(2)**: 236-243.

Kodali,V. P. and Sen, V. 2008. Antioxidant and free radical scavenging activities of an exopolysaccharide from a probiotic bacterium. *Biotechnology Journal*, **3 (2)**: 245 - 251.

Ramiah, K., van Reenen, C. A. and Dicks, L. M. T. 2007. Expression of the mucus adhesion genes *Mub* and *Map*A, adhesion-like factor EF-Tu and bacteriocin gene *pla*A of *Lactobacillus plantarum* 423, monitored with real-time PCR. *International Journal of Food Microbiology*, **116**: 405–409

Sanchez, B., Reyes-Gavilan, C. G. de los and Margolles, A. 2006. The F_1F_0-ATPase of *Bifidobacterium animalis* is involved in bile tolerance. *Environmental Microbiology*; **8(10)**: 1825–1833.

□□□

Chapter 22

Designing of Novel Probiotics for Functional Foods

V. K. BATISH AND SUNITA GROVER
Dairy Microbiology Division, National Dairy Research Institute, Karnal

Probiotic Lactic acid bacteria bacteria have been frequently used as active ingredients in functional foods such as bio-yoghurts, dietary adjuncts and other health related products. The health benefits for consumers attributed to probiotic bacteria can be grouped into two categories viz. Nutritional benefits and therapeutic benefits. Nutritional benefits include their role in bio-availability of calcium, zinc, iron, manganese, copper and phosphorus, increasing the digestibility of proteins in yoghurt and synthesis of vitamins. The therapeutic benefits include treatment of gastrointestinal disorders, hyper-cholesterolaemia, lactose intolerance, suppression of pro-carcinogenic enzymes, immuno-modulation and treatment of food related allergies. Considerable advancements made in the field of genetic engineering of lactic acid bacteria especially the availability of genome sequences of probiotic strains of lactobacilli and bifidobacteria have opened myriad opportunities for the development of strains with safe and effective health promoting effects. However, food grade genetic modifications are required so that these bacteria can be directly used in dairy products. Food-grade genetic modifications are the limited types of modifications that would be considered acceptable for food microorganisms with safety as the chief concern. Different tools for food grade modifications include homologous recombination, food grade deletions, food grade gene replacement, food grade integration into the chromosome, food grade cloning vectors etc. Metabolic engineering has also been explored to suitably modify the probiotic strains for better functionality and also to rule out the possibility of any adverse effects in the immuno compromised subjects.

Recently, the concept of "inverse pathogenesis" used to describe the bioengineering of "generally recognized as safe" organisms with genes from more versatile pathogenic species has become a growing area of interest. However, one of the limitations of probiotics in clinical application is that the most effective probiotic strains often prove to be fragile during industrial processing such as drying or heating. Improving the stress tolerance of probiotic strains is thus an important biological and clinical goal. 'Patho-biotechnology' describes the exploitation of highly adapted pathogenic stress survival and host evasion strategies for the development of improved probiotic cultures. In addition to improving their physiological stress tolerance, recent studies have led to the development of 'designer probiotics' which specifically target enteric infections by blocking crucial ligand-receptor interactions between the pathogen and host cell. In this context, a genetically engineered probiotic *E. coli* strain that secretes an anti-HIV peptide derived from the C-terminal heptad repeat (HR-2) region of the transmembrane subunit of the HIV-1 envelope glycoprotein (Env), and functions by blocking HIV entry to the host cells has been constructed. When administered orally or as a rectal suppository, this 'live microbicide' colonizes the gut mucosa and secretes the peptide in situ, thereby providing protection in advance of HIV exposure for up to a month. Many of the pathogens responsible for the major enteric infections exploit oligosaccharides on the surface of host cells as receptors for toxins and or adhesins, enabling colonization of the mucosa and entry of the pathogen or secreted toxins into the host cell. Blocking this adherence prevents infection, while toxin neutralization ameliorates symptoms until the pathogen is eventually overcome by the immune system. When administered orally, these probiotics bind to and neutralize toxins in the gut lumen and interfere with pathogen adherence to the intestinal epithelium.

In addition to a therapeutic role, certain probiotics have also been shown to function as effective prophylactic agents, being specifically engineered to function as novel vaccine delivery vehicles. Stimulating both innate and acquired immunity, these strains lack the possibility of reversion to virulence, which exists with the more conventional pathogenic platforms currently in development. Furthermore, recent developments in synthetic and systems biology, based on the rapidly advancing 'omics' technologies, has and will continue to lead to the emergence of an ever-increasing number of novel genetic loci with defined additional functions. This coupled with computer aided bioinformatics and novel tools for genetic modification will ultimately lead to the development of artificial micro-organisms and eventually to a new class of probiotics assembled from the components of various origins and tailored to fulfil all the requirements of an ideal therapeutic agent.

With the advancements made in food grade genetic engineering tools and genomics of lactic acid bacteria, it has now become possible to design probiotics for improved as well as new beneficial functions to meet the requirements of consumers. However, acceptance of genetically engineered 'designer probiotics', however, will likely be less easily achieved. Nevertheless, this reluctance (both in the developed and developing world) should eventually be overcome by rigorous scientific controls such as adequate biological containment and proper risk-benefit analysis of the potential advantages of such a strategy. The various issues and strategies related to designing of probiotics will be discussed in this presentation.

Suggested Reading

Marco, M. L., Pavan, S. and Kleerebezem, M. 2006. Towards understadning molecular modes of probiotic action. *Current opinion in Biotechnology*. **17:** 204-210

Sleator, R. D. and Hill, C. 2006. Patho-biotechnology - using bad bugs to do good things. *Current topics in Biotechnology*. **17**:211-216.

Sleator, R. D. and Hill, C. 2008. New frontiers in probiotic research. *Letters in Applied Microbiology*. **46:** 143–147.

Sudhamani, M., Batish, V.K. and Heller, K.J. **2008.** Characterization of pSMA23, a 3.5 kbp plasmid of *Lactobacillus casei* and application for heterologous expression in *Lactobacillus*. *Plasmid*. **59:** 11-19

Vaughan, E. E., Mollet, B and de Vos W. M. 1999. *Current opinion in Biotechnology*. **10:** 505-510.

□□□

Chapter 23

Bio-Safety Principles and Containment Designs in Food Laboratory

NARESH KUMAR
Dairy Microbiology Division, NDRI, Karnal -132001

Introduction

A biohazard can be defined as any organism, or material produced by such an organism, that is known or suspected to cause human or animal disease. Exposure to bio-hazardous agents may occur via puncture wounds or as a result of absorption through the respiratory tract, digestive system, skin and mucous membranes. Such exposures may result while handling animals, cell cultures and tissues or diagnostic specimens. Activities which are those specifically addressed are involving work with recombinant DNA, various bacterial, fungal, and parasitic agents, live viruses, experimentally infected research animals , Human blood and tissues, receipt, handling, and disposal of biological materials. Exposure to bio-hazardous agents is intended to be prevented or limited by establishing and following the appropriate bio-safety level practices and conditions. Environmental Health & Safety must approve any activities carried out involving bio-hazardous agents.This information is applicable to all laboratories, research, service and support activities that may involve exposure to bio-hazardous agents or materials.

Classification of Pathogens

Essentially, microbiological pathogens are classified according to their impact upon the individuals who manipulate them. Agents that pose little or no risk are assigned to Risk Group 1, while those with the greatest hazardous potential are in Risk Group 4. Risk assessment is based upon several factors, including severity of induced disease, route (s) of infection ,virulence and infectivity of the microorganism, antibiotic resistance patterns ,availability of effective medical treatment or vaccine ,whether the pathogen is indigenous ,possible effects on other animals and plants. Before setting up experiments involving new biohazards, consideration should also be given to conditions under which the infectious agent is used. Work involving release of microbial aerosols, passage in animals also increases the hazard. In these cases, pathogens should be handled as if they were in the next highest risk group: i.e., if the experimental procedure is likely to generate large amounts of aerosolized Risk Group 2 agent, the physical and operational requirements applicable to Risk Group 3 agents should be observed.

Genetically Engineered Organisms

The term «biotechnology» describes a variety of techniques for manipulation of cells; biotechnology has long been used for purposes such as selective breeding of animals and food production. More recently, in vitro incorporation of segments of genetic material from one cell into another («recombinant DNA technology») has resulted in altered organisms that can manufacture products such as vaccines, hormones, interferon and enzymes. A genetically altered organism may be directly pathogenic or toxic or, if released into the environment, transfer undesirable genetic traits to wild species or mutate into a pathogenic form. When assessing the risk of, and containment level required by, a genetic engineering protocol, the following approach is recommended: if the components of a genetic manipulation are not hazardous, then the altered organism is unlikely to present a risk, and no restrictions are needed. However, if one of the components is potentially hazardous, a risk level appropriate for the known hazard is assigned and modified as required. Subsequent modifications depend on factors such as: expression of the transferred gene in the recombinant organism, ability of the vector to survive outside the laboratory, expected interactions between transferred gene, host and other factors

Risks and characteristics associated with pathogens from Risk Groups 1 to 4, and recommended containment level and class of biological safety cabinet

Risk group	Risk assessment	Characteristics	Examples	Contain ment level	Bio safety cabinet
1	Low individual; low community.	Unlikely to cause disease in animals or humans	*Bacillus subtilis, E. coli* K12	1	Not required
2	Moderate individual; low community.	Rarely cause serious human or animal disease; effective prevention and treatment available; limited risk of spreading.	*Salmonella typhimurium*	2	Class I *or* Class II
3	High individual; Low community	May cause serious disease in humans or animals; effective prevention and treatment available; unlikely to be spread by casual contact.	*Yesinia pestis, , Bacillus anthracis,* cultured isolates of HIV*	3	Class II
4	High individual; high community.	Likely to cause very serious disease in humans or animals; readily transmitted from one individual to another, or between animals and humans; preventative vaccines or effective treatment not available.	hemorrhagic fever	4	Class III

Containment Level

The term «containment» is used in describing measures used to provide a barrier between the infectious organism(s) being handled and the worker (and, ultimately, the community at large). Containment is achieved through the use of appropriate safety equipment, facility design and lab procedures and practices. Careful consideration must be given to both facility design and work practices to ensure protection of laboratory personnel, their colleagues and the community as a whole. Four containment levels are outlined in the Health guidelines: of the four containment levels, the highest safety standards (Level-4) are reserved for the most hazardous pathogens (Risk Group 4), and the least stringent (Level-1) for those which have minimal impact on health (Risk Group 1).

Level-1 containment: Level-1 containment is used when working with agents (Risk Group 1) that pose no risk to healthy adults: The laboratory may be near a public area but doors should be kept closed. Work may be carried out on an open bench top. Lab surfaces (walls, ceilings, furniture and floors) should be cleanable. Open windows should have insect screens. Eyewash stations and hand washing facilities should be available. Street clothes and lab coats should not be kept together. Disinfection should be carried out as required, using effective concentrations and contact times; solutions should be replaced regularly.

Level 2 containment: Level 2 containment is appropriate for work with Risk Group 2 agents. The following precautions, in addition to those for containment Level 1, are recommended: The facility should be away from public areas and should have self-closing doors. A biohazard sign with relevant information should be posted at the entrance. Service and custodial staff should be informed of the hazards; the latter should be expected to clean floors only. Items should be autoclaved or chemically decontaminated before removal from the facility. Use Class I or II biological safety cabinets for procedures that generate infectious particles. Procedures should be carried out such that aerosol generation is minimized. An emergency spill response plan should be in place and posted in a visible location. Vacuum lines should be equipped with HEPA filters. Lab coats may be front-closing, but should not be worn outside the lab. Wear gloves to prevent skin contamination.

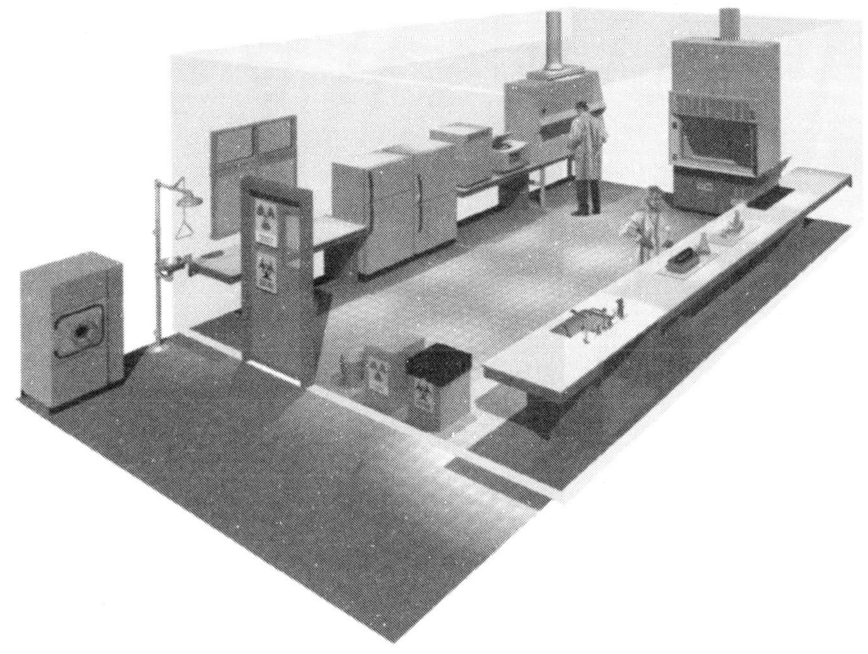

A typical Biosafety Level 2 laboratory

Level 3 containment: Level 3 containment is recommended for work with Risk Group 3 agents. Measures should include the recommendations outlined for levels 1 and 2, plus the following: The lab should be away from general work areas, with controlled access. There should be a change and shower area within the containment facility perimeter. The area should be kept at negative pressure relative to surrounding areas. Supply and exhaust air should be HEPA-filtered or provided by dedicated systems. A hands-free hand washing sink should be located near the exit. Lab windows should be unbreakable and sealed shut. Lab personnel should be trained in handling, disposal, and emergency procedures. Written protocols for these procedures should be developed and posted in a visible location. Personnel should wear solid-front lab clothing, which should be autoclaved before laundering or disposal. A medical surveillance program is recommended.

Level 4 containment: Level 4 containment is used for work with Risk Group 4 agents. Additional recommendations include: physical isolation of the laboratory, with an airlock for access ,entry restricted to authorized personnel and recorded in a log book: no one should work alone, use of class III biological safety cabinets and/or positive-pressure protective suits, additional safety measures for ventilation, waste treatment, and gas and water services.

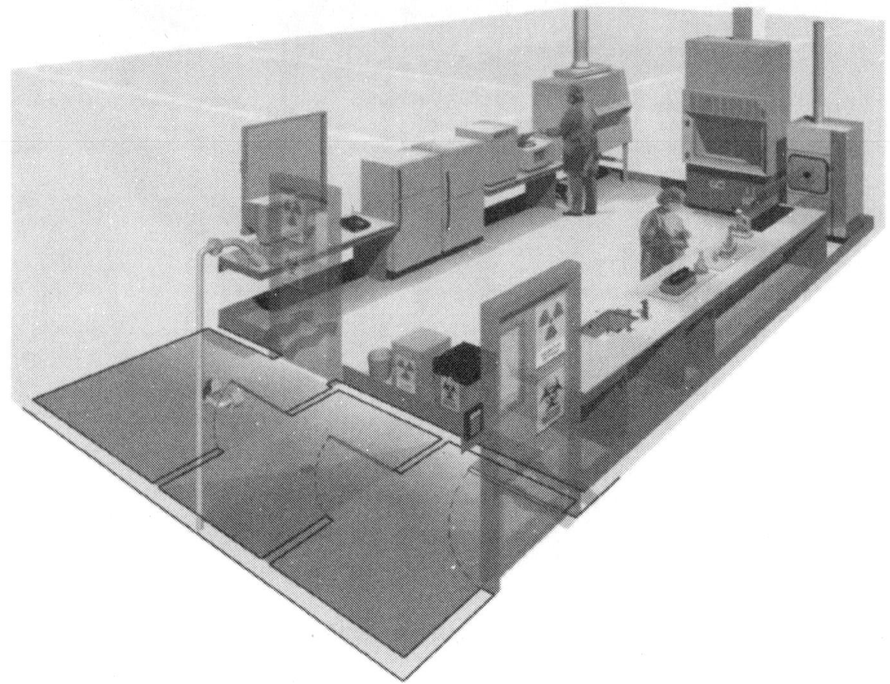

Biological Safety Cabinets

Biological safety cabinets reduce the risk of airborne infection by reducing the escape of aerosolized infectious agents into the laboratory environment. In addition to protecting workers, some biological safety cabinets protect the work inside the cabinet from airborne contamination (product protection). Biological safety cabinets minimize contact between the operator and pathogens through the use of directional airflow, HEPA filtration of supply and/or exhaust air, and, in some cases, a physical barrier such as a plastic or glass shield. Filters are an essential component of the biological safety cabinet, and have particle removal efficiencies of 99.97% or better for 0.3 micron diameter particles. This size particle is used as the basis for filter definition because it is considered the most difficult to remove. Thus, a filter that can trap 0.3-micron diameter particles can easily eliminate other sizes. HEPA filters consist of continuous sheets of glass fiber paper pleated over rigid corrugated separators and mounted in a wooden or metal frame. The filter medium is delicate and should never be touched. As well, the gaskets used to seal the filter frame to the cabinet must not be disturbed; thus the biological safety cabinet should not be moved without subsequently being tested and certified. While HEPA filters remove particulates from an airstreams, they are not effective at collecting chemical gases or vapours. Thus, it is inadvisable to use recirculation Class II cabinets with agents which have significant amounts of hazardous volatile or radioactive components. Although Class III and 100% exhaust Class II cabinets can be used in experiments which involve use of chemicals of moderate toxicity, it should be remembered that these cabinets are not explosion-proof. Use of flammable or explosive products is to be avoided unless the cabinet has been specifically designed for their use.

Classes of Biological Safety Cabinets: Horizontal and vertical clean benches are not biological safety cabinets: HEPA- filtered air is directed over the work surface and then discharged directly into the room. Thus, these units provide product protection, but do not protect the operator from exposure to the materials being handled; they must not be used for work with potentially infectious or toxic materials. There are three basic types of biological safety cabinet, each providing different levels of containment:

Class I: i) open-fronted ii) protects operator and environment iii) for work with low and moderate risk agents (Risk Groups 2 and 3) where product protection is not critical

General principle of operation: An inward flow of room air through the work opening, away from the operator, prevents the escape of airborne pathogens into the laboratory. Negative cabinet pressure is created by a blower that exhausts the air, either into the room or to the outside, through a HEPA filter. It is this HEPA filtration of exhaust air that provides environmental protection. A disadvantage of this type of cabinet is that the product is exposed to contaminants that are pulled in from the room environment. In addition, internal air turbulence may result in cross-contamination within the cabinet.

Class II: i) open-fronted ii) protects operator, product and environment from particulate contamination iii) for work with low to moderate risk agents (Risk Groups 2 and 3)

General principle of operation: Escape of pathogens into the worker's environment is prevented by an inward flow of room air which enters the front opening without crossing the work area and by HEPA filtration of exhaust air (this provides environmental protection), while downward flow of HEPA-filtered air through the work area removes work zone contaminants and protects the product. The amounts of room air drawn into the intake grille and the amount of air exhausted through the exhaust filter are equal. This balance is critical: positive pressure will allow the outflow of pathogens, while negative pressure will result in inflow of room contaminants. The different types of Class II cabinets (e.g., Type A, Type B or 100% exhaust) vary in: airflow velocities, Amount of cabinet air recirculated (from 0 to 70%) ,Amount of cabinet air exhausted (from 30 to 100%), destination of exhaust air (back to lab or outside) ,exhaust ducting (building system versus dedicated ducts). It should be kept in mind that toxic or radiolabel led chemicals must not be handled in cabinets that recirculate air within the cabinet or exhaust into the laboratory.

Class III : i) totally enclosed, gas tight, with glove ports for manipulation of pathogens ii) provides the greatest level of operator and product protection iii) for work with high risk pathogens (Risk Group 4)

General principle of operation: These cabinets form a physical barrier between the operator and microbiological agent. Internal negative pressure confines any leaks to the inside of the cabinet. Supply and exhaust air is HEPA-filtered; a dedicated exhaust fan, separate from that of the facility ventilation system, discharges directly to the outdoors. There is no recirculation of air within the cabinet. A Class III cabinet system must be designed to allow for safe introduction, handling and removal of all materials throughout the procedure. Equipment such as the incubator, refrigerator, centrifuge, autoclave and chemical dunk tank are connected to the cabinet system.

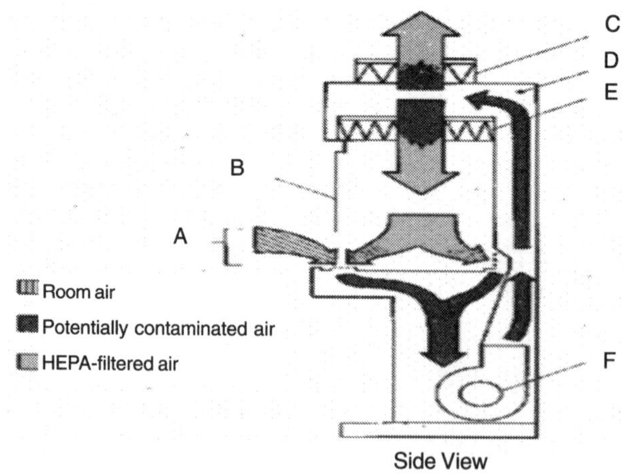

Working Safely in a Biological Safety Cabinet: Biological safety cabinets must be combined with good work practices for optimum safety and contamination control. Recommended practices when using a biological safety cabinet include the following: Movement of arms into and out of the cabinet can disrupt airflow, adversely affecting cabinet performance. Whenever possible, place all materials needed for a procedure inside the cabinet before starting. Avoid bringing non-essential equipment and supplies into the cabinet. Place supplies, equipment and absorbent towels so that air intake or exhaust grilles are not obstructed. Keep opening and closing of lab doors and other personnel activity to a minimum. If a burner is deemed to be indispensable (remember that burners contribute to the heat load, generate convection currents that interfere with airflow and may damage the filters), use one that has a pilot flame. Attach a HEPA filter cartridge between the vacuum trap and the source valve. Work at least 4-6 inches inside the cabinet window. Carry out work on an absorbent pad to contain small spills. Clean up spills as soon as they occur; remove and disinfect the grille if contaminated. Designate separate areas within the cabinet for contaminated and clean materials; place contaminated material at the rear of the work area.

Cabinet Start Up and Shut down Procedures: Before using the cabinet: Turn off the UV lamp; turn on the fluorescent lights. Disinfect the work surface. Place essential items inside the cabinet. Allow the blower to run for at least five minutes before starting work.

After completion of work: Leave blower on for at least five minutes to purge the cabinet. Remove and decontaminate equipment and materials, and disinfect cabinet surfaces. Turn off the blower and fluorescent lamp, and turn on the UV light.

Maintenance/Certification of Biological Safety Cabinets: Biological safety cabinets must be tested and certified annually. Cabinet performance must also be evaluated: upon initial installation in the laboratory,when moved from one building or laboratory to another, when moved from one area to another within the same room, whenever maintenance is carried out on internal parts, and whenever filters are changed, Annual certification is provided at no charge, and can be arranged by contacting Environmental Health & Safety. Costs of certification of new installations and relocated cabinets are the responsibility of the user.

Emergency Spill Response: All individuals who work in a lab where pathogens are used must know how to handle these agents safely and what to do in case of a spill. An emergency spill response protocol specific for the microorganisms in use should be prepared and posted in a visible location within the laboratory.

Prevention: An accident prevention plan should be the first priority. General safety precautions include: Limit access to rooms where microbiological agents are used, Wear appropriate protective clothing, Use the appropriate biological safety cabinet, Use plastics rather than breakable glassware to reduce likelihood of puncture wounds, cuts and generation of aerosols in the event of an accident, Transport materials on carts that have lipped shelves, using secondary containers (i.e. tubs) to catch spills, Disinfect waste.

The spill Response Plan: Response procedures should be established before a spill occurs. Assessment of the hazards presented by the pathogen(s) in use should be based upon: virulence and infectivity of the agent, viability - e.g., does the organism become inactive when dried? Route of entry - e.g., can the organism enter the body via aerosols or splash to the eye? Quantity and location of possible spill, immune status of the individuals at risk .The necessary clean-up materials should be available on site. In preparing a spill response kit, ascertain that it contains the appropriate clean-up materials, protective clothing and equipment. The kit should be stored in a visible and accessible location immediately outside the facility and should include: disposable protective clothing (e.g., long-sleeved coat or gown, mask, gloves), absorbent paper, autoclavable container and bags, disinfectant appropriate for the pathogen(s) handled: be sure to replace the disinfectant before it expires, autoclavable squeegee or forceps.

Spill Response Procedures: The appropriate spill response depends on the nature of the spilled organism and on the size of the spill. Small spills can be cleaned up immediately by lab personnel, provided that the organism does not pose a health risk (i.e., if the spill consists of low to moderate risk agents). Cover with a disinfectant-soaked towel (using a spray bottle for distributing the disinfectant generates aerosols and is to be avoided). Autoclave or discard contaminated material in a biomedical waste container. For spills of large volumes of moderate risk agents or small volumes of high risk agents, proceed as follows:

- Treat serious injuries before attempting to contain the spill.
- Evacuate the area immediately if exposure to the aerosolized microorganism presents a potential health hazard; close the

facility door(s) and allow aerosols to settle for 30 minutes. Remove contaminated clothing and place it in an autoclave bag or other sealed container; disinfect and wash exposed skin.

- Don the appropriate protective clothing and cover the spill with absorbent material such as paper towels to reduce splashing. Pour disinfectant around the perimeter of the spill rather than directly onto it to minimize creation of aerosols. Work the disinfectant toward the centre and let it sit for at least 20 minutes.
- If the spilled material has leaked through the grilles of a biological safety cabinet, leave the cabinet running and pour in enough disinfectant (avoid alcohol due to explosion hazard) to dilute the spill tenfold. Drain the catch tray after the time interval appropriate for the disinfectant.
- Wipe down any adjacent walls, cabinets, furniture and equipment that may have been splashed.
- Use forceps/squeegee and dustpan to pick up and transfer the contaminated material into an autoclavable bag or biomedical waste container.
- Decontaminate the waste and cleaning utensils.

Safe Handling of Laboratory Equipment: In addition to avoiding obvious sources of infection such as splashes, cuts, accidental inoculation or ingestion, laboratory workers should also be aware that some pathogens, when airborne, may cause infection if inhaled. Aerosolized microorganisms are generated during most routine laboratory procedures involving manipulation of liquid suspensions: another potential source of aerosolized pathogens is from animal cages where bedding is contaminated with feces and urine (Refer to Section 7 for a discussion of some of the biohazards associated with animal handling). Not all pathogens can infect via the aerosol route, but for those that do, the risk of infection depends on the type and concentration of agent and on the health status of the exposed individual. The degree of penetration and retention of airborne pathogens in the respiratory tract is determined primarily by size: particles which are 10 m m in diameter or smaller are most efficiently inhaled, deposited and retained in the upper respiratory tract or in lung alveoli. Larger particles (140 m m or greater diameter) are also of concern because they can settle and contaminate work surfaces, equipment and personnel.

Selection and Use of Laboratory Equipment: Whenever lab equipment is purchased, preference should be given to equipment that limits contact between the operator and the infectious agent .is corrosion-resistant, easy to decontaminate and impermeable to liquids .has no sharp edges. Every effort should be made to prevent equipment from becoming contaminated. To reduce the likelihood of equipment malfunction that could result in leakage, spill or unnecessary generation of aerosolized pathogens: Review the manufacturer's documentation. Keep for future reference. Ensure that anyone who uses a specific instrument or piece of equipment is properly trained in setup, use and cleaning of the item. Decontaminate equipment before it is sent out for repairs or discarded. The following sections outline some of the precautions and procedures to be observed with some commonly used laboratory equipment.

Centrifuges: Improperly used or maintained centrifuges can present significant hazards to users. Failed mechanical parts can result in release of flying objects, hazardous chemicals and bio-hazardous aerosols. The high-speed spins generated by centrifuges can create large amounts of aerosol if a spill, leak or tube breakage occurs.

When using high-speed or ultra centrifuges, additional practices should include: Connect the vacuum pump exhaust to a disinfectant trap. Record each run in a log book: keep a record of speed and run time for each rotor. Install a HEPA filter between the centrifuge and the vacuum pump. Never exceed the specified speed limitations of the rotor.

Lyophilizers (Freeze-Driers): Aerosols may be produced during operation of a freeze drier and when material is being removed from the chamber. When lyophilizing bio-hazardous materials: Load samples in a biological safety cabinet. Check glass vacuum containers for nicks and scratches. Use only glassware that was designed for high vacuum use. Use a disinfectant-containing trap for the vacuum pump exhaust. After completion of the run, decontaminate all accessible surfaces.

Mixing Apparatus: Homogenizers, shakers and sonicators can release significant amounts of aerosols during their operation and should be operated in a biological safety cabinet. When using any mixing equipment, remember to: Check condition of gaskets, caps and bottles before using. Allow aerosols to settle for at least one minute after use before opening containers, opening in a biological safety cabinet if possible. Cover tops of blenders with a disinfectant-soaked towel during operation. Immerse sonicator tip into solution to a depth sufficient to avoid creation of aerosols. Disinfect all exposed surfaces after use.

Freezing Apparatus: Spills inside freezing equipment may place laboratory and maintenance personnel at risk; for safe use of such equipment: Periodically check freezers, liquid nitrogen tanks and dry ice chests for broken ampoules, tubes etc. To minimize breakage and leaks, place primary containers such as test tubes inside secondary containers prior to storage in freezing units. For electrical safety, remember to shut down units before proceeding with decontamination.

Vacuum/Aspirating Equipment: Glass vacuum vessels may rupture and shower laboratory personnel with glass fragments and flask contents. To reduce these risks: Use metal flasks and vacuum traps whenever possible. Tape glass containers with duct or adhesive tape to contain glass shards in case of rupture or, use a secondary metal container that is at least as tall as the vacuum flask. To prevent exposure of lab personnel or maintenance employees who may be required to repair the central vacuum system, vacuum line connections that draw bio-hazardous aerosols or fluids should be fitted with: a HEPA filter in the line leading into the vacuum line: cartridge-type in-line filters provide an effective barrier to escape of aerosols into vacuum systems, and are commercially available for this purpose (discard used filters as biomedical waste) .an overflow flask in case of accidental aspiration of liquids out of the collection vessel.

Pipettes: Improper handling of pipettes can lead to contamination of the user and/or to generation of hazardous aerosols. Mechanical pipetting aids should be used for all pipetting procedures: never pipette by mouth. Selection of a pipetting device should be based upon: intended use, ease of handling, delivery accuracy, user preference, quality of seal formed with pipettes to be used; liquid should not leak from the pipette tip, whether the pipetting aid can be sterilized

Autoclave: Autoclaves are ideal for decontaminating bio-hazardous waste and for sterilizing surgical dressings, glassware and microbiological media and liquids. They must be loaded carefully to allow for steam penetration, since steam must contact pathogens in order to destroy them. Longer times are needed for larger loads, large volumes of liquid and denser materials. Proper loading and packing procedures include the following precautions: Wrap packages to allow for steam penetration; aluminum foil does not allow steam penetration, and should not be used for wrapping. Do not overload the chamber. Avoid over packing of autoclave bags. Do not seal bags or close bottles and other containers tightly. Do not stack containers.

Waste Disposal: To protect individuals and the community from unnecessary exposure to bio-hazardous agents, biomedical waste must not be disposed of with regular waste.

Laboratory Safety Protocols: Basic requirements for a laboratory using infectious materials are:

- Ensure that all laboratory personnel, including service and custodial staff and visitors, understand the chemical and biological dangers associated with the lab. Affix biohazard signs to Laboratory Information Cards on doors outside laboratories where bio-hazardous material is handled or stored. Post the spill response protocol in a visible location in the laboratory.
- Restrict laboratory access and keep doors locked when the laboratory is unattended.
- Keep the facility clean and free of clutter. Ensure that emergency safety devices (e.g., fire extinguishers, eyewashes, etc.) are easily accessible and in working order.
- Ensure that all personnel, students and visitors adhere to University policies for eye and face protection and for protective clothing. Remove lab coats or gowns and gloves before leaving the laboratory; never wear lab clothing in eating facilities.
- Avoid eating, drinking, smoking, storage of food and food utensils, application of cosmetics or lip balm and insertion or removal of contact lenses in the laboratory.
- Restrain long hair. Avoid wearing loose clothing or jewelry, shorts and open-toed shoes or sandals.
- Carry out procedures so as to minimize risks of splashes, spills and generation of aerosols.
- Refrain from pipetting by mouth.
- Wash hands after handling infectious material (even when gloves have been worn) and before leaving the laboratory.
- Decontaminate all contaminated materials before disposal or reuse.

- Decontaminate laboratory surfaces following any spill of bio-hazardous materials and at the end of each workday.
- Report all spills and accidents/incidents

Suggested Reading

Biological Safety: Principals and practices, 3rd Edition. Fleming DO and Hunt DL, eds. American Society for Microbiology, ASM Press, Washington DC, 2000.

Bio-safety in microbiological and biomedical laboratories (BMBL), 4th edition. U.S. Department of Health and Human Services, Public Health Service, Centers for Disease Control and Prevention, and National Institutes of Health. U.S. Government Printing Office, Washington, 1999.

Laboratory Bio-safety Guidelines, 3rd edition. Public Health Agency of Canada, Office of Laboratory Security, Ottawa, Canada, 2004.

Laboratory Bio-safety Manual, 3rd edition. World Health Organization. Geneva, 2004.

□□□

Chapter 24

Approaches to Develop Low - Cholesterol Milk Products

VIVEK SHARMA, DARSHAN LAL AND AMIT KUMAR
Dairy Chemistry Division, National Dairy Research Institute, Karnal -132001

Introduction

Milk is considered to be a complete food as it contains almost all essential nutrients required for human health and growth. Lipids, the most important constituent of milk, play significant role in the nutrition, flavour and physico-chemical properties of milk and milk products. They are also rich source of fat-soluble vitamins (A, D, E & K) and essential fatty acids, apart from having pleasant sensory attributes. Milk fat is easily digestible than other oils and fats. It contains number of components which show anticarcinogenic activity, e.g. sphingomyeline, conjugated linoleic acid, b-carotene etc. So one (especially vegetarians) cannot avoid it in one's diet. But recent trend, in the society, is against fat-rich dairy products due to the presence of saturated fat & cholesterol as these are known to increase the incidence of coronary heart disease (CHD). CHD is one of the common causes of heart attack. Through a period of time, many researchers have shown that dietary cholesterol, serum cholesterol and occurrence of coronary heart disease (CHD) have positive correlation. Milk fat contains about 0.25 to 0.40% cholesterol. Consumption of ghee and other fat-rich dairy products makes appreciable contribution to cholesterol intake. Furthermore, some cholesterol oxidation products (COPs) have been reported to be more harmful than cholesterol itself as they are cytotoxic, atherogenic, mutagenic and carcinogenic. In recent years, demand of cholesterol-free foods has increased tremendously. This has led to increase in market of margarine, vegetable fat filled dairy products, milk fat replaced dairy products, etc. Owing to the adverse affects of cholesterol on human health, various physical, chemical and biological methods have been developed for reducing cholesterol in foods including milk.

Cholesterol Content in Milk

Animal food products like milk and milk products, meat and meat products and eggs are the major sources of cholesterol in our diet. Among these, chicken egg contains highest amount (about 215 mg/egg) of cholesterol. Cholesterol accounts for 0.25-0.45% of the total lipids in milk. Cholesterol concentrates in the milk fat globule membrane (MFGM). In milk, 80% of the cholesterol is associated with the milk fat globules and the remaining 20% is partitioned into the skim milk phase where it is associated with fragments of cell membrane (Patton & Jensen, 1975).

Pantulu and Murthy (1982) observed 8-10 times higher content of cholesterol in whey than in whole milk. Srinivasan (1984) reported the average cholesterol content of cow and buffalo milk as 2.8 and 1.9 mg/g fat, respectively. Bindal and Jain (1972) estimated free and esterified cholesterol in Desi ghee, using TLC method and reported their contents as 0.288 and 0.038% and 0.214 and 0.056 % in cow and buffalo ghee, respectively. Prasad and Pandita (1987) observed cholesterol content of ghee prepared from milk of Haryana, Sahiwal and Sahiwal X Friesian cows and from Murrah buffaloes, to be 303, 310, 328 and 240 mg/100 g fat, respectively.

Approaches to Reduce Cholesterol in Milk

Since dairy products contain significant amounts of cholesterol, a number of processes for removal of cholesterol have been developed to produce low-cholesterol dairy products. These include steam stripping, molecular distillation, solvent or super-critical extraction, reaction with cyclic anhydride, enzymatic method and treatments with adsorbents like saponin, activated charcoal and cyclodextrin. These are briefly discussed below.

a) Steam Stripping

This process is similar to that used in the deodorization of vegetable oils and removal of unsaponifiable matter. To remove cholesterol by steam stripping, the fat is first deareted under vacuum after which it is heated with steam upto 232°C and then subjected to steam at low pressure in cylindrical tall chamber. The anhydrous milk fat (AMF) passing over a series of plates is spread in many thin layers, which increases the stripping efficiency. The steam rises and carries with it the evaporated cholesterol to be condensed and collected with other volatiles. This process can remove upto 93% of cholesterol though with 5% fat losses. The major disadvantage to the process is that it removes flavouring compounds also (Schlimme & Kiel, 1989).

b) Molecular Distillation

In this process, AMF is molecularly distilled at temperature 190 and 210°C at a vacuum of 10^{-4} Torr. Fractions distilled at 190 and 210°C represented 3.43 and 3.99 % of the initial mass and contained more than 93 % of the total cholesterol (Lanzani *et al*, 1994 and Sharma *et al.*, 1999). Arul *et al.* (1988) fractionated AMF into four fractions at temperatures of 245 and 265°C and pressure of 220 and 100 mm Hg. Two low melting point fractions were blended together to yield a total of three fractions (liquid, intermediate and solid). About 78% of the total cholesterol was found in the liquid fraction while the remaining was found in the intermediate (18%) and solid (4%) fractions in the esterified form. But, because of the high heat used in the process, the quality of the end product is adversely affected.

c) Solvent Extraction

In this process butter oil is mixed with propane and ethanol in the mixing vessel. The low viscous mixture of butter fat, ethanol and propane is fed into the extraction column. A mixture of ethanol and water, containing a small amount of propane is used as extractant. The extract, a solution of cholesterol and butter fat in a mixture of ethanol, water and some propane is withdrawn at the bottom of the extraction column, which is splitted into two phases. The upper phase consists of fat and cholesterol, which are subsequently separated, in a further processing step. Around 90 to 95% of the cholesterol is extracted in this counter-current procedure operated at 30°C and 10 bar (Czech *et al.*, 1993).

d) Supercritical Carbon Dioxide Extraction

Some studies have shown that supercritical carbon dioxide (SC-CO_2) can be used to fractionate AMF with evidence that cholesterol can be concentrated into selected fractions. Kaufmann *et al.*, (1982) obtained two fractions of milk fat by SC-CO_2 extraction at a pressure of 200 bars and temperature of 80°C. In this process, the liquid fractions were enriched in total cholesterol. However, Huber *et al.* (1996) observed that direct supercritical extraction of cholesterol from AMF is not feasible because of the low selectivity of cholesterol and poor solubility of AMF. Moreover, under these conditions, important milk flavours also get separated with the cholesterol. Therefore, they proposed another process for cholesterol removal from AMF, dissolved in SC-CO_2 under high solubility conditions for AMF (40 MPa at 70°C) to achieve rapid extraction. In this process, the dissolved AMF in SC-CO_2 is passed isobarically and isothermally through a high-pressure column, filled with a suitable adsorbent (e.g. silica gel) to eliminate cholesterol. Finally, the supercritical mixture is fractionated by either descending or ascending temperature profile in separators connected in series. Karkare and Alkio (1993) found that over 99% of cholesterol from milk fat could be removed using an SC-CO_2 extraction system equipped with a silica gel column.

e) Reaction with Cyclic Anhydride

Gu *et al.* (1994) developed a method for cholesterol removal from milk fat based on the reaction between the hydroxyl group of cholesterol and a cyclic anhydride such as succinic anhydride. The conversion of cholesterol into an acid derivative makes it possible to remove these from fats by extraction with aqueous alkali. Addition of acetic acid increases the rate of reaction and prevents the distillation of cyclic anhydride from reaction mixture. They removed 50% cholesterol from animal fats but alongwith it a- tocopherol (50%), g- and d- lactones also get removed.

f) Enzymatic Method

McDonald *et al.* (1983) have described an enzymatic process using cholesterol reductase for conversion of cholesterol to biologically inactive, e.g., non-toxic, non-absorbable products like coprosterol, which is either not or is only poorly adsorbed by the body. This approach, which is theoretically suitable for reducing the cholesterol content of milk fat, has been verified biologically at least in part, by the finding that a portion of the intestinal cholesterol is reduced to coprosterol by intestinal bacteria and subsequently eliminated.

g) Adsorption Methods

Cholesterol can be removed by its adsorption on certain material. Adsorbents, which are used to remove cholesterol, are activated charcoal, saponins and b-cyclodextrin.

i. Activated charcoal

Bindal *et al.*, (1994) could remove half of the cholesterol present in milk fat through treatment of liquid fat with activated charcoal. Another activated charcoal method claimed 95% of cholesterol removal from AMF but many other compounds including yellow pigments were also removed simultaneously (Sharma *et al.*, 1999).

ii. Saponins

Saponins are naturally occurring plant compounds that can be used to selectively bind to cholesterol and precipitate it out. 80% and 90% cholesterol reduction in cream and anhydrous milk fat was obtained by using this method (Riccomini et al., 1990). Oh et al. (1998) found 70.5% of the cholesterol removal when milk was treated with 1.5% saponin at 45°C for 30 min. Further, addition of 0.25% celite increased cholesterol removal to 72%. However, the methods using activated charcoal or saponins are relatively non-selective and remove flavour and nutritional components also when cholesterol is removed (Lee *et al.*, 1999; Sharma *et al.*, 1999).

iii. β-cyclodextrin

Beta cyclodextrin, one of the well known members of cyclodextrin family, is a cyclic oligosaccharide of seven glucose units joined 'head to tail' by a-1, 4 linkage and is produced by the action of enzyme cyclodextrin glycosyl transferase (CGT) on hydrolyzed starch syrup. Beta cyclodextrin has torus like structure (Figure B). The central cavity is hydrophobic, giving the molecule its affinity for non-polar molecules such as cholesterol (Szejtli, 2004). The radius of the cavity can accommodate a cholesterol molecule almost exactly, explaining the highly specific nature of b-cyclodextrin's ability to form an inclusion complex with cholesterol (Hettinga, 1996).

Seth and Singh (1994) observed that buffalo milk samples on treatment with beta-cyclodextrin at 1, 2 and 4%, the cholesterol level was reduced from 36 mg to 25 mg, 19.5 mg and 8 mg per 100 ml, respectively, corresponding to 30.5, 45.8 and 77.6% reduction, respectively. Lee *et al.*, (1999) studied the cholesterol removal from homogenized milk and reported that 92 – 95% cholesterol removal from milk by using b-cyclodextrin at 0.5-1.5% level. Ahn and Kwak (1999) used b-cyclodextrin for cholesterol removal in cream. Average cholesterol removal increased as -cyclodextrin concentration increased, regardless of other factors. At 10% of b-cyclodextrin level, the effect of stirring speed was more noticeable than that of stirring time. When stirring was 1600 rpm and stirring was 20 min, the cholesterol was reduced by 97.99% using 15% b-cyclodextrin. Recently NDRI has developed a process to remove cholesterol from ghee, where removal of about 90% cholesterol has been reported by using b-cyclodextrin in the range of 1 to 7.

Protocol for cholesterol estimation in milk fat after saponification using enzymatic diagnostic kit:

Milk fat (0.1-0.15 g) in test tube with teflon lined screw cap

↓

Add 5 ml 5% methanolic KOH and mix

↓

Incubate capped tubes in water bath for 90 °C/ 20 min with shaking every 5 min.

↓

Cool contents by tap water

↓

Add 1 ml distilled water

↓

Add 5 ml hexane

↓

Vortex the contents for 1 min

↓

Centrifuge at 2000 rpm/ 2 min

↓

Pipette out upper hexane layer

↓

Take 0.2 ml of aliquot in dry test tube

↓

Evaporate solvent under nitrogen at 60-70°C

↓

Add 10 μl of absolute ethanol to dissolve dried residue

↓

Add 1.0 ml cholesterol reagent provided in kit and incubate at 37°C/10min

↓

Cool to room temp (28 – 30°C).

↓

Measure colour (pink) intensity at 505 nm

Calculation

$$\text{Cholesterol (mg/100 g)} = \frac{0.02 \times \text{OD of sample} \times \text{ml of hexane (5 ml)} \times 100}{\text{OD of standard} \times \text{ml of hexane aliquot (0.2 ml)} \times \text{Weight of sample (gm)}}$$

where,

0.02 is the concentration (mg) of cholesterol in 10 μl of standard solution provided in the kit.

Suggested Reading

Ahn, J. and Kwak, H. S. (1999) Optimizing cholesterol removal in cream using beta-cyclodextrin and response surface methodology. *J. Food Sci.* **64**(4): 629-632.

Arul, J. A., Boudreau, A., Makhlouf, J., Tardif, R. and Grenier, B. (1988). Distribution of cholesterol in milk fat fractions. *J. Dairy Res.* **55**: 361-371.

Bindal, M. P., Wadhwa, B. K., Lal, D., Rai, T. and Aggarwal, P. K. (1994) Removal of cholesterol from milk and milk products: Application of biotechnical processes. *NDRI Annual Report.* pp. 98-99.

Czech, B., Peter, S. and Weidner E. (1993) Effective removal of cholesterol from butter fat. *Scandinavian Dairy Information,* **7**(4): 56-58.

Gu, Y. F., Chen, Y. and Hammond, E. G. (1994) Use of cyclic anhydrides to remove cholesterol and other hydroxy compounds from animal fats and oils. *J. Am. Oil Chem. Soc.* **71**: 1205-1207.

Hettinga, D. (1996) Butter. in *Bailey's industrial oil and fat products,* Vol. 3, 5th edn, (eds Y. H. Hui), John Wiley & Sons, INC. New York, pp. 1-23.

Huber, W., Molero, A., Pereyra, C. and Martinez de la Ossa E. (1996) Dynamic supercritical carbon dioxide extraction for removal of cholesterol from anhydrous milk fat. *Int. J. Food Sci. Tech.* **31**: 143-151.

Karkare, V. and Alkio, M. (1993) Removal of cholesterol during milk fat fractionation by supercritical carbon dioxide. *Agric. Sci. Finland.* **2**(5): 387-393. (Cited in DSA, 56: 1522).

Kaufmann, W. , Biernoth, G., Frede, E., Merk, W., Precht, D. and Timmens, W. (1982) Fractionation of butter fat by extraction with supercritical carbon dioxide. *Milchwissenschaft.* **37**: 92-96.

Lanzani, A., Bondioli, P. Mariai, C., Folegatti, L., Venturii, S., Fedeli, E. and Barreteau, P. (1994) A new short-path distillation system applied to the reduction of cholesterol in butter and lard. *J. Am. Oil Chem. Soc.* **71**(6): 609-614.

Lee, D. K., Ahn, J. and Kwak, H. S. (1999) Cholesterol removal from homogenized milk with beta-cyclodextrin. *J. Dairy Sci.* **82**: 2327-2330.

Oh, H. I., Chang, E. J. and Kwak, H. S. (1998) Conditions of the removal of cholesterol from milk by treatment with saponin. *Korean J. Dairy Sci.* **20**: 253-260.

Pantulu, P. C. and Murthy, M. K. R. (1982) Lipid composition of skimmed milk and whey. *Asian. J. Dairy Res.* **1**(1): 17-20.

Riccomini, M. A., Wick, C., Peterson, A, Jimenez-Flores, R. and Richardson, T. (1990) Cholesterol removal from cream and anhydrous milk fat using saponins. *J. Dairy Sci.* **73**(1): 107.

Schlimme, E. and Kiel, D. (1989) Removal of cholesterol from milk fat. *European Dairy Magazine,* 12-21.

Seth, R. and Singh, A. (1994-95) Removal of cholesterol from milk using b-cyclodextrin and preparation of milk products from such treated milk. *NDRI Annual Report.* pp. 95.

Sharma, R., Nath, B. S. and Lal D. (1999) Approaches for cholesterol removal from milk fat: An overview. *Indian J. Dairy and Biosciences.* **10**:138-146.

Szejtli, J. (2004) Cyclodextrins. In Chemical and Functional Properties of Food Saccharides (eds Piotr Tomasik), CRC Press, Boca Raton, pp. 271-289.

□□□

Chapter 25

Cosmeceuticals : A Healthy Way of Personal Care

SHILPA VIJ
Dairy Microbiology Division, National Dairy Research Institute, Karnal

Introduction

Cosmeceuticals, "performance cosmetics", "functional cosmetics", "dermoceuticals", "active cosmetics".... these are the buzz words in twenty first century personal care. The revolution they triggered is apparent from the fact that cosmetics are no longer visualized as products that cover up or camouflage imperfections in personal appearance. Today's healthful cosmetics offer protective, healing and rejuvenative attributes as well. A cosmeceutical is an ingredient with medicinal properties that manifests beneficial topical actions and provides protection against degenerative skin conditions. The word "Cosmeceutical" was popularized by Albert M. Kligman, a professor at the University of Pennsylvania Medical School in the late 1970s. It encompasses cosmetic actives with therapeutic, disease-fighting, or healing properties, serving as a bridge between personal care products and pharmaceuticals. Originally appearing in global markets in the 1990's as an off shoot of the nutraceuticals revolution, cosmeceuticals are now recognized as a rapidly growing segment in health and personal care in the U.S., Europe and Japan. The emerging trend of "beauty inside and out" is becoming increasingly popular, wherein orally consumed nutritional supplements and topically applied cosmeceuticals work in harmony to promote physical appearance and well-being. A radiant appearance is only a reflection of optimal health and well-being. The difference between a cosmeceutical and cosmetic is that cosmetics don't cause the skin to actually change, but cosmeceuticals do. Cosmetics don't penetrate the skin and they are not active ingredients. Cosmetics can only add moisture and help the skin look temporally better. Cosmeceuticals truly change the skin for the better and skin learns to be healthier and younger.

Like "Nutraceuticals," the term "Cosmeceuticals" is not recognized by the Food, Drug, and Cosmetic Act (as amended). Cosmetics and cosmeceuticals are tested only for safety. Efficacy testing is not mandatory. Bioactive food components such as milk peptides, certain vitamins and minerals, phytonutrients from herbs, including various oils, and botanical extracts are all cosmeceuticals. Several botanicals with a history of use in traditional cultures have entered the growing cosmeceuticals market. With fewer adverse side effects and the added advantage of multifunctionality, cosmeceuticals are increasingly being used in mainstream cosmetic products. Spice extractives, for instance, have a long history of use in personal care products as aroma constituents, facilitating brand identity and aesthetic appeal. It is only recently that mainstream personal care products focus on the "cosmeceutical" benefits of such ingredients. The goodness of ginger, mint, cinnamon, pepper and other spices is no longer limited to their sensory characteristics alone. Such spices represent the new image of the "healthy and natural" in personal care.

Advantage of Combined Nutraceuticals and Cosmeceuticals

The important aspect of effective use of an ingested or skin-applied nutrient is its absorption (permeation of epidermal barrier or crossing gastrointestinal barrier by the compound) and ultimately bioavailability of the compound to the target tissues and/or receptors. Natural cosmetics (cosmeceuticals) or ingestible supplements (nutraceuticals) e.g. vitamins, minerals and botanicals may have different rates of absorption and bioavailability, depending on their chemical and biochemical forms. For example, free isoflavones, like genistein, are considered more absorbable/bioavailable than their glycosylated, or bound form, genistin; organic forms of minerals, such as selenium in complex with the amino acid methionine - selenomethionine - is shown to be better absorbed/bioavailable than inorganic sodium selenite. In addition to a more absorbable/bioavailable form of a compound, recent research shows that there are specialized biochemicals that may facilitate nutrient crossing through the biological barriers. For example, alkaloids responsible for the pungent taste of fruits from black pepper (*Piper nigrum*, Linn. fam. Piperaceae), piperine and its derivative tetrahydropiperine (THP), are considered a new class of oral and topical absorption/bioavailability enhancing compounds.

More recently, orally consumed nutritional supplements (nutricosmetics) and topically applied cosmeceuticals work in harmony to promote physical appearance and well being. Several private companies are active in cosmeceuticals, at all levels, with some leading niche markets, targeted towards specific personal care needs. Anti-aging cosmetics are therefore the most popular category in cosmeceuticals. The Anti-aging cosmetics market encompasses a range of functionalities afforded by natural ingredients that heal, soothe, replenish nutrients, and rejuvenate tired skin, hair and nails. Historically, natural materials including botanicals, milk, honey and other nutrients have been used in ancient civilizations to achieve such effects.

Type of Cosmeceuticals

Three types of cosmeceuticals are most popular: Skin Cosmeceuticals, Hair cosmeceuticals and others

a) Skin Cosmecuticals Aging and Nutricosmetics

Skin Cosmecuticals Aging and Nutricosmetics

A radiant appearance is only reflection of optimal health and well being. This forms the underlying principle of current trends in the personal care industry. Cosmetic products no longer seek to cover up signs of aging. The root causes of skin, hair and nail damage are addressed externally by using cosmeceuticals and internally with nutraceutical supplements. Vitamin and mineral deficiencies, poor antioxidant status, impaired digestion and compromised immune functions are all reflected in a sallow complexion, lifeless hair and brittle or discolored nails. Oral intake of antioxidants such as carotenoids, selenium, pro anthocyanidins (found in grape seed extract, apples and other plant sources) and vitamin E is reported to reduce the risk of DNA damages by ultraviolet radiation that leads to skin aging and skin cancers. A recent study reports that curcumin, the antioxidant pigment from turmeric is useful in the oral treatment of melanoma. In other studies, orally administerd phytonutrients such as soy isoflavones and green tea polyphenols were found to offer protection against photoaging through inhibiting the action of enzymes that degrade connective tissues.

Skin Aging and Cosmeceuticals

The skin actively participates in the body's homeostasis through its abundant microcirculatory and sensory functions. Its appendages participate in the excretion of sweat and metabolic by-products, and by possibly participating in the regulation of melatonin - a hormone that regulates circadian rhythms in the body. Healthy looking skin therefore reflects overall health and warrants optimal well-being. Cosmoceuticals not only prevent skin damage such as wrinkles, dry skin, loss of collagen and elasticity, they can also repair already damaged skin and help to reprogram it.

Natural antioxidants that quench free radicals are an essential component of antiaging formulations. They potentially offer protection against damage to the tissues against the detrimental effects of environmental and other agents. Biochemical reactions that accelerate the progression of skin aging have their roots in inflammatory processes, as inflammation generates micro-scars that mature into blemishes or wrinkles. Various types of inflammatory mediators such as leukotrienes and prostaglandins, cytokines and growth factors may influence melanin synthesis by affecting the proliferation and functioning of melanocytes. Protein kinase C, the enzyme, that phosphorylates proteins, may also influence the growth and differentiation of melanocytes. Cytokines such as endothelins (also known as vasoconstrictive peptides) are also reported to accelerate melanogenesis. Natural anti-inflammatory agents are therefore included in anti-aging formulations, and serve to soothe, heal and protect skin tone and integrity. Lipids play an important role in maintaining the barrier functions of the skin. In general, unsaturated fatty acids reinforce the skin's barrier function, prevent moisture loss through the epidermis, provide structural integrity to the skin damaged by external influences and are anti-inflammatory. They also help to soften and smoothen the skin by inhibiting the formation of corneous cells. Lipid compounds that provide an occlusive effect to prevent water loss, repair lipid layers and restore barrier functions are therefore an integral part of anti-aging formulations. Bioactive peptide technology inherent in ingredients such as Matrixyl (a branded palmitoyl oligopeptide) has made a mark in anti-aging skin care products. Runaway successes based on bioactive peptides include branded products such as "Regenerist". Nutrients such as Vitamin A and derivatives, Vitamin B-5, Vitamin C and derivatives, Vitamin E and derivatives also find versatile uses in personal care products. Photoaging is accelerated by exposure the ultraviolet rays of the sun, and prolonged sun exposure over a period of several years is reported to increase the risk of skin cancer. Sunscreens serve to protect the skin form ultraviolet radiation. The toxicity of effective dose levels of synthetic sunscreens in terms of potential damage to the body is a matter of great concern. A recent study reports the possible endrocrine disruption effects of certain sunscreens. Approved sunscreens such as octyl methoxycinnamate (OMC) and 4-methyl benzylidene camphor (4-MBC) have been found to show estrogenic activity in *in vitro* and *in vivo* studies.

Skin Lightening

A significant claim made for cosmetic products is the skin tone lightening functionality. Hydroquinone the chemical agent used to lighten the skin is now subject to regulatory restrictions in several markets on account of its potential carcinogenicity. Cosmeceuticals such as kojic acid and derivatives, bearberry extract (arbutin), paper mulberry extract, Ascorbic acid derivatives, azelaic acid, licorice extract, and other cosmeceutical actives are therefore being increasingly used in cosmetic formulations that claim skin lightening. These actives are safer and effective alternatives to hydroquinone and related materials. However several natural materials including kojic acid may not be as effective in skin lightening as hydroquinone. In this context, the efficacy of ultra pure tetrahydrocurcumin as skin tone lightener was found to be several folds greater than that of kojic acid.

Natural Antimicrobials and Preservatives

Antimicrobials in cosmetics serve to address skin, hair and nail infections as well as to improve the shelf life of cosmetic formulations. In personal care formulations that target skin conditions such as acne, there is an increasing need for economical active ingredients with negligible side effects and a long history of topical use. With the increased occurrence of antibiotic resistant microbial strains and our expanding knowledge of deleterious side effects associated with prolonged antibiotic use, natural ingredients including essential oils, probiotics and botanical extracts present attractive alternatives for use as topical antimicrobials. Innovative long chain alcohols, natural phenolic compounds and other natural extractives that inhibit microbial growth or possess bactericidal/fungicidal properties are potential options to parabens, and other synthetic preservatives in cosmetic formulations. Natural antimicrobials such as rosemary extract, sage extract, olive leaf extract, certain mushroom extracts, spice essential oils, and probiotics are effective deodorants as well

Natural Moisturizers and Conditioners

Moisturizers function to smooth out the age lines, help brighten, and tone the delicate skin. Moisturizers usually incorporate emollients to smoothen the skin surface by working their way into the nonliving outer layers of the skin, filling spaces between the layers and lubricating, and humectants to help skin cells absorb and retain moisture in these layers. Healthy Remedies Balancing Lotion has been created for women containing ingredients, which diminish the appearance of fine lines, and wrinkles, uplift the neck area and moisturize the dry, sagging skin. Some of those ingredients include black cohosh, soy extract, and vitamins A and E. Augmenting the skin's natural moisture balance are a nourishing complex containing hyaluronic acid and a revival complex containing green tea leaf extract, and glutathione Natural topical moisturizers that nourish and tone the skin represent another innovative application of natural extractives in personal care products. One example is Coriander seed oil, a rich source of petroselinic acid, linoleic acid and related fatty acids. These fatty acids are constituents of ceramides that are inherently present in the stratum corneum and prevent moisture loss from the skin surface. Coriander seed oil therefore has a healthful role in personal care. Polysaccharides such as Chitosan and derivatives, low molecular weight glycans for example, tamarind seed polysaccharides, tissue components such as hyaluronic acid and complexes, and other actives are popular as natural moisturizers. Fats such as shea butter, cocoa butter and coconut oil derivatives are other naturals that support skin texture and hydration.

Functions of Some Natural Components in Skin Cosmeceuticals

- Retinol-is used to help reverse sun damage and to inhibit collagen and elastin breakdown. It reverses photoaging and stimulates collagen synthesis, it is good for acne too.

- Vitamin C-can lighten skin, because it inhibits melanin production. It also stimulates formation of collagen and protects skin from free radical damage.
- Glycolic acid-exfoliates the skin and speeds up the renewal process. GA used in combination with Vitamin C and Retinol triggers the formation of collagen
- Hyaluronic acid-is a natural moisturizer that attracts and binds water, helping to support and lubricate the skin. It transports the essential nutrients to the skin's viable cells.
- Soy-stimulates collagen production, improves elasticity and moisture in the skin due to estrogen-type and antioxidant effects of its metabolites: genistein and daidzen.
- DMAE-is a powerful antioxidant which occurs naturally in fish oil. People 2who regularly take DMAE get improvement in the facial musculature, resulting in a leaner look. It stimulates muscles to contract and tighten under your skin. It is also known as a " Brain" or smart drug, increasing the chemicals in the brain that are needed for short-term memory and concentration.
- MSM-Nature's Beauty Mineral, which plays an essential role in the health and maintaining of the collagen and elastin of the skin and is vital for the formation of healthy cells.
- Acethyl L-carnitine and R-alpha Lipoic acid-this combination slows the cell aging process.
- N-Acetyl Cistein-helps the formation of karnitin, enhancing skin's strength and provides powerful immune support
- Hyaluronic acid-promotes water retention and cellular rejuvenation of skin cells, increasing elasticity and adverse the formation of wrinkles.
- Lycopene-life-long protection against harmful UV radiation
- Green Tea extract-slowing down skin aging process
- Resveratrol-is a stimulator of collagen synthesis, has powerful antioxidant properties

b) Hair cosmeceuticals

Hair care, color, and style play an important role in people's physical appearance and self-perception. Among the earliest forms of hair cosmetic procedures in ancient Egypt were hair setting by the use of mud and hair coloring with henna. In ancient Greece and Rome, countless ointments and tonics were recommended for the beautification of the hair, as well as remedies for the treatment of scalp diseases. Henry de Mondeville was the first to make a distinction between medicinal therapies intended to treat diseases and cosmetic agents for the purpose of beautification. But today's delineation of cosmetics from pharmaceuticals has become more complex through the development of cosmetics with physiologically active ingredients, i.e. cosmeceuticals. Shampooing is by far the most frequent form of cosmetic hair treatment. While shampoos have primarily been products aimed at cleaning the hair and scalp, current formulations are adapted to the variations associated with hair quality, hair care habit, and specific problems such as treatment of oily hairs, dandruff and for androgenic alopecia related to the superficial condition of the scalp. Cosmetics for the treatment of hair are applied topically to the scalp and hair. While they can never be used for therapeutic purposes, they must be harmless to the skin and scalp, to the hair, and to the mucous membranes and should not have any toxic effect, general or local, in normal conditions of their use. Mausner has patented a shampoo composition, which cleans the hair and scalp without doing any damage to the fragile biological equilibrium of the scalp and hair. A hair care cosmetic compositions comprising iodopropynyl butylcarbamate and/or a solution of zinc pyrithione in N-acyl ethylenediamine triacetate has been patented, which includes an appropriate carrier and a nonallergenic dry extract of yarrow (Achillea millefolium L.), obtained by oxidation of a water- alcohol solution extract of flower tops of yarrow. The extract contains less than 0.5% by weight of polyphenolic derivatives, is used for the treatment of hair, in particular oily hair, based on extract of yarrow. Buck has patented a method for treatment for androgenic alopecia wherein Liquor Carbonic Detergents are topically administered. It is generally accepted that genetic hair loss arises from the activation of an inherited predisposition to circulating androgenic hormones. A hair cosmeceutical product includes - conditioning agents, special care ingredients, and hair growth stimulants. Conditioning agents are intended to impart softness and gloss, to reduce flyaway and to enhance disentangling facility. A number of ingredients may be used, mostly fatty ingredients, hydrolyzed proteins, quaternized cationic derivatives, cationic polymers, and silicons. Special care ingredients are aimed at modifying specific problems relating to the superficial scalp. These shampoos are formulated around one or more specific ingredients selected for their clinical effectiveness in these conditions. Accordingly, current antidandruff ingredients are virtually all-effective antifungal agents - zinc pyrithione, octopirox, and ketoconazole. Hair growth stimulants cannot be expected to have any impact on hair growth due to short contact time and water dilution. A minoxidil-related compound (2,4-diamino-pyrimidine-3-oxide) is a cosmetic agent with claim of acting as a topical hair growth stimulant. Its target of action has been proposed to be the prevention of inflammation and perifollicular fibrosis. Some degree of efficacy of 2,4-diamino-pyrimidine-3-oxide has been claimed in the prevention of seasonal alopecia. Recent approval in the United States of two new products, Propecia and Rogaine Extra Strength (Minoxidil) 5%, indicated in men to promote scalp hair growth, have added a new dimension to treatment options offered by physicians in treating androgenetic alopecia.

c) Other cosmeceuticals - Eye and Dental

The skin beneath the eye lacks subcutaneous fat and has virtually no oil glands. This delicate skin needs protection and plenty of moisture to replenish and repair, which helps to reduce the signs of premature aging. As the skin ages, it becomes thinner, drier, and rougher. Over-exposure to the elements and to environmental pollution aggravates this condition. Many topical skin-soothing products intervene in this process, but products for this area need to be particularly gentle and specially formulated with ingredients that work from the inside out by interacting with the cells under the skin's surface – without irritating the eyes. There are numerous cosmeceutical eye creams that nourish the skin with natural emollients and beneficial nutrients. The other functional ingredients include butcher's broom, chamomile, and vitamin E, antioxidants - vitamins A, C and E, green tea and tiare flower, Ginkgo biloba and also cucumber, calendula and bisabolol, an active constituent of chamomile, to calm irritated skin. A key ingredient in the eye lifting moisture cream – that treats puffiness, irritation, and also protects against future skin damage is yeast which helps to plump up the wrinkles. The eye wrinkle cream helps forestall the signs of aging and generally contains wheat germ and corn oil, squalene and carrot extract. Eye firming fluid has aosain, an algae extract from seaweed that helps the skin to maintain elasticity. Lawlor had developed dental care compositions, which are useful for providing a substantive composition on the surfaces of oral cavity, which can provide prophylactic, therapeutic, and cosmetic benefits. Lutein and Zeaxanthine make up the yellow pigment of ratina.

Source of Cosmeceutical Ingredients

Botanicals, "*Bioprotection*", Formulation

Botanical extracts that support the health, texture and integrity of skin and hair, are widely used in cosmetic formulations. Plant materials from which these extracts are prepared have a long history of traditional "cosmeceutical" use, although the term itself is of recent origin. Among the more popular functional natural ingredients, several antioxidants used in cosmetics, are scientifically proven to offer additional benefits in supporting skin texture, appearance and tone. However, while in traditional cultures, plant materials were used in crushed or dried and powdered form, their incorporation into contemporary cosmetic formulations presents unique challenges. Highly colored or gritty plant extracts need to be blended seamlessly into "milky" or clear creams lotions and gels. This is where a judicious blend of art and science comes in handy. The goodness of herbs and botanicals is extracted out and the actives are tailored for use in conventional formulations with their biological activity remaining intact.

One example is the curcuminoids-rich turmeric extract well known for its antioxidant properties, antimicrobial effects, and beneficial effects on inflammation. Turmeric has been traditionally used by South Asian women in skin care since ancient times. However, its yellow color may be unattractive to contemporary formulators. An innovative patented colorless (white to very light tan) derivative, Tetra hydrocurcuminoids, addresses this drawback, and offers effective protection against sun damage. Its antioxidant action is of a comprehensive "bioprotectant" nature, efficiently preventing the formation of free radicals, while quenching pre-formed ones as well, thereby protecting the skin cells from damage by UV radiation and the resultant inflammation and injury. This in turn has far reaching beneficial effects on overall health and well being, rendering a healthy glow to the skin. Additionally the composition efficiently lightens skin tone.

Milk Proteins

Milk proteins or their hydrolysates have already been used in compositions for holding the hair. They endow the hair with good shape-retention and, contrary to other proteins and keratins, they do not give rise to a phenomenon of making the hair sticky. These compositions possess, however, the drawback of causing a phenomenon of powdering on brushing the hair, which gives the hair an unattractive appearance. Moreover, milk proteins or their hydrolysates, when used in styling mousses, generate foams whose stability and expansion are insufficient., the combinations of some keratin hydrolysates with milk proteins and/or their hydrolysates endowed the hair with good shape-retention without causing stickiness or powdering of the hair.

Some Commercially available cosmeceutical products

1. Epicutin™ TT (INCI Name: Cyclodextrin (and) Melaleuca Alternifolia (Tea Tree) Leaf Oil Tea Tree Oil encapsulated in cyclodextrins.

   Application: Emulsified and gel-type cosmetics. To be used e.g. in facial care for applications of oily skin with a tendency of acne. Effective long term regulation and control of sebum. Due to its antiseptic, anti-bacterial and anti-irritating effects, Epicutin-TT is used in shampoos, soaps, creams and deodorants.

2. Follicusan™ (INCI Name: Water (and) Alcohol Denat. (and) Panthenyl Ethyl Ether (and) Milk Protein (and) Lactose (and) Inositol (and) Acetyl Cysteine (and) Acetyl Methionine (and) Sodium Citrate (and) Citric Acid) Bioactive signaling proteins combined with ethyl panthenol, inositol as well as acetyl cysteine and acetyl methionine, in aqueous-alcoholic medium.

   Application : Aqueous and aqueous-alcoholic hair care formulas. Designed for preventive care of scalp and hair by vitalizing the cells of the scalp including hair follicles. Counteracting premature, accelerated loss of hair (Alopecia). Normalizes excess sebum secretion, thus counteracting formation of dandruff

3. Glycoderm™ (P) (INCI Name: Water (and) Honey (and) Phospholipids (and) Sphingolipids (and) Hyaluronic Acid)Delivery system built with plant-derived phospholipids and ceramides, loaded with hyaluronic acid and natural blossom honey.

   Application : Aqueous and emulsified skin care preparations. Restores the lipid barrier of the stratum corneum of dry and fissured skin and simultaneously builds a moisturising network inside the stratum corneum, fixes water to the lipid bilayers.

4. Lactokine™ Fluid (INCI Name: Milk Protein) Network of activated and stabilized signaling molecules derived from milk which promote vitalization of the skin protection system and reduction of melanin synthesis. Also available as LACTOKINE AC, suitable for formulations at low pH.

   Application : Recommended for aqueous and emulsified skin care products. For regenerative care of stressed and aged skin. For prevention of premature skin aging. Skin protection through skin cell vitalizing.

5. MPC™ - Milk Peptide Complex (INCI Name: Whey Protein)Bioactive polypeptides (cytokines) with functionally valuable milk components as carrier and stabilizers.

   Application : Emulsified and gel-type skin regenerating products. Designed for mature skin to rejuvenate and smooth wrinkles. Prevention of environmentally induced, accelerated aging of skin (biological and environmental skin aging). Increases thickness of epidermis, stimulates cell-cell communication between epidermis and dermis. Provides also: Slimming activity, body contouring, reduces occurrence of Cellulite.

6. Phytosan™ (INCI Name: Water (and) Glycerin (and) Glycine Soja (Soybean) Seed Extract) Contains proteins, glycoproteins and polysaccharides isolated from soybean seeds.

   Application : Emulsified and gel-type skin care products. Designed to counteract the signs of biological and UV-induced premature skin aging. As a "biological sun protection factor" it helps to reduce UV-induced DNA damage when applied as pretreatment and prevents release of ECM-decomposing proteases. By stimulating skin regenerative functions it improves firmness and micro-relief of the skin.

7. Modukine™ (INCI Name: Lactose (and) Milk Protein) Biologically active milk polypeptides for regenerative care of sensitive and extremely dry, atopic skin.

   Application : Acts against inflammatory skin conditions and helps to restore an imbalanced lipid barrier.

Conclusion

The global trend in the cosmetic industry towards developing 'medicinally' active cosmetics, and in the Pharmaceutical industry towards 'cosmetically' oriented medicinal products as part of a current 'life-style' ideology. The future promises increasingly sophisticated formulations for cosmetics and skin-care products. Cosmetic companies are finding ways to deliver small-dose ingredients that do not require medical regulations and to introduce steroids and hormones into lip balms, which would result in production of cosmeceuticals that could help to improve body mass, nail, and hair growth. New challenges will also be presented to government regulatory agencies as more chemicals with true biological activity are invented and tested. Claim substantiation and premarketing testing must also evolve to accurately assess efficacy and safety issues with important implications for total body health. The new vehicles and delivery systems combined with established ingredients will alter percutaneous absorption, requiring re-evaluation of substances with an assumed good safety profile. Biotechnology will also compete directly with the pharmaceuticals and cosmetic businesses. The most influential angle over the coming 5 years will be the links between internal health, beauty, and anti-aging.

Table 1: List of Cosmeceuticals

Ingredient	Purported action	Source
Vitamins	Antioxidant	Vitamins A,C and E
α-Hydroxy acids (AHAs)	Exfoliates and improves circulation	Fruit acids (glycolic acid, lactic acid, citric acid, tartaric acid, pyruvic acid, maleic acid etc)
β- Hydroxy acids (BHAs)	Antibacterial	Salicylic acid
Essential fatty acids	Smoothens, moisturizes and protects	Linoleic, linolenic and arachidonic acid
Coenzyme Q10 (Ubiquinone)	Cellular antioxidant	Naturally occurring in skin
Allantoin	Soothes	Comfrey
Aloe vera	Softens skin	*Aloe vera*
Amica	Astringent and soothes	*Amica montana*
Calendula	Soothes, softens and promotes skin cell formation	*Calendula officinalis*
β -Bisaboiol	Anti-inflammatory, antibacterial and calms irritated skin	Chamomile flower
Cucumber	Cools, refreshes and tightens pores	Cucumber

contd...

Lupeol	Antioxidant and skin conditioning agent	*Cratacva nurvula*
Ginkgo	Antioxidant that smoothes, rejuvenates and promotes youthful appearance	*Ginkgo biloba*
Ivy	Stimulates circulation and helps other ingrdients penetrate skin	*Hedera spp. (Ivy family)*
Panthenol	Builds moisture and soothes irritation	Provitamin B_6
Witch hazel	Tones	*Hamamelis virginiana*
Green tea extract	Antioxidant	Green teas
Neen oil limonoids	Antimicrobial	*Azadirachta indica*
Pycnogenol	Antiaging effect	Grape seed extract
α-Lipoic acids, resveratrol, polydatins	Potent free radical scavengers and antioxidant	Fruits and vegetables
Furfuryladenine	Improves hydration and texture of skin	Plant growth hormone
Kinetin	Free radical scavenger and antioxidant	Plant and yeast
Sodium hyaluronate	Lubricant between skin tissues and maintains natural moisture	Natural protein
β-carotene	Minimizes lipd peroxidation and cellular antioxidant	Carrots and tomatoes
Retinoic acid	Smoothes skin, promotes cell renewal and improves circulation to skin	Vitamin A
Tetrahydrocurcuminoides	Antioxidant and antiaging	*Curcuma longa*
Centella	Skin conditioning agent, increases collagen production, improves texture and integrity of skin and reduces appearance of stretch marks	*Centella asciatica*
Boswellia	Antiinflammatory and antiaging	*Boswellia serrata*

contd...

Coriander seed oil	Anti-inflammatory and antiirritant, skin lightening properties	*Coriandrum sativa*
Turmeric oil	Antibacterial and antiinflammatory	*Curcuma longa*
Coleus forskofii oil	Antimicrobial, aroma therapyl perfumer	*Coleus forskoflii*
Arjunolic extract	Antioxidant and anti-inflammatory	*Terminalia arjuna*
Ursolic acid	Anti-inflammatory, collagen build-up	*Rosemarinus officinalis*
Oleanolic extract	Antioxidant, antifungal, improves texture and integrity of skin	Olive leaf
Rosemary extract	Antioxidant, antimicrobial and anti-inflammatory	*Rosemarinus officinalis*
Dry extract from yarrow	Treatment of oily hair	*Achillea millefolium*
Licorice extract	Skin whitening properties, antioxidant, antimicrobial and anti-inflammatory	*Glycyrrhiza glabra*
Horse chestnut extract	Supports blood circulation, wound healing effect and anti-inflammatory	*Aesculus hippocastanum*

□□□

Chapter 26

Identification, Analysis and Characterization of Biological Macromolecules using Proteomic Approaches for Future Application in Food Science

ASHOK K. MOHANTY
Animal Biotechnology Centre, National Dairy Research Institute, Karnal

Introduction

Proteomics is the study of the proteome, the protein complement of the genome. Proteomics or, more appropriately functional proteomics refers to the branch of discovery of science focusing on proteins. The term 'proteome' is used to describe the complete set of proteins that is expressed, and modified following expression, by the entire genome in the lifetime of a cell. The term "proteomics" and "proteome" were coined by Marc Wilkins and colleagues in the early 1990s and mirror the terms "genomics" and "genome", which describe the entire collection of genes in an organism. Today, proteomics is a scientific discipline that promises to bridge the gap between our understanding of genome sequence and cellular behaviour. It can be viewed as more of a biological assay or tool for determining gene function.

Initially the term was used to describe the study of the expressed proteins of a genome using two-dimensional (2D) gel electrophoresis, and massspectrometry (MS) to separate and identify proteins and sophisticated informatics approaches for deconvoluting and interrogating data. This approach is now referred to as "expression" or "global profiling" proteomics. The scope of proteomics has now broadened to embrace the study of "protein-protein" interactions (protein complexes), referred to as cell-mapping proteomics (Blackstock and Weir, 1999).

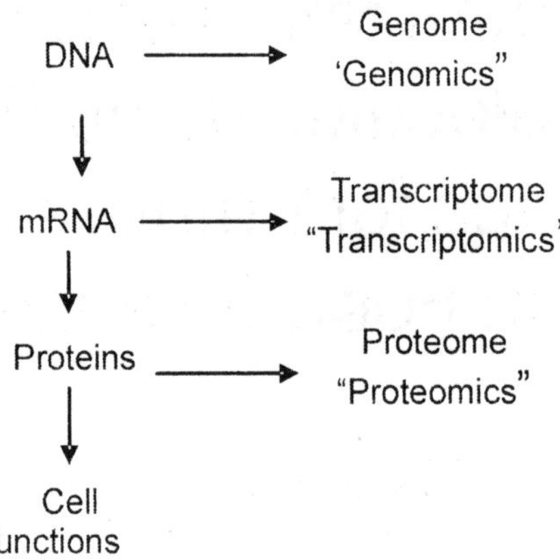

Biological context of Genomics and Proteomics

The many Faces of Proteomics

Proteomic analysis (or analytical protein chemistry). The large-scale identification and characterization of proteins, including their posttranslational modifications, such as phosphorylation and glycosylation. Analysis is done with the aid of mass spectrometry or Edman degradation.

Expression proteomics (or differential display proteomics). Two-dimensional gels are used for global profiling of expressed proteins in cell lysates and tissues. This conventional approach is being challenged by non-2D gel methods, such as liquid-based isoelectric focusing (IEF) or ion-exchange chromatography / reversed-phage high-performance liquid chromatography (RP-HPLC). Proteins are typically identified by massspectrometry (MS). In many situations, these methods are complemented by DNA-based array methods.

Cell-mapping proteomics (or cataloging of protein-protein interactions). Protein-protein interactions and intracellular signaling are determined by identification of protein complexes (obtained by affinity purification and protein identification by MS) or by direct DNA readout (e.g. yeast two-two hybrid, phage display, ribosome display, and RNA-peptide fusions).

Proteomics *vis-a vis* Genomics

Large-scale genome sequencing

One of the most biological achievements to emerge during the last 40 years has been the completion of draft DNA sequences of the human genome, published by the International Human Genome Sequencing Consortium (a publicly) funded project and by Celera Genomics (a commercial effort). This has provided a blue print of the information needed to create a human being and revealed for the first time the organization of a vertebrate's DNA. The public project estimates that there are 31, 000 protein-encoding genes, where as Celera finds ~26,000, with many more still to be found (a current estimate suggests that the number of protein-encoding genes may be on the order of 60,000).

Interestingly, the number of coding genes in the human sequence is not dramatically different from the numbers reported for phylogenetically remote organisms: 6, 000 for a yeast cell, 13,000 for a fly, 18,000 for a worm, and 26,000 for a plant (Genomes Online Databases at http://wit.integratedgenomics.com/GOLD). The number of genes reported for multicellular organisms is not highly accurate because of limitations of existing abinitio gene prediction methods used to identify genes. The existence of an open reading frame (ORF) in genomic data does not necessarily imply the existence of functional gene. In human DNA, gene prediction by abinitio methods is difficult because of extensive alternative splicing, lower density exons, and high proportions of interspersed repetitive sequences. Given the unreliability of abinitio gene prediction software, all genes will need to be experimentally identified and annotated. Hence, verification of a gene product by proteomic analysis is an important first step in annotating the genome.

Disparity between mRNA Profiling and Protein Profiling

There is no simple correlation between changes in mRNA expression levels (transcriptomics) and those in protein levels (proteomics). The link between transcript levels and protein levels in a given cell or tissue is difficult. It is also understood that array-based gene expression monitoring or other gene expression methods for measuring mRNA abundances, alone are insufficient for analyzing the cell's protein complement. There is a marked disparity between the relative expression levels of mRNAs and those of their corresponding proteins. Differing stability of mRNAs and different efficiencies in translation can affect the generation of new proteins. Once formed, proteins differ significantly in stability and turnover rates. Many proteins involved in signal transduction, transcription factor regulation, and cell-cycle control are rapidly turned over as a means of regulating their activities. Also mRNA levels tell us nothing about the regulatory status of the corresponding proteins, whose activities and functions are subject to many endogenous posttranslational modifications and other modifications by environmental agents. Therefore, the complication arises when considering the complementarity of genomics and proteomics. Despite the notion that one gene gives rise to one protein, the situation in eukaryotic cells is more likely six to eight proteins per gene. Thus, there may be several hundred thousand human proteins after splice variants and essential posttranslational modifications are included. For example, 22 different forms of human a-1-antitrypsin have been observed in human plasma. Such biological complexities can be unraveled using proteomic studies to understand how cells modulate and integrate signals.

Identification and Analysis of Proteins

Four key platform technologies are crucial to any proteomics strategy aimed at elucidating the function of an unknown gene.

- Sample preparation & handling
- Determination of partial amino acid sequence information
- Protein identification and quantification
- Cell mapping

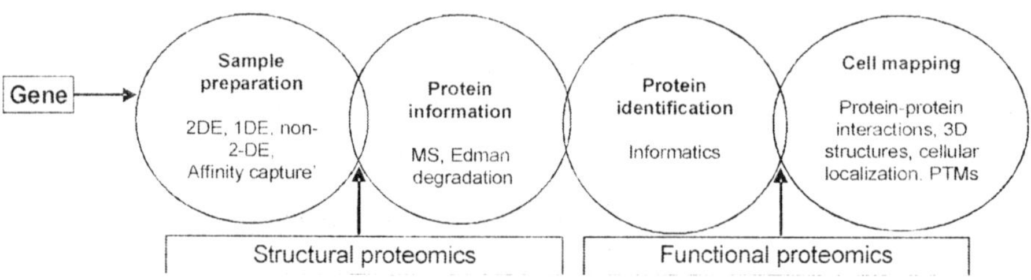

Technology platforms for Proteomics

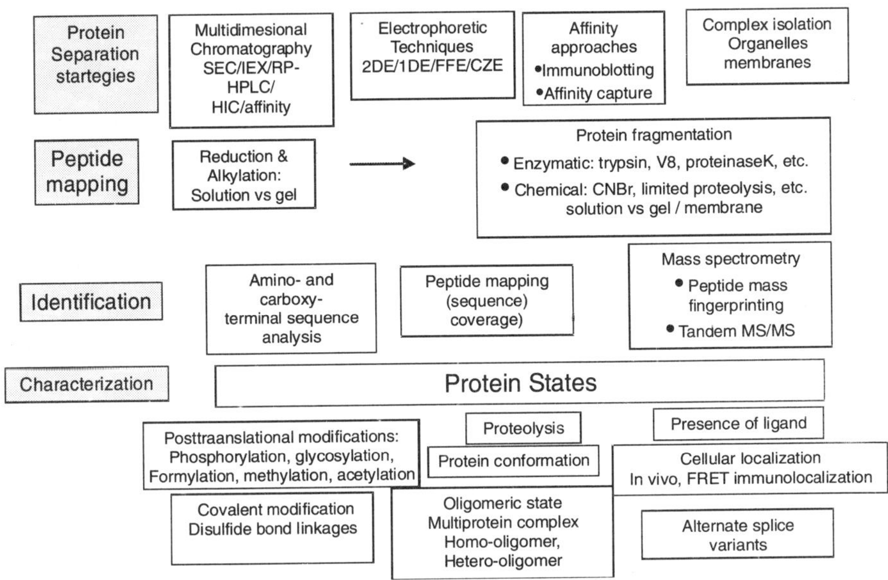

Proteomics Strategies for the Identification and Analysis of Proteins

SEC: size exclusion chromatography; IEX: ion-exchange chromatography; RP-HPLC: reversed -phase high-performance liquid chromatography; HIC: hydrophobic interaction chromatography; 2DE: two-dimensional gel electrophoresis; 1DE: one dimensional gel electrophoresis; FFE: free-flow electrophoresis; CZE: capillary zone electrophoresis; FRET: fluorescence resonanace energy transfer

Protein Separation Strategies

One of the rate-limiting steps in any proteomic analysis study is obtaining, and then handling, sufficient quantities of target protein (s) from its original biological source. The classical method of quantitative and qualitative expression proteomics combines protein separation by high-resolution 2D gel electrophoresis with MS or MS/MS identification of selected protein spots. Because, even the best 2D gels can routinely separate no more than 1500 proteins, this technique is limited to the most abundant proteins if a crude protein mixture (whole-cell lysate) is used. 2D electrophoresis is limited by the amount of material that can be applied to the first-dimension immobilized pH gradient gel (~150 ug to low milligram quantities). Hence 2D gels have limited 'scale up' capability. For this reason, it is often desirable to "trace enrich" for a particular subclass of proteins. By analyzing proteins in a cellular compartment or organelle, it is possible to reduce the complexity and differences in abundance of a subset of proteins within a cell.

Two-Dimesional SDS-PAGE: This separation method has become synonymous with proteomics and remains the single best method for resolving highly complex protein mixtures. Two-dimensional SDS-PAGE is actually a combination of two different types of separations. In the first, the proteins are resolved on the basis of isoelectric point by IEF. In the second, focused proteins are then are further resolved by electrophoresis on a polyacrylamide gel. Thus 2D-SDS-PAGE resolves proteins in the first dimension by isoelectric point and in the second dimension by molecular weight. Dedicated 2D-SDS-PAGE systems are available that use immobilized pH gradient (IPG) strips and relatively foolproof hardware to facilitate the transfer of proteins from the IPG strip into the SDS-PAGE slab gel. The IPG strip is based on the use of immobilized pH gradients, in which polycarboxylic acid ampholytes are immobilized on supports to reproducibly create stable pH gradient. The use of narrow pH ranges facilitates the separation of proteins with highly similar isoelectric points. Proteins separated by 2D gels are visualized by conventional staining techniques, including silver, Coomassie, amido black stains and fluorescent staining. Silver-staining and newer fluorescent dyes are the most sensitive.

Cell Mapping and Identification of Proteins in Complexes

One way to observe interacting proteins involved in a given biological process is to specifically enrich for these proteins. Typically, this requires knowledge of the activity of atleast one protein in the multiprotein complex. Under nondenaturing conditions, interacting proteins can be enriched from complex protein mixtures (e.g., cell lysates) using methods such as:

- *Coimmunoprecipitation* or "pull-down" techniques using antibodies directed against one of the component proteins.
- *Coprecipitation* using affinity-tagged recombinant proteins and antibodies directed against the "tag" epitope
- *Protein-affinity-interaction chromatography* (e.g using recombinant glutathione S-transferase (GST)-fusion proteins and glutathione-affinity chromatography).
- *Isolation of intact multiprotein complexes* (e.g., nuclear pore complex, ribosome complexes, spliceosomes).

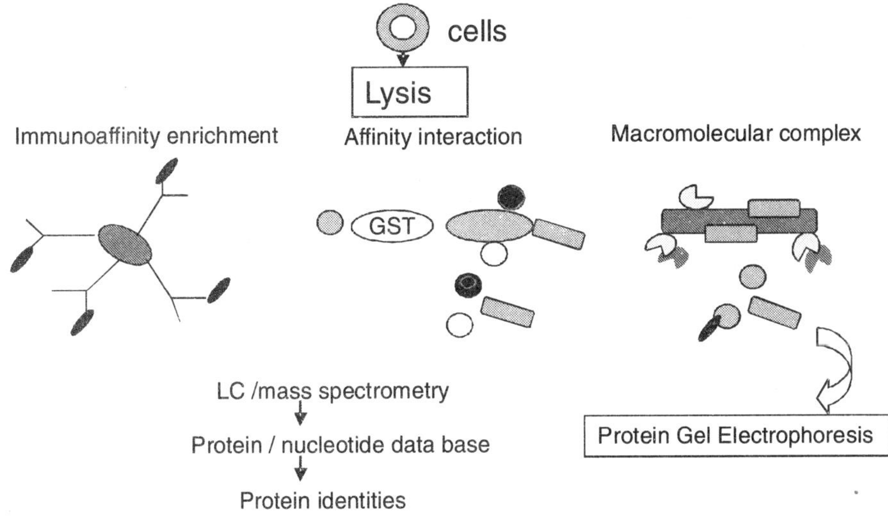

Cell mapping: affinity capture methods

Determination of Partial Amino Acid Sequence

Usually, the final step of most proteomic studies, independent of the purification method employed, utilizes either SDS-PAGE or 2D acrylamide gels to separate the proteins for identification and characterization. Following electroblot and transfer to an inert membrane, such as polyvinylidine difluoride (PVDF), intact proteins can be identified directly by amino- or carboxy-terminal amino acid sequence analysis or indirectly from peptides generated by *in-gel* or *on-membrane* digestion of the protein with a protease usually trypsin. MS-based methods usually identify a protein, not by analyzing it directly, but by analyzing the peptides derived from proteolytic digestion. Usually, a small number of peptides yield sufficient information to permit protein identification (by peptide mass finger printing (PMF) and/or MS/MS of individual peptides. In contrast to peptides, the molecular mass of intact proteins is usually insufficient to allow database identification.

MALDI-MS is used to determine the accurate mass of a group of peptides derived from a protein by digestion with a sequence-specific protease, usually trypsin, thus generating a peptide mass map or peptide mass fingerprint. Because trypsin cleaves proteins at the amino acids arginine and lysine, the masses of tryptic peptides can be predicted theoretically for any entry in a protein sequence database.

Electrospray ionization and tandem mass spectrometry is used to sequence the isolated peptides from a peptide mixture.

Differential Display Proteomics

A fundamental aspect of proteomic research is the determination of protein expression levels between two different states of a biological system (e.g. relative quantification of protein levels), such as that encountered between a normal and diseased cells or tissues. This is referred to as differential display or comparative proteomics. This can be done in two ways such as running and comparing samples in ***2D-SDS-PAGE*** or with ***LC-MS and Isotope tags.***

For differences in the protein-expression profiling, we compare 2D-gels from two different samples for differences in the occurrence or intensity of protein spots. This approach provides a useful means of comparing proteomes. However, identification of protein is cumbersome and difficult by this procedure. Application of peptide mass fingerprinting and LC-MS-MS analysis now makes it possible to identify essentially any protein one can detect by staining the gel. Therefore, the critical task in comparative proteomics with 2D gels is identifying the features that differ between the gels.

The LC-MS approach to proteome comparisions is conceptually the opposite of the 2D gel approach. Whereas the 2D gel approach separates proteins and begins with an image comparision, the LC-MS approach separates peptides and ends with data mining to assess differences between samples. Two protein samples are treated with reagents to "tag" them. The tags are chemically identical, except that one contains heavy isotopes (e.g. ^{2}H, ^{15}N, ^{13}C, ^{18}O etc....) and the other contains light isotopes. The samples are digested and the peptides are analyzed by LC-MS-MS. Analysis of the MS-MS data allows identification of the protein present. Examination of the full-scan spectra corresponding to each MS-MS scan then allows measurement of the ratio of the light- and heavy- isotope tagged peptides. This ratio corresponds to the ratio of that protein in the two samples. This approach provides a relative quantification of the level of a particular protein in two samples.

Mapping Protein Modifications

Vast majority of all eukaryotic proteins are posttranslationally modified and more than 200 posttranslational modifications (PTMs) of amino acids have been reported so far. Practically all PTMs are associated with either an increase or a decrease in molecular mass. The two major PTMs of proteins are phosphorylation and glycosylation. Phosphorylation of proteins is a ubiquitous regulatory mechanism in both eukaryotes and prokaryotes. Intracellular phosphorylation is regulated by protein kinases (dephosphorylation is regulated by protein phosphatases), which are activated in response to extracellular signals and trigger cells to switch on or off many diverse processes such as metabolic pathways, kinase cascade activation, membrane transport, gene transcription, and motor mechanisms. Protein phosphorylation can be examined in several ways such as 'phosphopeptide mapping of ^{32}P-labelled proteins and peptides', amino-terminal sequencing using Edman degradation procedure and Mass spectrometry. While all the three methods are equally good, MS is an ideal tool in proteomics studies of PTM identification and characterization because of its high sensitivity.

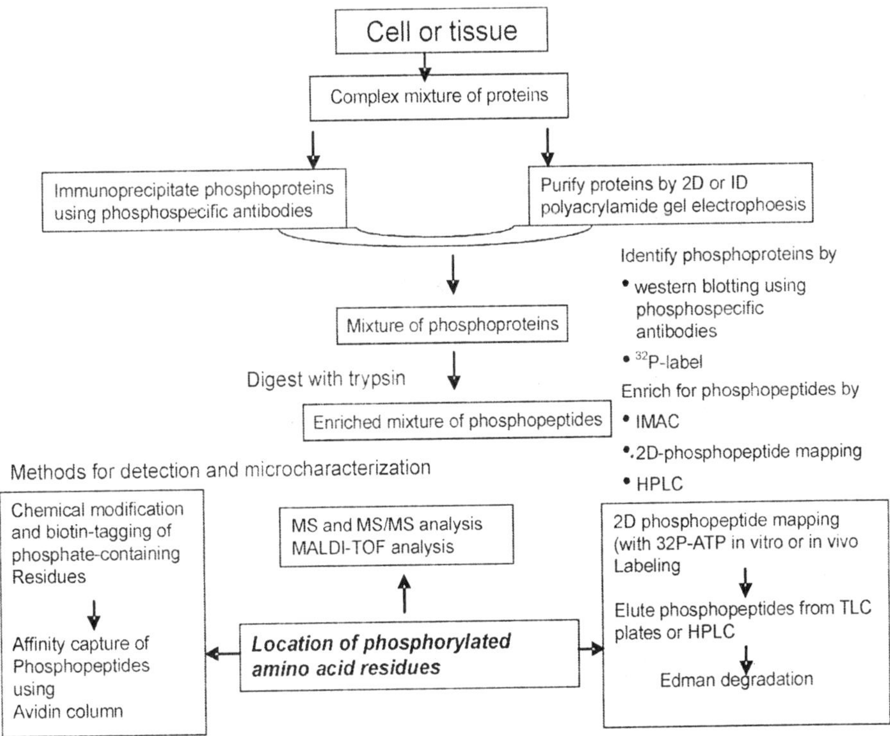

Summary of various analytical methods for phosphopeptide / phosphoprotein enrichment, detection and microanalysis

Protein 3D-structure characterization by X-ray crystallography

Proteins are polymers of amino acids, which perform specific biological functions in the body. The functional properties of proteins depend upon their three-dimensional structures. The three-dimensional structure arises because particular sequences of amino acids in polypeptide chains fold to generate, from linear chains, compact domains with specific three-dimensional structures. The folded domains can either serve as modules for building up large assemblies such as virus particles or muscle fibers or can specify specific catalytic or binding sites as found in enzymes or proteins that carry oxygen or that regulate the function of DNA. To understand the biological function of proteins, we should be able to predict the three-dimensional structures from amino acid sequences. However, it is not easy to do so merely from amino acid sequence information. Recently biotechniques such as X-Ray crystallography, NMR (nuclear magnetic resonance) or homology modeling have been developed to determine the three-dimensional structure of proteins. Among these techniques, X-ray crystallography is most widely used method to determine the three-dimensional structure of proteins. X-ray crystallography involves both wet biochemistry, physics and bioinformatics tool to determine the structure of biomolecules. For this, the candidate protein, whose structure is to be determined is, purified to homogeneity from a source and crystallized using different crystallization conditions. Once the crystals are produced, it is subjected to X-ray exposure in an X-ray crystallography set up. The diffracted X-ray data are merged and processed using various crystallographic softwares. The processed data is used to solve the structure of proteins by either molecular replacement or anomalous dispersion methods. Once the structure of the protein is solved, the structure factors generated from the processed data are used to create electron density maps of the protein. These electron density maps represent in continuity the electron densities for individual amino acids as observed in a protein chain. To determine the three-dimensional structure of a protein, we should have prior information of its amino acid sequence. Individual amino acids are modeled into the electron density map using different molecular graphics programmes to get a final model of the protein. Once the model of the protein is built, the structure is refined either by CCP4 programme suite (Collaborative Computational Project) or CNS (Crystallography and NMR system).

Molecular graphics play an important role in modeling of protein structure using X-ray crystallographic data. Functions such as side chain placement, loop, ligand, fragment fitting, structure comparison, analysis and validation are routinely performed using molecular graphics. Model building of protein structure are done using various molecular graphics programs such as 'O', 'COOT' and 'FRODO'". These tools have allowed protein crystallographers to enjoy what is widely held to be the most thrilling part of their work; giving birth (as it were) to a new protein structure. The CCP4 program suite is an integrated collection of software for macromolecular crystallography, with a scope ranging from data processing to structure refinement and validation.

Once the three-dimensional structure of the intended protein has been determined, it is analysed in detail regarding its functionality by analyzing its active site, substrate interaction sites, motifs, domains and protein environment influenced by different critical amino acids.

Application in Food Science

Various food grade bacteria with specific reference to lactic acid bacteria are used for production of fermented foods. Different lactic acid bacteria produce different proteins in the system during fermentation. Global analysis of proteome of useful lactic acid bacteria using proteomic approaches will help to identify various proteins expressed in the system. Systematic analysis of the proteins expressed will give an idea about the importance of various proteins during fermentation. Differential expression proteomics in between various useful microorganisms will help us to identify useful biomolecules specific for a particular microorganism. This will help us to identify organism specific cellular markers for future application in food product development.

Milk constitutes an important ingredient of food system. This is constituted of number of growth promoting proteins, enzymes and signaling molecules. Global and differential expression analysis of proteins in milk of various animal species will help us to identify various new proteins of health importance. Until now, most of the abundant proteins have been studied in milk. However, many low abundant proteins which are secreted in milk are generally ignored and have not been studied in detail. Proteomic approaches will help to identify useful low abundant proteins, which can be studied further for understanding their beneficial properties. Thus proteomic techniques are extremely useful for discovery of novel biomolecules for future applications in food science.

❑❑❑

Chapter 27

Potential Applications of Nanotechnology in Food Industry

GAUTAM KAUL AND THIPPESWAMY, H.S
Animal Biochemistry Division, National Dairy Research Institute Karnal

Introduction

Nanotechnology deals with the capability to image, measure, model, control, and manipulate matter at dimensions of roughly 1–100 nanometers, where novel interfacial phenomena introduce new functionalities. This exceptional capability has led to a vast array of new technologies that have an impact on virtually every aspect of science and technology, industry, economy, the environment and human lives. All organisms represent a consolidation of various nanoscale-size objects. Atoms and molecules combine to form dynamic structures and systems that are the building blocks of every organism's existence. For humans, cell membranes, hormones, and DNA are examples of vital structures that measure in the nanometer range. In fact, every living organism on earth exists because of the presence and interaction of various nanostructures. Even food molecules such as carbohydrates, proteins and fats are the results of nano scale-level mergers between sugars, amino acids, and fatty acids. The electron microscope and, more recently, the development of tools such as probe microscopes, has provided unparalleled opportunities for understanding heterogeneous food structure at the sub-molecular level. This has provided new solutions to previously intractable problems in food science and offers new approaches to the rational selection of raw materials, or the processing of such materials to enhance the quality of food products. As it applies to the food industry, nanotechnology involves using biological molecules such as sugars or proteins as target-recognition groups for nanostructures that could be used, for example, as biosensors on foods. Such biosensors could serve as detectors of food pathogens and other contaminants and as devices to track food products. Nanotechnology may also be useful in encapsulation systems for protection against environmental factors. In addition, it can be used in the design of food ingredients such as flavors and antioxidants. The goal is to improve the functionality of such ingredients while minimizing their concentration.

The recent explosion in the general availability of nanoproducts makes it almost certain that nanotechnology will have both direct and indirect impacts on the food industry. Some nanoscale phenomena have been utilized in nutraceutical and functional food formulation, manufacturing, and processes. New concepts based on nanotechnology are being explored to improve product functionality and delivery efficiency. Some of these nano-based technologies are outlined below.

Nanoparticulate Delivery Systems for Foods

Systems containing large interfacial areas such as emulsion, dispersion, and bicontinuous structured fluid are a rich source of new knowledge. Newly developed capabilities in nanoscale characterization offer a better visualization of these structures in nanometer resolution, and further a better understanding of their functionality. When amphiphilic molecules like surfactants, lipids, and copolymers that have both polar and nonpolar characteristics are dispersed in a polar solvent, hydrophobic interactions cause them to spontaneously self-assemble into a rich array of thermodynamically stable, lyotropic, liquid crystalline phases with characteristic length scales in the nanometers. These include micelles, hexagonal (tubular) structures, lamellar structures, and cubosomes, which possess a high degree of molecular orientation order despite the fact that they exist in a liquid state.

Micelles: These are submicron spherical particles, typically 5–100 nm in diameter, that are formed spontaneously upon dissolution of surfactants in water at concentrations that exceed a critical level, known as the "critical micelle concentration" (CMC). This self-assembly process is thermodynamically driven; i.e., interactions of the hydrophobic tail group of surfactants with water are minimized, while interactions of the hydrophilic surfactant head groups with water are maximized. Because of this, micelle integrity under a given set of environmental conditions (pH, temperature, salt concentration) is often maintained for many years. A remarkable property of micelles is that they have the ability to encapsulate nonpolar molecules such as lipids, flavorants, antimicrobials, antioxidants, and vitamins. Compounds that ordinarily are not water soluble or are only sparingly soluble can, with the help of micelles, be made water soluble. Micelles containing solubilized materials are referred to as microemulsions or swollen micelles. While micelles have been used as a delivery system for pharmaceutical compounds for quite a long time, their use as carrier systems for functional food components has only recently attracted increased attention. Reports of successful application of microemulsions include encapsulation of limonene, lycopene, lutein, and omega-3 fatty acids using a variety of food-grade emulsifiers, although in some cases addition of ethanol as a co-surfactant was required. Patent applications have been filed for the use of microemulsions to incorporate essential oils in flavoured carbonated beverages and to encapsulate alpha-tocopherol to reduce lipid oxidation in fish oil.

Liposomes : Liposomes or lipid vesicles are formed from polar lipids that are available in abundance in nature, mainly phospholipids from soya and egg. Like micelles, liposomes can incorporate a wide variety of functional components in their interior. However, in contrast to micelles, they can be used to encapsulate both water and lipid-soluble compounds. Liposomes are spherical, polymolecular aggregates with a bilayer shell configuration. Depending on the method of preparation, lipid vesicles can be unilamellar or multilamellar, containing one or many bilayer shells, respectively. Liposomes typically vary in size between 20 nm and a few hundred micrometers. Their core is aqueous in nature, its chemical composition corresponding to that of the aqueous solution in which the vesicles are prepared. Because of the charge of the polar lipids used in the preparation of liposomes, charged but water-soluble ionic species can be trapped inside the liposomes. The pH and ionic strength of the liposomal core can thus differ from those of the continuous phase in which the liposomes are later dispersed.

Liposomes have been successfully used to encapsulate proteins and provide a microenvironment in which proteins can continue to function regardless of external environmental conditions. On the other hand, the interior of the bilayer has properties resembling those of an organic solvent. Consequently, lipid compounds can be encapsulated inside the bilayer, a process known as adsolubilization. Liposomes have been shown to increase shelf life of dairy products by encapsulating lactoferrin, a bacteriostatic glycoprotein as well as nisin Z, an antimicrobial polypeptide. Antimicrobial efficiency of other ingredients in the encapsulated form has also been reported. Liposomal entrapped phosvitin was used to inhibit lipid oxidation in a variety of dairy products and ground pork. Recent research has demonstrated that, liposome-encapsulated vitamin C retained 50% activity after 50 days of refrigerated storage, whereas free ascorbic acid lost all activity after 19 days.

Nanoemulsions : These are simply very fine oil-in-water (o/w) emulsions with mean droplet diameter of 50–200 nm. An emulsion is defined as a mixture of two completely or partially immiscible liquids, such as oil and water, with one liquid being dispersed in the other in the form of droplets. Examples of emulsified food products are mayonnaise, milk, sauces, and salad dressings. In contrast to these well-known o/w emulsions, nanoemulsions are small enough not to scatter light in the visible region of the spectra; thus, they appear clear instead of being optically opaque. Because of their small size, they also do not cream within an appreciable time. Creaming is the process whereby oil droplets move to the top of the emulsion to form a concentrated oil-droplet layer. This is often followed by a complete breakdown of the emulsion, yielding a clearly visible oil layer on top of the emulsion. Nanoemulsions and macroemulsions can be manufactured in a similar fashion using high-pressure homogenizers, or membrane and microfluidic channels. It should be noted that the proper choice of surfactants and/or polymers is critical in the production of nanoemulsions. Because of their small size, nanoparticles have excellent penetration properties to ensure rapid delivery of high concentrations of active ingredients to cell membranes. Bioavailability of lipophilic active ingredients can be substantially improved by delivery in Nanoemulsions. For example, nanoemulsions have been used in parenteral nutrition for quite some time. Also because of their small size, they may also exhibit some interesting textural properties that differ from those of an emulsion containing larger droplets. For example, they may behave like a viscous cream even at low oil droplet concentrations, a fact that has attracted attention in the development of low-fat products.

Biopolymeric nanoparticles : These consist of a matrix of biopolymers that may be linked through intermolecular attractive forces or through chemical covalent bonds to form solid particles. Nanoparticles may consist of a single biopolymer or may have a core-shell structure. Because of the versatility in terms of compounds that can be encapsulated and the degree to which these particles can be engineered and surface properties can be tailored, they have rapidly become the most promising nanoscale delivery systems in the pharmaceutical and cosmetics industries. Food-grade biopolymers such as proteins or polysaccharides can be used to produce nanometer-sized particles. Using aggregative (net attraction) or segregative (net repulsion) interactions, a single biopolymer separates into smaller nanoparticles. The nanoparticles can then be used to encapsulate functional ingredients and release them in response to distinct environmental triggers. One of the most common components of many biodegradable biopolymeric nanoparticle is polylactic acid. But its high cost and susceptibility to hydrolytic breakdown were believed to make it unsuitable for use in biomedical or agricultural applications or sparingly used in research. However, the use of this polymer as an ideal material for sutures was discovered in the 1970s, and a process was developed in the 1980s to produce the polymer via bacterial fermentation, greatly reducing costs and increasing production rates. Today, a wide variety of natural and synthetic polymers have been used to encapsulate and deliver compounds. Among these are chitosan, a natural antimicrobial and anti-oxidative polymer obtained from crustacean shells and the synthetic polymers L-, D-, and D,L-polylactic acid (PLA), polyglycolic acid (PGA), and polycaprolactic acid (PCL). Copolymers created using combinations of the monomers lactide, galactide, and caprolactone are also increasingly used.

Cubosomes : These are bicontinuous cubic phases which consist of two separate, continuous, but non-intersecting hydrophilic regions divided by a lipid layer that is contorted into a periodic minimal surface with zero average curvature. The continuous and periodic structure results in a very high viscosity of the bulk cubic phase. However, cubosomes prepared in dispersion maintain a nanometer structure identical to that of the bulk cubic phase but yield a much lower, water-like viscosity. Its tortuosity can be useful for slowing diffusion in controlled transport applications. Its isotropic optical property permits uses in many different products. Compared to liposomes, cubosomes have much higher bilayer area-to-particle volume ratios. The cubosome structure can be changed by modifying the environmental conditions, such as pH, ionic strength, or temperature, thus achieving controlled release of the carried compound. Cubosomes may be used in controlled release of solubilized bioactives in food matrices as a result of their nanoporous structure (approximately 5–10 nm); their ability to solubilize hydrophobic, hydrophilic, and amphiphilic molecules; and their biodegradability and digestibility by simple enzyme action. The cubic phase is strongly bioadhesive, so it may find applications in flavor release via its mucosal deposition and delivery of effective compounds. Yet, its tortuous structure may lead to applications where masking unpleasant taste or flavor is desirable, because of the slow effective diffusivity. The rate of release appears tunable through system optimization or ideal formulation of products for specific purposes.

Nanolaminates : Besides nanodispersions and nanocapsules, another nanoscale technique that is commercially viable for the food industry is nanolaminates. Consisting of two or more layers of material with nanometer dimensions, a nanolaminate is an extremely thin food-grade film (1–100 nm/ layer) that has physically bonded or chemically bonded dimensions. Because of its advantages in the preparation of edible films, a nanolaminate has a number of important food-industry applications. Edible films are present on a wide variety of foods: fruits, vegetables, meats, chocolate, candies, baked goods, and French fries. Such films protect foods from moisture, lipids, and gases, or they can improve the textural properties of foods and serve as carriers of colors, flavors, antioxidants, nutrients, and antimicrobials. Currently, edible nanolaminates are constructed from polysaccharides, proteins, and lipids. Although polysaccharide- and protein-based films are good barriers against oxygen and carbon dioxide, they are poor at protecting against moisture. On the other hand, lipid-based nanolaminates are good at protecting food from moisture, but they offer limited resistance to gases and have poor mechanical strength. Because neither polysaccharides and proteins, nor lipids provide all of the desired properties in an edible coating, researchers are trying to identify additives that can improve them, such as polyols. For now, coating foods with nanolaminates involves either dipping them into a series of solutions containing substances that would adsorb to a food's surface or spraying substances onto the food surface. While there are various methods that can cause adsorption, it is commonly a result of an electrostatic attraction between substances that have opposite charges. The degree of a substance's adsorption depends on the nature of the food's surface as well as the nature of the adsorbing substance. Different adsorbing substances can constitute different layers of a nanolaminate like polyelectrolytes (proteins and polysaccharides), charged lipids, and colloidal particles. Consequently, different nanolaminates could include various functional agents such as antimicrobials, anti-browning agents, antioxidants, enzymes, flavors, and colours.

Conclusion

Undoubtedly nanotechnology is having potential applications in all areas of food production and processing, however many of the methods are either too expensive or too impractical to implement on a commercial scale. There is an urgent need for nanoscale techniques that are most cost-effective in development of new functional materials, food formulations, food processing at microscale and nanoscale levels, product development, and storage. Although the products of nanotechnology intended for food consumption are likely to be classified as novel products, they require testing and clearance, and there are concerns, particularly in the area of food contact materials, that there could be inadvertent release and ingestion of nanoparticles of undetermined toxicity. Such concerns need to be addressed because the ultimate success of products based on nanotechnology will depend on consumer acceptance. Consideration should be given to the consequences of the use of nanotechnology to enhance the bioavailability of nutrients. This should consider the safety of the products, the consequences of enhanced or altered metabolism, and also the need for labelling, regulation and testing of health claims for such food supplements.

Suggested Readings

Chen, H, Weiss, J and Shahidi, F., Nanotechnology in nutraceuticals and functional foods. *Food Technol.*, 2006. **60 (3):** 30-36.

Flanagan, J. and Singh, H., Microemulsions: A potential delivery system for bioactives in food. *Crit. Rev. Food Sci. Nutr.*, 2006. **46:** 221-237.

Moraru, C.I., Panchapakesan, C.P., Huang, Q., Takhistove, P., Liu, S., and Kokini, J.L., Nanotechnology: A new frontier in food science. *Food Technol.*, 2003. **57(12):** 24-29.

McClements, D.J., Decker, E.A., and Weiss, J., inventors; University of Massachussetts, assignee. Novel procedure for creating nanolaminated edible films and coatings, U.S. patent application. 2005.

Nakajima, M., Development of Nanotechnology and Materials for Innovative Utilization of Biological Functions, Proceedings of the 34th United States and Japan Natural Resources (UJNR) Food and Agriculture Panel, Susono, Japan. 2005.

Scott, N. and Chen, H., Nanoscale science and engineering for agriculture and food systems, Report submitted to Cooperative State Research, Education and Extension Service (CSREES), U.S. Dept. of Agriculture. 2003.

Taylor, T.M., Davidson, P.M., Bruce, B.D., and Weiss, J., Liposomal nanocapsules in food science and agriculture. *Crit. Rev. Food Sci. Nutr.*, 2005. **45:** 1-19.

Worldnutra., 6th International Conference and Exhibition on Nutraceuticals and Functional Foods. Anaheim, Calif., October 2005.

□□□

Chapter 28

Role of Probiotic Bacteria Against Cancer

MANOJ KUMAR,PRAMOD KUMAR AND P.K AGGARWAL
Dairy Microbiology Division, National Dairy Research Institute, Karnal

Intoduction

Several lactic acid bacteria may help prevent initiation of colon cancer. It has also been demonstrated that lactic acid bacteria slow the growth of experimental cancers, although the results are not long-term. It appears that lactic acid bacteria can reduce the levels of colon enzymes that convert procarcinogens to carcinogens. Specifically, lactic acid bacteria can reduce levels of the enzymes β-glucuronidase, nitroreductase, and azoreductase (Fernandes *et al.*,1990). Lactic acid bacteria may also be involved in the direct reduction of procarcinogens, for example, by taking up nitrites and by reducing the levels of secondary bile salts (Fernandes *et al.*,1990). In most reports, these effects only occur during the period of time that the bacteria are consumed (Marteau *et al.*, 1990). Changes in enzyme activity in humans have been observed with *L. acidophilus* and *B. bifidum* (Marteau *et al.*,1990), and LGG (Goldin *et al.*,19984). Animal studies show fewer tumors in those exposed to a carcinogen, in the presence of LGG, compared to the animals exposed to the carcinogen without the benefit of LGG

(Goldin *et al.*, 1996). In humans, epidemiological reports show that populations eating fermented dairy products have a decreased risk of colon cancer (Kampman *et al.*, 1994). However, there is not yet a clear relationship between lactic acid bacteria intake and cancer prevention. An antitumour effect has been reported by oral intake of LAB in *in vitro* studies but, since colon carcinogenesis is a multistage process the mechanisms, if validated, remain to be elucidated (Hirayama and Rafter, 1999). A preventive effect on malignant development could be mediated by production of antimutagens and LAB binding of mutagens, and this has been reviewed (Lankaputhra and Shah, 1998) (Table 1). In general, live cells of probiotic bacteria showed higher anti-mutagenic activity, and this was permanent, in contrast to killed cells. In this *in vitro* study, butyric acid, and to a lesser extent, acetic acid inhibited mutagens. Strains of *B. lactis* were shown to express antimutagenic properties, probably linked to cell wall constituents. The antimutagenic effect was active also after acid and bile treatment, mimicking the GI transport, and interestingly, enhanced in the presence of whole milk (Lo *et al.*, 2004). One mechanism for this effect can be binding of mutagens, and heterocyclic aromatic amines were shown to be bound to the cell wall of certain bacteria, such as *B. longum* and other LAB, and thereby be detoxified (Orrhage *et al.*, 1994; Knasmüller *et al.*, 2001). In *L. plantarum* KLAB21, however, the antimutagenic effect was mediated by three glycoproteins which are secreted extracellularly (Rhee and Park, 2001).

Antigenotoxic activity against 4-nitroquinoline-1- oxide was shown in in vitro tests for strains of several Lactobacillus species whereas only one L. acidophilus strain inhibited MNNG (N-methyl N-nitro N-nitrosoguanidine). All strains also showed antimutagenic properties and were viable after the tests (Caldini *et al.*, 2005). Intake of LAB in fermented milk or other products influence gut flora enzymes, like β-glucuronidase, β-glucosidase and nitroreductase (Wollowski *et al.*, 2001). LAB and Bifidobacteria have lower activities of these enzymes than Enterobacteriaceae, Clostridia and Bacteroides. A daily intake of *L. acidophilus* and *B. bifidum* for 3 weeks decreased the activity of nitroreductase but increased the activity of β-glucosidase (Marteau *et al.*, 1990). This could be an advantage since β-glucosidase may release flavonoids which have antimutagenic, antioxidative and immune stimulatory effects (Stoner and Mukhtar, 1995). Daily intake of fermented vegetables for 3 weeks decreased the faecal levels of β-glucuronidase and nitroreductase. Most animal and human studies do indicate that feeding certain LAB decrease faecal enzyme levels which may be involved in formation of carcinogens. Also, beneficial effects can be attributed to immunpotentiating effects by LAB strains. One specific effect was shown by heat-killed L. plantarum L-137 which restored the inhibited IL-12 production in DBA/2 mice with tumors (Murosaki et al., 2000). Conjugated linoleic acid has been highlighted recently because of its effect to reduce carcinogenesis, atherosclerosis and body fats (Chin et al., 1992; Nicolosi et al., 1997). One LAB strain, L. plantarum JCM 1551 was shown to efficiently produce conjugated linoleic acid from castor oil in the presence of lipase which is an interesting "side effect" of anti-tumor activities of LAB (Ando et al., 2004). Mycotoxins are well known contaminants of grains and other food raw materials, and e.g. ochratoxin A may accumulate in domestic animals and remain in food products of animal origin (Petzinger and Weidenbach, 2002). Ochratoxin A is carcinogenic, genotoxic, immunosuppressive and nephrotoxic (Petzinger and Ziegler, 2000). In a recent study, strains of Lactobacillus and *Lactococcus* were shown to be resistant to Ochratoxin A (5–10 μg/disc), and just one strain of *L. helveticus* was sensitive (0.1 μg/disc) (Piotrowska and Zakowska, 2005). Furthermore, all strains tested reduced the amount of Ochratoxin A in the growth medium as measured after 120 h. *L. acidophilus* CH-5, *L. rhamnosus* GG and a couple of other strains reduced the initial level of Ochratoxin A by more than 50% (Piotrowska and Zakowska, 2005).Intake of such LAB may potentially reduce carcinogenic/ genotoxic effects of consumption of food contaminated by Ochratoxin A. AFB1 a mycotoxin produced by the common fungi *Aspergillus flavus* and *Aspergillus parasiticus,* is an established human hepatocarcinogen (Vamio *et al.,* 1993; Eaton *et al,* 1994). Various strains of lactic acid bacteria isolated from either dairy products or healthy human microbiota showed that two *Lactobacillus rhamnosus* strains, LGG and LC705 (both possess probiotic properties) were found to be most efficient strains in binding a range of mycotoxins, including aflatoxin B1 (El-Nezami *et al.,* 2006). A single viable bacterium has been found to bind > 10^7 aflatoxin B1 molecules (Haskard *et al.,* 2000).

Protective Roles of Probiotics on Cancer.

Probiotics are defined as live microbial food ingredients that have a beneficial effect on human health (Salminen *et al.,* 1998).The concept of probiotics evolved at the turn of the 20th century from a hypothesis first proposed by Nobel Prize winning Russian scientist Elie Metchnikoff, who suggested that the long, healthy life of Bulgarian peasants resulted from their consumption of fermented milk products (Bibel, 1998). He believed that when consumed, the fermenting bacillus (*Lactobacillus*) positively influenced the microflora of the colon, decreasing toxic microbial activities.

Some probiotic bacterial strains successfully bind mycotoxins; *Lactobacillus rhamnosus* binds away AFB1 *in vivo* and thus reducing bio-absorption of the toxin from the gut while *L. acidophillus* and *Bifidobacterium longum* neutralize AFB1 and AFM1 by similar binding mechanism (Haskard *et al.*, 2001). Epidemiological studies in Africa and Asia, where high levels of exposure occur, link aflatoxin B1 exposure and hepatocellular carcinomas (Qian *et al.*, 1994, Bosch *et al.* 1988, Bulatao-Jayme *et al.*,1982), and the risk is synergistic with hepatitis viral infections (Qian *et al.*, 1994). The mechanism of action for aflatoxin B1 mutagenicity begins with metabolic activation by CYP3A4, CYP3A5 and/or CYP1A2 (Garner *et al.*,1972 ,Reynolds *et al.*, 1987) that forms an exo-8,9- epoxide and subsequent adduct formation and DNA damage (Essigmann *et al.*, 1977, Iyer *et al.,1994*). This damage has been shown in vitro to cause guanine nucleotide substitutions (Lilleberg *et al.*, 1992) specifically to codon 249 of the p53 gene (Aguilar *et al.,1993*).

Colorectal cancer represents a major public health problem accounting for over 1 million cases and about half a million deaths worldwide (Chau *et al. ,2006)*. Survival from colon cancer at 5 years has been found to vary demographically and estimated to be 65% in North America, 54% in Western Europe, 34% in Eastern Europe, and 30% in India (Parkin *et al.*, 2005). Although chemotherapy and radiotherapy have been applied as the surgical adjuvant treatments of colon cancer, they vary in success rates for local recurrence, disease-free survival, and overall survival (Martenson *et al.*, 2004*)*. In addition, these treatments may present some side effects such as an increased risk for infections, hair loss, fatigue, lip sores, nausea, vomiting, diarrhea, and bloody stools. Diet interventions and natural bioactive supplements have now been extensively studied to reduce the risks of colon cancer, as a cause of prevention instead of cure. Postulated mechanisms include reduction in the activity of several cancer causing bacteria, desmutagenic and anticarcinogenic properties (Collins *et al.*,1999, Chau *et al.*, 2006).

There are fifty-seven cytochrome P450 encoded in the human genome, mainly catalyzing the metabolism of steroids, bile acids, eicosanoids, drugs and xenobiotic chemicals (Guengerich *et al.,2003)* However, some of the P450s are also active carcinogens(Table 2). Past epidemiological researches have shown increased risk of colon cancer in individuals with high P4501A2 activity. The metabolic activation of food-borne heterocyclic amines to colon carcinogens in humans is hypothesized to occur via N-oxidation followed by O-acetylation to form the N-acetoxy arylamine that binds to DNA to give carcinogen-DNA adducts. These steps are catalyzed by hepatic cytochrome P4501A2 and acetyltransferase-2 (NAT-2), respectively (Lang *et al.,1994)*. It has been postulated that probiotics such as *Bifidobacterium* could lower the risks of colon cancer, by producing metabolites that could affect the mixed-function of P450s and subsequently affect the conversion of azoxymethane from proximate to ultimate carcinogen (Campbell *et al.,1976)*. These had led to the suggestion that probiotic could suppress colon cancer.

In-depth investigations have also showed that cultured milk possess desmutagenicity and this activity increase with increasing numbers of viable cells, indicating that probiotic play an important role in the inhibition of mutagenicity (Usman *et al.*,1998). Thyagaraja and Hosono (Thyagaraja *et al.*, 1993) found that probiotic isolated from "idly", a traditional cereal pulse product of India could exert desmutagenicity on various spice mutagens, heterocyclic amines and aflatoxins. Subsequent studies on the desmutagenicity properties of probiotic suggested that the desmutagenic substances may reside in the cellular envelope of the bacterial cell wall (Singh *et al.*, 1997). Cell wall preparation of *B. infantis* was found to inhibit tumor activity in mouse peritoneal cells in-vitro (Sekine *et al.*,1995), while cell wall preparation of heat-killed *L. casei* (LC9018) was found to induce immunity against tumor induction in a randomized, controlled and comparative study involving 223 patients with stage III cervical cancer. The antitumor effects were found to be due to the activation of macrophage by LC9018 (Okawa *et al., 1993)*. Also, mutagens were suggested to be bound to the cell wall of probiotics.

Evidence that probiotics can influence carcinogenesis is derived from a variety of biological effects :

- Effects on bacterial enzyme activities.
- Antigenotoxic effects in vitro and in vivo.
- Effects on pre-cancerous lesions in laboratory animals.
- Effects on tumour incidence in laboratory animals

Potential mechanisms of anti-carcinogenicity and anti-genotoxicity as exhibited by many probiotics are described below:-

1. Binding/ adsorption of carcinogens - Several carcinogens such as heterocyclic amines and AFB1 are reportedly adsorbed or bound in *vitro* by LAB and other intestinal bacteria. A concomitant decrease in mutagenicity is often reported. (Morotomi and Mutai, 1986; Orrhage *et al* 1994; Zhang *et al* 1990; Zhang and Ohta 1991; Bolognani *et al.*, 1997)

2. Effect on bacterial enzymes, metabolite production - Oral administration of some LAB/fermented milks to animals and humans leads to decrease in certain metabolites and enzymes purported to be involved in synthesis or activation of carcinogens, genotoxins and tumour promoters such as â-glucuronidase, â-glucosidase, nitrate reductase and ammonia. Such changes in enzyme activities concomitantly suppressed tumours. This would appear to be due to the low specific activity of these enzymes in LAB (Saito *et al.*, 1992). Such changes in enzyme activity or metabolite concentration have been suggested to be responsible for the decreased level of preneoplastic lesions or tumours seen in carcinogen-treated rats given pre and probiotics (Reddy and Rivenson 1993; Rowland *et al.*, 1998).

3. Stimulation of host enzymes involved in carcinogen inactivation - There are isolated reports citing that administration of LAB results in increased activity of antioxidative enzymes or process that protect cells against carcinogen induced damage. These include glutathione-s-transferase, glutathione, glutathione reductase, glutathione peroxidase which consequently reduce the apoptotic rate and the necrobiotic changes in liver. Such mechanisms of protection could be effective against wide range of dietary carcinogens possibly influencing several cancer sites. Lowering of hepatic GSH levels and enhanced lipid peroxidation was studied in animals dosed with AFB1and other carcinogens. It may be useful to employ hepatic lipid peroxidation as a biochemical marker during AFB1 initiated hepato-carcinogenesis. (Aguilar- Delfin *et al.*, 1996; Corrales *et al.*, 1992; Gasso *et al.*, 1996). Many food-borne carcinogens such as heterocyclic amines and polycyclic aromatic hydrocarbons are known to be conjugated to glutathione, which appears to result in inactivation. The enzyme involved, glutathione transferase (GST), is found in

the liver and in other tissues including the gut. In a study of the effect of *B. longum* and lactulose on AOM (azoxymethane)-induced ACF (aberrant crypt foci) in the colon showed that the activity of GSH in the colonic mucosa was inversely related to the ACF numbers (Challa *et al.*, 1997).Such a mechanism of protection would be effective against a wide range of dietary carcinogens that the various LAB can inhibit genotoxicity of dietary carcinogens. The degree of inhibition was strongly species dependent. For example Pool-Zobel *et al* (1993) demonstrated that *Lactobacillus casei* and *L. lactis* inhibited the mutagenic activity of nitrosated beef by over 85% whereas *Lactobacillus confusus* and *Lactobacillus sake* had no effect. Pre-treatment of mice with *L. casei* 1 h before the administration of genotoxins can lead to a moderate reduction ($P < 0.05$) in the genotoxicity of PCB (procarbazine) and URE (urethane) (Abraham S.K and Paul Jaishree, 2001). Using the technique of single cell microgel electrophoresis (Comet assay), Pool-Zobel *et al.*, (1996) investigated the ability of range of species of LAB to inhibit DNA damage in the colon mucosa of rats treated with the carcinogens MNNG (N-methyl-N-nitro-N-nitrosoguanidine) or 1,2-dimethyl-hydrazine (DMH). All the strains of lactobacilli and bifidobacteria tested - *L. acidophilus* (isolated from a yoghurt sample), *Lactobacillus gasseri, L. confusus, B. breve* and *Bifidobacterium longum,* prevented MNNG-induced DNA damage when given at a dose of 10^{10} cells/kg body weight eight hours before feeding the carcinogen. In most cases the DNA damage was reduced to a level similar to that in untreated rats. *Streptococcus thermophilus* was not as effective as the other LAB strains. The Comet assay has also been used to evaluate the effect of a prebiotic, lactulose, on DNA damage in the colonic mucosa. Rats that were fed a diet containing 3% lactulose and given DMH, exhibited less DNA damage in colon cells than similarly treated animals fed a sucrose diet. In the latter animals, the percentage of cells with severe DNA damage comprised 33% of the total compared with only 12.6% in the lactulose-fed rats (Rowland *et al.*, 1996).

4. Modulation of immune/ inflammatory response - Suppression of colon tumour growth by yoghurt in DMH (dimethyl hydrazine) treated mice was associated with suppression of the inflammatory immune response (Wollowski, 2001). Many case studies have shown that consumption of fermented milk resulted in reduction of risk of both breast, Liver and colon cancer.

Table 1: Factors which may exert beneficial effects on malignancies.

Binding of mutagens
Production of antimutagenic substances
Inhibition of procarcinogenic enzymes like nitroreductase and β-glucuronidase
Increased production of β-glucosidase which release flavonoids
Deconjugation of bile salts
Binding of heterocyclic amines
Production of antioxidants and scavenging of ROS
Production of conjugated linoleic acid from castor oil
Immunopotentiation

Probiotics and Apoptosis

There are s reports that probiotic, *Lactobacillus rhamnosus* GG (LGG), prevents cytokine induced apoptosis in intestinal epithelial cell models. Culture of LGG with either mouse or human colon cells activates the anti-apoptotic Akt/protein kinase B. This model probiotic also inhibits activation of the pro-apoptotic p38/mitogen-activated protein kinase by tumor necrosis factor, interleukin-1á or ò-interferon Furthermore, products recovered from LGG culture broth supernatant show concentration-dependent activation of Akt and inhibition of cytokine-induced apoptosis. These observations suggest a novel mechanism of communication between probiotic microorganisms and epithelia that increases survival of intestinal cells normally found in an environment of pro-apoptoticcytokines (Fang *et al.,*2002)

Role of Bifidobacterium longum in the induction of apoptotic deletion in the human enterocyte-like Caco-2 cell line Bifidobacterium longum is a probiotic, known for its beneficial effects to the human gut and even for its immunomodulatory and antitumor activities. Recently, many studies have stressed out the intimate relation between probiotic bacteria and the GIT mucosa and their influence on human cellular homeostasis . Bacteria were incubated with Caco-2 cells to investigate apoptotic deletion. Cocultures of Caco-2 cells with B. longum resulted positive in DNA fragmentation test, only when adherent strains were used (B12 and B18). These results indicate that the interaction with adherent B. longum can induce apoptotic deletion of Caco-2 cells, suggesting a role in cellular homeostasis of the gastrointestinal tract and in restoring the ecology of damaged colon tissues (Nissen *et al.,* 2006).

Conclusion

In conclusion, cancer, which in a high proportion of the population is due to somatic mutations occurring during the lifetime of an individual, could be retarded or prevented by preventing these mutations. LAB and prebiotics that enhance LAB have been shown to deactivate genotoxic carcinogens. In model systems in vitro and vivo they have been shown to prevent mutations. DNA damage has been prevented and chemopreventive systems may be stimulated in vivo in colon tissues. From a mechanistic point of view, LAB offer potential as chemoprotective agents and thus further research is clearly needed to quantify the beneficial effects for prevention of human cancer.

Table 2: Examples of biomarkers of exposure and susceptibility

Carcinogen	Target organ	Biomarker Genetic susceptibility		Phenotypic biomarker	
		Activating enzyme with genetic polymorphisms	Detoxifying enzyme with genetic polymorphisms	Adduct	p53 Mutation
Aflatoxin B1	Liver	CYP3A4,[1] CYP3A5, CYP1A2	Glutathione transferases, glucuronyl transferases	Adducts in DNA, albumin, haemo-globin and urine; urinary metabolites	Codon 249 AGG to AGT
Polycyclic aromatic hydrocarbons	Lung, oropha-ryngeal, breast, gastrointestinal and genitourinary tracts	CYP1A1, CYP1B1, epoxide hydroxylase	Glutathione transferases, glucuronyl transferases	Adducts in DNA, albumin, hemoglobin; urinary metabolites	Increased G-to-T mutations at codons 157, 248, 273
N- nitrosa-mines	Gastrointestinal, lung, oropharyngeal	CYP2A6, CYP2E1	-	O6-methyl-guanine, 7-meth-ylguanine	-
Hetero-cyclic amines	Colon, breast	CYP1A2, NAT1[2]	NAT2	Adducts in albumin, and DNA adducts	-

References

Abraham S.K and Paul, J. 2001. Protective effects of lactic acid bacteria against genotoxicity in mice. *Current Science*. Vol. **80:** NO. 10.

Aguilar, F., Hussain, S. P. & Cerutti, P. (1993) Aflatoxin B1 induces the transversion of G!T in codon 249 of the p53 tumor suppressor gene in human hepatocytes. *Proc. Natl. Acad. Sci.* U.S.A. **90:** 8586–8590.

Aguilar-Delfin, I., Lopez-Barrera, F. and Hernandez-Munoz, R. (1996) Selective enhancement of lipid peroxidation in plasma membrane in two experimental models of liver regeneration: partial hepatectomy and acute CCl4 administration. *Hepatology* **24:** 657- 662

Ando, A., Ogawa, J., Kishino, S. and Shimizu, S. (2004). Conjugated linoleic acid production from castor oil by *Lactobacillus plantarum* JCM 1551. Enzyme Microb. *Technol.* **35:** 40–45.

Aso Y, Akazan H. Prophylactic effect of a Lactobacillus casei preparation on the recurrence of superficial bladder cancer. *Urol Int* 1992; **49:** 125-9.

Bibel, D.J. 1988. Elie Metchnikoff's bacillus of long life. *ASM News* **54:** 661–665.

Bolognani, F., Rumney, C.J., and Rowland, I.R. 1997. Influence of carcinogen binding by lactic acid producing bacteria on tissue distribution and *in vivo* mutagenicity of dietary carcinogens. Food Chem. *Toxicol.* **35:** 535-545.

Bosch, F. X. & Munoz, N. (1988) Prospects for epidemiological studies on hepatocellular cancer as a model for assessing viral and chemical interactions. *IARC Sci. Publ.* **89:** 427–438.

Bulatao-Jayme, J., Almero, E. M., Castro, M. C., Jardeleza, M. T. & Salamat, L. A. (1982) A case-control dietary study of primary liver cancer risk from aflatoxin exposure. *Int. J. Epidemiol.* **11:** 112–119.

Caldini, G., Trotta, F., Villarini, M., Moretti, M., Pasquini, R., Scassellati-Sforzolini,G., Cenci, G. (2005). Screening of potential lactonacilli antigenotoxicity by microbial and mammalian cell-based tests. *Int. J. Food Microbiol.* **102:** 37–47.

Campbell, T.C.; Hayes, J.R. The effect of quantity and quality of dietary protein on drug metabolism. *Federation Proceedings* 1976, **35:** 2470–2474.

Challa, A., Rao, D.R., Chawan, C.B., and Shackelford, L. 1997. Bifidobacterium longum and lactulose suppress azoxymethane induced aberrant crypt foci in rats. *Carcinogenesis.* **18:** 517-521.

Chau, I.; Cunningham, D. Adjuvant therapy in colon cancer—what, when and how? *Annals of Oncology* 2006, **17:** 1347-1359.

Chin, S.E., Liu, W., Storkson, J.M., Ha, Y.L. and Pariza, M.W. (1992). Dietary sources of conjugated dienoic isomers of linoleic acid, a newly recognized class of anticarcinogens. *J. Food Comp. Anal.* **5:** 185–197.

Collins, M.D.; Gibson, G.R. Probiotics, prebiotics and synbiotics: Approaches for modulating the microbial ecology of the gut. *American Journal of Clinical Nutrition* 1999, **69:** 1052S-1057S.

Corrales, F., Gimenez, A., Alvarez, L., Caballeria, J., Pajares, M. A., Andreu, H.,Pares, A.,Mato, J. M. and Rodes, J. (1992) S-adenoslymethionine treatment prevents carbon tetrachloride-induced S-adenosylmethionine synthetase inactivation and attenuates liver injury. *Hepatology* **16:** 1022-1027

El-Nezami, H.S., Nektaria N Polychronaki, Jing Ma, Huilian Zhu, Wenhua Ling, Eeva K Salminen, Risto O Juvonen, Seppo J Salminen, Tuija Poussa and Hannu M Qian, G. S., Ross, R. K., Yu, M. C., Yuan, J. M., Gao, Y. T., Henderson, B. E., Wogan, G. N. & Groopman, J. D. (1994) A follow-up study of urinary markers of aflatoxin exposure and liver cancer risk in Shanghai, People's Republic of China. *Cancer Epidemiol. Biomarkers Prev.* **3:** 3–10

Essigmann, J. M., Croy, R. G., Nadzan, A. M., Busby, W. F., Jr., Reinhold, V. N., Buchi, G. & Wogan, G. N. (1977) Structural identification of the major DNA adduct formed by aflatoxin B1 in vitro. *Proc. Natl. Acad. Sci. U.S.A.* **74:** 1870- 1874.

Fernandes C.F., K.M. Shahani. 1990. Anticarcinogenic and immunological properties of dietary lactobacilli. *Journal of Food Protection* **53:** 704 - 710.

Garner, R. C., Miller, E. C. & Miller, J. A. (1972) Liver microsomal metabolism of aflatoxin B1 to a reactive derivative toxic to Salmonella typhimurium TA 1530. *Cancer Res.* **32:** 2058–2066.

Gasso, M., Rubio, M., Varela, G., Cabre, M., Caballeria, J., Alonso, E., Deulofem, R., Camps, J., Gimenez, A., Pajares, M., Pares, A., Mato, J. M. and Rodes, J. (1996) Effects of S-adenosylmethionine on lipid peroxi-dation and liver fibrogenesis in carbon tetrachloride-induced cirrhosis. *J. Hepatology* **25:** 200-205

Goldin B.1996.The metabolic activity of the intestinal microflora and its role in colon cancer. *Nutrition Today* **31**(6)(suppl):24S - 27S.

Goldin B.R., S.L. Gorbach. 1984. The effect of milk and lactobacillus feeding on human intestinal bacterial enzyme activity. *Am. J. Clin. Nutr.* **39:** 756 - 61.

Guengerich, F.P. Cytochromes P450, drugs, and diseases. *Molecular Interventions.* 2003, **3:** 194-204.

Haskard, C., C. Binnion, and J. Ahokas. 2000. Factors affecting the sequestration of aflatoxin by Lactobacillus rhamnosus strain GG. Chem. *Biol. Interact.* **128:** 39–49.

Hirayama, K. and Rafter, J. (1999). The role of lactic acid bacteria in colon cancer prevention: mechanistic considerations. *Ant. v. Loewenhoek* **76:** 391–394.

Iyer, R. S., Coles, B. F., Raney, K. D., Thier, R., Guengerich, F. P. & Harris, T. M. (1994) DNA adduction by the potent carcinogen aflatoxin B-1- mechanistic studies. *J. Am. Chem. Soc.* **116:** 1603–1609.

Kampman E., R.A. Goldbohm, P.A. van den Brandt, et al. 1994. Fermented dairy products, calcium and colorectal cancer in the Netherlands cohort study. *Cancer Research* **54:** 3186 – 3190.

Lang, N.P.; Butler, M.A.; Massengill, J.; Lawson, M.; Stotts, R.C.; Maurer-Jensen, M.; Kadlubar, F.F. Rapid metabolic phenotypes for acetyltransferase and cytochrome P4501A2 and putative exposure to food-borne heterocyclic amines increase the risk for colorectal cancer or polyps. *Cancer Epidemiology Biomarkers and Preventions* 1994, **3:** 675–682.

Lankaputhra, W.E.V. and Shah, N.P. (1998). Antimutagenic properties of probiotic bacteria and of organic acids. *Mut. Res.* **397:** 169–182..

Lilleberg, S. L., Cabonce, M. A., Raju, N. R., Wagner, L. M. & Kier, L. D. (1992) Alterations in the p53 tumor suppressor gene in rat liver tumors induced by aflatoxin B1. Prog. *Clin. Biol. Res.* **376:** 203–222.

Lo, P.-R., Yu, R.-C., Chou, C.-C. and Huang, E.-C. (2004). Determinations of the antimutagenic activities of several probiotic bifidobacteria under acidic and bile conditions against benzo[*a*]pyrene by a modified Ames test. *Int. J. Food Microbiol.* **93:** 249–257.

Marteau, P., Rochart, P., Flourie, B., Pellier, P., Santos, L., Desjeux, J.F., and Rambaud, J.C. (1990). Effect of chronic ingestion of a fermented dairy product containing *Lactobacillus acidophilus* and *Bifidobacterium bifidum* on metabolic activities of the colonic flora in humans. *Am. J. Clin. Nutr.* **52:** 685–688.

Martenson, J.A.; Willett, C.G.; Sargent, D.J.; Mailliard, J.A.; Donohue, J.H., Gunderson, L.L.; Thomas, C.R.; Fisher, B.; Benson, A.B.; Myerson, R.; Goldberg, R.M. Phase III study of adjuvant chemotherapy and radiation therapy compared with chemotherapy alone in the surgical adjuvant treatment of colon cancer: Results of intergroup protocol 0130. *Journal of Clinical Oncology* 2004, **22:** 3277–3283.

Morotomi, M., and Mutai, M. 1986. In vitro binding of potent mutagenic pyrolyzates to intestinal bacteria. *J.Natl.Cancer. Inst.* **77:** 195-201.

Murosaki, S., Muroyama, K., Yamamoto, Y. and Yoshikai, Y. (2000). Antitumor effect of heat-killed *Lactobacillus plantarum* L-137 through restoration of impaired.

Okawa, T.; Niibe, H.; Arai, T.; Sekiba, K.; Noda, K.; Takeuchi, S.; Hashimoto, S.; Ogawa, N. Effect of LC9018 combined with radiation therapy on carcinoma of the uterine cervix. *Cancer* 1993, **72:** 1949–1954.

Orrhage, K., Sillerström, E., Gustafsson, J.-Å., Nord, C.-E. and Rafter, J. (1994). Binding of mutagenic heterocyclic amines by intestinal and lactic acid bacteria. *Mut. Res.* **311:** 239–248.

Parkin, D.M.; Bray, F.; Ferlay, J.; Pisani, P. Global cancer statistics 2002. *CA Cancer Journal for Clinicians* 2005, **55:** 74–108.

Petzinger, E. and Weidenbach, A. (2002). Mycotoxins in the food chain: the role of ochratoxins. *Livestock Prod. Sci.* **76:** 245–250.

Petzinger, E. and Ziegler, K. (2000).Ochratoxin A from a toxicological perspective. *J. Vet. Pharmacol. Therap.* **23:** 91–98.

Piotrowska, M. and Zakowska, Z. (2005). The elimination of Ochratoxin A by Lactic Acid Bacteria strains. *Polish J. Microbiol.* **54:** 279–286.

Pool-Zobel, B.L., Munzner, R., and Holzaapfel, H. 1993. Antigenotoxic properties of lactic acid bacteria in the S. typhimurium Mutagenicity assay. *Nutr Cancer.* **20:** 261-270

Pool-Zobel, B.L., Neudecker, C., Domizlaff, I., Ji, S., Schillinger, U., Rumney, C., Moretti, M., Vilarini, I., Scasellati-Sforzolini, R., and Rowland, I.R. 1996. Lactobacillus and Bifidobacterium mediated antigenotoxicity in the colon of rats. *Nutr Cancer.* **26:** 365-380.

Reddy, B.S., and Rivenson, A. 1993. Inhibitory effect of Bifidobacterium longum on colon, mammary, and liver carcinogenesis induced by 2-amino-3-methylimidazo[4,5,-f]quinoline, a food mutagen. *Cancer Research.* **53:** 3914-3918.

Reynolds, S. H., Stowers, S. J., Patterson, R. M., Maronpot, R. R., Aaronson, S. A. & Anderson, M. W. (1987) Activated oncogenes in B6C3F1 mouse liver tumors: implications for risk assessment. *Science* **237:** 1309–1316.

Rhee, C.-H. and Park, H.-D. (2001). Three glycoproteins with antimutagenic activity identified in *Lactobacillus plantarum* KLAB21. *Appl. Environm. Microbiol.* **67:** 3445–3449.

Rowland, I.R., Rumney, C.J., Coutts, J.T., and Lievense, L.C. 1998. Effect of Bifidobacterium longum and inulin on gut bacterial metabolism and carcinogen-induced aberrant crypt foci in rats. *Carcinogenesis.* **19:** 281-285.

Saito, Y,Takano and Rowland, I.R. 1992. Effects of soybean oligosaccharides on the human gut microflora in in vitro culture. Microb. Ecol. Health Dis. **5:** 105-110.

Salminen, S. and von Wright, A. 1998. Current probiotics—safety assured? *Microbial Ecol. Health Dis.* **10:** 68–77.

Sekine, K.; Ohta, J.; Onishi, M.; Tatsuki, T.; Shimokawa, Y.; Toida, T.; Kawashima, T.; Hashimoto, Y. Analysis of antitumor properties of effector cells stimulated with a cell wall preparation (WPG) of *Bifidobacterium infantis. Biological and Pharmaceutical Bulletin.* 1995, **18:** 148–153.

Singh, J.; Rivenson, A; Tomita, M.; Shimamura, S.; Ishibashi, N.; Reddy, B.S. *Bifidobacterium longum*, a lactic acid-producing intestinal bacterium inhibits colon cancer and modulates the intermediate biomarkers of colon carcinogenesis. *Carcinogenesis*, 1997, **18:** 833–841.

Stoner, G.D. and Mukhtar, H. (1995). Polyphenols as cancer chemopreventive agents. *J. Cell. Biochem.* **22:** 169–180.

Thyagaraja, N.; Hosono, A. Antimutagenicity of lactic acid bacteria from "Idly" against foodrelated mutagens. *Journal of Food Protection* 1993, **56:** 1061–1066.

Usman; Hosono, A. Desmutagenicity of milk cultured with *Lactobacillus acidophilus* strains against mutagenic heated tauco. *Food Chemistry and Toxicology* 1998, **36:** 805-810.

Vamio H, Heseltine E, Wilbourn J. Report of an IARC Working Group Meeting on Some Naturally Occurring Substances. *Int J Cancer* 1993; **53:** 535–537.

Wollowski, I., Rechkemmer, G. and Pool-Zobel, B.L. (2001). Protective role of probiotics and prebiotics in colon cancer. *Am. J. Clin. Nutr.* **73**(suppl), 451S-455S.

Wollowski, I., Rechkemmer, G. and Pool-Zobel, B.L. 2001.Protective role of probiotics and probiotics in colon cancer. *Am. J. Clin. Nutr.*, **73:** 451-455.

Zhang, X.B., and Ohta, Y. 1991. *In vitro* binding of mutagenic pyrolyzates to lactic acid bacterial cells in human gastric juice. *J. Dairy. Sci.* **74:** 752-757

Zhang, X.B., Ohta, Y., and Hosono, A. 1990. Antimutagenicity and binding of lactic acid bacteria from a Chinese cheese to mutagenic pyrolyzates. J. Dairy. Sci. 73: 2702-2710.

□□□

Chapter 29

High Hydrostatic Pressure (HHP) Technology in Dairy Processing

REKHA CHAWLA AND A.K. SINGH
Dairy Technology Division, National Dairy Research Institute, Karnal-132001

Introduction

In the era of health consciousness, consumer of the date is getting much and much aware regarding his health and so with the food components. Consumer demands foods, which are natural, nutritionally good; free from chemical preservatives, microbiologically safe with extended shelf-life. Today many processed foods we buy from a supermarket are superheated to kill bacteria, e.g. Juice, milk, and canned products. Processing at high temperature though makes the food risk free but at the same time lowers the nutritional quality of foods because most of the nutrients present in foods are heat labile in nature. To overcome these problems, now days, several non- thermal food processing technologies are coming into picture and high hydrostatic pressure technology (HHP) is one of them. It is also known as High pressure processing (HPP) and Ultra high pressure processing that is UHP, & considered as "Cold processing" technique since temperature employed in most of the processes is of ambient range (Farr, 1990). Also, energetic optimization and heat recovery in the food industry has been a focus in the past decades for conventional processes, but their replacement by novel techniques for food preservation or modification may still provide a potential to reduce energy consumption and costs of operation, as well as to improve sustainability of production. This newer concept of processing is of immense importance and is gaining acceptance, owing to its ability to inactivate pathogenic micro-organisms with minimal heat treatment, along with almost complete retention of nutritional and sensory characteristics of fresh food without scarifying its shelf life. Introducing newer techniques of cold pasteurization or related techniques of non-thermal processing prior to drying can provide good opportunities to continue improving energy efficiency of food processing. High-pressure technology is increasingly being used in the food industry particularly to produce high-value-added products.

Hite (1899) was the first researcher who reported the effect of HHP on food borne micro-organisms by subjecting milk to pressure of 650 MPa and obtained a significant reduction in the viable number of microbes. Though the technology was evolved in Japan but now a days being promising, number of countries are adopting this technology, and till date variety of food products like jams, fruit juice, meat have been processed. Nevertheless, interest in HHP derives from its ability to deliver foods with fresh-like taste without added preservatives. In high pressure food is subjected to pressures as high as 6000 times the atmospheric pressure usually within range of 300-700 MPa & is effective in inactivating most of the vegetative bacteria at pressure above 400 MPa. The most attractive feature is the uniform processing ability as the pressure is applied uniformly throughout the food material, independent of its mass and time. HHP can be used to process both liquid and solid (water-containing) foods and adds advantages such as bacterial free food with extended shelf life, additive free food, manipulated texture with other enhanced desirable food attributes (Makhal *et al*, 2003).

Principle

The operating principles behind this technology are as follows

Le Chatelier's Principle : Any phenomenon in equilibrium chemical reaction, phase transition, change in molecular configuration is accompanied by decrease in volume can be enhanced by pressure.

Isostatic Principle: The transmittance of pressure is uniform & instantaneous (independent of size & geometry of food).

In a high pressure processing, the pressure vessel is filled with a food product and pressurized for a desired time, following which it is depressurized. A simplified flow- sheet as below-

Pack Food in Sterilized Container
↓
Load in a Pressure Chamber
↓
Fill Chamber with Water
↓
Pressurize Chamber
↓
Hold under Pressure
↓
De-Pressurize the Chamber
↓
Remove Processed Food

The time required to pressurize the vessel is influenced by the compressibility of the pressure medium and the nature of the food material. Generally in almost all the cases, water is used as the pressure - transmitting medium. Presence of air in the food increases the pressurization time; since air is considerably more compressible than water. Food is then kept under high pressure for the required process time and upon completion of the exposure in pressurized vessel, depressurization is done quite rapidly. Normally, process pressure of 680 MPa will result in a 15% compression of the liquid treated.

Mechanism of HHP

According to Le chatelier's principle, HHP affects any those phenomenon in food systems where volume changes are involved and results in decrease in volume. The process affects non-covalent bonds (hydrogen, ionic, hydrophobic bonds) substantially, owing to their sensitivity towards pressure and compounds with low molecular weight (responsible for nutritional and sensory characteristics) are not affected, but high molecular weight components (whose tertiary structure is important for determining its functionality) are sensitive (Carlez *et al*, 1994).

Process can be broadly classified into three main categories and are:

- Batch process
- Semi continuous process
- Continuous process

Effect of HHP on Microorganisms

Milk being a perishable commodity is usually heat- treated using a range of time temperature combination treatments to provide acceptable safety and shelf life. Though heat treatment extends the shelf life of the product but at the same time destroys its natural nutrients and to preserve its nutritional value alternatively HHP treatment can be given. The resistance of microorganisms to pressure in food is very variable depending on HP processing conditions (pressure, time, temperature, cycles, etc.), food constituents and the properties and the physiological state of the microorganism (Smelt, 1998). Cells at their exponential growing stage are more sensitive to pressure than cells in the stationary phase. The bacterial spores are always more resistant than vegetative cells and they can survive at pressure of 1000 MPa (Cheftel, 1992). However, it has been found that the pressurization along with mild heat treatment triggers spores to germinate and after germination, microorganisms lose their resistance towards pressure & heat and gets killed (Gould and Sale, 1970). Gram-positive microorganisms are more resistant to pressure than gram-negative. For ex. Gram-positive microorganisms need the application of 500–600 MPa at 25 °C for 10 min to achieve inactivation, while gram-negative microorganisms can be inactivated with treatments of 300–400 MPa at same time temperature combination.

Control of microorganisms in milk by High pressure application

Foods	Pressure (MPa)	Time (min)	Organism (s)
Whole milk	680	10	*Natural microflora* *Listeria monocytogens*
Skim milk	400	30	*Aerobic bacteria Psychrotrophs, Pseudomonas, Enterobacteriaceae*
Ewes milk	200-500	5-15	*Listeria innocula 910 CECT*

Many studies have been conducted on raw milk using this technology and have been proved that HPP treatment gives raw milk quality (pressurized at 400–600 MPa) comparable to that of pasteurized milk, as it is equally effective in destroying pathogenic and spoilage microorganisms but not that of sterilized milk due to presence of HP resistant spores.

Advantages of HHP

- Rapid and uniform distribution of pressure throughout the sample, which helps to produce less thermal degradation.
- It is suitable to handle both particulate and pumpable foods.
- Process time is less dependence of product shape and size.
- Opportunities for new process product development.
- Comparison to thermal process, food may not undergo significant chemical changes.
- Pressure accelerates traditional thermal inactivation kinetics.
- High pressure processing helps to eliminate the risk of various food borne pathogens such as Escherichia coli, Salmonella and Listeria inside packaged food products without additives.
- High pressure can extend the shelf life by 2-3 folds over a non-pasteurized counterpart and improve food safety.
- High pressure processing cycle takes no longer than six minutes compared to traditional high-temperature processing that takes an hour or longer, without causing chemical changes that degrade food quality and thus is a rapid method.
- High pressure processing is that it does not affect the flavour, nutritional value or texture.
- Though installation cost is high but processing cost is low.
- Flexible system for no. of products like *glucamole, juices, oysters, ham, fruit jellies and jams, pourable salad dressings, salsa, poultry and rice products*

HPP in Dairy Industry

Effect of HHP on Different milk constituents

HP influences the physico-chemical and technological properties of milk. When milk is subjected to HP, the **Casein Micelles** are disintegrated into smaller particles which in turn are accompanied by an increase of caseins and calcium phosphate levels in the serum phase of milk and by a decrease in the both non-casein nitrogen and serum nitrogen fractions (Law *et al.*, 1998). Pressure above than 3000 atmosphere tends towards irreversible denaturation as compared to reversible denaturation within range of 1000-3000 atmospheres (Jaenicke, 1981).

Pressurization of milk causes conformational changes of Milk Proteins. With the application of HP treatment, the size and number of casein micelles increases as spherical particles changes to form chains or clusters of sub-micelles. Thus, reduces RCT (Rennet Coagulation Time).

Liu *et al.* (2005) studied the effect of HHP on hydrophobicity of whey protein concentrate (WPC) and recorded that HHP treatment of WPC yields increase in the number of binding sites and led to certain modifications of proteins that enhance hydrophobicity and showed promising results for improving functional properties of foods.

Studies carried out by Gervilla *et al.* (2001) on Free fatty acids (FFA) content (lipolysis of milk fat) in ewe's milk have showed that HP treatments between 100–500 MPa at 4, 25 and 50° C did not increase FFA content, even some treatments at 50° C showed lower FFA content than fresh raw milk, Thus this phenomenon of great interest to avoid production of off flavours, which otherwise gets develop because of lipolytic rancidity in milk. However, no damage on the milk fat globule membrane occurred, being proved by the lack of increase in lipolysis. These modifications on distribution of milk fat globules could be due to phenomenon of aggregation and disaggregation. This provides some advantages for HP-treated milk, because HP treatment increases the stability of milk treated at 25 and 50° C, whereas at 4 °C (low temperatures) increases the creaming-off. This aspect could improve cream separation in the elaboration of butter.

Lactose in milk and milk products may isomerise to lactulose by heating and then degrade to form acids and other sugars. No changes in these compounds are observed after pressurization (100–400 MPa for 10–60 min at 25 °C), suggesting that no Maillard reaction or lactose isomerization occur in milk after pressure treatment (López Fandino *et al.*, 1996).

Micelle disintegration induced by HP treatment also affects milk Colour. A study was carried out by Harte *et al.* (2003) to observe the series of changes during combine treatment of thermal and HHP and it was observed that milk subjected to HHP or thermal treatment and then HHP, lost its white colour and turns into yellowish colour and may be due to reduction in size of casein micelles (Needs *et al.*, 2000), whereas milk when first subjected to HP and then to thermal treatment regained its whitish colour and is attributed to reversible nature of casein micelles (or reaggregation of disrupted micelles) towards HHP treatment when applied in the range of 300-676 MPa followed by thermal treatment. Similar observations had been recorded in HP treated ewe's milk by Gervilla *et al.,* (2001).

Contrary to thermal treatments, where covalent as well as non-covalent bonds are affected, HP treatment at room and mild temperatures only disrupts relatively weak chemical bonds (hydrogen bonds, hydrophobic bonds, ionic bonds). Thus, small molecules such as vitamins, amino acids, simple sugars and flavour compounds remain unaffected by the HP treatment. HP treatment of milk at 400 MPa (@ 2.5 MPa/s for 30 min at 25 °C) results in no significant loss of vitamins B_1 and B_6 (Sierra *et al.,* 2000).

Cheese production from HP-treated milk

Milk pasteurization (heating at 72–74 °C for 15 s or equivalent treatments) destroys pathogenic and almost all, but not all, spoilage microorganisms, and it is the most important heat treatment applied to cheese milk to provide acceptable safety and quality. However, milk pasteurisation is known for its adverse affects with respect to many sensory characteristics of cheese, leading to alterations in texture and often delayed maturation (Grappin & Beuvier, 1997). HP technology can be used to increase the microbiological safety and quality of milk to produce high quality cheeses. It has been observed that denaturation of whey proteins due to applied pressure results in interaction between denatured whey protein and casein, which in turn increases the retention of former within casein matrix in cheese. Thus, these changes results in modifying the technological aptitude of milk to make cheese, improving the rennet coagulation and yield properties of cheese milk (Gonzalez *et al.*, 2004; Trujillo *et al.,* 1999). Lopez *et al* (2006), evaluated the combined effect of ultrahigh pressure homogenization (UHPH) followed by HHP treatment on the count of *Staphylococcus aureus* and found complete inactivation of *S. aureus* after 15 days of storage in UHPH+HHP treated cheese. However, the application of HP technology to cheese milk causes differences in cheese composition and ripening in comparison to pasteurised milk cheese. The HP-treated milk cheeses have higher moisture, salt and total free amino acids contents than raw or pasteurised milk cheeses. On the other hand, cheeses made from HP-treated milk showed a similar level of lipolysis to cheeses made from raw milk, whereas the level of lipolysis in cheese made from pasteurised milk was lower and this behaviour was explained by heat-sensitive but partial pressure-resistant characteristics of the indigenous milk lipase.

Cheese Ripening Acceleration

Being cheese ripening a quite expensive process, acceleration of ripening is highly desirable. Most of the work in this field has been done by elevation of ripening temperature, addition of cheese slurries or exogenous enzymes or by the use of adjunct starters, either as such or in modified form. The potential use of HP to accelerate cheese ripening was first shown in a patent by Yokoyama, Sawamura and Motobayashi (1992). Experimental Cheddar cheese samples were exposed to pressure from 0.1 to 300 MPa for 3 days at 25°C after cheese making. Best results were obtained at 50 MPa, where cheese with free amino acid and taste comparable to that of 6 months old commercial cheese was obtained. In particular, the kind of starter bacteria added to the cheese milk was highly proteolytic and at least 10-fold higher than conventional inoculation rates. In certain cheese varieties such as Mozzarella and Gouda, pressurization increases rate of proteolysis on exposure to pressure treatment of 400-600 MPa for 5-15 min.

Many HP conditions have been tested for accelerating cheese ripening that involved 'high' HP treatments (400–600 MPa) short times (5–15 min) or an initial 'high' HP treatment (400–600) short times (5–15 min) followed by a 'low' HP treatment (50 MPa) long times (72 h) and different cheese varieties While long treatments at moderate pressure produce an increase in proteolysis while the treatment is applied, short and intense treatments produce a permanent effect on proteolysis rates. Saldo *et al.* (2001) evaluated the effect of (HHP) treatment (50 MPa for 72 h) on Garrotxa goat cheese and found that HHP produced changes in primary proteolysis during the treatment and along with this, solute diffusion, water holding capacity, and salt distribution was improved by HHP.

Yoghurt, and Ice- Cream

Yoghurt, a very popular dairy product suffers from common defect of syneresis and low viscosity or rigidity. Yoghurt quality can be improved in terms of its preservation and improved rheological properties by pressurization treatment. Skim milk treated with combined treatments of high hydrostatic pressure (400-500 MPa) and thermal treatment (85°C for 30 min) showed increased yield stress, resistance to normal penetration, elastic modulus and reduced syneresis (Harte *et al.*, 2003).

Reps *et al.* (1999) investigated the effect of pressurization on inactivation of microflora present in yogurt and found that HP treatment of 400 MPa completely inactivates *Lactobacillus bulgaricus* but *Streptococcus thermophilus* has been found to be more resistance towards pressure and its resistance varies from strain to strain & is in the range of 35.3 to 99.9 % which gives an idea that shelf life of yogurt can be enhanced by HHP treatment.

HPP treatment Induces fat crystallization, Shortens the time required to achieve a desirable solid fat content, & thereby thus reduces the ageing time of Ice Cream, and also enhances the physical ripening of cream for making butter (Buchheim and Frede, 1996). In a study it has been observed by Eberhard *et al.*, 1999 that pressurization treatment improves whipping ability of cream when treated for 2 minutes at 600 MPa and is possibly due to better crystallization properties of milk fat.

Limitations of HHP

- Food must contain water as whole phenomenon is based upon compression.
- High hydrostatic pressure also can cause structural changes in structurally fragile foods such as strawberries.
- High pressure may not inactivate spores.
- Some enzymes are very pressure resistant.
- Low acid shelf-stable products such as soups are not commercially available yet because of the limitations in killing spores with HHP.

Epilogue

HPP products are becoming choice of a new consumer in terms of health and safety aspects. Being one of the emerging technologies in developing countries, High-pressure technology offers the Food Technologists an opportunity to develop novel products with enhanced shelf life and higher safety with better sensory and nutritional aspects of food intact within and being applicable to a wide range of products, this technology offers food processors to manufacture minimally processed shelf stable products. In coming future, the day is not far off when we will have these products available in local markets and presence of such products will eradicate the products made by the obsolescing technologies available that only help in preserving the food but destroys its nutritive value.

Suggested Reading

Buchheim W. and Frede E. (1996). Use of high-pressure treatment to influence the crystallisation of emulsified fats. *DMZ Lebensmittel industrie and Milchwirtschaft* 117(**5**): 228–237.

Carlez A.; Rasec J.P.; Richard N. and Cheftel J.C. (1994). Bacterial growth during chilled storage of pressure treated minced meat. *Lebensm . Wiss. U. Technol.* 27: 48-54.

Cheftel J.C. (1992). Effect of high hydrostatic pressure on food constituents- An overview. *In: High Pressure and Biotechnology* (Ed. By: Balny R Hayashi; Heremans H.; and Masson K), Colloque INSERM, John Libbey and Co. Ltd., London, England.

Eberhard P.; Strahm W. and Eyer H. (1999). High pressure treatment of whipped cream. *Agrarforschung* 6 (**9**): 352–354

Farr D. (1990). High Pressure Technology in the Food Industry. *Trends Food Sci. Technology*. **1:** 4-16.

Gervilla R.; Ferragut V. and Guamis B. (2001). High hydrostatic pressure effects on colour and milk-fat globule of ewe's milk. *J. Food Sci.* **66**(**6**): 880–885.

Gonzalez-Martin M. F. San; Chanes- Welti J. S. and Barbosa- Canovas G. V. (2004). Cheese manufacturing assisted by ultra-high pressure. IFT Meeting, July 12-16, Las Vegas, NV.

Gould G.W. and Sale A.J.H. (1970). Initiation of germination of bacterial spores by hydrostatic pressure. *J. Gen. Microbiol.* **60:**335-346.

Grappin R. and Beuvier E. (1997). Possible implications of milk pasteurisation on the manufacture and sensory quality of ripened cheese: A review. *International Dairy Journal* **7**(**12**): 751–761.

Harte F; Luedecke L; Swanson B and Barbosa-Canovas .G.V. (2003) Low fat set yogurt made from milk subjected to combinations of High hydrostatic pressure and thermal processing. *J. Dairy Sci.* **86:** 1074-1082.

Jaenicke R. (1981). Enzymes under extreme conditions. *Ann. Rev. Biophy. Bioeng.* **10:**1

Law A.J.R.; Leaver J.; Felipe X.; Ferragut V.; Pla R. and Guamis B. (1998). Comparison of the effects of high pressure and thermal treatments on the casein micelles in goat's milk. *J. Agricultural and Food Chemistry.* **46** (**7**): 2523–2530.

Liu X. ; Powers J.R. ; Swanson B.G.; Hill H.H.; Clark S. (2005). High hydrostatic pressure affects flavor-binding properties of whey protein concentrate. *J. Food Sci.*. **70** (**9**):C581-584.

López Fandiño; Carrascosa A.V. and. Olano A. (1996). The effects of high pressure on whey protein denaturation and cheese-making properties of raw milk. *J. Dairy Sci.* **79**(**6**): 929–1126.

Lopez. Pedemonte.T; Brinez.W.J; Roig-Sagues .A.X; Guamis. B. (2006). Fate of Staphylococcus aureus in cheese treated by ultrahigh pressure homogenization and high hydrostatic pressure. *J. of Dairy Science*. **89**(**12**): 4536-4544.

Makhal S; Bhavana Vashishtha ; Mandal S; and Kanawjia S. K. (2003). High Hydrostatic Pressure in Food Preservation: Philosophy and Development. *Indian food Industry* **22**(**1**): 38-45

Needs E.C.; Capellas M.; Bland P.; Manoj P.; MacDougal D. B. and Gopal P. (2000). Comparison of heat and pressure treatments of skimmed milk, fortified with whey protein concentrate, for set yoghurt preparation: effects on milk proteins and gel structure. *J. Dairy Research* **67**: 329–348.

Reps A.; Warminska Radyko I. and Dajnowiec F. (1999). Effect of high pressure on yoghurt. In: H. Ludwig, Editor, *Advances in High Pressure Bioscience and Biotechnology*, Springer, Heidelberg, Germany. Pp: 453–456.

Saldo J.; Sendra E.; Guamis B. (2001). Hard cheese structure after a high hydrostatic pressure treatment at 50 MPa for 72 h applied to cheese after brining. *Lait.* **81**(**5**): 625-635.

Sierra I.; Vidal Valverde C. and López Fandiño R. (2000). Effect of high pressure on the Vitamin B_1 and B_6 content in milk. *Milchwissenschaft* **55**(**7**): 365–367.

Smelt J.M. (1998). Recent advances in the microbiology of high pressure processing. *Trends in Food Science & Technology*. **9**: 152–158

Trujillo A.J.; Royo C.; Ferragut V.; and Guamis B. (1999). Influence of pressurization on goat milk and cheese composition and yield. In: *Advances in High Pressure Bioscience and Biotechnology*. Ed. By Ludwig Horst, Proceedings of International Conference on High Pressure Bioscience and Biotechnology, Heidelberg, Aug 30-Sept 3, 1998. Springer -Verlag Berlin, New York . Pp: 457-460.

Yokoyama H.; Sawamura N. and Motobayashi N. (1992). Method for accelerating cheese ripening. *European Patent application EP* 469 857.

❑❑❑

Chapter 30

Use of DNA Microarray Technology in Food Microbial Pathogen Detection, Identification and Characterization

DINESH KUMAR & B.K. JOSHI
National Bureau of Animal Genetic Resources, Karnal-132 001

Introduction

It would be ideal to develop a method of pathogen detection in food samples that minimizes enrichment steps, is sensitive enough to detect few microorganisms, and possesses the specificity to identify a pathogen among a complex sample with mixed micro flora present. The detection and identification of pathogens on live food animals and plants as well as in processed food and water is essential for ensuring the safety of food products for human consumption.

The difficulty in implementing pathogen detection techniques in food science involves the complexity of food samples that are usually exposed to large numbers of various microorganisms from the environment, making it difficult to ensure the sensitivity and specificity of these methods. Traditional methods of pathogen detection and identification include microbiological culturing techniques, where the pathogen is identified based on biochemical characteristics, and immunological techniques to detect specific antigens of the pathogen. However, these detection methods are very time consuming, as some microorganisms are difficult to culture and may grow slowly. As well, immunological methods can result in false-positive results because of cross-reactivity of antibodies. In addition, routinely used biochemical and immunological tests do not provide information about the potential pathogenicity or virulence of identified microorganisms. Molecular methods offer several potential advantages over conventional microbiological techniques. Methods based on polymerase chain reaction (PCR) are rapid and sensitive enough to detect very small quantities of microorganisms. In addition, DNA microarray technology provides a means of detecting thousands of specific PCR products simultaneously and is thus a useful method of analysis of amplified pathogen DNA (Call *et al.* 2001). Resolving PCR amplified DNA fragments via agarose gel electrophoresis is limited to a certain number of primer sets and to the length of the PCR product. Chizhikov *et al.* (2001) compared gel electrophoresis to oligonucleotide microarray analysis of multiplex PCR products of six genes encoding bacterial virulence factors and antigenic determinants. The microarray was able to detect the bacterial factors in fifteen different strains of *Salmonella* spp., *Shigella* spp., and *Escherichia coli*, whereas the gel electrophoresis analysis resulted in unexpected non-specifically amplified products that were difficult to characterize, leading to ambiguous results. In addition to improving detection and analysis of PCR products by hybridization, DNA microarrays can improve the detection of microorganisms via direct genomic DNA without PCR amplification, although this is less sensitive. Most importantly, DNA microarray can be an excellent tool for characterizing microorganisms by discriminating among sequence variations among isolates and by identifying the presence or absence of genes.

DNA Microarray Technology

DNA microarrays consist of a solid support, most commonly glass or silicon, upon which DNA fragments or oligonucleotides are immobilized. In the Affymetrix approach to produce high-density tiled oligonucleotide arrays, probes (usually 25 nucleotides long) are synthesized directly on the wafer surface using a photolithographic manufacturing process. In genomic profiling experiments, DNA extracted from a biological sample is fluorescently labelled and hybridized to the microarray. Alternatively, sample DNA is amplified by PCR, labelled by fluorescence, and hybridized to the array. In gene expression profiling experiments, mRNA extracted from a test sample is applied to the array. DNA microarrays allow for rapid, simultaneous detection of thousands of specific DNA sequences or reveal the level of expression of thousands of genes. Although this technology may be expensive for laboratory use and requires specially trained staff to effectively analyze the abundant information obtained from this technique, it has the advantages of being a more rapid, efficient method of pathogen detection when compared with classical detection methods involving culturing, while providing specific identification of pathogens and their characterization.

Two main types of arrays are produced: DNA arrays, such as genomic microarrays, and oligonucleotide arrays.

a) *Genomic microarrays*: These comprise whole genome, genomic fragments from a cDNA library, or open reading frames from a strain of microorganism. These DNA fragments are attached to the glass or silicon support and are then hybridized with labelled nucleic acid from the test sample. Genomic arrays can also be adapted to include genomic DNA from several related strains of a microorganism, referred to as a mixed genomic microarray. Genomic microarrays are useful for pathogen identification based on fingerprinting genomic isolates, classifying them into phylogenetic groups, and studying their genetic diversity by determining constant and variable genes among the isolates, providing clues regarding the acquisition of virulence genes or genes required for survival in specific niches (Taboada *et al.* 2004). Oligonucleotide arrays generally involve oligonucleotides between 18 and 70 nucleotides long (Call *et al.* 2001) and are often used for pathogen detection and genomic analysis because these arrays can consist of multiple sequence variants of a target gene, allowing for more sensitive detection of polymorphisms and for differentiation among strains.

b) *Oligonucleotide microarrays*: These are capable of detecting single-point mutations and are also very useful for epidemiological or phylogenetic studies, as they can detect mutational evolutionary changes in response to novel environments. Often, these genetic variations can be traced back to the gene, identifying virulence or antibiotic resistance determinants, suggesting the altered nucleotides are a result of adaptation. Therefore, DNA microarrays can be adapted to combine effective strategies for detecting and identifying pathogenic isolates from food samples, while providing information for the characterization and potential determination of the pathogenicity of the isolate.

Application of DNA Microarrays for Pathogen Detection

DNA microarray technology is a powerful new tool for pathogen detection, most commonly in environmental or clinical diagnostic specimens (Lemarchand *et al.* 2004). However, the methods applied to obtain detectable levels of pathogenic DNA via microarray analysis may potentially be applied in agriculture for food safety practices. Several challenges for microarray use as a detection tool exist, such as obtaining a suitable level of sensitivity to detect low levels of pathogen, the possibility of quantifying detected pathogens, and the capacity to differentiate among several strains in a mixed sample (Wu *et al.* 2001). Sample preparation will continue to be significant challenge for the detection of pathogens in food systems. Particularly, DNA or RNA extraction of pathogens present in complex food samples as well as detection of food-borne pathogens present in low numbers in a complex mixture of microorganisms could be extremely challenging. Therefore, interpreting negative results can pose a real challenge, and assay validation is necessary to confirm the suitability of microarrays to detect and identify pathogens in food systems. In addition, to achieve specific hybridization, the major challenges are to select optimal probes for each gene or group of genes and to adjust the hybridization and washing conditions. However, this could be difficult to achieve, since one set of hybridization and wash conditions does not provide optimal target discrimination for all probes on the array. Research has focused on overcoming these obstacles to improve the functionality of DNA microarray as a tool to detect and identify bacterial and viral pathogenic organisms in the food chain.

a) Detection of pathogenic bacteria:Microarray analysis can be conducted on genomic DNA from food-borne pathogenic bacteria to detect the presence of pathogens and identify them. For instance, Wu *et al.*(2001) detected a minimum of 25 ng of genomic DNA, corresponding to approximately 5.6 × 106 cells of pathogenic bacteria from environmental samples. However, there may be applications where the use of genomic DNA may not be sensitive enough because of the variability of complexity of samples collected from food. As well, this method could pose great challenges if pathogen populations are small.

   Commonly, DNA microarrays are used for detecting hybridized PCR products amplified from the genomic DNA of pathogenic microorganisms. This method takes advantage of the microarray's ability to serve as a series of dot-blots for rapidly detecting polymorphisms, while detecting only the sequences specific to the probes, thereby eliminating the nonspecific amplified products often produced by PCR. Thus, combining microarray with PCR increases the specificity of PCR by not detecting nonspecific PCR products while improving the sensitivity of microarray by amplifying nucleic acid sequences from small quantities of microorganisms. When compared, DNA microarray detection of amplified PCR products was 32 times more sensitive than detection with gel electrophoresis (Call *et al.* 2001). There are several examples where research has demonstrated that an oligonucleotide microarray combined with PCR amplification could be a sensitive method for detecting food-borne pathogens. For the purpose of food safety surveillance, researchers were able to detect 55 CFU/mL of enterohemorrhagic *E. coli* that were isolated from chicken rinsate, by immunomagnetic cell capture (Call *et al.* 2001). The pathogens detected were then genotyped with the array using multiplex PCR of four virulence determinants. Yoshida *et al.* (2004) found that hybridization of antigen-specific PCR products improved specificity compared with hybridization of genomic DNA to an oligonucleotide microarray designed to detect 100 different *Salmonella* serotypes based on *Salmonella* O and H antigens. In a similar effort to apply microarray use in food safety, Delaquis *et al.* (2002) have designed a microarray with DNA probes specific for virulence factors of *E. coli* O157:H7 to differentiate this microorganism from nonpathogenic *E. coli* strains isolated from food samples such as milk. In addition, this array is being used to genetically characterize *E. coli* strains isolated at various stages of the ground beef production system. Delaquis *et al.* (2002) have also designed microarrays for detecting *Listeria monocytogenes* in milk samples using 16S rRNA-based oligonucleotides that are specific for the pathogen and plan to use this microarray in conjunction with the *E. coli* specific array for simultaneous detection and genetic characterization of both pathogens.

Multi-Pathogen Identification (MPID) Microarray

Often, food samples are contaminated with various pathogenic organisms or several closely related species, thus it is essential that a microarray be able to discriminate among these pathogens or species of pathogens while specifically identifying all organisms. Wilson *et al.* (2002) have developed a Multi-Pathogen Identification (MPID) microarray for identification of 18 pathogens (11 bacteria, 5 RNA viruses, and 2 eukaryotes). MPID microarray could be used to screen for the presence of pathogens by examining specific diagnostic regions and creating specific fingerprints. This microarray assay offers an alternative to current methods for high throughput screening of bacterial and viral pathogens. In addition, a DNA microarray based on the gyrB genes that encode the subunit B protein of DNA gyrase (topoisomerase type II protein) was developed to identify closely related enteric bacteria at the species level (Kakinuma *et al.* 2003). Hong *et al.* (2004) developed an oligonucleotide microarray based on a mutation region of 23S rDNA for the detection of 14 bacterial pathogens isolated from food-borne infections. High sensitivity and specificity was achieved with this array for 10 pathogenic species; however, two species had weak cross-reactivity with *E. coli* but were still detected, and two strains did not result in successful hybridization. This microarray combined with the PCR amplification of the 23S rRNA resulted in a detection level of 10 CFU/mL and could be used as a rapid diagnostic test. Given the success in efficiently identifying and discriminating among several pathogenic species, this array could be successful if applied to food samples to assess food safety, although sensitivity could be affected by the complexity of the sample collected for PCR amplification.

Validation of Microarray Data by Real Time PCR

The ability of microarray technology to quantify detected pathogens is limited by relative comparisons of intensity of genomic DNA with standard controls. Very carefully controlled experimental conditions would be required to make useful relative comparisons, although ideal conditions are often difficult to achieve in food science applications. An alternative method to using genomic DNA hybridized to microarray to quantify pathogens is the use of real-time PCR coupled with microarray, although the use of these combined methods has not been well documented. More often real-time PCR is used as an alternative to low density microarrays or to validate microarray data. Real-time PCR allows the monitoring of PCR products as they are produced, alleviating some of the biases associated with conventional PCR (Hanna *et al.* 2005). As such, real-time PCR is fast becoming the method of choice for detection and quantitative gene expression of a limited number of genes; however, microarray technology provides unparalleled opportunities for the simultaneous detection of thousands of genes or target DNA sequences.

b) Detection of parasites

Oligonucleotide microarray based detection of parasites

There have been a few studies where DNA microarray has been used to identify parasitic pathogens in water samples. Wang *et al.* (2004) developed an oligonucleotide microarray based on genus-, species-, and subtype-specific probes. This array was used in conjunction with multiplex PCR to detect and genotype *Entamoeba histolytica, Entamoeba dispar,Giardia lamblia,* and *Cryptosporidium parvum,* all of which are protozoan parasites associated with waterborne disease outbreaks.

SNP microarray based detection of parasites

Straub *et al.* (2002) have used a microarray based on single nucleotide polymorphisms within a region of the hsp70 gene of *Cryptosporidium parvum* to detect *Cryptosporidium* species, to differentiate between closely related species and isolates of *Cryptosporidium parvum,* and to identify the genotypes commonly associated with human disease. The application of DNA microarrays to identify and study contaminating protozoan organisms could be an effective measure in quality control and assurance of human safety in water sources.

c) *Detection of viral pathogens:*The MPID microarray developed by Wilson *et al.* (2002) could be used for the detection of viruses identified as candidates for biological terrorism. In addition, the development of DNA microarrays for detecting viral pathogens for clinical diagnostics may have prospects if applied to agriculture and food science. Wang *et al.* (2002) developed a microarray with 140 genome sequences to screen a diverse range of human respiratory viruses, using the most conserved sequences within a viral family as the oligonucleotide probes on the microarray. The use of conserved sequences allowed for the detection of unsequenced or uncharacterized members of a viral family and displayed unique hybridization patterns for discerning viral subtypes. The microarray was efficient at detecting a wide variety of viruses from respiratory tract specimens that were amplified by random PCR to detect pathogens of unknown sequence. These results indicate that PCR amplification of similarly complex food specimens may likewise be used with this microarray system and random PCR to detect known viruses as well as those that are uncharacterized.

Use of DNA Microarrays in Characterizing Isolates

a) Whole-genome and mixed-genome arrays:An advantage of using DNA microarray to detect pathogens is the potential to use this tool to further characterize the genomes. Whole-genome, mixed-genome, and PCR amplified probe microarrays have been developed to genotype pathogens, to describe genetic diversity, to classify based on phylogeny, and to study the epidemiology by identifying polymorphisms of conserved or variable genes among isolates. Obtaining this information can provide clues to the pathogenicity, evolution, host specificity, virulence, and antibiotic resistance of an isolate. Several species of food pathogen have been subjected to these characterization studies, such as *Listeria monocytogenes, E. coli* O157:H7, *Salmonella, Campylobacter jejuni*, and *Helicobacter pylori*. Ideally, the ability to characterize pathogenic organisms will be useful in food safety by providing information to allow for specific identification of isolates, to understand the pathogenesis based on presence of virulence genes, and to indicate how new pathogenic strains evolved epidemiologically and phylogenetically. Obtaining this information on an isolate will allow researchers to direct efforts at the control and prevention of human infection or disease as well as the proper identification of the pathogenic organism before it reaches the food chain.

   Whole-genome microarray of Campylobacter jejuni for strain differentiation:DNA microarray technology can be useful for characterizing microbial isolates from food sources. Genotyping methods or fingerprinting typically involve electrophoresis of DNA fragments to identify fragments based on differences in length. The banding pattern would be compared among different isolates as an estimate of genetic relatedness. However, if sequence information is available for the organism of interest, genome microarray can be a useful alternative for genotyping isolates and will provide pertinent information about the genome compared with the electrophoresis method, where it is difficult to identify the sequences encoded by the fragments. Dorrell *et al.* (2001) constructed a whole-genome microarray from a *Campylobacter jejuni* reference strain that was then hybridized with 11 different strains and analyzed for differential spotting patterns. Researchers were subsequently able to identify genes in the reference strain that were absent or divergent in some of the strains tested, such as genes encoding surface structures and genes that function in acquiring iron and DNA restriction or modification. Thus, genomic array can be used to provide information about genetic diversity and classes of genes that may be dispensable or variable within a species and about virulence factors that may be conserved. Similarly, Taboada *et al.* (2004) used a full-genome DNA microarray of a *Campylobacter jejuni* strain to compare the genomes of 51strains from around the globe that are phenotypically and epidemiologically diverse. This research encompassed the use of diverse strains to study the genetic relationships among them and to identify several variable genes that can be used for genotyping. Genotyping via genomic DNA microarray can allow researchers to gain an understanding of the pathogenicity and epidemiology of dangerous food pathogens to develop approaches to reduce the occurrence of these microorganisms in the food chain.

Mixed-genome microarray of L. monocytogenes for strain differentiation

A mixed-genome DNA microarray was developed by Borucki *et al.* (2003) to differentiate related isolates of *L. monocytogenes*. The array was constructed using 600 base pair fragments from a random shotgun library created with 10 different strains. When the array was hybridized to *L. monocytogenes* strains, phylogenetic groupings previously devised were confirmed by probe hybridization patterns. Additional experiments with this mixed genome microarray identified four probes that could differentiate among 50 strains (Call *et al.* 2003b). Furthermore, several polymorphic probes were recognized and sequenced, identifying divergent strains related to virulence and niche specific survival of the pathogen (Call *et al.* 2003b). The mixed genome DNA microarray for *L. monocytogenes* proved useful in differentiating among strains, characterizing strains phylogenetically, and identifying potential virulent genes and, thus, could be helpful in detecting this food borne pathogen and providing information to prevent and treat infection and disease.

Whole-genome microarray of E.coli for strain differentiation

Comparisons of strains to expose genetic diversity have also been conducted on the food pathogen *E. coli* O157:H7.Whole genome microarrays of non pathogenic *E. coli* K12 have been used for comparative genomic hybridization to demonstrate the genetic diversity of pathogenic isolates, to identify virulence genes present in these strains (Dobrindt etal. 2003), and to demonstrate the use of the microarray as a tool for phylogenetic analysis in addition to genome comparison (Fukiya *et al.* 2004). More recently, oligonucleotide microarray was used to characterize genomic differences between two lineages of *E. coli* O157:H7, both of which could be isolated from cattle (Zhang *et al.* 2004).

Whole-genome microarray of H. pylori for strain differentiation

The potential food-borne pathogen *H. pylori* has also been studied using DNA microarrays to characterize genetic diversity among strains as an indication of pathogenicity. Helicobacter pylori strains exhibit variability in the extent of virulence, possibly associated with a pathogenicity island, which manifest as gastritis, ulceration, and gastric cancer caused by the most virulent strains. The variance in pathogenicity prompted researchers to investigate genetic diversity among strains with a range of virulence (Salama *et al.* 2000).A whole-genome microarray was constructed based on the known sequences of two strains and was hybridized to genomic DNA from several strains with unknown sequences. The study identified conserved genes as well as species specific genes comprising approximately 18% of a strain's genome. The researchers proposed that the species-specific genes represent adaptations to various hosts or are related to the extent of virulence. The researchers were also able to identify possible virulence genes that are co-inherited with the pathogenicity island. The information regarding H. *pylori* virulence genes and the method of using DNA microarray to identify genes associated with extent of virulence could be useful for screening food samples, not only for identification of this organism but also to suggest the degree of pathogenicity.

Microarray based on annotated ORF of Salmonella for strain differentiation

DNA microarray has been used to construct a phylogenetic tree based on the relationship among various strains of *Salmonella* as an indication of evolutionary adaptation to new environments (Chan *et al.* 2003).The microarray was constructed based on annotated open reading frames (ORFs) of *Salmonella enterica* subsp.enterica serovar Typhimurium LT2, a serovar from the subspecies responsible for most infections of mammals, and was hybridized to genomic DNA from members of each subspecies of *Salmonella enterica* and *Salmonella bongori*. The phylogenetic tree constructed based on Typhimurium LT2 indicated that there are several genes that are homologous in other *Salmonella* subspecies and are likely acquired by horizontal transfer between strains and lineages. This study, and a study conducted by Chan *et al.* (2003) with a very similar microarray, indicates that there is a high level of genetic exchange among *Salmonellae* during the evolution of the bacteria and that tracing the exchange of various genes during evolution according to phylogenicity may provide clues as to the adaptation of *Salmonellae* to mammalian species.

b) Virulence factors

The use of DNA microarray in food safety can be extended beyond pathogen detection and studies of relatedness to the determination of virulence factors. Virulence genes can be detected by microarrays to identify pathotypes of microorganisms present in live food animals or plants or processed food samples. This information would be useful to determine the disease-causing potential of isolated organisms. Bekal *et al.* (2003) developed a DNA microarray with PCR-amplified products representing *E. coli* virulence genes for use as a diagnostic tool. This array allowed for rapid classification of *E. coli* isolates into various pathogenic groups based on virulence factors and enabled identification of relatedness to phylogenetic groups. Bekal *et al.* (2003) have evidence to suggest that this microarray has the potential to detect emerging new pathotypes arising from the horizontal transfer of virulence genes.

c) Antibiotic resistance

Antibiotic resistance is a characteristic of pathogenic microorganisms that should be traced and identified to ensure food safety. Infections caused by antibiotic-resistant pathogens are very difficult to treat. In addition, it is difficult to remove contaminating resistant organisms from food products or live food animals or plants, since the agricultural tools of choice are antibiotics. Booth *et al.* (2003) conducted an oligonucleotide microarray to detect point mutations of pathogens that are associated with antibiotic resistance. Although this experiment was not designed for food-borne organisms, Booth *et al.* (2003) successfully discriminated among five point mutations associated with resistance in genes from *Neisseria gonorrhoeae*.

Feasibility in Food Science

DNA microarray is a promising tool for future use in food safety applications. To make microarrays more practical for use with food samples, efforts have been made to improve convenience for use in agriculture and to maximize performance with the complexity of samples collected from food. Research has also been focused on making the microarray process more rapid and specific for application to food safety protocols. Factors that limit the utility of DNA microarray for field detection of food-borne pathogens are the time and effort required to isolate DNA from microorganisms, the labelling and hybridization of the DNA to the microarray, and the expensive laboratory equipment required to analyze the hybridization profile. Bavykin *et al.* (2001) developed a portable system for preparing bacterial samples for oligonucleotide analysis to eliminate many of these limiting steps that are usually encountered. The portable system involved a syringe-operated silica mini column for the 25 min isolation, fractionation, fragmentation, and fluorescent labelling of DNA as well as the removal of excess label. Bavykin *et al.* (2001) then hybridized DNA isolated from *Bacillus thuringiensis, E. coli,* and *Bacillus subtilis* to oligonucleotide probes for 16S ribosomal RNA identification and analyzed the hybridization using a portable battery-powered microchip imager to obtain an image on Polaroid film. The researchers recommended that for rapid detection and approximate estimates of quantities of targeted microorganisms, a comparison could be made between the image obtained from the field sample and an image obtained from hybridization with a known number of bacterial cells to compare fluorescence intensity and hybridization pattern. Otherwise, the portable device could be attached to a cooled charge-coupled device camera and a personal computer to conduct more accurate, yet stationary quantification. This research illustrates the potential to modify DNA isolation, labelling, and hybridization procedures to maximize the efficiency of DNA microarray for use in agriculture or with processed food and water samples, although this particular technology may not be effective if the samples are complex and contain many microorganisms.

Conclusions and Future Prospects

In the future, the development of microarray as a tool in food safety will need to focus on the challenge of identifying possibly very few pathogenic organisms that are among many normal florae in large volumes of sample that could have substances that interfere with isolation. Once this challenge is overcome, DNA microarray can be an excellent tool for simultaneously detecting a wide variety of pathogens by developing specific molecular signatures for certain forms of contamination, using specific microarray probes. The use of specific molecular signatures for microarray analysis would be very useful to rapidly determine the presence of a pathogen from food samples in a field setting. After initial pathogen identification, further studies involving DNA microarray could be conducted to characterize isolates, identifying particular strains and presence of virulence factors or microbial resistance genes, thereby providing information that could be valuable in determining the pathogenicity of the isolated organism. Thus, DNA microarray has excellent potential for use in food safety provided that future research improves the feasibility of applying this tool to complex food samples, making microarray amenable to the field samples collected from agriculture and processed food and water.

Suggested Reading

Bavykin, S.G. *et al.*, 2001. Portable system for microbial sample preparation and oligonucleotide microarray analysis. *Appl. Environ. Micobiol.* **67:** 922–928.

Booth, S.A. *et al.*, 2003. Design of oligonucleotide arrays to detect point mutations: molecular typing of antibiotic resistant strains of *Neisseria gonorrhoeae* and hantavirus infected deer mice. *Mol. Cell. Probes,* **17:** 77–84.

Borucki, M.K. *et al.*, 2003. Discrimination among *Listeria monocytogenes* isolates using a mixed genome DNA microarray. *Vet. Microbiol.* **92:** 351–362.

Bryant, P.A. *et al.*, 2004.Chips with everything: DNA microarrays in infectious diseases. *Lancet,* **4:** 100–111.

Call, D.R. *et al.*, 2001. Detecting and genotyping *Escherichia coli* O157:H7 using multiplexed PCR and nucleic acid microarrays. *Int. J. Food Microbiol.* **67:** 71–80.

Call, D.R. *et al.*, 2003. Mixed-genome microarrays reveal multiple serotype and lineage-specific differences among strains of *Listeria monocytogenes. J. Clin.Microbiol.* **41:** 632–639.

Chan, K. *et al.*, 2003. Genomic comparison of *Salmonella* enteric serovars and *Salmonella bongori* by use of an *S. enteric* serovar Typhimurium DNA microarray. *J. Bacteriol.* **185:** 553- 563.

Chizhikov, V. , *et al.*, 2001. Microarray analysis of microbial virulence factors. Appl. Environ. *Microbiol.* **67:** 3258–3263.

Delaquis, P. *et al.*, 2002. Application of DNA microarrays for the identification and tracking of *E. coli* O157:H7 and *Listeria monocytogenes* in food production systems. 2002 Meeting of Agriculture and Agri-Food Canada Food Network, Lacombe, Alberta, 30–31 May 2002. p. 23.

Dobrindt, U. *et al.*, 2003. Analysis of genome plasticity in pathogenic and commensal *Escherichia coli* isolates by use of DNA arrays. *J. Bacteriol.* **185:** 1831–1840.

Dorrell, N. *et al.*, 2001. Whole genome comparison of *Campylobacter jejuni* human isolates using a low-cost microarray reveals extensive genetic diversity. *Genome Res.* **11:** 1706- 1715.

Fukiya, S. *et al.*, 2004. Extensive genomic diversity in pathogenic *Escherichia coli* and *Shigella* strains revealed by comparative genomic hybridization microarray. *J. Bacteriol.* **186:** 3911–3921.

Hanna, S.E., *et al.*, 2005. Real-time polymerase chain reaction for the food microbiologist: technologies, applications and limitations. *J. Food Sci.* **70:** R49–R53.

Hong, B. *et al.*, 2004. Application of oligonucleotide array technology for the rapid detection of pathogenic bacteria of food borne infections. *J. Microbiol. Methods,* **58:** 403–411.

Kakinuma, K. *et al.*, 2003. Detection and identification of *Escherichia coli, Shigella* , and *Salmonella* by microarrays using gyrB gene. *Biotechnol. Bioeng.* **83:** 721–728.

Lemarchand, K. *et al.*, 2004. Molecular biology and DNA microarray technology for microbial quality monitoring of water. *Crit. Rev. Microbiol.* **30:** 145–172.

Porwollik, S. *et al.*, 2003. A non-redundant microarray of genes for two related bacteria. *Nucleic Acids Res.* **31:** 1869–1876.

Salama, N. *et al.*, 2000. A whole-genome microarray reveals genetic diversity among Helicobacter pylori strains. *Proc. Natl. Acad. Sci. U.S.A.* **97:** 14668–14673.

Straub, T.M. , *et al.*, 2002. Genotyping *Cryptosporidium parvum* with an hsp 70 single-nucleotide polymorphism microarrray. *Appl. Environ. Microbiol.* **68:** 1817–1826.

Taboada, E.N. *et al.,* 2004. Large-scale comparative genomics meta-analysis of *Campylobacter jejuni* isolates reveals low level of genome placticity. *J. Clin. Microbiol.* **42:** 4566- 4576.

Wang, D. , *et al.,*. 2002. Microarray-based detection and genotyping of viral pathogens. *Proc. Natl. Acad. Sci. U.S.A.* **99:** 15687-15692.

Wang, Z. *et al.,* 2004. Detection and genotyping of *Entamoeba histolytica, Entamoeba dispar, Giardia lamblia,* and *Cryptosporidium parvum* by oligonucleotide microarray. *J. Clin. Microbiol.* **42:** 3262-3271.

Wilson, W.J. *et al.,* 2002. Sequence-specific identification of 18 pathogenic microorganisms using microarray technology. *Mol. Cell. Probes,* **16:** 119-127.

Wu, L. *et al.,* 2001. Development and evaluation of functional gene arrays for detection of selected genes in the environment. *Appl. Environ. Microbiol.* **67:** 5780-5790.

Yoshida, C. 2004. *Salmonella* antigen detection using microarray analysis. 104th General Meeting of American Society for Microbiology, 2004.

Zhang, Y. *et al.,* 2004. Characterization of genomic diversity of *Escherichia coli* O157:H7 using oligonucleotide microarray. Federal Food Safety and Nutrition Research Meeting. Kanata, Ontario, Canada, 3-5 October 2004. p. 85.

□□□

Chapter 31

Prebiotics : The Novel Functional Food Ingredients for Health

R.K. SHARMA AND TULIKA
Animal Biochemistry Division, National Dairy Research Institute, Karnal

Introduction

There is an increasing awareness and demand among consumers around the world for the *functional foods* and *functional food ingredients,* which provide some health-benefits beyond traditional nutritional value. Probiotics, Prebiotics and Synbiotics are gaining lot of importance in this respect. Probiotics are living microorganisms, which upon ingestion exert health benefits to their host. *Lactobacillus* and *Bifidobacterium* are important probiotic bacteria generally added to fermented foods and other foodstuffs. The benefits associated with probiotics are dependent on their viability during storage, and passage through the gastrointestinal tract, their growth and metabolic activity, which can be maximized by incorporation of prebiotics. *'A prebiotic is a nondigestible food ingredient that beneficially affects the host by selectively stimulating the growth and/or activity of one or a limited number of bacteria in the colon, and thus improves host health'* (Gibson and Roberfroid, 1995). According to more recent definition, 'A prebiotic is a selectively fermented ingredient that allows specific changes, both in the composition and/or activity in the gastrointestinal microflora, that confer benefits upon host wellbeing and health' (Gibson et al., 2004).

Not all dietary carbohydrates are prebiotics, and clear criterion has been established for classifying a food ingredient as prebiotic (Roberfroid, 2007). These are:

1) Resistance to gastric acidity, to hydrolysis by mammalian enzymes and to gastrointestinal absorption
2) Fermentation by intestinal microflora
3) Selective stimulation of the growth and/or activity of those intestinal bacteria that contribute to health and well being.

Resistance does not necessarily mean that prebiotic is completely indigestible but it should guarantee that a significant amount of it must reach intestine to serve as a fermentation substrate.

Since its origin, prebiotics concept has gained a lot of momentum and a vast number of non-digestible carbohydrates are being evaluated for their prebiotic ability. A number of studies are available on inulin and FOS which have health promoting potential that primarily includes positive modulation of microbiota, immune modulation, lowering of hyperglycemia and stimulation of mineral absorption. Galactooligosaccharides (GOS) are fairly bifidogenic and reported to reduce microbial enzyme activities like nitrate reductase, b-glucuronidase involved in production of carcinogens. Other probable candidates for the prebiotic status are glucooligosaccharides, isomaltooligosaccharides, lactosucrose, polydextrose, soybean oligosaccharides, galactomannans and xylooligosaccharides for which preliminary or even promising data already exist. However, the evidence for prebiotic status is still not sufficient, and they cannot presently be classified as prebiotics.

Prebiotics as Functional Fiber

The prebiotics have also been termed as 'functional fiber', 'soluble fiber', 'bioactive carbohydrates'. These serve as sources of energy for the beneficial microorganisms. However, all types of fibers are not prebiotics. The combination of prebiotic and probiotic is termed as synbiotic, which can provide synergistic effects. Intestinal flora in human comprises of approx. 10^{14} bacteria, a complex, mixed population of different types of bacteria. These are most abundant in the colon which is recognized as a highly metabolic organ with eventually more and also more complex functions than the rest of gastrointestinal tract. Linking diet with health is an old, well-recognized concept in Indian system of medicine. The type of diet consumed determines the intestinal physiology, metabolism of an individual. Among the different food ingredients certain non-digestible carbohydrates (dietary fiber) have emerged as useful prebiotics. These generally constitute the soluble form of fiber and are fermentable. A major part of research on prebiotic activity has been conducted around inulin and fructooligosaccharides. Oat and barley beta-glucan, galactooligosaccharides (GOS), soybean oligosaccharides, resistant starch, are among others, gaining importance for the various suggested beneficial physiological effects. Honey has also been reported to contain a number of fermentable carbohydrates including oligosaccharides which show prebiotic properties, hence a natural health ingredient. Interest in the probiotic and prebiotic foods has risen very fast in Japan followed by European countries and the US.

In India, no clear data/recommendations are available regarding dietary fiber, their incorporation for desired physiological benefits. The manufacturers/producers of prebiotic carbohydrates (functional fiber) are also not much known. With changing lifestyle, dietary habits the inclusion of health promoting fiber in diet, processed foods, appears to be an important consideration in near future. In India, some companies have already entered the market to introduce probiotic dairy foods. There is increased interest worldwide in adding prebiotic carbohydrates /non-digestible oligosaccharides in different types of foods. The emergence of Prebiotic carbohydrate (functional fiber) containing value-added foods, bakery products, beverages, dairy drinks may be the next scenario in the Indian food market also.

Enzymatic Synthesis of Non-Digestible Galactooligosaccharides

Non-digestible oligosaccharides (NDOs) are claimed to behave as dietary fibers and prebiotics. Incorporation of non-digestible oligosaccharides in diet provides an opportunity to improve the gut microecology including bacterial populations and physiological effects. -galactosidase, commonly known as lactase, catalyzes the hydrolysis of lactose to its constituent monosaccharides, glucose and galactose. In addition, they are also known to catalyze a series of transferase reactions, using lactose and its hydrolysis products resulting in the formation of galactose rich oligosaccharides. Earlier, the formation of these oligosaccharides has been seen in a negative light, frequently described as unwanted products. However, in the recent past, these galactooligosaccharides have been recognized as Prebiotics. These have been suggested to have a number of beneficial effects on human health, which include bifidogenic activity, stimulation of mineral absorption, hypolipidemic effect, prevention of colon cancer. Many food grade microorganisms, including the lactic acid bacteria, are well-recognized source of b-galactosidase. Boon *et al.* (2000) studied oligosaccharide formation by enzyme from *Bacillus circulans, A. oryzae, K.lactis and K. fragilis*. Differences concerning amount, size and type of oligosaccharides were reported. Rabiu *et al.* (2001) extracted -galactosidase from different *Bifidobacterium* sp. and used the enzyme in glycosyl transfers reactions to synthesize oligosaccharides from lactose. Albayrak and Shang (2002) immobilized b-galactosidase of *A. oryzae* on cotton cloth and used for production of galactooligosaccharides from lactose. They reported maximum galactooligosaccharides production of 20 per cent (w/v) of the initial lactose (50% conversion) at an initial lactose concentration of 500 g/lit.

Whey, a byproduct of paneer/cheese industry, contains considerable lactose, which can be used for production of galactooligosaccharides. In our laboratory, b-galactosidase of a strain of *Streptococcus thermophilus* was found to be a better producer of galactooligosaccharides in comparison to enzyme from *Lactobacillus delbrueckii* susp. *bulgaricus* using lactose as substrate as well as lactose in whey (Kumar, 2005).

Health Benefits

Composition of the large-intestinal microbiota is affected by various factors which include age, susceptibility to infections, nutritional status, immunologic status of the host, pH, transit time, interactions between flora components and presence and availability of fermentable material in the gut. Diet exerts a major influence on gut bacterial populations, of which, prebiotics or fermentable fibers play an important role (Collins and Gibson, 1999). Whelan *et al.* (2005) reported a large increase in fecal bifidobacteria by fortification of the enteral formula with only 5.1 g/L of FOS in human subjects with apparent reduction in clostridia.

Short chain fatty acids are the major end products of bacterial fermentation of polysaccharide, oligosaccharide, protein, peptide and glycoprotein precursors by anaerobic micro-organisms. However, in quantitative terms carbohydrates are the most important SCFA progenitors. Some prebiotics are well known for their SCFA production profile: fructooligosaaccharides and resistant starch are considered butyrogenic; L-rhamnose is known to produce propionate while lactulose raises acetate: propionate ratio. Butyrate has been reported to act as a preferred energy fuel for colonic epithelial cells. Other than serving as the principal energy source for colonic epithelial cells, butyrate also exhibits anti-tumorigenic effects on colonocytes and plays a significant role in the regulation of a number of genes associated with the key processes of apoptosis, proliferation and differentiation (Daly *et al.*, 2005), induces glutathione-s-transferase (GST) expression that detoxifies many carcinogens (Sauer *et al.*, 2007). SCFAs also prevent overgrowth of pH-sensitive pathogenic bacteria by lowering pH values, enhance mineral absorption in colon and modulate nitrogen balance by protonating ammonia to produce ammonium ions.

Different types of indigestible carbohydrates such as galactooligosaccharides, resistant starch, fructooligosaccharides, inulin have been reported to stimulate the absorption and retention of several minerals particularly calcium and magnesium (Sako *et al.*, 1999; Scholz-Ahrens *et al.*, 2001). Effect of feeding indigestible carbohydrates on mineral absorption has been attributed to organic acids produced as a result of fermentation by the intestinal bacteria. Most of the findings were obtained in animal experiments. Chonan *et al.* (2001) concluded that the action of intestinal bacteria is necessary for the effects of GOS feeding.

Several bacterial enzymes e.g. b-glucuronidase, -glucosidase and nitroreductase may play a significant role in colon carcinogenesis by converting procarcinogens to proximal carcinogens. Rowland and Tanaka (1993) administered GOS or GOS plus *B. breve* to human flora-associated rats and found increases in the cecal concentration of total anaerobic bacteria, lactobacilli and bifidobacteria. Through feeding of GOS the cecal pH, activities of β-glucuronidase and nitrate reductase were reduced. Van Doccum *et al.* (1999) showed that daily intake of 15 g of GOS in young healthy volunteers resulted in a significant increase in fecal acetate and a significant decrease in fecal b-glucuronidase activity.

Lipid and glucose metabolism can also be modulated by prebiotics. Experimental data are available from animal and human studies on hypotriglyceridemic and hypocholesterolemic effects of prebiotics.

When added in the diet of rats and other experimental animals, inulin-type fructans modulate hepatic lipid metabolism with consequences on triglyceride accumulation in the liver and/or serum lipids. In rats, reduced triglyceridemia is linked to a decrease in *de novo* lipogenesis in hepatic tissue and to a decrease in the expression of key hepatic lipogenic enzymes, as reflected by a decrease in fatty acid synthase mRNA (Delzene *et al.*, 2002). Yamashita *et al.* (1984) suggested that ingestion of fructo-oligosaccharides in diabetic patients led to lower serum cholesterol levels. Gut peptides are the focus of attention to explain different effects on glucose and lipid metabolism. Glucagon like peptide-1 (GLP-1) is a key hormone released from enteroendocrine L cells in response to nutrient ingestion. It is produced by tissue specific posttranslational processing of its precursor proglucagon peptide by prohormone convertase-1. It promotes satiety, secretion of insulin and proliferation of b-cells in the pancreas. In rats fed on diets enriched with oligofructose significant increase in colonic and/or portal plasma GLP-1 have been reported (Cani *et al.*, 2004; Cani *et al.*, 2005a,b) The high content of intestinal GLP-1 was corroborated by increased colonic proglucagon mRNA level. Oligofructose is also able to modulate other gastrointestinal peptides (such as PYY and ghrelin) which could be involved in control of food intake. However, humans studies are needed in relation to the role of prebiotics, probiotics, synbiotics, in modulation of energy metabolism, body mass, gut peptides, insulin sensitivity. The complex interactions between the different functional food ingredients and host still needs to be elucidated.

Suggested Reading

Albayrak, N. and Yang, S.T. 2002. Production of galactooligosaccharides from lactose by *Aspergillus oryzae* b-galactosidase immobilized on cotton cloth. *Biotechnol. Bioengg.*, **77(1):** 8-19.

Boon, M.A., Janssen, A.E.M. and Riet, K.V. 2000. Effect of temperature and enzyme origin on the enzymatic synthesis of oligosaccharides. *Enz. Microb. Technol.*, **26**:271-281.

Cani, P. D., Dewever, C. and Delzenne, N. M. 2004. Inulin-type fructans modulate gastro-intestinal peptides involved in appetite regulation (glucagon-like peptide-1 and ghrelin) in rats. *Br. J. Nutr.* **92:** 521–526.

Cani, P. D., Daubioul, C. A., Reusens, B., Remacle, C., Catillon, G. and Delzenne, N. M. 2005a. Involvement of endogenous glucagon-like peptide-1(7–36) amide on glycaemia-lowering effect of oligofructose in streptozotocin-treated rats. *J Endocrinol* **185:** 457–465.

Cani, P. D., Neyrinck, A. M., Maton, N. and Delzenne, N. M. Oligofructose promotes satiety in rats fed a high-fat diet: involvement of glucagon-like peptide-1. *Obesity Res.* 2005b; **13:**1000-1007.

Chonan, O., Takahashi, R. and Watanuki, M. 2001. role of activity of gastrointestinal microflora in absorption of calcium and magnesium in rats fed b 1-4 linked galactooligosaccharides. Biosci. *Biotechnol. Biochem.,* **65** (8): 1872-1875.

Collins, M. D. and Gibson, G. R. 1999. Probiotics, prebiotics, and synbiotics: approaches for modulating the microbial ecology of the gut. *Am. J. Clin. Nutr.,* **69:** 1052S-1057S.

Daly, K., Cuff, M. A., Fung, F. and Shirazi-Beechey, S. P. 2005. The importance of colonic butyrate transport to the regulation of genes associated with colonic tissue homoeostasis. *Biochem Soc Trans.,* **33:** 733-735.

Delzenne, N. M., Daubioul, C., Neyrinck, A., Lasa. M. and Taper, H. S. 2002. Inulin and oligofructose modulate lipid metabolism in animals: review of biochemical events and future prospects. *Br. J. Nutr.,87: Suppl* **2:**S255-S259.

Gibson, G.R. and Roberfroid, M.B. 1995. Dietary modulation of the human colonic microbiota: Introducing theconcept of prebiotics. *J. Nutr.* **125:**1401-1413.

Gibson, G.R., Probert, H.M., Van Loo, J., Rastall, R.A. and Roberfroid, M. 2004. Dietary modulation of the human colonic microbiota: updating the concept of prebiotics. *Nutr. Res. Rev.,* **17:** 259-275.

Kumar, P. 2005. Study on oligosaccharide formation by b-galactosidase of lactic acid bacteria. M.Sc. Thesis, N.D.R.I., Karnal, India.

Rabiu, A.B., Jay, J.A., Gibson, R.G. and Rastall, A.R. 2001. Synthesis and fermentation properties of novel galactooligsaccharides by b-galactosidases from Bifodobacterium species. *Appl. Environ. Microb.,* **67** (6) : 2526-2530.

Roberfroid, M. 2007. Prebiotics: the concept revisisted. *J Nutr.* **137:** 830S- 837S.

Rowland, I. and Tanaka, R. 1993. The effects of transgalactosylated oligosaccharides on gut flora metabolism in rats associated with a human fecal microflora. *J. Appl. Bacteriol.,* **74:** 667-674.

Sako, T., Matsumoto, K. and Tanaka, R. 1999.Recent progress on research and applications of non-digestible galactooligosaccharides. *Int. Dairy J.,* **9:**69-80.

Sauer, J., Richter, K.K. and Pool-Zobel, B.L. 2007. Products formed during fermentation of the prebiotic inulin with human gut flora enhance expression of biotransformation genes in human primary colon cells. *Br J Nutr.,* **97:** 928-937.

Scholz-Ahrens, K.E., Schaafsma, G., van den Heuvel, E.G.H.M., and Schrezenmeir, J. 2001. Effects of prebiotics on mineral metabolism. *Am. J. Clin. Nutr.*, **73** (suppl): 459S-464S.

Van Dokkum, W., Wezendonk, B., Srikumar, T.S. and van den Heuvei, E.G.H.M. 1999. Effect of nondigestible oligosaccharides on large-bowel functions, blood lipid concentrations and glucose absorption in young healthy male subjects. *Eur. J. Clin. Nutr.*, **53**:1-7.

Whelan, K., Judd P.A., Preedy, V.R., Simmering, R., Jann, A. and Taylor, M. A. 2005. Fructooligosaccharides and fiber partially prevent the alterations in fecal microbiota and short-chain fatty acid concentrations caused by standard enteral formula in healthy humans. *J Nutr.*, **135:** 1896–1902.

Yamashita, K., Kawai, K. and Itakura, k. 1984. Effect of fructooligosaccharides on blood glucose and serum lipids in diabetic subjects. *Nutr. Res.*, **4**:961-966.

□□□

Chapter 32

Regulatory and Voluntary Guidelines for Probiotics

J B PRAJAPATI AND SUJA SENAN
Department of Dairy Microbiology, SMC College of Dairy Science, AAU, Anand- 388110

Introduction

Worldwide, we witness today an attitudinal change among people, leading them to acknowledge the value of peace over pace of life and health over wealth. Giving an impetus to this evolution in personal health care, is the revolutionary scientific researches that have reiterated the irrefutable role, the daily diet plays in human wellness. As a result, the development and consumption of functional foods, or foods that promote health beyond providing basic nutrition are on the rise. Striking an analogy to the quotation on a man's heart and his stomach, it won't be wrong to state that the fastest way to a consumer's pocket is through his troubled stomach. This has been the ideology behind companies like Arla, Danone, Danisco, Fonterra, Nestle, Yakult and Amul along with a large number of small food and dietary supplement companies, jumping on to the bandwagon of functional fermented dairy based foods to get a bite on health with the aid of edible bacteria with goodness, the probiotics and hence flourishing in the health food market. Unfortunately, the other side to this booming optimistic probiotic market is a growing sceptism. The aware consumers today are taking the probiotic mantra with a pinch of salt, metaphorically. Many so-called 'probiotic' products are unreliable in content and unproven clinically (Reid, 2005). This sceptism associated with the unproved beneficial claims of bacteria explains the delay of almost a century between the first established report on beneficial bacteria and today's probiotic wave. To understand the reasons for such sceptism it would be worthwhile to have a glance at the historical development of probiotics.

The tradition of taking milk fermented with beneficial microbes, runs very deep into the Indian culture. The treatise of ayurvedic medicine *Charaka Samhita* in 1000 BC speaks of *Jatharagni* (fire in the stomach) as the sustaining force of all living beings, and states that the first line of treatment for disease related to the stomach is "*Takra*" or fermented milk. There is the mention of *karambha* (blended curd-rice dish) in *Rigveda*. But the birth of the probiotic (pro – life microbes) concept occurred in 1900 when Nobel Laureate Elie Metchnikoff, at the Pasteur Institute, linked health and longevity of the Bulgarian peasants to the ingestion of bacteria present in their traditional fermented milks (Metchnikoff, 2004). He correctly assumed that, "Systematic investigations should be made on the relation of gut microbes to precocious old age, and on the influence of diets which prevent intestinal putrefaction in prolonging life and maintaining the forces of the body". Thus historically, the use of microbes was essential in food fermentations and their selection process since ages was totally based on back slopping using bacteria which we know today as Generally Regarded as Safe (GRAS). Moreover unlike drugs, probiotics lacked reports of adverse events associated with their use. Thus with such a safety record, it was comparatively easy for the present probiotic product manufacturers to piggyback on traditional use and information in public domain to advertise and market their health care products without any legislative body governing the product safety or efficacy. It was recently that probiotics were subjected to scientific scrutiny to examine their purported health benefits. Reports on problems associated with commercial probiotic products are growing parallel to the large variety of probiotics being launched in the market. The documentation is often based upon case reports, animal studies or uncontrolled small clinical trials, and few products declare the content of microorganisms (Ljung and Wadström, 2005). Peer-reviewed papers have been published worldwide, evaluating commercial probiotic-containing products. These reviews and reports have identified the following problems (1) The number of live microbes of each strain delivered through the end of shelf life is often not accurately reflected on the label; (2) Microbes contained in the product are not always named in accordance with scientifically valid nomenclature (3) Claims of efficacy are not always adequately substantiated; and (4) Use of the term 'probiotic' on labels of products with no established record of a physiological (health) benefit in humans. (Berman and Spicer, 2003; Elliot and Teversham, 2004; Huff, 2004). Furthermore, rare cases of infection, such as bacteraemia, endocarditis, pneumoniae, septic arthritis and meningitis, associated with certain *Lactobacillus* strains have been reported, mostly in immunocompromised patients (Cannon et al., 2005; Salminen et al., 2006). The risk that probiotic bacteria carrying acquired antimicrobial resistance traits may be horizontally transferable to autochthonous gut microbiota (Danielsen & Wind, 2003) and issues such as virulence and transfer of antibiotic resistance (Agostoni et al., 2004) have eaten into the confidence of consumers and the sale of manufacturers. The solution to maintain the confidence of the consumers and the sale of the manufacturers lies in the proper framework of guidelines for the selection of strains, scientific validation of claims and tests to check the safety, toxicity, efficacy and labeling of the product based on the functional properties of probiotic products. It is the onus on the scientific community and the government to establish just legislative framework, and communicate with the probiotic industry regarding use of health claims for proper labeling, marketing and advertising of wholesome probiotic. Surprisingly, till date there is neither a legally recognized definition of, nor a standard of identity for, the term "probiotic" although a number of definitions have appeared in the scientific literature since the term was coined in the 1960's.

Definitions of Probiotics

The term probiotic, meaning "for life," is derived from the Greek language. It was first defined by Lilly and Stillwell in 1965 and as the understanding increased, various definitions have come in (Table 1). Finally, a Joint FAO/WHO Expert Consultation on Evaluation of Health and Nutritional Properties of Probiotics in Food including Powder Milk with Live Lactic Acid Bacteria, redefined probiotics for the purpose of this meeting as 'Live microorganisms which when administered in adequate amounts confer a health benefit on the host'.

It is the first definition that was framed as to provide an international consensus.

Table 1: List of Definitions of probiotics given by various scientists and agencies.

Published Definition	Reference
Substances produced by one microorganism, that stimulate the growth of another.	Lilly and Stillwell, 1965
Organisms and substances that contribute to intestinal balance.	Parker, 1974
Probiotics are live microbial feed supplements which beneficially affect the host animal by improving its intestinal microbial balance.	Fuller, 1989
A viable mono- or mixed-culture of microorganisms which applied to animal or man, beneficially affects the host by improving the properties of the indigenous microflora.	Havenaar and Huis In't Veld, 1992
Living microorganisms, which upon ingestion in certain numbers, exert health benefits beyond inherent basic nutrition.	Schaafsma, 1996
Probiotics are live microbial food ingredients that have a beneficial effect on human health	Salminen et al., 1998
Probiotics are microbial cell preparations or components of microbial cells that have a beneficial effect on the health and well-being of the host.	Salminen et al., 1999
A microbial dietary adjuvant that beneficially affects the host physiology by modulating mucosal and systemic immunity, as well as improving nutritional and microbial balance in the intestinal tract.	Naidu et al. 1999
A preparation of or a product containing viable, defined microorganisms in sufficient numbers, which alter the microflora (by implantation or colonization) in a compartment of the host and by that, exert beneficial health effects in this host.	Schrezenmeir and Vrese, (2001)
Probiotics are live microorganisms which when administered in adequate amounts confer a health benefit on the host	FAO/WHO, 2001

Apart from the lack of a universally accepted legal definition of probiotics, the lack of a global legislation to make probiotic health claims and develop a suitable legislative framework to systematize the claiming procedure is creating barriers to trade. Precise and validated health claims can tell the consumers in an unambiguous way the benefits the product shall have on their health and thus help the buyers to take a rational decision. Functional foods, handicapped with a lack of a legal definition, often fall into the grey area of food and drug. This has complicated the measures to launch a probiotic product with proper health claims and emphasizes the need for secure framework for making health claims. We shall now look into the laws governing probiotic products and their health claims in the various parts of the world. Before going to the various laws, it is interesting to first read the arguments laid down by Przyrembel in 2001 (Table 2) questioning the need of legislation for probiotics.

Table 2 : Arguments against and for a special legislation on prebiotics and probiotics

Against	For
Fermented foods with live microbes do not essentially differ from probiotics	Probiotic foods differ from other fermented foods with live microbes.
The term *probiotic* need not be defined.	The term *probiotic* can be used inappropriately and mislead consumers.
The use of microbes in food is covered in existing rules and laws on food production.	Microbes are selected for their special properties, in addition to technologic considerations.
The labeling of microbes used is covered in product standards and general labeling directives.	The consumer should be informed about the probiotic properties of selected microbe strains.
Indigestible substances (ie, ingredients) are covered by existing laws on food and food labeling	Criteria defining probiotic bacterial strains and products allow control by authorities
	A prebiotic or probiotic claim offers an advantage in marketing.
	Indigestible substances (ie, ingredients) are selected for their effects on special colonic bacteria, not for technologic or nutritional purposes
	Safety

Health Claims – Definition and Types

Food labels are important tools to educate consumers about the healthfulness of specific foods and the benefits of following a nutritious diet.

Health claims describe the relationship between a substance (food, food component or dietary supplement) and a disease or a health related condition.

Nutritional Claims

1. Nutrient Content Claim - Characterizes the level of nutrient; e.g. Milk is a good source of calcium.
2. Structure/Function Claim - Describes nutrients role in health maintenance; e.g. Calcium helps build strong bones.
3. Health Claim - Describes nutrient's role in disease risk reduction; e.g. Calcium may reduce the risk of osteoporosis.

Health Claims: Any representation that states, suggests or implies that a relationship exists between a food or constituent of that food and health.

Nutrient function claim: A form of claim that refers to the physiological role of the nutrient in growth, development and normal functions of the body.

Enhanced function claim: Claims that concern specific beneficial effects of foods or their constituents in the context of the total diet on physiological or psychological functions or biological activities. Such claims relate to a positive contribution to health, improvement of a function, or modification or preservation of health.

Reduction of risk of disease: Claims relating to the consumption of a food or food constituent in the context of the total diet that may help reduce the risk of a specific disease or health related condition.

FAO Guidelines

In view of this growing popularity of probiotic foods, and the lack of international consensus on the methodology to assess their efficacy and the safety, FAO (Food and Agricultural Organization) and WHO (World Health Organization) initiated work to examine the scientific evidence on the functional and safety aspects of probiotics in food. In particular, an expert consultation on the health and nutritional properties of powder milk with live lactic acid bacteria was convened by FAO and WHO in Cordoba, Argentina in 2001, and an expert working group organized in 2002 to develop guidelines for the evaluation of probiotics in food. The resultant report not only established that there were indeed excellent data on probiotics, but that a set of Guidelines or Operating Standards was required. The guidelines constitute a set of parameters required in order for a product or strain to be termed 'probiotic'. The reports also suggest that food and small dietary supplement companies who currently produce most of the available probiotic products find the standards stringent but it is essential that they follow the guidelines (Reid, 2005). The consultation confined its discussion to the following:

- Properties of probiotic strains and their assessment
- Probiotic product specifications, quality assurance and regulatory issues
- Safety and beneficial human health effects.

Standard methods for clinical evaluations are comprised of Phase 1 (safety), Phase 2 (efficacy), Phase 3 (effectiveness) and Phase 4 (surveillance). Important guidelines for the Assessment of Probiotic Microorganisms are given below.

1) Selection of probiotic strains for human use

The group have emphasized on the specificity of the action of the microbes and not the source of the probiotic organism that is more important. The ability to remain viable at the target site and to be effective at that site should be verified for each potentially probiotic strain. The report stresses on the refinement of *in vitro* tests to predict the ability of probiotics to function in humans.

2) Classification and identification of individual strains

The Consultation recommended that probiotics be named according to the International Code of Nomenclature to ensure understanding on an international basis. The Consultation strongly urged that for the sake of full disclosure, probiotic strains be deposited in an internationally recognized culture collection. Since probiotic properties are strain related, it is suggested that strain identification (genetic typing) be performed with reliable methods.

3) Defining and measuring the health benefits of probiotics

The use of probiotic microorganisms to confer health benefits on the host must indicate the dosage regimens and duration of use. The Consultation strongly recommended that each product should indicate the minimum daily amount required for it to confer specific health benefit(s). Such evidence should, where possible result from *in vitro,* animal (where appropriate) and human studies.

4) Testing methods for establishing health benefits conferred by probiotic microorganisms

Proper in vitro studies should establish the potential health benefits of probiotics prior to undertaking *in vivo* trials. Tests such as acid and bile tolerance, antimicrobial production and adherence ability to human intestinal cells should be performed depending on the proposed health benefit (Collins et al., 1998; Havenaar and Huis in't Veld, 1992). For *in vivo* testing, randomized double blind, placebo controlled human trials should be undertaken to establish the efficacy of the probiotic product and human studies in which adequate numbers of subjects should be enrolled to achieve statistical significance.

5) Safety considerations

The FAO/ WHO Guidelines recommend that, at a minimum, probiotic strains be characterized using a series of specific tests: (a) Determination of antibiotic resistance patterns; (b) Assessment of certain metabolic activities (e.g., D-lactate production, bile salt deconjugation); (c) Assessment of side-effects during human studies; (d) Epidemiological surveillance of adverse incidents in consumers (post-market); (e) If the strain under evaluation belongs to a species that is a known mammalian toxin producer, it must be tested for toxin production (f) If the strain under evaluation belongs to a species with known hemolytic potential, determination of hemolytic activity is required; and (g) Assessment of lack of infectivity by a probiotic strain in immunocompromized animals (adding a measure of confidence in the safety of a probiotic).

6) Probiotic product specifications, quality assurance and regulatory issues

Regulatory issues: Government regulations differ among countries; however the status of probiotics as a component in food is currently not established on an international basis. For the most part, probiotics come under food and dietary supplements because most are delivered by mouth as foods. The Consultation recommends that disease reduction claims be permitted for specific probiotics if these have been demonstrated after following the proposed guidelines outlined in the report.

Appropriate labeling: To clarify the identity of a probiotic present in the food, the consultation recommends that the microbial species be stated on the label. There is a need to accurately enumerate the probiotic bacteria in food products. The label should state the viable concentration of each probiotic present at the end of shelf life

Manufacturing and handling procedures: To ensure that any given culture maintains the beneficial properties, the stock culture should be maintained under appropriate conditions and be checked periodically for strain identity and probiotic properties. Furthermore, viability and probiotic activity must be maintained throughout processing, handling and storage of the food product containing the probiotic, and verified at the end of shelf-life. Adequate quality assurance programmes should be in place.

7) Post Market Surveillance

The Consultation recommends that probiotic producers, medical professionals and public health officers consider some form of system to monitor the health outcomes of long-term probiotic administration. This is suggested as a means to gain insight into potential side effects as well as assess long-term benefits. A necessary prerequisite for surveillance is a proper trace-back system.

Situation in European Union

Sales of functional food and drink in Europe have experienced considerable growth, with a doubling of sales between 2000 and 2005. European markets for functional foods generally are less developed, compared to the US and the Japanese markets. This has been attributed to restrictive and inconsistent health claim legislation in and between the European countries. In the European Union, there is no harmonized legislation on health claims, which means that they are dealt with at a national level. Because of increasing interest in the concept of "Functional Foods" and "Health Claims", the European Union set up a European Commission Concerted Action on Functional Food Science in Europe (FUFOSE) in 1995. The programme was coordinated by the

International Life Sciences Institute (ILSI) Europe and the aim was to develop and establish a science-based approach to the evidence needed to support the development of food products that can have a beneficial effect on an identified physiological function in the body, that can improve an individual's state of health and well-being and/or reduce the risk of disease.

The main thrust of the consensus document was to suggest a scheme to link claims for functional foods to solid scientific evidence. FUFOSE, works with the following goals.

- To assess critically the science base required to provide evidence that specific nutrients and food components positively affect target functions in the body;
- To examine the available science from a function-driven perspective rather than a product-driven one ;
- To reach consensus on targeted modifications of food and food constituents, and options for their application.

The work generated by FUFOSE served as basis for the Process for the Assessment of Scientific Support for Claims on Foods project. The PASSCLAIM document delivers criteria to assess the scientific support for claims on foods. It is a consensus document that has been borne of wide and intensive consultation among diverse stakeholders including academic experts, representatives of public interest groups, regulators, and the food industry. The set of criteria developed were derived through an iterative continuous improvement process that involved developing and evaluating criteria against key health claim areas of importance in Europe today. These areas included diet-related cardiovascular disease, bone health, physical performance and fitness, body-weight regulation, insulin sensitivity and diabetes risk, diet-related cancer, mental state and performance and gut health and immunity. Expert groups reviewed the availability of indicators of health and disease states within their respective areas of expertise. The PASSCLAIM approach to claims on foods and food components addresses consumers' concerns about misleading and unsubstantiated claims on foods and, when implemented, can contribute to a higher level of confidence among European consumers

Focusing in the area of interest, gut health and immunity, PASSCLAIM have acknowledged that to substantiate a claim that a food or nutrient maintains healthy digestion or strengthens the body's natural resistance is difficult due to the very large number of commensal bacteria and the lack of proper definition of the optimal function of the integrated systems that shape, monitor and maintain the body's defenses. Hence as a first step to optimize health, the project has defined normal bowel habit and transit time, identified their role as risk factors for disease and described how they may be measured. Similarly they have tried to define what a healthy gut flora is in terms of the dominant genera and their metabolism and listed the many varied and novel methods for determining these parameters. They have further put boundaries on what constitutes optimal or improved gastric emptying, gut motility, nutrient and water absorption and the function of organs such as the liver, gall bladder and pancreas. Similarly the gastrointestinal well beings have also been discussed. Although the characteristics of well being are ill defined and merge imperceptibly from acceptable to unacceptable. Nevertheless, this is an important area of future work and method development.

There are three main reasons for assessing the scientific substantiation of claims: 1) to provide truthful information and to support consumer confidence in foods with claims, 2) to satisfy regulatory requirements, and 3) to allow fair market competition. The decision on the justification of a claim should give due weight to current knowledge and the evidence-base for outcomes, and the primary basis for allowing a claim should be the soundness of the science or evidence-base. A claim should reflect its scientific basis and, at the same time, should be understandable, and not be misleading for the intended consumer.

Through an iterative process of discussion in expert groups and workshops, a set of criteria for the scientific substantiation of claims are developed, which are given below.

Criteria for the scientific substantiation of claims

1. The food or food component to which the claimed effect is attributed should be characterised.
2. Substantiation of a claim should be based on human data, primarily from intervention studies the design of which should include the following considerations:

(a) Study groups that are representative of the target group.

(b) Appropriate controls.

(c) An adequate duration of exposure and follow up to demonstrate the intended effect.

(d) Characterisation of the study groups' background diet and other relevant aspects of lifestyle.

(e) An amount of the food or food component consistent with its intended pattern of consumption.

(f) The influence of the food matrix and dietary context on the functional effect of the component.

(g) Monitoring of subjects' compliance concerning intake of food or food component under test.

(h) The statistical power to test the hypothesis.

3. When the true endpoint of a claimed benefit cannot be measured directly, studies should use markers.

4. Markers should be: - biologically valid in that they have a known relationship to the final outcome and their variability within the target population is known; methodologically valid with respect to their analytical characteristics.

5. Within a study the target variable should change in a statistically significant way and the change should be biologically meaningful for the target group consistent with the claim to be supported.

6. A claim should be scientifically substantiated by taking into account the totality of the available data and by weighing of the evidence.

In the Europena Union food and feed area, the most relevant scientific committees have been the Scientific Committee on Food (SCF), the Scientific Committee on Animal Nutrition (SCAN) and the Scientific Committee on Plants (SCP). With the establishment of the European Food Safety Authority (EFSA) by Regulation 178/2002, system went through a radical change. The aim of EFSA is "to form a common basis of measures governing food and feed taken in the Member States and at Community level".

A regulation on nutrition and health claims made on foods was introduced in the European Union in 2007. This Regulation provides opportunities for the use of health claims on foods in Europe, including reduction of disease risk claims. The Regulation will be fully implemented by January 2010. A community list of permitted and rejected claims will be established and made available to the public. The newly adopted EU legislation on nutrition and health claims made on foods will specify in detail the conditions for their use and establish a system for scientific evaluation of claims. The claims on foods will be broadly categorized into four groups: (I) 'content' claims listed in the Annex including nutrient content claims, comparative claims and claims for 'other substances' with a nutritional or physiological effect; (II) health claims based on generally accepted scientific evidence under Article 13.1; (III) health claims based on newly developed scientific data and for those requiring IPR protection under Article 13.5; (IV) and reduction of disease risk claims and claims referring to children's development and health under Article 14. The substantiation of health claims will be accomplished by considering the totality of the available data and by weighing of the evidence. The emphasis will be on the benefit conferred, and the methods used to obtain and measure the claimed benefits will need to be consistent with the FAO/WHO Guidelines 2002 and with the PASSCLAIM criteria for the scientific substantiation of claims.

With the health and wellness mega trend driving growth in probiotics and other health-promoting dairy cultures, the EU's impending Nutrition and Health Claims Regulation is a mixed blessing for the industry. The major players are already working to get their health claims onto the approved list under Article 13 of the legislation. But some of their more cautious dairy customers are delaying the development of new products until they know for sure what they will and will not be able to claim after the regulation comes into force. The legislation will also help to boost consumer confidence that any health claims are fully justified. There could be fewer products, due to the over stringency but any claims will be based on scientifically and clinically proven evidence. This seems to be the popular consensus among the experts, manufacturers and educated consumers of the European Union.

United Kingdom

The Joint Health Claims Initiative (JHCI) is the UK's response to the growing need for clarity about the use of health claims and was instigated by representatives from the food industry, consumer and enforcement groups. A voluntary Code of Practice (Joint Health Claims Initiative,2000) was launched in December 2000 and since then it has provided pre-market, product-specific advice to the food industry, maximizing the opportunity for food manufacturers to convey health messages whilst minimizing their risk of being challenged. The JHCI does not seek to replace current regulatory and self regulatory controls; rather, it aims to enhance these systems, by clarifying and strengthening the requirements for evidence to substantiate health claims on food, beverages and dietary supplements. A Code Administration Body (CAB) has been established to supervise the operation of the Code and provide practical guidance to interested parties. JHCI has produced step-by-step guidelines based on its own practical experience and designed to meet the requirements of the claim assessment process. JHCI recommends that the Commission refer to "disease risk-reduction claims" rather than the "reduction of disease risk factor claims", as the latter tends to suggest the reduction of a biomarker rather than the reduction of risk of developing disease as a result of making wise dietary choices.

Swede Code for Probiotics

Sweden was the first country to allow certain health claims within a Code of Practice introduced in 1990 (before entry into the EU). The Swedish Nutrition Foundation (SNF) plays an advisory role and has been the coordinating body. Principals of the Code are the Swedish Food Federation and the Swedish Food Retailers Federation. Sweden developed a self-regulating programme, which originally allowed mention of eight diet-related diseases or risk factors in connection with the relevant nutrient or dietary fibre content of a food. The eight connections were (1) Obesity – energy; (2) Cholesterol level in the blood - saturated fatty acids and some types of dietary fibre; (3) Blood pressure – salt; (4) Atherosclerosis - factors affecting blood cholesterol and blood pressure, and omega-3- fatty acids; (5) Constipation - dietary fibre; (6) Osteoporosis - calcium; (7) Caries - easily fermentable carbohydrates; (8) Iron deficiency – iron; (Coronary) heart disease – wholegrain.

Since 2001, product-specific health claims have also been permitted on food products, which, through their composition, contribute positively to a nutritionally adequate diet. The Code of Practice, Rules for Health Claims in the Labeling and Marketing of Food Products, is implemented by a coalition of farmers, food trade and retail organizations. However, by 1 July 2007 the EC-Regulation (1924/2006) on nutrition claims and health claims on foods was applied in all EC Member States. However, during the implementation of the Regulation, parts of the Code may be useful as guidelines on how to apply health claims in a responsible manner on a national level (Bryngelsson and Nils-Georg, 2007).

The Code is applicable to regular foods, but explicitly not to supplements and other foods appearing as medicinal products, e.g. in the form of capsules and pills. The Regulation, in contrast, is applicable also to food supplements. Three categories of health claims are defined by the Code : (i) generic nutrient function claims (ii) generic reduction of disease risk claims and (iii) product-specific physiological claims (PFP).

The Swedish Code stresses that wordings of generic claims should ensure that they are not perceived as product specific, i.e. they must not indicate that the particular product in question has a certain effect on disease risk reduction. It would be especially important to put such claims in the context of a balanced diet. For this purpose, the two-step principle was introduced in the Code, and required for disease risk reduction claims. The two-step principle means that (i) the connection between diet and a physiological function/reduced risk of disease and (ii) the product's composition are given separately, but in connection to each other. Within the Code, probiotic claims have therefore been considered as PFP, requiring scientific documentation on the final product. Those evaluated and approved within the Code will be suggested by SNF for the national Article 13 list. The Code includes a generic nutrient function claim for dietary fibre referring to "bowel regularity". More specific claims on prebiotic effects, e.g. "stimulates bifidobacteria" or "increases calcium absorption" have not been scientifically evaluated within the Code, and are therefore not included in the SNF list.

Japanese Situation

Japanese have a long history and recognition in eating nutritious foods to which they credited their excellent health record and longevity. Ministry of Health and Family Welfare (MHLW) conducted studies on functional foods, led by the Special Study Group on the Systematic Analysis and Development of Food Function. This project was focused on investigating the benefits of functional foods in controlling the physiological function of a living body (Arai, 1996). The reports stressed on the tertiary function of foods, directly involved in the modulation of the human physiological systems such as the immune, endocrine, nerve, circulatory as well as digestive systems, while the primary and secondary functions related to nutrition and sensory satisfaction, respectively. The concept of functional foods grew from the tertiary function and Japan originated the concept of "functional foods" in the eighties and is the only nation that has legally defined functional foods. The Government, however, prohibited the use of the term 'functional' because of its implication as a drug-like effect. As a result, the term 'health foods' was widely used and recognized by consumers and even became part of their diets, replacing the term 'functional foods'. The concept of the 'functional foods' was ultimately integrated into the 'foods for specified health uses' (FOSHU) system. (Ohama *et al.,* 2006). The positive progress of Japan's FOSHU program has created food concepts that may eventually be applied in other industrialized nations. FOSHU has a number of unique features:

1. It is a voluntary procedure but has government approval
2. Specific health claims are approved prior to use
3. Approval is based on documented scientific safety and effective considerations

Approved products may use the FOSHU label on the products.

The Ministry of Health, Labor and Welfare (MHLW) decided to frame a new regulatory system of 'health foods' called 'foods with health claims' (FHC) and 'foods with nutrient function claims' (FNFC) in April 2001. In addition, MHLW has changed the existing FOSHU, FNFC and other systems in 2005. Such changes include the new subsystems of FOSHU such as (1) standardized FOSHU, (2) qualified FOSHU and (3) disease risk reduction claims for FOSHU.

The leading category of FOSHU products is made up of strain-specific (15 different bacteria are now FOSHU functional ingredients) probiotic products with on-label claims for improved gastrointestinal health. Yakult Honsha has converted nearly its entire line of yogurts and cultured lactic acid bacteria beverages to FOSHU status in recent years. 336 products like Soft drink, yogurt, biscuit cookie, table sugar, soyabean curd, vinegar, chocolate, powdered soup, fermented milk, yoghurt, miso soup, cereal have been approved by FOSHU till date with prebiotics and probiotics giving the functionality with the health claim stated as "Food that improve gastrointestinal conditions" The Japaneese FOSHU can act as a role model to other developing countries in developing their own system of regulating the functional food market especially to keep in check the large number of manufacturers who would tend to make exaggerated claims to cash on the consumer's health concerns.

To acquire a FOSHU label, applications are submitted to local regional authorities and to the Ministry of Health and Welfare (MHW). The application process takes about 1 year. Companies making an application are required to provide the following.

1. Scientific documentation demonstrating the medical or nutritional basis for a health claim
2. The basis of the recommended dose of the functional food of the functional ingredient
3. Information demonstrating the safety of the ingredient
4. Information on physical and chemical characteristics
5. Relevant test methods
6. Compositional analysis
7. Scientific papers in the fields of medicine/ nutrition that substantiate the health claims.

Although the MHW is responsible for overseeing the FOSHU regulations, the regulatory process is administered by a private industry group, the Japan Health Food and Nutritional Food Association (JHFNFA). Because the FOSHU system is voluntary, companies still market more than 1,000 functional foods that are not FOSHU-approved. Some companies, generally smaller ones, simply make express health claims for unapproved products in violation of the law. Functional foods sold outside the FOSHU system are simply marketed as "health foods." There is no law defining health foods, and manufacturers are free to market any food as a "health food" as long as they do not violate the Food Hygiene Law or Pharmaceutical Affairs Law.

The US Situation

In both food and drug areas, the US regulatory authority is the FDA. In the USA, food products containing probiotic organisms may not systematically claim health benefits. FDA has defined four categories of foods and the intended use determines how the food is regulated. The categories are:

- *Conventional foods*:
- *Dietary supplements*: Foods intended to supplement the diet and are marketed as conventional foods but they are to be labeled as dietary supplements.
- *Foods for special dietary use*: foods intended to supply special; dietary need for physiological conditions like weight gain, infant formulas, food allergy etc.
- *Medicinal foods*: Foods intended for dietary management of specific disease, under the supervision of doctor or health professional.

Probiotic products, as other nutraceuticals, can fall under several Statutes and it is not always clear where the legislation draws the line between the categories. Considering the above categories probiotics could fit into any of the above catergories; but they are seldom seen in the category of medicinal foods or foods for special dietary use. A number of conventional food in US contain probiotics especially yoghurt, cultured milk and cottage cheese. The main market for probiotics in the USA is under the dietary supplements category in the form of pills, powders and drinks. The reason is that prior approval of structure function claims is not required for this category. FDA's post-marketing responsibilities include monitoring safety, e.g. voluntary dietary supplement adverse event reporting, and product information, such as labeling, claims, package inserts, and accompanying literature. The Federal Trade Commission regulates dietary supplement advertising. Regarding claims and labeling Nutrition Labeling and Education Act - NLEA (1990) allowed health and disease preventional claims on food labels. The Dietary Supplement Health and Education Act - DSHEA (1994) created another category of statements generally referred to as structure/functional claims that can be made for dietary supplements. These statements may claim a benefit related to a nutrient deficiency disease or describe the role of a nutrient or dietary ingredient to maintain such structure or function in humans. The manufacturer is responsible for ensuring the accuracy and truthfulness of these claims and are also expected to include a disclaimer along with the claim on a dietary supplement that FDA has not evaluated the claim. Many probiotics sold as dietary supplements may make structure and function claims like " when taken daily, helps fortify your body's natural balance and help keep your body at its best" (Actimel, Danone). Food and Drug Administration Modernization Act -

FDAMA (1997) provided a second way to use health claims in foods. This allows certain health claims to be made on account of the submission of a notification based on the authoritative statement from a scientific body of the US government or the National Academy of Sciences. This method of managing health claims cannot be applied for dietary supplements. As can be seen from the list of FDA approved health claims none of the claims can be applied to probiotic products. Statements such as probiotics help maintain proper digestion is more of a structure function statement than a health claim.

FDA's Center for Food Safety and Applied Nutrition (CFSAN) regulates probiotics and prebiotics marketed as dietary supplements or food ingredients. A probiotic used to diagnosis, cure, mitigate, treat, or prevent disease is a drug and a biological product. FDA's Center for Biologics Evaluation and Research (CBER) regulates probiotic products when used for clinical indications. CBER's Office of Vaccines Research and Review has regulatory jurisdiction over most probiotic products for clinical use. CFSAN, in conjunction with the Agency's field staff, is responsible for promoting and protecting the public's health by ensuring that the nation's food supply is safe, sanitary, wholesome, and honestly labeled, and that cosmetic products are safe and properly labeled.

Canada

In Canada, probiotics are defined by the Natural Health Products Regulations as a monoculture or mixed culture of live microorganisms that benefit the microbiota indigenous to humans. A probiotic is limited to non-pathogenic microorganisms and are regulated by the Natural Health Products Directorate (NHPD). Probiotics are also sold in foods (almost exclusively dairy products) regulated by Food Product Directorate.

Products on the Canadian market can carry three types of health claims: structure/function, non-specific risk reduction or treatment. These are assessed by the NHPD based on the credibility, strength and quality of evidence provided. In the NHPD, products can be considered as either 'traditional' or 'non-traditional', Traditional use claims encompass the sum total of knowledge, skills and practices based on the theories, beliefs and experiences indigenous to different cultures, used in the maintenance of health. They require a product to have a history of at least 50 consecutive years of use. Non-traditional claims occur when the evidence based on scientific data and all three types of claims are considered.

Australia and New Zealand

Food Standards Australia New Zealand (FSANZ) is an independent statutory agency established by the Food Standards Australia New Zealand Act 1991. Working within an integrated food regulatory system involving the governments of Australia and the New Zealand, set food standards for the two countries. Food standards provide the minimum regulatory burden necessary to maintain a safe food supply and informed consumers. The regulations currently restrict the use of claims on food to nutrient content and health maintenance claims. Claims about the prevention or reduction of disease risk are currently prohibited (with one exception about folate) although this may soon change (Food Standards Australia New Zealand, 2007).

Nonetheless, health claims are already present in the Australian food market on labels, in advertising and the internet (Williams et al., 2007). The system is divided between high and general-level claims. General-level claims consist of claims relating to the ability of foods to affect performance, well-being and non-serious disease. There is a pre-approved central list of claims. "Calcium is good for strong bones and teeth, when consumed as part of a healthy diet containing a variety of foods" or "Gives you energy, when consumed as part of a healthy diet with a variety of foods" are examples. A probiotic claim in this field may state something like: "Yoghurt high in strain X and Y may reduce your risk of stomach upset, when consumed as part of a healthy diet with a variety of foods."

Higher level claims require pre-market approval and refer to the ability of a food to reduce the risk of disease, control disease or improve health. Scientific substantiation is required for each along with dosage information. There is a split in regulations governing dietary supplements because supplements in Australia are regulated under the Therapeutic Goods Act, which very strictly controls health claims under a centralized registry. New Zealand's regulations, while operating under a different structure, are similar in effect. *Bifidobacterium* species and *Lactobacillus* species are listed as approved active ingredients for complementary medicines.

India

The foods are covered by PFA and drugs are covered by FDA in India. However, laws for functional foods and probiotic are not clear. The guidelines are being set by Ministry of Health, Ministry of Food Processing Industries, ICMR and some other agencies.

India's regulatory environment for nutraceuticals, dietary supplements and functional foods is in transition. Historically, these sectors were governed under multiple laws, but in 2006, India passed the Food Safety and Standards Act to integrate and streamline the many regulations covering nutraceuticals, foods and dietary supplements. The act calls for the creation of the Food Safety and Standards Authority (FSSA, 2005). Once established, the FSSA will be charged with drafting rules and regulations for companies in the food sector to be licensed by local authorities, and a system of checks and balances, including product-recall procedures, enforcement and penalties. FSSA defines foods for special dietary uses or functional foods, nutraceuticals or health supplements. It is a relatively liberal regulation that permits claims as long as they are not misleading. Supplements are more strictly controlled, with most considered as drugs and pre-market approval is required for probiotic health claims.

Conclusion

As the awareness towards 'food for health' increases, all the nations of the world have become active to safeguard consumers from fake health claims by making regulations for probiotics and functional foods. However, due to dynamic nature of livings systems (the host as well as the product), it is difficult to generalize effects and make hard and fast rules/standards for the products containing live probiotic bacteria. As the understanding is increasing, the regulations are getting evolved and standardized. Taking the role model from Japan and guidelines proposed by FAO/WHO, many countries are now ready to offer legal and voluntary guidelines for probiotics.

References

Agarwal S, Hordvik S, Morar S. (2006). Nutritional claims for functional foods and supplements. *Toxicology*, **221:** 44–49.

Aggett J, Antoine JM, Asp NG, Bellisle F, Contor L, CummingsJH, Howlett J, Müller DJG, Persin C, Pijls LTJ, Rechkemmer G, Tuijtelaars S and Verhagen H.(2005). Passclaim Process For The Assessment of Scientific Support for Claims on Foods Consensus on Criteria. *Eur. J. Nutr.* **44:** Supplement 1.

Agostoni, C., Axelsson, I., Braegger, C., Goulet, O., Koletzko, B. and Michaelsen, K. F. (2004). Probiotic bacteria in dietetic products for infants: a commentary by the ESPGHAN Committee on Nutrition. *Journal of Pediatric Gastroenterology and Nutrition*, **38**(4), 365-374.

Berman S and Spicer D. (2003). Safety and Reliability of Lactobacillus Supplements in Seattle, Washington (A Pilot Study). Internet Journal of Alternative Medicine. Volume 1 Number 2.

Bryngelsson S and Asp NG.(2007). Health claims according to Article 13 of the EC Regulation: suggested priorities with reference to the Swedish Code on health claims and emphasis on relevance. *Scandinavian Journal of Food and Nutrition*; **51**(3): 127-136

Claims That Can Be Made for Conventional Foods and Dietary Supplements http://www.cfsan.fda.gov/~dms/hclaims.html

Codex Guidelines: Food Labelling - Complete Texts (2005) http://www.codexalimentarius.net/web/publications

Collins JK, Thornton G and O'Sullivan GO (1998). Selection of probiotic strains for human applications. *Int Dairy J*, 8: 487-490.

Council Regulation (EEC) No. 2092/91 on organic production of agricultural products and indications referring thereto on agricultural products and foodstuffs. Brussels: EEC, 1991.

Danielsen, M., & Wind, A. (2003). Susceptibility of *Lactobacillus* spp. to antimicrobial agents. *International Journal of Food Microbiology*, **82**(1), 1-11.

Elliot E and Teversham K. (2004). An evaluation of nine probiotics available in South Africa. *S Afr Med J*., **94**(2):121-4.

FAO/WHO Food Standards Programme. Codex General Guidelines on Claims Joint Codex Alimentarius. Vol 1a. Rome: FAO/WHO, 1998. (CAC/GL 1-1979, Rev 1-1991.)

FDA, 2004. Code of Federal Regulation, Title 21, Part 101 Food Labeling, http://www.access.gpo.gov/nara/cfr/waisidx 04/21cfr101 04. html.

FDA/CFSAN Issued Documents – "Overview of Dietary Supplements" Issued by the Center for Food Safety and Applied Nutrition (CFSAN), US FDA, January 3, 2001 http://www.cfsan.fda.gov/~dms/ds-oview.html

Fuller, R. (1989). Probiotics in Man and Animals. *J. Appl. Bacteriol.* **66:** 365-378

GRAS • http://www.cfsan.fda.gov/~dms/grasguid.html

Guidance for Industry Structure/Function Claims Small Entity Compliance Guide. http://www.cfsan.fda.gov/~dms/sclmguid.html

Havenaar R and Huis in't Veld JHJ (1992). Probiotics: A general view. In: Wood BJB: The Lactic Acid Bacteria, Vol. 1: The Lactic Acid Bacteria in Health and Disease, Chapman & Hall, New York, NY: 209–224.

Health Claims in the Labelling and Marketing of Food Products – The Food Sector Code of Practice Revised Version September 2004 http://www.snf.ideon.se/snf/en/rh/Healh_claims.

Huff BA. (2004). "Probiotics" might not be what they seem. *Can Fam Physician.* **50:**583-7.

Joint FAO/WHO Food Standards Programme. Report of the 26th Session of the Codex Committee on Food Labeling. Ottawa: FAO/WHO, 1998 May 26–29. (Alinorm 99/22.)

Lilly, D.M. and Stillwell, R.H. (1965). Probiotics: Growth Promoting Factors Produced by micro- organisms. *Science* **147:** 747-748.

Ljung, A., and Wasdström, T. (2005). Lactic acid bacteria as probiotics. Curr. *Issues Intestinal Microbiol.* **7:** 73-90.

Metchnikoff, E. (2004). The prolongation of life: Optimistic studies (reprinted edition 1907). New York, USA: Springer Publishing Company.

Ohama H, Ikeda H and Moriyama H. (2006). Health foods and foods with health claims in Japan. *Toxicology* **221:** 95–111.

Parker, R.B. (1974) Probiotics: the Other Half of the Antibiotic Story. *Anim. Nutr. Health* **29:** 4-8.

Przyrembel H. (2001). Consideration of possible legislation within existing regulatory frameworks. American *J. Clin. Nutr.,* **73**(2) 471S-475s.

Reid G. (2005). The Importance of Guidelines in the Development and Application of Probiotics. *Current Pharmaceutical Design,* **11:** 11-16.

Salminen, M. K., Rautelin, H., Tynkkynen, S., Poussa, T., Saxelin, M. and Valtonen, V. (2006). Lactobacillus bacteremia, species identification, and antimicrobial susceptibility of 85 blood isolates. *Clinical Infectious Diseases,* **42:** 35-44.

Salminen, S., von Wright, A., Morelli, L., Marteau, P., Brassard, D., de Vos, W., Fonden, R., Saxelin, M., Collins, K., Mogensen, G., Birkeland, S.-E. and Mattila-Sandholm, T. (1998) 'Demonstration of Safety of Probiotics - a Review. *Int. J. Food Microbiol.* **44:** 93-106.

Schrezenmeir J, Vrese M. Probiotics, prebiotics, and synbiotics-approaching a definition. *Am J Clin Nutr* 2001; **73**(suppl):361S-364S.

Tamime Adnan (2005), Current legislation of probiotic products. In: Probiotic dairy products. Blackwell Publishing, pp. 73-89.

□□□

Chapter 33

Calcium Fortified Milk and Products

SUMIT ARORA
Dairy Chemistry Division, National Dairy Research Institute, Karnal

Introduction

Food manufacturers add vitamins and minerals to food products to attract the attention of health-conscious consumers (Weaver, 1998) and to respond to public health needs. Interest in fortifying foods with calcium is growing due to the difficulty for many individuals to meet calcium requirements with unfortified foods; the large incidence of osteoporosis, especially in elderly women; and the marketing advantage of using a health claim linking calcium to osteoporosis. Calcium is an essential mineral for human health (e.g., bone strength). Indians are more prone to osteoporosis, a debilitating bone disorder resulting from porous bones, with an estimated five crore people suffering from life threatening disease. Major factors responsible for high prevalence of osteoporosis in India are inadequate intake of calcium and Vitamin D. As a part of preventive measure, the Indian society for bone and mineral research (ISBMR) suggested to the authorities and industry to go for calcium fortification of food products (Hindustan Times, 2002). Calcium requirements vary throughout an individual's life and for different population groups. Thus it is imperative that dietary intake of calcium be increased in population groups that have sub - optimal calcium nutritional status. Poor dietary habits are also seen to be responsible for this situation, especially when consumption of fast food is gaining acceptance. If a calcium supplement is given to a population which already has an adequate supply, this does not affect growth in terms of height, but increases bone density (Danone, 1999).

Cow and buffalo milk contains 125 to 150 mg/100 g and 180 to 200 mg/100 g, calcium, respectively. ICMR (1989) recommends RDA for calcium as 400 to 600 mg (1000 mg in pregnancy/lactation) whereas NIH (1994) recommends 1200 to 1500 mg of calcium per day. High intake of calcium will only be possible through formulating calcium enriched dairy products. Addition of calcium to milk resulted in a new distribution of ions between milk serum and casein micelles, leading to physicochemical changes in casein micelles.

Fortification with Calcium

Calcium is one of the most essential nutrient and must be provided regularly in the diet. This mineral alone constitutes about 2 per cent of the total body weight and most of this is distributed in the bones, therefore, an adequate daily intake of calcium is vital for maintaining structural integrity of the body skeleton and strength of the teeth. Besides its structural role, calcium is equally important in regulating various metabolic processes in the human body. It also plays a role in the transmission of impulse in nerve cell and in case of injury forms an important component of the blood clotting. More recently, central role of calcium in the prevention and treatment of a number of diseases i.e. osteoporosis, hypertension, colon cancer, kidney stones and lead absorption has also been acknowledged (National Dairy Council, 1992).

The average rate of calcium accumulation in the skeleton between birth and maturity is of the order of 60 mg per day but the most rapid deposition occurs during adolescence when the positive balance required rises to about 300 mg or more daily. There is general agreement that extra dietary calcium is required for optimum skeletal growth at this time, with recommended intakes at puberty around 1200 mg or more (Vital news, 2000)

Selecting the Calcium Fortificant

The ideal calcium source for fortification of foods should be highly absorbable and available to enhance bone mass, inexpensive and safe for human consumption. The ultimate nutritional benefit of the calcium source must aim at increasing peak bone mass during growth and to reduce bone mass later in life (Weaver, 1998). Several commercial calcium salts are available to the food manufacturers for use in food and dietary supplements e.g. inorganic salts like calcium carbonate, calcium chloride, calcium phosphate and organic salts like calcium citrate, calcium lactate, calcium gluconate. The salt best suited for fortification should possess a high nutritional value and low interference with the absorption of other nutrients besides its cost effectiveness and minimal effects on consistency, mouthfeel and taste of the product. In general, organic salts of calcium are more bioavailable than the inorganic salts (Goldscher and Edelstein, 1996). The inorganic salts (Calcium carbonate 40 %, phosphate 31% and chloride 36%) contain more calcium than organic salts (lactate 13 %, citrate 21% and gluconate 9%) but their application in food and beverages remain limited due to their poor solubility and the way in which they influence taste in final applications.

Table 1: Solubility and Absorbability of Calcium salts

Source	Solubility mM/L	Fractional Absorption
Calcium Oxalate	0.04	0.102±0.040
Calcium carbonate	0.14	0.296±0.054
Tri calcium phosphate	0.97	0.252±0.130
Calcium citrate	7.3	0.242±0.049
Calcium citrate Malate	80	0.363±0.076
Bisglycin calcium	1500	0.440±0.104

According to Weaver (1998) there is no significant difference **(Table 1)** in the bioavailability of most of the calcium compounds available for fortification of food and the manufacture of nutritional supplements, so for most applications there is no need to use an expensive calcium source. Using milk calcium, as a fortificant is also gaining momentum because it has beneficial sensoric properties such as neutral taste and bland odour. It also offers marketing benefits in terms of labelling and claims as it is derived from milk (Witte, 2000). Evidence suggests that salts such as calcium oxalate and calcium carbonate with molecular weights under 100 Kd can be absorbed intact by passive diffusion rather than by vitamin D dependent active mechanism which necessitates calcium dissociation from any complex (Benway, 1993).

Soy milk has also been fortified (Valente Alvarez, 1996) using a mixture of calcium gluconate and calcium-D-saccharate as the calcium source and protecting the stability of the soymilk product by using sodium hexametaphosphate as a calcium sequestering agent. LactaidÒ low fat, which is a lactose free milk marketed by McNeil consumer healthcare, Ft.Washington, PA., fortified with tribasic calcium phosphate, has 500 mg of calcium per serving, 200 mg more than standard milk products (Klahorst, 2001). In addition to traditional dairy products such as milk and Yoghurt drinks, new concept and ideas are being developed for calcium and other micronutrient fortified beverages to help consumers meet the recommended daily intake.

Calcium Sources and Bioavailability

Calcium source is only present in significant amounts in a limited range of foods. Some vegetable, e.g., green leafy one, such as broccoli provide quite large amounts of calcium, others include cabbage, beans, watercess. Milk and milk products are a major source of calcium. Calcium levels are not reduced when milk is skimmed. The adsorption of calcium is affected by its solubility and chemical form and by the presence of inhibitors in the food or total diet (Allen, 1982).

The bioavailability of calcium has been shown to be reduced by a variety of dietary components capable of forming insoluble complexes with calcium in the intestine. These include phytate, oxalate and uronic acid and other components of non-starch polysaccharide (NSP) (dietary fibre). Calcium in spinach is less well utilized than calcium from milk or from other green vegetables such as broccoli. The high oxalate content of spinach has been assumed to reduce the absorption of calcium by forming insoluble calcium complexes in the intestine (Heaney, 1988). Calcium in milk is easily absorbed by the intestine and is considered to have higher bioavailability than calcium in vegetables or cereals. Two factors in milk - lactose and casein, have been suggested to be responsible for promoting calcium absorption and leading to apparent higher bioavailability (Miller, 1989).

Lactose delays the precipitation of calcium in vitro by forming calcium-lactose complexes (Charley and Saltman, 1963) and it may also do so in the intestinal lumen. According to Kansal(1998) Lactose stimulates passive component of calcium absorption in the illeum and the colon in rats. Lactose enhances calcium absorption in Vitamin D deficient as well as Vitamin D adequate animals. Casein may also act by increasing the proportion of soluble calcium in the intestinal lumen (Lee *et al.*, 1979). It is possible that the higher bioavailability of milk calcium may merely be due to the absence of factors, which normally inhibit the absorption of calcium such as phytate and oxalate. The type of calcium source did not affect calcium absorption, but milk components increased the apparent calcium absorption.

Milk and other dairy products are the most important sources of calcium in readily available form. There is an increasing trend in producing calcium and vitamin D enriched milk and other drinks. A part of calcium in milk is associated with the casein fraction, where it associated with phosphorylated serine. The caseinophosphopeptides (CPP) released in the gastrointestinal tract during normal digestion can improve passive transport of calcium in the lower portion of small intestine. These CPP forms soluble complexes with calcium phosphate salts and thus improve calcium absorption (Kansal, 1998). Sodium intake raises the urinary excretion of calcium thus increasing the amount of calcium that has to be absorbed from diet to maintain calcium balance, otherwise calcium would be resorbed from the skeleton to compensate the reduced bioavailability (Goulding and Mclntosh, 1986).

In infants, the protein and calcium requirements are generally met from milk. In fact, the micellar form of calcium found in milk is best absorbed and has high (almost 2 times) bioavailability compared to that from nuts and vegetables (like sesame seeds, almonds, whole wheat bread, spinach etc). The reason for the poorer bioavailability of calcium through vegetarian source is the presence of chelators like phytate and oxalate in vegetables, which makes calcium insoluble and unavailable for absorption. However, regular intake of milk will help in meeting the daily protein and calcium requirements (Neelam and Hemalatha, 2001).

Interaction between Calcium and Other Minerals

It has been suggested that a high intake of calcium may impair the absorption of a number of other minerals including zinc, copper, magnesium and iron. However, the results of studies in human indicate that under normal dietary conditions, the absorption of zinc, copper and magnesium is not impaired to any great extent (Sandstrom and Lonnerdal, 1989; Lonnerdal, 1989). Jackson and Lee (1992) concluded from a review of the effect of dairy products on iron absorption that their nutritional benefits outweigh any small inhibitory effect they might have on iron availability.

Dietary Factors Affecting Calcium Bio Availability

Optimal bone mineralization obviously requires calcium, but phosphorous must be present for the formation of hydroxyapatite. The dissociation of calcium intake from that of phosphorous may restrict calcium and phosphorus retention. A calcium source with a ratio Ca/P > 1 is a plus. Lactose is reported to stimulate calcium passive absorption (Vital news, 2000).

Because it is bound to peptides and proteins, milk calcium is more likely to remain in solution when the pH is unfavourable. Milk calcium may be absorbed in distant intestine in the absence of vitamin D under the influence of lactose. Dairy products do not contain anything likely to inhibit the intestinal absorption of calcium like phytates, oxalates, uronic acids or the polyphenols of certain foods. The hyper calciuric effect of sulfates from the sulphur containing amino acids of milk is offset by the hypocalciuric effect of phosphorous (Vital news, 2000). The uptake of calcium is increased when it is ingested during a meal and combined with protein and lactose as in case of dairy products (Danone World News Letter, 1999).

The choice of calcium fortification compound may be made more on technological grounds, since there is little variation in the bioavailability of the calcium from different compounds (Couzy, 1996). Only a fraction (25 to 35%) of the calcium in the diet is absorbed, most of the dietary calcium stays in the intestinal lumen and contributes to the daily loss of calcium.

Mechanism Involved in Fortification

Calcium fortification of milk and milk based beverages has been carried out. Calcium fortificant preparations including stabiliser and emulsifiers have been used for this purpose to maintain calcium in suspension so as to improve mouthfeel and appearance of products (FAO, 1995). An important relationship exists between calcium addition and protein concentration, which is one of the determining factors for successful heat stability (Valente Alvarez, 1996). The ratio of calcium-to-protein may not exceed 40 mg calcium per gram of protein in order not to swamp the protein with calcium. It is the overload of calcium on the protein that is the first determining factor. Another important processing step involves the careful control of final pH followed the calcium salt addition. Finally, the use of a calcium-sequestering agent must be a controlled addition, dependant upon the actual calcium load of the system. Upon addition of calcium salt, there is a marked drop in the pH, which will render the system unstable unless it is corrected to near neutrality by the addition of a food grade base (Arora, 2005).

Conclusion

The ongoing success of calcium-fortified dairy products speaks for itself. Consumers increasingly prefer food with added nutritional value rather than taking nutritional supplements to fill up their calcium needs. Milk products are seen to be one of the best means to help fight calcium deficiency; they are further enriched with calcium in order to achieve higher calcium intakes per serving. However, the challenge for Dairy and food manufacturers is to provide a product having high calcium content with good taste and appealing properties.

Suggested Reading

Arora, S; Ranjan, P; Sharma, G.S; Sindhu, J.S; Singh, G and Kansal, V.K. (2005). Effect of calcium fortification on heat stability and physico - chemical properties of mixed milk (cow and buffalo 1:1). *Indian J. Dairy Sci.*, **58**(4): 242-246.

Allen, L.H (1982). Calcium bioavailability and absorption: a review. *Am. J.Clin. Nutr.* **35:** 783–808.

Benway, D.A; Weaver, C.M and Magnusen, H.M (1993). Intestinal absorption of calcium oxalate. IFT book of abstracts, 443 abstract.p114.

Charley, P and Saltman, P (1963). Chelation of calcium by lactose: its role in transport mechanism. *Science* **139:** 1205–1206.

Couzy.F (1996) Foods naturally rich in calcium or fortified with calcium. Abst no 073 published in 1st World Congress on calcium and vitamin D in human life. 8–12 Oct, Rome, Italy.

Danone World News Letter (1999) Nutritional and Health benefits of Yoghurt and Fermented milks. Calcium and bone growth. No. 19. Centre International de Recherche Daniel Carasso. France.

FAO (1995). Micronutrient fortification of food: Technology and quality control. Technical consultation on food fortification. Rome, Italy. 20–23 Nov.

Goldscher, R.L and Edelstein, S (1996). Calcium citrate: A revised look at calcium fortification. *Food Technol.,* 96-98.

Goulding, Aand Mclntosh, J (1986). Effects of sodium chloride on calcium balance, parathyroid function and hydroxyproline excretion in prednisolone-treated rats consuming low calcium diet. *J. Nutr.* **116:** 1037 - 1044.

Heaney, R.P (1988). Calcium absorbability from Spinach. *Am. J. Clin. Nutr.* **47:** 707 - 709.

Hindustan Times (2002). Indians more prone to osteoporosis: PTI. Hyderabad, oct 21st .

ICMR (1989). Nutrient requirements and recommended dietary allowances for Indian. Indian council of medical research, National Institute of Nutrition, Hyderabad.

Jackson, L.S and Lee, K (1992). The effect of dairy products on iron availability. *Crit. Revi.Food.Sci.Nutr.* **31:** 259 - 270.

Kansal, V.K. (1998). Milk offers dietary calcium in best available form. *Ind. Dairyman.* **50**(8) 23–27.

Klahorst, S (2001). Calcium's role. Food Product design: Design Elements. www.foodproductdesign.com.

Lee, Y.S (1979) An enhanced intestinal absorption of calcium in the rat directly attributed to dietary casein. *Agri. Bio. Chem.* **47:** 2009 - 2011.

Lonnerdal, B (1989). Food and dietary factors influencing levels and bioavailability of trace elements. In southgate, D; Johnson, I and Fenwick, G.R Ed Nutrient availability: chemical and biological aspects. *London, Royal Society of Chemistry.* 131- 139.

Miller, D.D (1989). Calcium in the diet: food sources, recommended intakes and nutritional bioavailability: *Adv. food nutr. Res.* **33:** 103 - 156.

National Dairy Council (1992). Calcium and Health. Fact file no 1. Nutrition service, 5-7 John Princes street, London.

Neelam and Hemalatha, R (2001). Some frequently asked questions on milk. *Nutrition.* **35**(2): 27-29.

NIH (1994). National Institute of Health Consensus Conference. Optimum Calcium Intake. *J.Am.Med Assoc.* **272**(24):1942-1948.

Sandstrom, B and Lonnerdal, B (1989). Promoters and antagonists of Zinc absorption, in Mills, C.F Ed Zinc in human Biology, London: Springer - Verlag: 57–78.

Valente Alvarez (1996) Improved method for calcium fortification of Soy milk. Dept of Food Sci, College of food, agriculture and environmental Science. Ohio State Univ.

Vital news (2000). News about milk proteins and Health Science. Dairy calcium and Health. No 5. Armor proteins. Saint- Brice-en-cogles. France. www.armor-proteines.com

Weaver, C.M (1998). Calcium in food fortification Strategies. *Int. Dairy. J.* **8:** 443- 449.

□□□

Chapter 34

Production of Functional Biomolecules by Propionibacteria

PRASHANT, SARANG DILIP POPHALY & S. K. TOMAR
Dairy Microbiology Division, National Dairy Research Institute, Karnal

Introduction

Propionibacterium is gram positive, non motile anaerobic to aerotolerant bacteria producing propionic acid, acetic acid and carbon dioxide. Glucose is commonly used as carbon source, but other substrates, i.e. fructose, mannose, galactose, glycerol and milk, can be used for growth. Besides propionic acid, acetic acid is produced under anaerobic conditions, with a ratio of 2:1 for propionic acid: acetic acid. The production of propionic acid is a clear advantage over other species as this compound is toxic in low levels for many other organisms, like other lactic acid bacteria, acetic acid bacteria and yeasts. As a result there is little chance of contamination with other microorganisms during fermentation.

All species of dairy propionibacteria are found in wide variety of habitats such as dairy products, silage, fermented olives, orange juice, maize and honey. Dairy Propionibacterium such as *P. freudenreichii* are traditionally used as starter culture in cheese production. They are also considered probiotic as they are acid tolerant and have ability to inhibit growth of undesirable microflora, stimulate growth of bifidobacterium, beneficially modify the enzymatic activities within the gut and it also produces vitamin B_{12}, trehalose and exopolysaccharides.

Vitamin B_{12}

The term vitamin B_{12} is used as a broad term which includes all the cobalt corrinoids of the cobalamin group, which include in particular cyanocobalamin, hydroxocobalamin, methylcobalamin and 5′desoxyadenosylcobalamin, characterized by a cyano, hydroxyl, methyl or 5'-desoxyadenosyl radical respectively. The methylcobalamin and 5′desoxyadenosylcobalamin compounds are known to be unstable to light in isolated form and are easily transformed to hydroxocobalamin in aqueous solution. For this reason, almost all commercial vitamin B_{12} preparations consist of the stable cyanocobalamin, which as such is not the chemical form in which vitamin B_{12} can be found in nature. The term natural vitamin B_{12} is defined so as to comprise all chemical forms of vitamin B_{12} naturally occurring in nature (Hugenholtz et al. 2002).Vitamin B_{12} is an essential vitamin in the human diet. It is used totreat pernicious anaemia and peripheral neuritis, and is used as a dietary supplement. Vitamin B_{12} is also an important animal feed supplement as growth enhancer. B12 serves as cofactor in many different carbon rearrangement reactions as occuring during breakdown of amino acids or metabolism of small carbon molecules such as glycerol, ethanolamine and propionic acid. The daily requirement of this vitamin for an average adult is about 1.2-1.5 μg/day, 0.3-0.4 μg/day for infants, 0.5-1.0 μg/day for age group 1-10 years and 2.0 μg/day for nursing mothers. Early symptoms of vitamin B12 malnutrition are unusual fatigue, faulty digestion and loss of menstruation and other symptoms are nervousness, numbness and tangling of hands and feet, mild depression, striking behavioral changes, paranoia, hyperactive reflexes and most notably anemia. Propionibacteria have already been known, for quite a long time, as sources of vitamin B_{12}. *Pseudomonas denitrificans, Bacillus megaterium* and *Propionibacterium* are three microorganisms used for commercial production of vitamin B_{12}. The dairy *Propionibacterium* has added advantage of being food-grade and vitamin B_{12} can thus be produced *in-situ* or added to the food, without need for elaborate processing.

Various studies on production of Vitamin B_{12} have been carried out varying the nutrient composition of the growth medium such as amino acid or mineral composition. Only cobalt-ions were found to affect B_{12}-production. Two experimental findings lead to major improvements in the production yields of vitamin B_{12}: (1) addition of the precursor dimethylbenzimidazole increased $_{B12}$-production, and (2) aerobic incubation in the latter phase of fermentation resulted in increased B_{12}-production (Roth *et al.* 1996). Using either or both of these strategies, production levels of 10 to 35 mg/L could be reached during fermentation of *Propionibacterium* in cheese whey as cheap growth medium.

Riboflavin

Riboflavin (Vitamin B+) is a water-soluble vitamin belonging to the B-complex group that is important for optimal body growth, red blood cell production, and release of energy from carbohydrates and fatty acids. In the body, riboflavin is found primarily as an integral component of the coenzymes flavin adenine dinucleotide (FAD) and flavin mononucleotide. These flavin-containing coenzymes participate in redox reactions in numerous metabolic pathways such as metabolism of carbohydrates, fats, and proteins. In addition, flavin-containing coenzymes are associated with the metabolism of folate, cobalamin, vitamin B6, and other vitamins, which is the reason plasma riboflavin is one of the determinants of plasma homocysteine levels, a factor known to influence the risk of cardiovascular disease, pregnancy complications, and cognitive impairment Hustad *et al,* 2002. The deficiency of this vitamin can cause severe symptoms in animals such as stunted growth.

Although riboflavin is found in a wide variety of foods (i.e., lean meats, poultry, dairy products, fish, grains, broccoli, turnip greens, asparagus, spinach, and enriched products), vitamin B2 deficiency is common in many parts of the world, particularly in developing countries, but can also be found in industrialized countries in the elderly and young adults. The use of riboflavin-producing strains in the production of dairy products such as fermented milk, yogurt, and cheese is feasible and economically attractive because it would decrease the costs involved during conventional vitamin fortification and satisfy consumer demands for healthier foods. In a study done by LeBlanc *et al,* 2006 inoculation of milk with *P. freudenreichii* B2336 (a spontaneous roseoflavin-resistant mutant derived from *P.freudenreichii* B374) has shown increased riboflavin production before inoculation with commercial yogurt cultures significantly increased the riboflavin content to almost double. It also prevented hepatomegaly, an abnormal enlargement of the liver due to riboflavin deficiency and further reverted morphologic changes observed in riboflavin-deficient rats.

Folate

"Folate" is a non-specific term referring to any folate compound with vitamin activity. The term folic acid is used for the chemically synthesised vitamin. The term folate is used as the generic descriptor for all derivatives of pteroic acid that demonstrate vitamin activity in humans. All folate compounds exist predominantly as polyglutamates containing from five to seven glutamate residues in γ-peptide linkage.

Folate is an essential nutrition component involved in many metabolic pathways, mainly in carbon transfer reactions such as purine and pyrimidine biosynthesis and amino acid interconversion. The daily recommended intake for an adult varies between 200µg and during pregnancy, the double dose 400µg is recommended for women since folate deficiency has been associated with the incidence of neural tube defects during the embryo development. A low folate intake has been associated with the risk of cardiovascular diseases. Low plasma folate concentrations correlate with elevated levels of homocystein which has been recognized as a risk factor in the coronary heart disease. Furthermore, there seems to exist some relation between low folate status and the incidence of certain forms of cancer.

Milk is a well known source of folates. The concentration of folate in milk has been reported to in the range of 20-50µg/lt. Most HPLC studies indicate 5-methyltetrahydrofolate (5-MTHF) as the major form of folates in milk. Fermented milk products are reported to contain even higher amounts of folate. The folate content in commercially available yoghurts varies between 2-10 µg/100 g. Folate is conspicuously absent in many food products and is considered an essential additive to the general diet.

Folate is an essential cofactor in bacterial metabolism and, as such, will be present in most, if not all, bacteria. Folate is produced in many forms in nature: methylated, hydrated and containing various sizes of polyglutamate tails. Uptake of folate by auxotrophic bacteria and possibly also by human cells, however, is claimed to be restricted to specific folate molecules with polyglutamate tails of threemoieties or less. Many bacteria are able to synthesize this cofactor by themselves from simple precursors, such as GTP, *p*-aminobenzoic acid and glutamate but some auxotrophic bacteria, including many lactic acid bacteria, have a strict growth requirement for folic acid.

Some strains of *Propionibacteria,* well known vitamin B12 producer, can produce substantial amounts of folate. In general, *Lactobacillus* strains do not produce folate with the exception of *Lactobacillus plantarum* where as *Streptococcus thermophilus* are the best known folate producers. Some strains, clearly, retain all the folate intracellularly while in other strains almost complete excretion or leakage of the folate is observed. This statement is supported by various studies where a substantial increase in folate production has been obtained by cell disruption. In a study by Holasova *et al.* 2004; *Propionibacterium freudenreichii* strains with butter starter culture has shown to produce approximately 1.12 μg/100 g of 5-MTHF (5-Methyltetrahydrofolate) in UHT milk.

Bacteriocin

Bacteriocins are classically defined as bactericidal proteins which show activity against species closely related to the producer culture (Tagg *et al.,* 1976). Bacteriocin provides a competitive advantage to the producing strains in fermented foods and also inhibits pathogenic microorganisms in the food milieu. An advantage of bacteriocins over classical antibiotics is that digestive enzymes destroy them. Bacteriocin producing strains can be used as part of starter cultures or as adjuncts in fermented foods in order to improve safety and quality. Only a few bacteriocins have been reported for propionibacteria. Among the dairy propionibacteria, bacteriocin production was first studied by Grinstead (1989). Till date only six bacteriocins have been described, viz. propionicin PLG-1, isolated from *Propionibacterium thoenii* P127; jenseniin G, isolated from *P. thoenii* P126, previously *P. jensenii*; propionicin T1, isolated from *P. thoenii* 419; propionicin SM1, isolated from *P. jensenii* DF1 and thoeniicin 447 isolated from P. thoenii 447 (Holo *et al.*2002).

- Propionicin PLG-1 is a bacteriocin isolated from *Propionibacterium thoenii* P-127 which is active against *P. thoenii, P. jensenii* and *P. acidipropionici*. It is also inhibitory to several LAB and some Gram negative bacteria, yeasts and molds. The viable count of sensitive strain *P. thoenii* P-5 was reduced by 99.6% by the extract. This bacteriocin is moderately thermostable (80°C, 30 min) and is proteinase sensitive. Several typical food borne pathogens and spoilage organisms belonging to genera *Bacillus, Staphylococcus, Clostridium, Yersinia* and *Salmonella* were however, found to be insensitive (Lyon and Glatz, 1991).

- Jenseniin G is a bacteriocin that is produced by *Propionibacterium jensenii* P126 (ATCC 4872), is stable during heat treatment (for 15 min at 100^{0}C) and from pH 3 to 12. It inhibits *Lactobacillus delbrueckii* ssp. *Bulgaricus* NCDO 1489, outgrowth of Clostridial spores and it is bacteriostatic to *P. acidipropionici* P-5. Jenseniin G inhibited a wide range of yogurt starter cultures. Lactobacilli are 2 to 16 times more sensitive than streptococci. Addition of jenseniin G (20 to 60 AU/ml) to stationary phase cultures of lactobacilli and streptococci in simulated yogurt limited acid formation and maintained acidity at pH 4.15 to 4.42 for 30 d at 5°C. These data suggest the usefulness of jenseniin G in preventing over acidification or post acidification of yogurt.
- Bacteriocin isolated from *P. jensenii* B1264 is a heat stable, negatively charged bacteriocin produced by *P. jensenii* B1264. It was inhibitory to PAB and *Lactobacillus delbrueckii* strains. Interestingly, mild protease treatment stimulated its activity and extended the inhibitory spectrum to more lactobacilli as well as *Lactococcus lacti*. The narrow spectrum of inhibition of this bacteriocin limits its usefulness in food preservation.
- Propionicin T1 is a bacteriocin produced by *P. thoenii* strains 419 and is active against all *P. thoenii, P. jensenii* and *P. acidipropionici* but not against *P. freudenreichii* (Faye *et al.*, 2000).
- Bacteriocin Thoeniin 447 is produced from *P. thoenii* 447. It possesses bactericidal activity against *Lb. delbrueckii* ssp. *bulgaricus* and bacteriostatic against *P. acnes*.

Propionic Acid

Propionic acid is well known to be inhibitory to a variety of organisms, ranging from bacteria to fungi. It is used for grain preservation, as antifungal agent and as herbicide (Luck, 1985). In the food industry, propionic acid and its salts are also accepted as preservatives for industrial use in bread manufacture. The concentrations that are required for inhibition are between 0.1% and 5%. Propionibacteria can produce a good amount of propionic acid, it can, therefore be used for the preservation of different foods. Propionic acid prevents the growth of fungi and bacteria by its accumulation within microbial cells and subsequent inhibition of enzymes involved in different metabolic processes. Addition of *Propionibacterium* spp. to silage has been shown to prevent the growth of undesirable organisms and increase the shelf life due to production of propionic acid; and the lowering of pH of silage. *Propionibacterium* spp. themselves are also very sensitive to propionic acid, and particularly to the undissociated form of this weak acid. The use of whey-based media for propionate production by propionibacteria has been patented (T. M. Anderson, U.S. patent 4,497,833, Feb. 1985; J. M. Sherman, U.S. patent 1,865,146, June 1932), and commercial products are available to the baking industry. Growth of the propionibacteria is already severely hampered by concentrations of undissociated propionic acid above 5 mmol.L–1 and growth stops completely at 10 mmol L^{-1}.

Trehalose

Trehalose is a well-known non-reducing disaccharide synthesized by a wide variety of organisms. New food products containing low calorie sugars and fat are in constant demand in the market. Trehalose is very promising as a sugar substitute in food: it is reported as being anti-carcinogenic and can be considered as a dietetic sugar since it is only partially digested in the human intestine. These properties place trehalose in the category of compounds known as nutraceuticals, defined as foods and food components with benefits for human and animal health. Given the beneficial properties of trehalose, the development of food products in which this sugar is produced in situ at the expense of glucose, lactose, or lactate is highly desirable.

By feeding rats with trehalose-containing diets, instead of sucrose, almost complete suppression of dental decay (caries) was observed. Other (biological) activities that can be attributed to trehalose are protection of proteins and whole cells against denaturation under different stress conditions such as heating, drying and freezing. The preserving and stabilising properties of trehalose on different biological systems have been widely demonstrated with different enzymes, membranes, human cells, tissues & organs it also protects cells and cell proteins from damage by oxygen radicals (Cardosoa *et al*, 2004). More recently, it has been shown that trehalose has important stabilizing effects on human proteins, preventing protein aggregation, facilitates protein refolding as well as formation of pathological conformational forms, and its application against illnesses, such as the Creutzfeld-Jakob disease has been disclosed.

Accumulation of trehàlose in *Propionibacterium* occurs as a result of different stress conditions, such as low temperature, high osmolarity and other adverse growth conditions. The enzymes specific for trehalose production are trehalose-6-phosphate synthase and trehalose 6- phosphate phosphatase. The intracellular concentrations that are reached vary enormously between different strains and are also strongly dependent on the growth conditions applied. In a study done by Cardosoa *et al* 2004, trehalose production as high as 450 mg per gram of cell protein of *Propionibacterium* was obtained. This demonstrates the actual potential of the use of propionibacteria for production of the low-calorie sugar trehalose in fermented foods.

Conclusion

Propionibacteria can produce several components that actually contribute to the health-status of (fermented) foods. These components-vitamin B12, riboflavin and folate as essential nutrients, propionic acid and bacteriocins as food preservatives and trehalose as low-calorie sugar and protein stabilizer, can be formed through fermentation with selected propionibacteria, and, as such, can enhance the both nutritional and functional values of fermented foods.

References

Cardosoa F.S., Gaspara P., Hugenholtzb J., Ramosa A., Santos H. (2004). Enhancement of trehalose production in dairy propionibacteria through manipulation of environmental conditions. *Int. J. of Food Microbiol* **91:** 195–204.

Faye T., Langsrud T., Nes I.F., Holo H., (2000) Biochemical and genetic characterization of propionicin T1, a new bacteriocin from *Propionibacterium thoenii, Appl. Environ. Microbiol.* **66:** 4230–4236.

Grinstead D.A., (1989.) The detection of bacteriophages and bacteriocins among the classical propionibacteria, M.S thesis, Iowa State University, Ames,

Holasova M., Fiedlerova V., Roubal P., Pechacova M. (2004) Biosynthesis of Folates by Lactic Acid Bacteria and Propionibacteria in Fermented Milk. *Czech J. Food Sci.* **22:** 175–181.

Holo H., Faye T., Brede D.A., Nilsen T., Odegard I., Langsrud T., Brendehaug J., Nes I F.(2002) Bacteriocins of propionic acid bacteria. Lait, INRA, EDP *Sciences,* **82:** 59-68

Hugenholtz J., Hunik J., Santos H., Smid E. (2002) Nutraceutical production by propionibacteria: Review. Lait, INRA, EDP. *Sciences* **82:** 103-112.

Hustad S, McKinley MC, McNulty H, Schneede J, Strain JJ, Scott JM, Ueland PM. (2002) Riboflavin, flavin mononucleotide, and flavin adenine dinucleotide in human plasma and erythrocytes at baseline and after low-dose riboflavin supplementation. *Clin Chem* **48:** 1571–7.

LeBlanc J. G., Rutten .G., Bruinenberg .P., Sesma .F., Giori .G .S de., Smid .E .J. (2006) A novel dairy product fermented with Propionibacterium freudenreichii improves the riboflavin status of deficient rats. *Nutrition* **22:** 645–651.

Luck H., (1985) Chemical preservation of food, Zent. bl. Bakteriol. *Mikrobiol. Hyg.* **180:** 311–318.

Lyon G.A., Glatz B.A., (1991) Partial purification and characterization of a bacteriocin produced by *Propionibacterium thoenii,* Appl. Environ. *Microbiol.* **57:** 701–706.

Roth J.R., Lawrence J.G., Bobik T.A., (1996) Cobalamin (coenzyme B12): Synthesis and biological significance, Ann. Rev. *Microbiol.* **50:** 137–181.

Weinbrenner D.R., Barefoots. F., Grinstead D.A. (1997) Inhibition of Yogurt Starter Cultures by Jenseniin G, a Propionibacterium Bacteriocin. *J Dairy Sci* **80:** 1246–1253.

□□□

Chapter 35

Combating Anti-Nutritional Factors in Food through Microbial Intervention

SUDHIR KUMAR TOMAR & SANDIP BASU
Dairy Microbiology Division, National Dairy Research Institute, Karnal

Introduction

Low levels of dietary fibre in our diet contribute to a long list of diseases, ranging in severity from dental caries through constipation to obesity, colorectal cancer, coronary heart disease and type 2 diabetes. In contrast, diets rich in insoluble cereal fibres are associated with a reduced risk of such metabolic disorders. Whole meal flour is rich in fibre, minerals, vitamins and many phytochemicals (phenolic compounds, sterols, tocopherols, tochotrienols etc.), which may partly account for its beneficial effects on human health. Cereals are used world-wide as staple foods. A large proportion of the world cereals production is processed by fermentation prior to consumption. The enhancement of attractive flavour and texture, and the improved shelf-life and digestibility is attributed chiefly to fermentation. In Africa and Asia, cereals of major importance are maize (*Zea mays*), sorghum (*Sorghum vulgare*), rice (*Oryza sativa*), and several minor grains such as the millets, especially pearl millet (*Pennisetum glaucum*) and finger millet (*Eleusine coracana*). Millets are cereals that grow under difficult ecological conditions and tolerate poor soils and a certain degree of drought better than any other cereal crops. In many countries especially in African countries, millet is often the main component of many meals and is essentially consumed as steam-cooked products (*Couscous*), thick porridges (*To^*) and thin porridges (*Ogi*) which can be used as a complementary food for infants and young children and is also used in brewing beer. Pearl millet is nutritionally better than most other cereals; it has high levels of calcium, iron, zinc, lipids and high quality proteins. But, as in other cereal grains, nutritional quality is considerably lowered by the presence of anti-nutritional factors leading to poor digestibility of proteins, carbohydrates and minerals. In developing countries, the low bioavailability of minerals (especially iron and zinc) in cereal-based foods is a crucial problem for infants and young children. The main anti-nutritional factors acting on iron and zinc bioavailability are **phytates**, certain **phenolic compounds** and **fibres**. Similarly, pulses contain certain anti-nutritional constituents such as **trypsin inhibitors, phytic acid, tannins** and **oligosaccharides** (raffinose, stachyose and verbascose) that could limit their protein and carbohydrate utilization. Trypsin inhibitors are low molecular weight proteins capable of binding to and inactivating the digestive enzyme, trypsin and oligosaccharides are responsible for flatulence. Tannins are known to form complexes with proteins, which are reported to be responsible for low protein digestibility and decreased amino acid availability. Peas (*Pisum sativum*) are an excellent source of proteins, carbohydrates and essential minerals, but its utility to humans is limited by the presence of antinutrients, such as phytic acid, protease inhibitors and tannins. Several outbreaks of food poisoning due to consumption of foods containing lectins have also been recorded.

Anti Nutritional Factors

2.1 Phytate

Many plant seeds including cereals, oil seeds and legumes contain phytic acid (myo-inositol 1,2,3,4,5,6-hexakis dihydrogen phosphate) that constitutes approximately 1–3% of the dry weight (Fig.1). This accounts for 60–90% of seed phosphorus necessary during plant germination. Phytic acid may have beneficial and deleterious effects on human and animal nutrition. Some of the benefits reported are that it has anticancer properties and preventive effects against heart disease and diabetes. The typical negative effect known is the binding of divalent cations such as magnesium, calcium, zinc and iron forming insoluble complexes, hence reducing their bioavailability.

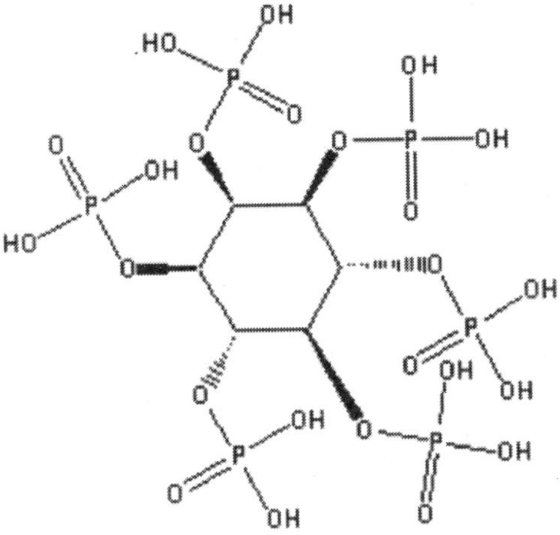

myo-inositolhexakisphosphate, **1**

Fig.1. Structure of Phytate

Humans and monogastric animals such as pigs and chickens have insufficient activity in their guts to have any substantial impact on the hydrolysis of phytate consumed in food/feed. Thus, unhydrolysed phytate phosphorus is excreted in manure and contributes to eutrophication of surface waters, which may result in toxic algal blooms and fish kills in areas of the world with intensive monogastric livestock production.

Phytase or myo-inositol hexaphosphate phosphohydrolase is an enzyme that hydrolyzes phytic acid (myo-inositol 1,2,3,4,5,6-hexakis dihydrogen phosphate) to myo-inositol and inorganic phosphate through intermediate myo-inositol phosphates to the mono-phosphate. Therefore, decreasing the phytate content in plant materials to increase their nutritional values and minimize the environmental problem like eutrophication is advisable. This helps explain the increased interest in phytate-degrading enzymes like phytase. Phytase has been reported in cereals, legumes, oilseeds and in animal tissues, but it is also produced by some microorganisms such as bacteria, yeast and fungi. Several phytase-producing bacteria have been examined— *Klebsiella aerogenes*, *Bacillus subtilis*, *Escherichia coli* and some ruminal bacteria. Some studies on phytase from *Saccharomyces cerevisiae* and

Schwanniomyces castellii have been reported. The most active producers of extracellular phytase are fungi belonging to the genus *Aspergillus*. The enzyme from *Aspergillus ficuum* is well studied. Phytases have also been characterized from other aerobic fungi– *Aspergillus carbonarius, Aspergillus oryzae* and *Aspergillus niger*. The principal function of phytase in seeds or grains is to produce inorganic phosphate from phytate during germination. The inorganic phosphate thus produced is then utilized for the purpose of plant growth. Phytase, sensitive to high temperature and pressure, is not heat stable and should be applied by avoiding excess heat during extrusion which may destroy the phytase effect. Heat treatment at 100°C for 10 min has been observed to result in loss of all phytase activity. Similarly high temperatures (>70° C) caused partial or total inactivation of native phytase. Most phytases have an optimal pH in the range of 4.5–6.0 and a temperature range of 45–60°C. Phytase activity is defined as *fytase* units (FTU), where one FTU is the amount of enzyme that liberates 1_mol inorganic orthophosphate/min from 0.0051 mol L^{-1} sodium phytate at pH 5.5 and a temperature of 37°C. This definition provides a useful measure of quantity of phytase activity and represents a simple bench mark measurement under well defined assay conditions. Phytase feed enzymes fall into two categories depending on the site where the hydrolysis of the phytate molecule is initiated. 3-Phytase preferentially liberates the P moiety at position C3, whereas 6-phytase commences at position C6 of the *myo*-inositol hexaphosphate ring. In theory, enzymic hydrolysis of phytate generates a series of lower *myo*-inositol phosphates esters, via a progression of step-wise dephosphorylation reactions, to yield inositol and six inorganic P moieties. The three commonly used phytase feed enzymes are derived from *A. niger*, which is a 3-phytase and *Peniophora lycii* and *Escherichia coli*, which are 6-phytases. However, there is evidence that *E. coli* derived phytase is more active in the small intestine than phytase derived from *P. lycii*, which may be attributable to the greater resistance of *E. coli* derived phytase to endogenous, proteolytic enzymes.

Although microbial phytases are considered of a great value in upgrading the nutritional quality of plant foods, very few studies have dealt with lactic acid bacteria. In one of these studies, It was evident that lactic acid bacteria used in milk and milk-based products such as strains of *L. acidophilus, L casei, Lactococcus lactis, L. delbrueckii* and *Streptococcus* spp. are very poor producers of phytase, whereas

L. amylovorus and *L. plantarum,* which are present in a variety of microbial systems of plant origin, resulted in the maximum yield of enzyme.

2.2 Polyphenols

The main sources of polyphenols (Fig.2) are fruits, tea and coffee, and to a lesser extent, these are present in vegetables, cereals and legume seeds. Tea (*Camellia senesis)* is the second most common beverage in the world next to water, and tea drinks are popular worldwide. Many epidemiological studies have shown that regular consumption of tea is associated with reduced risk of several forms of cancer. For example, regular drinking of green tea protects against lung and skin cancer and heart diseases. These protective effects are often attributed to the tea polyphenols (TP), in particular, the catechins. The primary catechins in TP include catechin (C), epigallocatechin gallate (EGCG), epicatechin gallate (ECG), epigallocatechin (EGC) and epicatechin (EC). Further studies demonstrated the antibacterial and antioxidative activities of TP, showing good prospects for their use as preservatives and antioxidants. TP exhibits strong free radical scavenging ability in both aqueous- and lipid-phase assay systems and in some cases tea polyphenols are up to five times more effective than vitamin C or vitamin E. In fact, TP has been employed as a natural antioxidant in the food industries of many countries. On the other hand, TP plays an important role in protein precipitation and enzyme inhibition, through forming various complexes. It is known that most polyphenols, such as tannic acid, gallotannin, catechin and proanthocyanidin, can react with proteins, which should result in the formation of sediment and haze. In food industry this phenomenon occurs frequently in the production of beverages, such as beers, fruit and vegetable juices, leading to low quality products. TP exhibits strong complexing abilities with enzymes. This should inevitably result in the change of enzyme molecular configuration and lead to the loss of catalytic activity. Many enzymes, such as tyrosinase, peroxidase, trypsin , decarboxylase, squalene epoxidase and ribonuclease, were found to be denatured by tea polyphenols. So it could be speculated that TP should bind and precipitate digestive enzymes and thereby reduce food digestibility, when excessively taken. Usually, TP is considered to be a safe natural product. However, it might act as an antinutritional factor, in terms of the inhibition of enzymes, when ingested in excess.

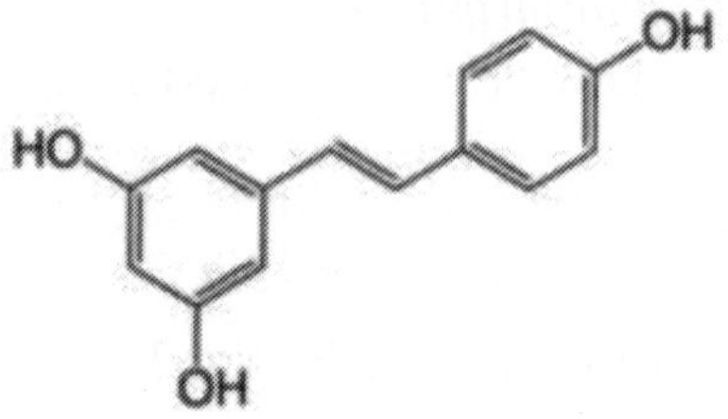

Fig 2. Structure of plolyphenol

Gut bacteria are known to act on polyphenols, producing metabolites with different physiological significance. For example, the soy isoflavone daidzein is microbially biotransformed to equol, which has a more potent estrogenic action than the precursor itself. There is also some evidence that the polyphenols in turn may influence colonic microflora. In one study carried out *in vitro*, the isoflavone genistein was shown to inhibit the internalisation of enteric bacteria by human Caco-2 and HT-29 enterocytes. Synergy between phenolics may play a role in mediating amylase inhibition and therefore have the potential to contribute to the management of type 2 Diabetes mellitus, which is characterized by high blood glucose levels. Polyphenols may act as inhibitors of amylase and glucosidase (similar to acarbose, miglitol and voglibose) leading to a decrease in post-prandial hyperglycemia.

Consumption of red wine might prevent coronary heart disease, because it contains polyphenols that protect against the oxidation of LDLcholesterol. Apples (*Rosaceae Malus* sp.) have been one of the human diets since ancient times and are one of the most commonly consumed fruits in worldwide. They are eaten both raw and in processed products such as juice, cider, brandy, jam and vinegar. Apples contain many types of phenolic acid derivatives and flavonoids (flavan-3-ols, flavonols, procyanidins, chalcones, and anthocyanins). Apple polyphenols have been reported to have various physiological functions including in vivo and clinical antiallergic activity, in vivo anti-caries activity and in vitro and in vivo inhibitory activity against some enzymes and receptors.

Post Harvesting Technology

Post-harvesting processes involved in reducing the anti-nutritional factors in food ingredients of plant origin. Phytase is frequently added to the feed as a supplement to improve phosphorus availability. Although microbial phytase from *Aspergillus niger* is commercially available, there is a need to explore other potential sources in an attempt to finding a more potent and stable form of the enzyme. Some evidence already suggests that plant phytase can also be effective in the digestion of phytate in pigs. Therefore, successful application of plant sources of phytase to enhance nutritional value of cereal-based diets will be an attractive alternative. Heat treatment significantly improves the protein quality in pulses by destruction or inactivation of the heat labile anti-nutritional factors. Cooking results in significant reduction in phytic acid and tannins in pulses. The seed coat (hull) of pulses is often indigestible and may have a bitter taste. Dehulling has been reported to improve the palatability and taste of chickpea, pigeonpea and lentil and reduce polyphenols or tannins, mostly present in the seed coat in pulses. Dehulling also reduces pulse cooking time by removing its impermeable seed coat which hindered water uptake during cooking.

Sprouting is the practice of soaking and leaving seeds until they germinate and begin to sprout. This practice is reported to be associated with improvements in the nutritive value of seeds. At the same time there are indications that germination is effective in reducing phytic acid and flatulence causing oligosaccharides (namely stachyose and raffinose), thereby increasing protein digestibility and improving sensory properties. In case of white kidney beans, faba beans and chickpeas; sprouting improved the protein/amino acid digestibility by decreasing anti-nutritional factors and increasing the true/apparent protein/amino acid digestibility.

Generally, in Africa and Asia, sorghum and millets are consumed after decortication; the grains are wetted and decorticated traditionally using a wooden mortar and pestle. Decortication is found to decrease antinutrients of pearl millet, decreasing the total polyphenols and phytic acid. It has been reported that the *E. coli* phytase is more resistant to pepsin activity than fungal phytases, which may explain the increased liberation of phytate-bound P. The ideal enzyme would have high specific catalytic activity (per unit of protein), good thermostability during feed processing, high activity under wide ranges of gut pH, resistance to proteolysis and good stability under ambient temperatures. Finally, it is important to recognise that the transit time and pH limitations within the digestive tract of poultry do not permit complete dephosphorylation of phytic acid to *myo*inosotol and inorganic P moieties. Exogenous phytase probably hydrolyses less than 0.35 of dietary phytate in broilers and clearly there is considerable scope to develop strategies that will increase phytate-P degradation in poultry. Thus, it appears that considerable advantage remains to be realized from further reductions of phytate concentrations either inherently in feed ingredients and/or in the gastrointestinal tract by enzymatic hydrolysis or other means.

Dephytinisation, or pre-treatment of feed ingredients with phytase, could become a practical alternative, particularly with oilseed meals. The use of low-phytate feed ingredients would also reduce dietary phytate contents. Presently, there is considerable interest in low-phytate maize, which has been developed following the identification of mutant genes that suppress phytate synthesis in the kernel without affecting total P concentrations. The use of low-phytate feed ingredients can also reduce dietary phytate contents. Future developments in molecular biology may increase phytase efficacy, reduce phytate accumulation in plants or increase endogenous phytase synthesis in both plants and animals.

Conclusion

The use of plant sources with low content of anti nutritional factors, introduction of post harvest treatments such as dehulling, decortication, introduction of processing methods such as cooking, malting, their reduction in food by in situ production of hydrolytic enzymes during fermentation, addition of microbial enzymes and improvement in their efficiency and production through molecular biological intervention are some of the strategies to redress the problem of anti nutritional constituents naturally present in plant material. in this context, it may be interesting to explore Lactic Acid Bacteria for their potential in combating anti nutritional factors through fermentation in cereal based fermented milk products e.g. *raabadi,* a traditional product of North India.

Suggested Reading

Badshah Khattak, A., Zeb, A., Bibi, N., Akbar Khalil, S. and Saeed Khattak, M. (2007). Influence of germination techniques on phytic acid and polyphenols content of chickpea (*Cicer arietinum* L.) sprouts. *Food Chemistry.* **104**: 1074-1079.

Bae, H. D., Yanke, L.J., Cheng, K.-J. and Selinger, L.B. (1999). A novel staining method for detecting phytase activity. *Journal of Microbiological Methods.* **39**: 17-22.

Carlson, D. and Damgaard Poulsen, H. (2003). Phytate degradation in soaked and fermented liquid feed – effect of diet, time of soaking, heat treatment, phytase activity, pH and temperature. *Animal Feed Science and Technology.* **103:** 141-154.

E. El Hag, M., H. El Tinay, A. and E. Yousif, N. (2002). Effect of fermentation and dehulling on starch, total polyphenols, phytic acid content and in vitro protein digestibility of pearl millet. *Food Chemistry.* **77:** 193-196.

El-Batal, A.I. and H. Abdel Karem, A.I. (2001). Phytase production and phytic acid reduction in rapeseed meal by *Aspergillus niger* during solid state fermentation. *Food Research International.* **34:** 715-720.

Faccia, M., Giovine, M., Gobbetti, M., Angelis, M. D., Gallo, G., Corboc, M. R., McSweeney, P. L.H. (2003). Phytase activity in sourdough lactic acid bacteria: purification and characterization of a phytase from *Lactobacillus sanfranciscensis* CB1. *International Journai of Food Microbiology.* **87:** 259-270..

Gargova, S. and Sariyska, M. (2003). Effect of culture conditions on the biosynthesis of *Aspergillus niger* phytase and acid phosphatase. *Enzyme and Microbial Technology.* **32:** 231-235.

H. Selle, P. and Ravindran, V. (2007). Microbial phytase in poultry nutrition. Animal *Feed Science and Technology.* **135:** 1-41.

H. Selle, P., J. Cowieson, A. and Ravindran, V. (2009). Consequences of calcium interactions with phytate and phytase for poultry and pigs. *Livestock Science.* **124:** 126-141.

He, Q., Lv, Y. and Yao, K. (2006). Effects of tea polyphenols on the activities of alpha-amylase, pepsin, trypsin and lipase. *Food Chemistry.* **101:** 1178-1182.

Lestienne, I., Buisson, M., Lullien-Pellerin, V., Picq, C. and Tre'che, S. (2007). Losses of nutrients and anti-nutritional factors during abrasive decortication of two pearl millet cultivars (*Pennisetum glaucum*). *Food Chemistry.* **100:** 1316-1323.

Okot-Kotber, M., Yong, K., Bagorogoza, K. and Liavoga, A. (2003). Phytase activity in extracts of flour and bran from wheat cultivars: enhanced extractability with b-glucanase and endo-xylanase. *Journal Cereal Science.* **38:** 307-315.

Richard, T., Lefeuvre, D., Descendit, A., Quideau, S. and Monti, J.P. (2006). Recognition characters in peptide-polyphenol complex formation. *Biochimica et Biophysica Acta.* **1760:** 951-958.

□□□

Chapter 36

Safety Aspects of Probiotics

S. D. POPHALY[*], PRASHANT SINGH[*], A.K. SINGH[**], S. K. TOMAR[*] AND RAMESHWAR SINGH[*]

[*]Dairy Microbiology Division, National Dairy Research Institute, Karnal

** Dairy Technology Division, N.D.R.I., Karnal

Introduction

Probiotics are defined as "Live microorganisms that when administered in adequate amounts confer a health benefit on the host". Probiotics are often regulated as dietary supplements rather than as pharmaceuticals or biological products. It is believed by many that the ideal probiotic should remain viable at the level of the intestine and should adhere to the intestinal epithelium to confer a significant health benefit. Probiotics must also be resistant to gastric acid digestion and to bile salts to reach the intestinal intact, and they should be non-pathogenic. Most probiotics are strains of *Bifidobacterium* or *Lactobacillus* species. Species from other bacterial genera such as *Streptococcus, Bacillus,* and *Enterococcus* have also been used as probiotics (Boyle *et al.* 2006), but there are concerns surrounding the safety of such probiotics because these genera contain many pathogenic species, particularly *Enterococcus*. Nonbacterial microorganisms such as yeasts from the genus *Saccharomyces* have also been used as probiotics for many years.

Mostly probiotics are formulated and marketed as foodstuffs or drugs thus generating a natural concern with regard to their safety for human consumption. Consideration of the safety of probiotics is therefore of utmost importance. The safety of the microbes that have been used traditionally in probiotics has been confirmed through a long period of experience. Probiotics may be implicated in 4 types of side-effects or adverse reactions: systemic infections, risk of deleterious metabolic activities, risk of adjuvant side-effects and of immunomodulation, and risk of gene transfer (Salminen and Wright 1998).

Proper characterization of probiotic organism is of supreme importance (Fig.1). This is essential as one organism from same genus may act as probiotic and other may be opportunistic pathogen. Even some established probiotic organism may cause adverse effects in susceptible individuals. The significance of human origin has been debated recently, but most if not all currently successful strains are reported to be of human origin. Bifidobacteria are naturally present as the predominant bacteria in the intestinal tract of breast-fed infants. Lactobacilli occur as normal flora of human intestine. Safety of these organisms was implicitly accepted till now, and reports of a harmful effect of these microbes to the host were rare (Ishibashi and Yamazaki 2001). However, recently, many species of the genera *Lactobacillus, Leuconostoc, Pediococcus, Enterococcus,* and *Bifidobacterium* were isolated frequently from various types of infective lesions. *L. rhamnosus, L. acidophilus, L. plantarum, L. casei, L paracasei, L salivarius, L. lactis,* and *Leuc. mesenteroides* are some examples of lactobacilli isolated from bacterial endocarditis; *L. rhamnosus, L. plantarum, Leuc. mesenteroides, Pediococcus acidilactici, B. eriksonii,* and *B. adolescentis* have been isolated from bloodstream infections and many have been isolated from local infections. Aguirre and Collins (14) similarly reported the isolation of *Lactobacillus, Pediococcus, Enterococcus,* and *Lactococcus* species from infection sites. Brook (15) reported the isolation of *Bifidobacterium (dentium* and *adolescentis)* and some species of *Lactobacillus* from pediatric infection sources.

Knowledge on survival of the probiotics within the GI-tract, their translocation and colonization properties, and the fate of metabolites is important for the evaluation of possible positive and negative effects of probiotic consumption. The survival of different probiotic strains in different parts of the GI-tract varies: Some strains are rapidly killed in the stomach while others can pass through the whole gut in high numbers(Martenau et al.1993). One theoretical concern with the safety of probiotics is that some have been designed or chosen to have good adherence to the intestinal mucosa, and this is considered important for their mechanism of action. Adherence to the intestinal mucosa may also increase bacterial translocation and virulence. The most potent probiotics, therefore, may have increased pathogenicity. Some enzymatic properties such as excessive deconjugation of bile salts or degradation of mucus have been postulated to be potentially detrimental (Marteau et al., 1995; Ruseler-van Embden *et al.*, 1995). Platelet aggregation properties (Korpela *et al.*, 1997), enzymes which seem to favour cardiac valve colonization (Pelletier *et al.*, 1996) and formation of unwanted metabolites can be studied *in vitro.*

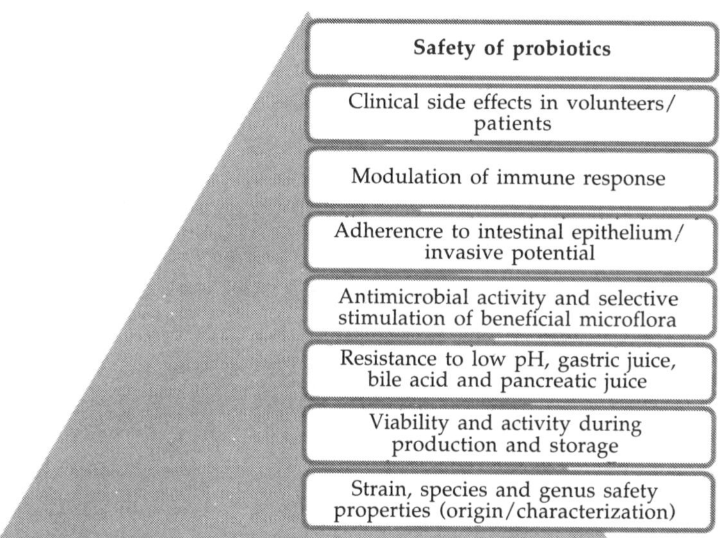

Fig 1: Safety aspects of Probiotics (Modified from Saarela *et al.*, 2000)

Assessing the risks of probiotic consumption could be a very expensive and time consuming task. A low risk may have to be accepted when recommended to immunocompromised individuals, but the risk to benefit ratio needs to be clearly established in such cases. This requires relevant information on the efficacy and safety of the products.

The factors that must be addressed in the evaluation of safety of probiotics include pathogenicity, infectivity, and virulence factors comprising toxicity, metabolic activity, and the intrinsic properties of the microbes. Donohue and Salminen (19) provided some methods for assessing the safety of lactic acid bacteria (LAB) through the use of in vitro studies, animal studies, and human clinical studies and indicated that some current probiotic strains are reported to fulfill the required safety standards.

Factors Determining Risk of Probiotics

Pathogenicity and Infectivity

The absence of pathogenicity and infectivity is a requisite of probiotic safety. Recently few incidences of clinical infections caused by probiotic organisms are reported but it is unlikely that they universally possess generalized infectivity. The isolation of (LAB) and Bifidobacteria from infections is likely to be the result of opportunistic infections. Although (LAB) or bifidobacteria may invade the host body by bacterial translocation or other routes, causing bacteremia, for these bacteria to actually cause systemic infections, from endocarditis and other infections to septicemia, both the bacterial factors and the host factors probably need to be involved Whether the probiotic bacteria are infective is difficult to prove, especially in anaerobes, which are generally considered to have no infectivity. Even if the bacteria are administered orally, infection does not normally occur in healthy animals; this is particularly so for bacteria with weak infectivity.

The causes of opportunistic infection include skin injury, chronic diseases, cancer, and drug-induced abnormality. Bacterial translocations induced by these and other factors are also considered to play an important role. Bacterial translocation is a phenomenon caused by a diminished intestinal barrier, resulting in the passage of bacteria (or bacterial components or products) across the mucous membrane and epithelium. The bacteria are then transported through the tunica propria to the mesenteric lymph nodes (MLN) and other organs. This results in bacteremia, which may progress to multiple organ failure and septicemia (Ishibashi and Yamazaki 2001).

Metabolic and Enzymatic Activity

Another requisite of probiotics is that the probiotic bacteria should not produce harmful substances by metabolic activities. One test is to determine whether the bacteria convert food components or biological secretions into secondary substances harmful to the host. For example, some intestinal bacteria act on proteins and their digested products to produce ammonia, indol, phenols, and amines. *Lactobacillus* and *Bifidobacterium* species have not been reported to produce very harmful compounds. Compared with other bacteria of the intestinal flora,

Bifidobacterium sp. have a lower deaminase activity involved in the production of ammonium from amino acids but a higher ammonia assimilation activity

Secondary bile acids are important harmful substances that are produced by intestinal bacterial actions on body secretions. They may exhibit carcinogenicity by acting on the mucous-secreting cells and promoting their proliferation, or they may act as promoters of carcinogenesis (Cheah, 1990). Many intestinal bacteria, including *Bifidobacterium* and *Lactobacillus* species, can deconjugate conjugated bile acids. However, some *Bifidobacterium* ,*Lactobacillus* and, and *S. thermophilus* strains have been reported to lack the 7-dehydroxylase activity that is related to the production of secondary bile acids (Takahashi and Morotomi, 1994).

Platelet-Aggregating Activity

Platelet aggregating activity has been considered to be a required test in the assessment of safety. Aggregation of platelets by bacteria is thought to contribute to the progression of infective endocarditis (53). The platelet-aggregating activity differs according to strains. The aggregation is thought to be associated with the proteins on the outer cell layer. The properties of the outer cell layer have been measured by hydrophobicity, hydroxyapatite adhesion, and salivary aggregation. *L. rhamnosus* strains isolated from infective endocarditis have higher activities than do laboratory strains of *L. rhamnosus* (Harty *et al.*, 1993). As for the other virulence factors, the activities of glycosidases and proteases (arylamidase), which might enable the breakdown of human glycoproteins and the synthesis and lysis of human fibrin clots, have been measured in *L. rhamnosus, L. paracasei* subsp. *paracasei,* and other strains. Some strains produce these enzymes, suggesting that they may have an infective property in causing endocarditis (Oakey *et al.*, 1995).

Antibiotic Resistance

Many probiotic organisms serve as ideal reservoir for a range of antibiotic resistance genes. This seems to be an important cause of concern. Especially in the case of *Enterococcus,* many strains, including those isolated from infection sites, have been shown to be multiply resistant to many antibiotics (Grey et al., 1991). These resistant bacteria may have acquired antibiotic resistance independently by contact with the antibiotics, or they may have acquired it by transformation. To prevent the undesirable spread of resistance or conferment of resistance to endogenous bacteria, probiotics should not carry resistance other than that required. Although special-purpose probiotics for use in combination with antibiotics have been developed through the introduction of multiple resistance to the bacteria), probiotics generally should not be designed to carry more resistance than is required for a specific purpose.

Mucus Degrading Effect

It has been reported that excessive degradation of the intestinal mucus layer by probiotics or by altered intestinal microflora may be detrimental. Some endogenous bacteria including several bacteroides species and even some lactobacilli and some strains of bifidobacteria have been reported to have the ability to degrade human intestinal mucus. For such effects Ruseler-van Embden and cowoerkers. In the Netherlands have studied the mucus degrading properties of three commonly used probiotic strains administered in fermented milks (*L. acidophilus, Bifidobacterium sp, L. rhamnosus* GG). No mucus degradation was observed *in vitro* or in gnotobiotic rats mono-associated with the test strains and thus the strains were considered safe for the mucus (Ishibashi and Yamazaki 2001).

Immunologic Effects

Several administration methods have been applied to probiotic bacteria. The most common is oral administration using either fermented foods or capsules. Other types of administration methods have also been used, especially in Japan. When administered *parentally,* bacterial cell wall such as peptidoglycan polysaccharides from different Gram-positive bacteria, including lactobacilli, can cause fever, arthritis, cardioangitis, hepatobiliary lesions or autoimmune diseases. Such side-effects may be mediated by different cytokines. Some probiotics have been shown to induce cytokine secretion and thus the relevance of such effects needs to be carefully assessed (Berg, 1983). Oral administration of high doses of LAB did not induce immunological side-effects in mice. However, a systemic uptake of cell wall polymers from intestinal lumen -hence the immunological side-effects- have been observed in rats with colonic injury (Deitch *et al.* 1991), and during small bowel bacterial overgrowth. No immunological side-effects caused by any orally administered probiotic has been reported in man.

Safety aspects include the following specifications (Saarela *et al.*, 2000):

1. Strains for human use are preferably of human origin.
2. They are isolated from healthy human GI tract.
3. They have a history of being non-pathogenic.
4. They have no history of association with diseases such as infective endocarditis or GI disorders.
5. They do not deconjugate bile salts (bile salt deconjucation or dehydroxylation would be a negative trait in the small bowel; (Marteau *et.al.,* 1995).
6. They do not carry transmissible antibiotic resistance genes.

Safety Assesment of Probiotics

In vitro Assessment of Virulence

Antibiotic Resistance and Transfer

The dissemination of antibiotic resistance genes can reduce the therapeutic possibilities in infectious diseases. Given the presence of such genes in LAB and indirect evidence that these genes could be transferred along the food chain it is therefore relevant to look for the presence of transferable antibiotic resistances in LAB strains that are or shall be used as probiotics for human consumption. Acquired resistances result either from mutations or accumulation of mutations in the organism's own DNA that finally leads to resistance against the corresponding antibiotic or from acquisition of transferable resistance genes from donors. Acquired antibiotic resistance due to mutations of housekeeping chromosomal genes most probably represents a low risk of horizontal dissemination in comparison with plasmid or transposons mediated transfer. Failure to demonstrate in vitro horizontal gene transfer does not exclude the risk of dissemination of genes. Standardised transfer methods are still required to estimate the probability for resistance transfer. On the other hand, several authors have recently shown that transfer of resistance genes occurs more frequently in vivo than in vitro (Jacobsen et al., 2007).

Metabolic Activity Assay

Glycodeoxycholic acid (GDCA) and taurodeoxycholic acid (TDCA) conjugated bile acids or primary bile salts are end products of the cholesterol metabolism, which emulsifies fat from daily food. LAB are able to hydrolyse (deconjugate) these bile salts, resulting in free (secondary) bile acids in the circulation. It has been suggested that the absence or limited extent of bile salt transformation capacity of bacteria added to food should be a prerequisite for labelling a product as GRAS. Indeed, these free bile acids can have harmful effects on the human body and there are some implications that free bile acids can act as mutagens. *In vitro* tests have shown a co-mutagenic effect of secondary bile acids, formed by 7-dehydroxylation and were also found to increase colonic cell proliferation in vitro. Furthermore, animal studies have demonstrated a co-carcinogenic effect of bile acids (Nagengast *et al.*, 1995). On the contrary, also positive effects have been ascribed to bile salt deconjugation activity. Indeed, a role in lowering of human serum cholesterol has been suggested. The field has not advanced much over the past 20 years and that the evidence for harmful effects is inconclusive. Moreover, for a large part of the commensal microbial community, including bifidobacteria, bile salt hydrolysis activity is a common trait. Finally, when probiotics are consumed only a small number of additional microbes are added compared to the total commensal microbial community present in the gut. Based on all these comments, it was concluded that bile salt deconjugation activity is irrelevant for safety assessment and is not recommended for in vitro safety testing of probiotics. (Vankerckhoven *et al.*, 2008).

Adhesion Assays

Adhesion and colonisation are essential features of most commensal bacteria. Adherence to the intestinal wall has been proposed as a selection criterion for probiotic bacteria. On the other hand, adhesion is also considered as a crucial step in the initiation of infections (Finlay & Falkow, 1997). Notably, lactobacilli isolated from infective endocarditis have been shown to possess some characteristics that could be associated with initial stages of colonisation of endocardial vegetation, such as binding to fibronectin, fibrinogen, and collagen. Nevertheless, adhesion capacities of probiotic bacteria have therefore raised some concern. But even tough there is general agreement that adhesion should not be considered as a prerequisite for translocation and that the currently available in vitro models for adhesion with cell lines are insufficient, whereas models with fibronectin and molecules typically used by pathogens could be used for validation.

Presence of virulence genes

Several putative virulence factors have been identified in enterococci. Aggregation substance and cytolysin are located on pheromone-responsive plasmids and contribute to the severity of endocarditis in animal models. Aggregation substance (encoded by *asa1*) increases adhesion to renal and heart cells. Cytolysin (***cylA*** needed for its expression) has been shown to contribute to the severity of enterococcal disease in humans. Gelatinase (encoded by ***gelE***) has been shown to exacerbate endocarditis in an animal model. Enterococcal surface protein (encoded by ***esp***) is associated with increased virulence, colonisation, and persistence in the urinary tract, biofilm formation, and a role in pathogenicity has been suggested. Hyaluronidase (encoded by the ***hyl*** gene) is believed to be associated with invasive disease. However, it was remarked that not all the virulence factors of pathogens are known yet and many genes have not yet been confirmed in virulence. Finally, it was decided that the use of *Enterococcus* probiotics that harbour several known virulence genes should be avoided.

In vitro Colon Model

Most of the Probiotic actions take place in dynamic conditions of stomach and gut. So these processes cannot be simulated in static in-vitro models. The TNO dynamic, computer controlled, *in vitro* gastro-intestinal models (nick-named TIM) simulate to a high degree these successive dynamic processes in the stomach and the small intestine and large intestine. The models are unique tools to study the stability, release, absorption and bioconversion of nutrients and nutraceuticals in the gastro-intestinal tract. Because the models can be sampled at different positions (stomach, duodenum, jejunum, ileum and colon) in time, the systems are suited to study the underlying mechanisms of functional foods. Various safety parameters may be studied in such models. The model may be useful for determining antibiotic resistance transfer from a probiotic to the commensal microbiota or for metabolite analysis of probiotics in the gut microbiota.

In vivo Assessment of Safety

Animal models are usually designed to mimic the pathogenesis of infectious or inflammatory diseases. Due to low pathogenicity for LAB, only a few models can be used for investigating their potential virulence. Based on the expertise of the PROSAFE partners (EU-funded project "Biosafety Evaluation of Probiotic Lactic Acid Bacteria used for Human Consumption"), the rat endocarditis and the immunocompromised mouse model were chosen for in vivo assessment of safety.

Experimental Endocarditis

The rat model of experimental endocarditis mimics the physiopathology of infective endocarditis in humans. Three main events occur during application of this model: (i) a pre-existing valve lesion (induced by catheterisation) becomes colonised within minutes during transient bacteraemia (intravenous bacterial inoculation), which involves valve tissue and bacterial factors; (ii) bacterial persistence and growth within the cardiac lesions; and (iii) dissemination of septic emboli to distant organs.

Immunocompromised Mouse Model

The immunocompromised mouse model was designed to mimic immunocompromised patients (e.g. cancer patients with chemotherapy). The model used within PROSAFE was derived from a pre-existing model evaluating the potential infectivity of enterococci. It was proven efficient with probiotic enterococci, since it detected differences between strains harbouring virulence genes and strains deprived in virulence genes.

The presence of bile acid deconjugation, adhesion to cell lines and extracellular matrix proteins, and survival of probiotics in the *in vitro* colon model were considered irrelevant in the biosafety assessment of probiotics. However, the in vitro colon model could be useful to study antibiotic transfer from a probiotic to the commensal microbiota or for metabolite analysis of probiotics in the gut microbiota among other applications. With regard to the presence of putative virulence genes, it was recommended that the use of *Enterococcus* probiotics or other LAB that harbour known and confirmed virulence genes should be avoided. If an *in vivo* assessment of potential pathogenicity in a specialised animal model is required, then the immunocompromised mouse model is considered too immature and not well validated, whereas the endocarditis model is the most reliable of the models tested in this study.

Clinical Relevance of Probiotic Organisms

Mostly all the established probiotic organisms commonly in use are gram positive. They do not have history of any major outbreak. There safety was largely believed on the basis of traditional knowledge of long safe use rather than proven scientific facts. But isolated cases of infection or adverse effects have been reported. *Lactobacillus, Bifidobacterium* are the most celebrated cultures used as probiotics and it also include yeast species *Saccharomyces*. A brief view of safety status of different probiotic organisms is given in the table 1.

Table 1: Safety status of Probiotic Organism (Adapted from Salminen and Wright 1998)

Organism	Infection Potential
Lactobacillus	Mainly non-pathogens, some opportunistic infections (usually in immunocompromised patients)
Lactococcus	Mainly Non-pathogenic
Leuconostoc	Mainly Non pathogenic, some isolated cases of infection
Streptococcus	Oral Streptococci mainly non-pathogenic; some may cause opportunistic infection
Enterococci	Some strains are opportunistic pathogens with haemolytic activity and antibiotic resistance
Bifidobacterium	Mainly non-pathogenic, some isolated cases of infection
Saccharomyces	Mainly non-pathogenic, some isolated cases of infection

Lactobacillus

Lactobacilli are Gram positive rods belonging to the group of LAB. Lactobacilli play a crucial role in the production of fermented foods: vegetables, meats and particularly fermented dairy products. Probiotic strains of lactobacilli are now widely used to give consumers a health benefit. Several recent reviews highlighted the benefits and limitations of their use in different medical and health-related areas: control of intestinal inflammation; maintenance of remission in Crohn's disease; improvement of human health; their value in treating infections during pregnancy; their therapeutic role in gastroenterology; management of allergic diseases; control of antibiotic-related diarrhoea; prevention of urinary tract infections and a number of other medical applications, as reviewed by Mercenier et al. (2003).

Lactobacilli and human health risks

Opportunistic pathogens: Despite their extremely widespread presence, there have been very few reported cases of infection due to lactobacilli. No cases of occupational disease or respiratory allergy implicating lactobacilli have been documented. Cannon et al. (2005) discussed the incrimination of this genus in endocarditis, bacteremia, localized infections and a variety of other infections including, abscess, cholecystitis meningitis and pneumonia. Recent infections involved patients with central venous catheter. *L. casei* (35.7%) and *L. rhamnosus* (22.9%) were the most commonly incriminated species.

Five cases of *Lactobacillus* associated infections have been reported in France since 1823: three cases of induced endocarditis; one induced liver abscess (Larvol *et al.*, 1996) and one case involving urinary tract infection. In fact, most of the rare cases of *Lactobacillus* infection occur in patients with underlying severe illnesses or immunocompromised patients. In four cases (three of endocarditis and one local Lactobacillus infection) out of a total of over 200 cases reported by Cannon *et al.* (2005), a possible risk factor could be "heavy dairy consumption" or "dairy drinks containing *L. rhamnosus* GG". One example was a patient suffering from inflammatory internal lesions with *L. rhamnosus* GG isolated and described as the causative agent, reported to have consumed large quantities of fermented milk containing the probiotic *L. rhamnosus* GG. (Bernardeau *et al.*, 2008)

It is worth noting that large amounts of lactobacilli have been safely administrated to immunocompromised patients. For example, two clinical studies have been conducted to assess the efficacy and safety of lactobacilli in small groups of specific immunocompromised patients (e.g. patients with HIV infection), premature babies, infants, children, the elderly, subjects with Crohn's disease or diarrhoeal diseases, and HIV-positive persons; no side-effects have been reported. Considering the way Lactobacillus taxonomy is being constantly updated and changed, it is probable that some previous reports of association between lactobacilli and pathogenic cases may have incriminated misidentified Lactobacillus (at genus and/or species level).

Fermentation and intrinsic properties: Several intrinsic properties of lactobacilli related to their metabolic activities may have an effect on human safety. For example, D-lactacte or biogenic amines (histamin, tyramin, etc.) can be produced and accumulated in fermented dairy products such as cheeses. Other lactobacilli properties of relevance for safety are tyrosine decarboxylation, bile salt deconjugase activity, enzymatic activities (azoreductase, nitroreductase, β-glucuronidase, glycosidase which may have undesirable toxicological effects); degradation of hyaluronic acid; platelet aggregation activity; colonization; and production of toxic metabolites.

Antibiotic resistance: Lactobacilli is the potential hosts for antibiotic resistance genes, with the risk of transferring these genes to many LAB and other pathogenic bacteria (Mathur and Singh. 2005). As a general rule, lactobacilli have a high natural resistance to bacitracin, cefoxitin, ciprofloxacin, fusidic acid, kanamycin, gentamicin, metronidazole, nitrofurantoin, norfloxacin, streptomycin, sulphadiazine, teicoplanin, trimethoprim, and vancomycin.

Data from various studies demonstrate the existence of interspecies differences of *Lactobacillus* spp. resistance to various antimicrobial agents. Despite the extremely widespread presence of lactobacilli, there have been very few reported cases of infections only in population having reduced immune function. These opportunistic infections may be complicated by the antibiotic resistance traits of some *Lactobacillus* strains. This is cause for caution and justifies performing antibiotic-susceptibility assays before technological and functional use. As adopted by EFSA (2005), intrinsic resistance and resistance due to mutation of chromosomal genes present a low risk of horizontal dissemination but acquired resistance mediated by added genes may present a risk for public health. To conclude, starter culture bacteria in dairy products do not seem to represent an important source for the spread of genes encoding resistance to antimicrobial agents. However, care must be taken regarding the antibiotic-resistance profiles of novel strains used as starters or probiotics in dairy products.

Bifidobacterium

To date, more than 30 species have been isolated and assigned to the genus *Bifidobacterium*, of which at least 11 species have been isolated in humans: *B. adolescentis, B. angulatum, B. bifidum, B. breve, B. catenulatum, B. dentium, B. gallicum, B. infantis, B. longum, B. pseudocatenulatum*, and *B. scardovii*. Contrary to starter cultures, bifidobacteria do not have a significant role either in the acidification of milk, or in the formation of texture and/or flavour, but are thought to play a health-promoting role in the gut of the consumer. *Bifidobacterium* strains that have been used as probiotics for some years and whose identities are documented to a certain extent include, among others, *B. breve* Yakult, *B. lactis* Bb12, *B. longum* BB-536, *B. bifidum* RO71 and *B. infantis* RO33. *Bifidobacterium lactis* DSM 10140, which is the most frequently used probiotic Bifidobacterium strain in dairy products worldwide.

There is no report to date of the recovery of probiotic bifidobacterial strains in infections following ingestion of dairy products, despite their widespread consumption in Europe and Japan. Bifidobacteria are generally considered safe. However, since most bifidobacterial strains used in dairy products are common members of the human intestinal microbiota, they may behave as opportunistic pathogens, like other commensal bacteria. Some commensal bifidobacteria have been connected with certain dental infections, pulmonary infections, bacteremia, abscesses and bloodstream infections. In infants aged 2 to 16, Brook (1996) described a prevalence of 57 *Bifidobacterium* isolates among 2033 specimens originating from chronic otitis media, abscesses, peritonitis, aspiration pneumonia and paronychia. In most cases, the non-probiotic *B. dentium* (formerly *B. eriksonii*) was implicated: this is the only bifidobacterial species classified in risk group 2 and regarded as a dental pathogen. A few cases also implicated *B. adolescentis,* which has been recovered pure from otitis media. Despite these few cases, the recovery of bifidobacteria in clinical infections is very rare in comparison to other commensals. Genome sequence analysis of Bifidobacterium longum and molecular biological analysis of other probiotic strains confirmed so far the absence of virulence and pathogenecity factors. However, tetracycline resistance genes tet(W), although probably not easy transferable, were found in Bifidobacterium strains, also in *Bifidobacterium animalis* subsp. *lactis,* the worldwide most used industrial strain (Meile *et al.,* 2008).

To our best knowledge, probiotic bifidobacterial strains have never been recovered in infections following ingestion of dairy products. *Bifidobacterium* genus is certainly among the safest genera.

Propionibacterium

The genus *Propionibacterium* comprises two distinct groups from different habitats: strains typically found on human skin, referred to as the "acnes group", and strains isolated from milk and dairy products, referred to as "dairy" or "classical" Propionibacteria strains. The dairy group of propionibacteria comprises four species *P. freudenreichii, P. acidipropionici, P. jensenii* and *P. thoenii. Propionibacterium freudenreichii* is the main species used as a ripening starter in Swiss and Emmental cheese. Dairy propionibacteria have also recently been proposed as probiotics.

All *Propionibacterium* isolates of human clinical relevance belong to the "acnes group". The species *P. freudenreichii* is classed as Generally

Recognized as Safe (GRAS) for use in cheese. Dairy propionibacteria do not possess any known virulence factors, although *P. thoenii* strains and some *P. jensenii* strains show â-haemolytic activity. Some *Propionibacterium* strains can form tyramine or histamine in vitro. To assess the safety of one *P. jensenii* probiotic strain in a rat model, a high dose (1010 cfu/rat/day for 81 days) was fed. The repeated dose toxicity and translocation of propionibacteria cells to blood and other tissues of the rats were evaluated (Huang et al., 2003). Results showed that the probiotic strain used had no adverse effect on general health status, body weight gain, or visceral organs. Concerning antibiotic resistance, dairy propionibacteria exhibit a natural resistance to oxacillin, aminoglycosides, 1st and 2nd generation quinolones, colistin, metronidazole and fosfomycin. () The antibiotic resistance in dairy propionibacteria is chromosomally encoded so less prone to horizontal gene transfer. In conclusion dairy propionibacteria have a long documented history of use in foods and are safe to be used as probiotic (Meile *et al.,* 2008).

Enterococci

The genus *Enterococcus* is the most controversial group of LAB. Traditional cheeses made from raw ewes' or goats' milk, in the Mediterranean countries, have *enterococci* microflora which play an important role in the ripening of these cheeses, probably through proteolysis, lipolysis, and citrate breakdown, hence contributing to their typical taste and flavour. Enterococci have the ability to produce bacteriocins, the so called enterocins, which are active against closely related Gram-positive bacteria such as *Listeria.* Enterococci are used in some countries as probiotics strains of *E. faecium* and *E. faecalis* species have been applied in human, probiotic supplements, *E. faecalis* strains have also been widely used as veterinary feed supplements. Since February 2004, 10 preparations (9 different strains of E. faecium) are authorized as additives in feeding stuffs in the European Union (European Commission, 2004). *E. faecium* SF 68, which is produced in Switzerland (Cylactin\ LBC SF68 ME10, F. Hoffmann-La Roche Ltd., Basel, Switzerland). It has been proposed to be clinically effective in the prevention of antibiotic-associated diarrhoea and in the treatment of diarrhea in children. In Denmark, a fermented milk product that contains *E. faecium* SF 68 (GAIO;MD Foods, Aarhus, Denmark) has been sold for several years because of its hypocholesterolemic effect on individuals. *E. faecium* CRL 183 for the development of a novel fermented soymilk product with potential probiotic properties.

Another probiotic product available on the market that contains an *E. faecium* strain is Walthers ECOFLOR (Walthers Health Care, Den Haag, Netherlands).

Enterococci are among the most common nosocomial pathogens and they have been implicated as an important cause of endocarditis, bacteremia, and infections of the urinary tract, central nervous system, intra-abdominal and pelvic infections, as well as of multiple antibiotic resistances. Several virulence factors have been described and the number of antibiotic resistant enterococci (**ARE**), especially vancomycin-resistant enterococci (**VRE**), is increasing. *E. faecalis* hemolysin/bacteriocin was a natural bifunctional component which posses both the characteristics. Cytolysin is unique among the lantibiotics in that it is able to lyse eukaryotic cells in addition to its bactericidal effect. Cytolysin is considered a virulence factor of *E. faecalis* strains (Foulquie et al., 2006).

Antibiotic resistance encompasses both natural (intrinsic) resistance and acquired (transferable) resistance. Enterococci possess a broad spectrum of antibiotic resistances within these two types. Examples of intrinsic resistance are vancomycin (VanC type) resistance in *E. gallinarum*, and resistance towards streptogramins in *E. faecalis*, as well as resistance to isoxazolylpenicillins, cephalosporins, monobactams, aminoglycosides (low level), lincosamides (mostly), and polymyxins. The resistance to ampicillin (especially in *E. faecium*), tetracyclines, macrolides, aminoglycosides (high level), chloramphenicol, trimethoprim/sulfamethozaxole, quinolones, and streptogramins (in *E. faecium* and related species) are acquired, as well as resistance towards glycopeptides (vancomycin). Unfortunately, VRE are also highly resistant to all standard anti-enterococcal drugs and therefore, VRE constitute a serious risk group. For safety reasons, a critical or important criterion to check is the absence of transferable antibiotic resistance in enterococci that could be used as co-cultures or starter cultures in foods, since it is strain-specific.

Aggregation substance (Agg) is a pheromone-inducible surface protein of E. faecalis, which promotes aggregate formation during bacterial conjugation. Agg mediates efficient enterococcal donor-recipient contact to facilitate plasmid transfer. Agg increases the hydrophobicity of the enterococcal surface, which may induce localization of cholesterol to phagosomes, and prevent or delay fusion with lysosomal vesicles. This trait may contribute to the pathogenesis of enterococcal infection through different mechanisms.

Gelatinase (Gel) is an extracellular metallo-endopeptidase involved in the hydrolysis of gelatin, collagen, haemoglobin, and other bioactive peptides. Extracellular surface protein (Esp) was first described in a clinical *E. faecalis* MMH594. Studies on the distribution of this surface protein revealed a significant enrichment in infection derived *E. faecalis* isolates. Esp is thought to play a role in adhesion and evasion of the immune response of the host. The incidence of Esp in food isolates differs for *E. faecium* and *E. faecalis*.

The use of enterococci as probiotics remains a controversial issue. While the probiotic benefits of some strains are well established, the emergence and the increased association of enterococci with human disease and multiple antibiotic resistances have raised concern regarding their use as probiotics. The fear that antimicrobial resistance genes or genes encoding virulence factors can be transferred to other bacteria in the gastrointestinal tract contributes to this controversy. (Foulquie et al., 2006) However, some Enterococci carry potential virulence factors and can display pathogenic traits. Furthermore, the number of VRE has been increasing during the last decade. Fortunately, these important features are strain-dependent. For these reasons, the selection of *Enterococcus* strains of interest in the food industry should be based on the absence of any possible pathogenic properties and transferable antibiotic resistance genes.

Lactococcus

Lactococci are Gram positive cocci and belong to the group of LAB. They are homofermentative and exclusively produce L(+) lactic acid. The *Lactococcus* genus includes five species, *L. garvieae, L. piscium, L. plantarum, L. raffinolactis* and *L. Lactis,* which is differentiated into subspecies *L. lactis* subsp. *cremoris, L. lactis* subsp. *hordniae* and *L. lactis* subsp. *lactis.* They are not â hemolytic and they are poorly á hemolytic. *L. lactis* subsp. *lactis* and *L. lactis* subsp. *cremoris. L. lactis* subsp. *lactis* and to a lesser extent *L. lactis* subsp. *cremoris* have long been extensively used in starter cultures for dairy fermentation (i.e. in cheeses, sour cream and butter). They can also be used for food preservation due to their ability to produce organic acids and bacteriocins, nisin being the best characterized and recognized. Their use as probiotics has been also considered.

Members of *Lactococcus* genus are most commonly granted GRAS status. *L. lactis* cannot be considered as an opportunist pathogen, as only two cases of endocarditis have been reported in the medical literature over a period of fifty years. *Lactococcus garvieae* has been associated with septicemic infections of fish, mainly in intensive modern aquaculture; it is the main risk factor for the Mediterranean European trout industry. It has been also isolated from bovine mastitis. Human infections due to *L. garviae* are scarce, since only six cases has been described, all of which concerned elderly or immunosuppressed patients (Casalta & Montel 2008). Thus *Lactococci* from dairy products can be generally considered safe, in view of their extensive, daily consumption by humans and their low incidence in human infections.

Streptococcus thermophilus

S. thermophilus is a thermophilic LAB widely used as starter in the manufacture of dairy products and may be regarded as the second most important industrial dairy starter after *L. lactis*. It is traditionally used in the production of yogurt and many cheeses, e.g. hard-cooked cheeses (Emmental, Gruyère, Parmigiano and Grana-types, etc.), Mozzarella and Cheddar.

Though *Streptococcus thermophilus* is of nonhuman origin and is known to suffer from exposure to gastric acidic conditions yet has a moderate ability to adhere to intestinal epithelial cells and is shown to survive gastrointestinal transit. Therefore, *S. thermophilus* could be considered as a transient probiotic. Also it has shown to improve intestinal microflora and lactose digestion in lactose intolerant individuals; stimulate the gut immune system, produce high quantity of folate extracellularly and alleviate the risk of certain cancers, ulcer and inflammation. It has also been reported to have been described as having positive effects on diarrhea in young children, enterocolitis in premature neonates and inflammatory gut disease. However, with time probiotic characteristics (deconjugation of bile salts, hydrophobicity and â-galactosidase activity) and the resistance to biological barriers (gastric juice and bile salts) were also found in *S. thermophilus* (Iyer et al., 2009).

Together, *S. thermophilus, S. salivarius* and *S. Vestibularis* form the salivarius group of viridans streptococci. *S. thermophilus* is the only streptococcal species to originate from dairy products and have a GRAS status. All other viridans streptococci are commensals in the oral cavity and gastrointestinal and genital tracts of mammals. Several cases of meningitis, endocarditis and bacteraemia due to *S. salivarius* and *S. vestibularis* oral streptococci have been reported. Recently, the genomes of three *S. thermophilus* strains were sequenced. A striking feature of these genomes is the absence or inactivation of the virulence related genes characterized in pathogenic *streptococci*. This supports the safe status of the dairy *S. thermophilus* strains (Delorme, 2008). Dissemination of tetracycline resistance among oral streptococci has been shown. Lastly, since the acquisition of new genes during cocultures in dairy fermentation has been suggested, transfers of antibiotic determinants or genes from undesirable species present in the same dairy environment to *S. thermophilus* cannot be excluded.

Saccharomyces boulardii

The genus *Saccharomyces* has 16 species, including *S. cerevisiae* and *S. boulardii*; of which two, *S.cerevisiae* and *S. boulardii*, are described in the literature as containing biotherapeutic agents. *Saccharomyces boulardii* are regarded as the most prominent representatives. Today, a considerable number of pharmaceutical preparations (capsules, powders, tablets, pellets) containing probiotic yeasts cells are commercially available. The beneficial properties of strains of some *Saccharomyces* spp are well documented, in addition to their nutritive value (e.g. provision of vitamins of the B group). Probiotic yeasts are generally resistant to gastrointestinal passage and most antibiotics. Yeast preparations have also been successfully applied, in combination with antibiotics, to treat *Clostridium difficile*-related diarrhoea. Probiotic *Saccharomyces* spp may also help to re-establish a normal gut function after long term antibiotic therapy. *S. cerevisiae* and *S. boulardii* release polyamines which help in repairing mucous membranes. These polyamines increase the activity of short chain fatty acids (SCFA) and disaccharide enzymes (lactase, maltase, sucrase). *S. boulardii* reduces the growth of *Clostridium albicans, Escherichia coli, Salmonella typhi, Shigella dysenteriae, Vibrio cholerae, Salmonella enteritidis* and *Clostridium difficile*. Some people use *S. boulardii* for lactose intolerance, urinary tract infections (UTIs), vaginal yeast infections, high cholesterol levels, hives, fever blisters, canker sores, and teen-age acne. *Saccharomyces* spp stimulate IgA production and the phagocytic system of gnotobiotic mice (Dixit *et al.*, 2009).

S. boulardii is likely safe for most adults when taken by mouth. It can cause gas in some people. Rarely, it might cause fungal infections that can spread through the bloodstream to the entire body (fungemia). *S. boulardii* is possibly safe for children. There is some concern that critically ill people, people who have a weakened immune system, and people who are taking medicines that alter the immune system might have an increased risk for developing a yeast infection that spreads to the bloodstream and the rest of the body (fungemia) if they take *S. boulardii* (Boyle et al., 2006). People with yeast allergy can be allergic to products containing *S. boulardii*, and are best advised to avoid these products.

Conclusion

Herein we reviewed the safety of probiotics and highlighted deficiencies in our understanding of their appropriate administration and their mechanisms of action. We learn that probiotics are safe for use in otherwise healthy persons, but should be used with caution in some persons because of the risk of sepsis. Newly developed probiotic strains should be thoroughly evaluated for safety before being marketed. Although much remains to be learned regarding the mechanisms of action and the appropriate administration of probiotic strains, it is clear that different strains can have very specific effects. Moreover their effects may vary in health and disease, in different disease states, and in different age groups. Thus, clinical trial results from one probiotic strain in one population cannot be automatically generalized to other strains or to different populations. Further studies are needed to explore mechanistic issues and probiotic interactions. In view of the increasing use of probiotics as health supplements and therapeutic agents, clinicians need to be aware of the risks and benefits of these treatments.

References

Aguirre, M., Collins, M.D., Lactic acid bacteria and human clinical infection. (1993). *Journal Applied Bacteriology*, **75:** 95–107.

Berg, R.D., Bacterial translocation from the gastrointestinal tracts of mice receiving immunosuppressive chemotherapeutic agents. (1983) *Current Microbiology*, 285–292.

Bernardeau, M., Vernoux, J. P., Henri-Dubernet, S., Guéguen, M. (2008) Safety assessment of dairy microorganisms: The *Lactobacillus* genus. *International Journal of Food Microbiology*, **126:** 278–285.

Boyle R.J., Roy M. R.B., Tang M. L. (2006). Probiotic use in clinical practice: what are the risks? *American Journal of Clinical Nutrition*, **83:** 1256-64.

Brook, I. (1996). Isolation of non-sporing anaerobic rods from infections in children. ***Journal of Medical Microbiology*, 45:** 21–26.

Brook, I., Isolation of non-sporing anaerobic rods from infections in children. (1996) *Journal of Clinical Microbiology*, **45:** 21–6.

Cannon, J.P., Lee, T.A., Bolanos, J.T., Danziger, L.H. (2005). Pathogenic relevance of Lactobacillus: a retrospective review of over 200 cases. *European Journal of Microbiological Infectious Disease*, **24:** 31–40.

Casalta, E., & Montel, M.,(2008). Safety assessment of dairy microorganisms: The *Lactococcus* genus. *International Journal of Food Microbiology*, **126:** 271-273.

Cheah, P.Y., Hypotheses for the etiology of colorectal cancer-an overview. (1990) *Nutrition and Cancer;* **14:** 5–13.

Deitch, E.A., Xu, D., Lu, Q., Berg, R.D., (1991) Bacterial translocation from the gut impairs systemic immunity. *Surgery*, **109:** 269–76.

Delorme, C. (2008). Safety assessment of dairy microorganisms: *Streptococcus thermophilus*. *International Journal of Food Microbiology*, **126:** 274–277.

Dixit ,K., Gandhi, D. N., Chauhan P. (2009). Selection and characterization of potential probiotic *saccharomyces* strains isolated from fruits, fermented Food and dairy products. *International Journal of Probiotics & Prebiotics*, **4:** 249-256.

Finlay, B. B., & Falkow, S. (1997). Common themes in microbial pathogenicity revisited. *Microbiology and Molecular Biology Reviews*, **61**(2): 136-169.

Foulquie Moreno, M.R., Sarantinopoulos, P., Tsakalidou, E., De, Vuyst, L. (2006). The role and application of enterococci in food and health. *International Journal of Food Microbiology*, **106:** 1-24.

Gasser, F., Safety of lactic acid bacteria and their occurrence in human clinical infections. (1994). *Bull Inst Pasteur;* **92:** 45–67.

Gray, J.W., Stewart, D., Pedler, S.J., (1991) Species identification and antibiotic susceptibility testing of enterococci isolated from hospitalized patients. *Antimicrobial Agents and Chemotherapy*, **35:** 1943–1945.

Harty, D.W.S., Patrikakis, M., Knox, K.W., Identification of *Lactobacillus* strains isolated with infective endocarditis and comparison of their surface-associated properties with those of other strains of the same species. (1993) *Microbiol Ecology and Health and Disease*, **6:** 191–201.

Huang, Y., Kotula, L., Adams, M.C. (2003). The in vivo assessment of safety and gastrointestinal survival of an orally administered novel probiotic, *Propionibacterium jensenii* 702, in a male Wistar rat model. *Food and Chemical Toxicology*, **41:** 1781–1787.

Ishibashi, N., Yamazaki, S., Probiotics and safety. (2001). *American Journal of Clinical Nutrition*, **73:** (suppl), 465S–470S.

Iyer, R., Tomar, S.K., Kapila, S., Mani, J., Singh, R. (2010). Probiotic properties of folate producing *Streptococcus thermophilus* strains. *Food Research International*, **43:** 103–110.

Jacobsen, L., Wilcks, A., Hammer, K., Huys, G., Gevers, D., Andersen, S. R. (2007). Horizontal transfer of tet(M) and erm(B) resistance plasmids from food strains of *Lactobacillus plantarum* to *Enterococcus faecalis* JH2-2 in the gastrointestinal tract of gnotobiotic rats. *FEMS Microbiology Ecology*, **59**(1): 158-166.

Marteau, P., Gerhardt, M.F., Myara, A., Bouvier, E., Trivin, F., Rambaud, J.C., (1995) Bifidobacteria and lactobacilli ingested in fermented dairy products can metabolize bile salts in the human small intestine. *Microbial Ecology in Health and Disease*, **8:** 151–157.

Mathur, S., & Singh, R., (2005). Antibiotic resistance in food lactic acid bacteria- A review. *International Journal of Food Microbiology*, **105**(3): 281–295.

Meile, L., Blay, G. L., Thierry, A. (2008) Safety assessment of dairy microorganisms: *Propionibacterium* and *Bifidobacterium*. *International Journal of Food Microbiology*, **126:** 316–320.

Mercenier, A., Pavan, S., Pot, B. (2003). Probiotics as biotherapeutic agents: present knowledge and future prospects. *Current Pharmaceutical Design*, **8:** 99–100.

Nagengast, F. M., Grubben, M. J., Van Munster, I. P. (1995). Role of bile acids in colorectal carcinogenesis. *European Journal of Cancer, 31A* **7-8:** 1067-1070.

Oakey, H.J., Harty, D.W.S., Knox, K.W., Enzyme production by lactobacilli and the potential link with infective endocarditis (1995). *Journal Applied Bacteriology*, **78:** 142–148.

Ruseler-van Embden, J., van Lieshout, L., Marteau, P., (1995) No degradation of intestinal mucus glycoproteins by *Lactobacillus* strain GG. *Microecol Ther*, **25:** 304–309.

Saarela, M., Mogensen, G., Fonden, R., Matto, J., Mattila-Sandholm, T., (2000) Probiotic bacteria: safety, functional and technological properties. *Journal of Biotechnology*, **84**(3): 197-215.

Salminen, S., Wright, A. Von., Current probiotics - safety assured? (1998) Microbiol Ecology and Health and Disease, **10:** 68-77.

Takahashi, T., Morotomi, M., Absence of cholic acid 7-dehydroxylase activity in the strains of Lactobacillus and Bifidobacterium. (1994). *Journal of Dairy Science* **77:** 3275–86.

Vankerckhoven, V., Huys, G., Vancanneyt, Marc., Vael, C., Klare, I., et al (2008) Biosafety assessment of probiotics used for human consumption: recommendations from the EU-PROSAFE project. *Trends in Food Science & Technology* **19:** 102 -114.

□□□

Chapter 37

Phytochemicals and their Utilization in Functional Foods

ASHISH KUMAR SINGH
Scientist (SS), D.T. Division, NDRI, Karnal

Introduction

Increasing awareness among consumers to know which specific molecules present in their food possess disease preventive or curative properties has led to the concept of "Functional Foods". Now the attention of scientific investigations has moved towards exploring the role of biologically active components on human health. Basic temptation in human being towards nature and the products that are natural, for every little disturbances related to health resulted in flourishing of market with products containing various therapeutic ingredients. Functional foods, pharma foods, designer foods and nutraceuticals are synonymous for foods that can prevent and treat diseases. Functional foods have been defined as foods that, by virtue of the presence of physiologically active components, provide a health benefit beyond basic nutrition. Epidemiological studies and randomized clinical trials carried out in different parts of the world have been demonstrated or at least suggested numerous health effects related to functional food consumption, such as reduction of cancer risk, improvement of heart health, enhancement of immune functions, lowering of menopause symptoms, improvement of gastrointestinal health, anti-inflammatory effects, reduction of blood pressure, antibacterial & antiviral activities, reduction of osteoporosis etc. Nutritional significance of plant molecules is well documented and increasing cases of cancers, coronary heart diseases, diabetes and many other chronic diseases, have been attributed to under consumption of fruits and vegetables in our diet. But beyond these known nutrients i.e. vitamins, fibers, plants have clearly more to offer and scientists are scurrying to discover exactly which plant components might fend off specific diseases. An ever-expanding array of previously unknown plant molecules with hard to pronounce names is being uncovered. But there exact metabolic role and how these can be utilized in designer food, need to be clarified.

All over world there has been growing demand for functional foods. Currently Japan leads the world in the production, with more than 100 production and consumption of such products.

Defining Functional Foods and Phytochemical

The term functional foods was first introduced in Japan in the mid-1980s and refers to processed foods containing ingredients that aid specifically bodily functions in addition to being nutritious. To date, Japan is the only country that has formulated a specific regulatory approval for functional foods, known as foods for specific health use (FOSHU). Currently 100 products are licensed as FOSHU foods in Japan. In US, the Functional foods category is not recognized legally. However, The Institute of Medicine's Food and Nutrition Board (IOM/FNB, 1994), defined functional foods as "any food or food ingredient that may provide a health benefit beyond the traditional nutrients it contains". To qualify as functional food, it should meet following three highlighted conditions:-

- It is a food (not capsule, tablet or powder) derived from naturally occurring ingredients.
- It ca and should be consumed as part of daily diet.
- It has particular when ingested, serving to regulate a particular body process. Such as
- Improvement of biological defence mechanisms
- Prevention of specific disease.

- Recovery from specific disease
- Control of mental and physical conditions
- Retarding the ageing procèss

"Phytochemicals"- phyto from Greek word for plants, denoting their plant origins. Most likely phytochemicals developed as a part of the plants own defense mechanism against environmental insult and only fortuitously provide benefits to man.

The number of identified physiologically has increased dramatically in the last decades and overwhelming evidence from epidemiological, in vivo, in vitro and clinical trial indicate that plant rich diet can reduce the risk of certain chronic diseases (Hasler, 2000) Health professionals are gradually recognizing the role of phytochemicals in health improvement.

Potential Phyto-Nutrients

Some phytochemicals with proven health benefits can be grouped under following categories.

Phytoestrogens

Phytoestrogens are a broad group of plant-derived compounds that are structural mimics of endogenous 17 beta-estradiol. Two major phytoestrogens, which are of great importance from a nutritional and health perspectives, include lignans (Flaxseed) and isoflavones (soy bean).

Isoflavones are flavonoid compounds and major flavonoids that have been identified in soyabeans are genistein, daidzein and glycitein. These compounds exist naturally in Soya in several glycoside forms, but it is the aglycone form of isoflavones that is biologically active.

These compounds either compete with or antagonize estrdiol action. Exact biochemical mechanism involving CYP3A monoxygenase activity in presence of phase I enzyme inducers such as dixamethane, on carcinogens was explained by Ronis et al. In another study dietary soy products, showed antitumer activity, by directly affecting the tumor cell proliferation and reduction in tumor angiogenesis (microvessle density). This activity may prevent prostrate cancer. Research has shown that diets rich in soya help to reduce blood levels of LDL (bad) cholesterol by an estimate of 12-15%. The isoflavones in soy foods are converted in the gut to phytoestrogens that may reduce LDL blood cholesterol. The health effect of these compounds depend upon the exposure level of phytoestrogens, the binding constant relative to estradiol and the selectivity of different tissue receptors.

Organosulfer Compouds

Termed as "promise of garlic" and most widely quoted herb in the literature. Besides its popularity as a recipe seasoning, garlic has long been promoted as a medicinal agent. Garlic and other allium vegetables - onions, chieves, leeks and scallions contain allylic sulphides.

Garlic has acquired a reputation in many cultures as a formidable prophylactic and therapeutic medical agents. Plant extracts containing organosulfur compounds have been shown to exhibit antimicrobial, hypolipidemic, hypoglycemic, antiothrombotic, antioxidant and anticarcinogenic (Shenoy and choughuley, 1992). The cardiovascular properties of garlic are less understood and it was suggested that antihypertensive properties of compounds present in garlic play a major role. Animal studies indicate that garlic supplementation in diet depressed the hepatic activities of lipogenic and cholesterogenic enzymes such as malic enzyme, fatty acid synthase, Glucose-6-phosphate dehydrogenase and 3-hydroxy-3-methyl-glutaryl CoA (HMG-CoA) reductase.

Active components identified is a family of thioallyll compounds and the intact garlic bulb contain an odorless amino acid called alliin, which is converted enzymatically by allinase into allicin, when the garlic cloves are crushed. Formation of allicin is responsible for characteristic garlic flavour. Allicin then immediately decomposes to form numerous sulpher-containing compounds. It has role in prevention of CVD possibly through antihypertensive properties. Animal feeding studies demonstrated the regulation of fatty acid and cholersterol synthesis via modulating. The role of certain enzymes, by garlic components. However, it is still unclear which component of garlic is responsible for cholesterol lowering effect.

Recent investigations also reveals the lowering of plasma concentration of cholesterol is mainly due to an inhibition of hepatic cholesterol synthesis. The human and animal studies to identify the active sulfur compound (s) of garlic responsible for hypocholsterolemic effect was carried out by Yeh and Liu (2001). On the basis of their studies with aged garlic extract (AGE) containing a wide range of organosulfur compounds, they concluded that cholesterol lowering properties of garlic/organosulfur compounds may be attributed in part from inhibition of hepatic cholesterol synthesis. They also observed the major role of water-soluble organosulfur compounds specially SAC as the active principle.

Epidemiologic studies in the past 15 years have established an inverse relationship between garlic consumption and the incidence of certain types of cancer include gastrointestinal, laryngeal, breast and colon cancer (Milner, 2001). The proposed mechanisms include, inhibition of N-Nitroso compounds formation, suppression of metabolic activation of carcinogen, enhanced DNA repair, reduced cell proliferations and /or induction of apoptosis. It is believed that many of these events occur simultaneously at cellular level and are responsible for chemoprotection role of garlic compounds. Milner (2001) has reviewed the role of garlic compounds in cancer prevention. Garlic used in various forms as food ingredient and processing involve may also affects the level of organosulfur compounds and efficacy of there. Heating has a nagative influence on beneficial effects of garlic.

Song and Milner (2001) demonstrated that heating of garlic for 30 Sec in microwave oven resulted in only 10% allinase activity of original. However, 60 Sec microwave heating not only completely inactivated allinase but also eliminated garlic's ability to suppress DMDA induced DNA-adduct formation.

Glucosinolates

About 100 glucosinolates have been identified in more than 450 plant species including 16 higher plant families, marine sponges and red algae. Cruciferous vegetables comprising cauliflower, broccoli, radish, horseradish cabbage, Brussels sprout contain relatively high content of glucosinolates, a potent anticarcinogen. Glucosinolates are a group of glycosides stored within cell vacuoles of all cruciferous vegetables. The functions of glucosinolates in plants are not very clear but their potent odour and taste suggests a role in herbivore feeding and microbial defense (Finley, 2005). Glucosinolates consist of a beta-D-thioglucose group; a sulfonated oxime group and a side chain derived from methionine, phenylalanine, tryptophan or branched chain amino acids. The sulfate group of a GS molecule is strongly acidic and plants accumulate. GS by sequestering them as potassium salts in plant vacuoles. Glucosinolates not bioactive in animals that consumes them until they have been enzymatically hydrolyzed to an associated isothiocynates.

Myrosinase, an enzyme found in plant cell, catalyze these GS to a variety of hydrolytic products, including isothiocynates in doles. Diindolylmethane (DIM), indole-3-carbinol (I3C), phenethly isothiocynate (PEITC) and sulphoraphane seems to be promising chemo preventive molecules, in brassica plants. In addition to the induction of phase I and II xenobiotic metabolizing enzymes resulting in the inhibition of the oxidative activation of carcinogens. I3C may reduce cancer risk by modulating estrogen metabolism. The C-16 and C-2 hydroxylations of estrogens involve competing cytochrome P-450-dependent pathways, each sharing a common estrogen substrate pool. Studies indicate that the increased formation of 2-hydroxylated estrogen metabolites relative to 16-hydroxylated forms, may protect against cancer. Another isothiocynate isolated from broccoli and termed as sulforphane has been shown to be the principal inducers of a particular type of phase-II enzyme, quinone reductage.

Carotenoids and Flavonoids

Consumption of B-carotene rich fruits and vegetables, has since long been known to prevent certain diseases like cataract. More recent studies have promoted their role in caner, and CVD prevention primarily owing to their anti-oxidative properties.

Carotenoids, C-40 polyisoprenoid structure with an extensive conjugated double bond system. Of the 600 or so carotenoids that have been identified, about 50 serve as precursors for vitamin A. The most researched among them are B-carotene & lycopene. These carotenoids inhibit the oxidation of LDL to its aterogenic form. Lycopene's ability to act as an anti-oxidant and scavenger of free radicals that are often associated with carcinogenesis is potentially a key for mechanism for its beneficial effects on human health. (Khachik et al., 1995). Lycopene may prevent carcinogenesis and athrogenesis by interfering passively with oxidative damage to DNA & lipoproteins. Lycopene is the most effective quencher of singlet oxygen in biological system.

Flavonoids are polyphenolic compounds and originally regarded as nutritionally inert, these is new increasing interest in the apparent anticarcinogenic properties of certain flavonoids, although exact biochemical mechanisms is not clear. A great deal of attention has been directed to the polyphenolic constituents of tea, particularly green tea. Catechins are the predominant and most significant of all tea polyphenols. The four major green tea catchins are epigallocatechin-3-gallate, epigallocatechin, epicatechin-3-gallate and epicatechin.

However, epidemiological studies are inconclusive, on the other side research findings in laboratory animals clearly support a cancer chemopreventive effect of tea components. Other flavonoids, with promising health stimulating effects include quercetin, kaempferol, myricetin, apigenin and luteolin. Source of these compounds are choclate, coca, fruit juices, nuts, red wine, raspberry, blackberry and citrus fruits.

Phytosterols

Phytosterols are another important terpene subclass. Two sterol molecules that are synthesized by plants are b - sitosterol and its glycoside. In animals, these two molecules exhibit anti-inflammatory, anti-neoplastic, anti-pyretic and immune-modulating activity. A proprietary mixture of b - sitosterol and its glycoside were tested *in vitro*, in animals and in human clinical trials. Phytosterols were reported to block inflammatory enzymes, for example by modifying the prostaglandin pathways in a way that protected platelets. Recently, the cytostatic activity of steroidal saponins from *Ruscus aculeatus* against leukemia HL60 cells has been reported.

In the body, phytosterols can compete with cholesterol in the intestine for uptake, and aid in the elimination of cholesterol from the body. Saturated phytosterols appear to be more effective than unsaturated ones in decreasing cholesterol concentrations in the body. These actions reduce serum or plasma total cholesterol and low-density lipoprotein (LDL) cholesterol. In mammals, concentrations of plasma phytosterol are low because of their poor absorption from the intestine and their faster excretion from liver, and metabolism to bile acids, compared to cholesterol.

Other Phytochemicals

There are certain other compounds present in plant foods, with significant health promoting effect include plant fatty acids, tocotrienols, phenolic derivatives and dietary fibers etc. Docosahexaenoic acid (DHA), which is one of the most important structural component of brain and ratina, and denovosynthesis of this compound is very rare. The decline in DHA intake could have serious implications for public health, since low plasma, DHA concentrations have been correlated with increased incidence of number of important chronic diseases such as depression, attention deficit disorders and Alzheimer's dementia. Crypthecodinium cohmii strain of marine algae is used for the commercial production of DHA rich oil. Spirulina, termed as wonder alga is one of riches source of omeg-3-fatty acids, quality protein and many other therapeutic molecule.

Plant polyphenols are secondary metabolites widely distributed in higher plants. Their distinguishing characteristics are: water solubility, molecular weights that range from 500 to 3000-4000 D, 12-16 phenolic groups and 5 – 7 aromatics rings per 1000 relative molecular mass, intermolecular complexation, and classification as condensed proanthocyanidins, galloyl and hexahydroxydiphenoyl esters and derivaties, or phlorotannins. Polyphenols historically have been considered as anti-nutrients by nutritionists, because some, eg. tannins have such adverse effects as decreasing the activites of digestive enzymes, energy, protein and amino acid availabilities, mineral uptake and having other toxic effects. Recognition of the antioxidant activities of many polyphenols has realigned thinking toward the health benefits provided by many of these compound.

Promising Functional Foods with Phytochemicals

Nutrients dense foods that provide benefits beyond basic nutrition have been already developed and commercially available. Such foods can be defined as "An accepted and tolerable food that has natural, naturally concentrated photochemical derived from fruits and vegetables or related food ingredients that a) indicated to be epidemiologically important in disease prevention, b) have shown to be useful in preventing cancer or other chronic diseases in animal bioassays, c)are unique in structure, class, pattern and metabolism. The concept of fortification is not new and some new formulations based on phytochemicals are discussed here:

Functional Drinks

The beverage market is on the cutting edge in functional food development. Energy drinks, isotonic (spare) beverages, herbal and green teas, fortified waters, smart drugs, caffeinated drinks and fringe. Beverages are relatively cheap to blend, fortified, bottle and distribute. These beverages are making inroads in the conventional drink market, currently dominated by aerated synthetic beverages. Soluble fiber containing beverages are already available. Tea and coffee, constituents can be incorporated into new formulations with other phytochemical to develop herbal drinks. Recently the anti-oxidiants or free radial scavenging ability of ascorbic acid, B-carotene lycopene, trocalnenols has been promoted. Free radicals are formed by autoxidation, photosensitization, enzymatic reactions and due to pollutants.

Some of the fruits and vegetables more important ones from beverage formulation standpoint include licorice, ginger, tea, citrus, carrots, tomatoes berries, mint and other herbs and spices. These have been much interest in developing beverages, which contain these types of compounds.

Diary Products

Probiotic cultures in fermented dairy products, have multifunctional role. In recent years, yogurts with oligosaccharides have been developed. These cultured dairy products may also incorporate, isoflavones of soy protein, B-glucan of oat, carotenoids and flavorioids of certain fruits in their formulation. Blackberry and raspberry as fruits or their extracts added in ice creams, dairy drinks, frozen desserts, not only as flavouring component, but also as source of anthocyanins and other phenotic derivatives. Similarly imitation dairy products may also include omega-3-fatty acids from flax seed oil and DHA from algae. Some infant formula have already available in market containing DHA. Dairy analogue, primarily based on soymilk, may be modified to increase the level of phytoestrogens, particularly isoflavones.

Confectionary Items

Bars either based on fruits or on cereals may serve as vehicle of phytonutrients. Commercially wide variety of nutritional bars has been developed for specific purpose. These bars can be enriched with fibers, anti-oxidants, phenolic substances, ant proteins, glucosinolates, in them. Oat bar, based on soluble oat fibers. Also contain amega-3-fatty acids to further strengthen its heart health image. Fruit preserves, fruit spreads, have been developed, using traditional herbs and products were further enriched by adding other phytonutrients. Encapsulation technology offers the product developer a new tool in protecting sensitive and expensive nutrients. A coated candy confectionary was patented in USA. The product has a soft candy center with 5-40% (w/w) water. Two coating was applied to the product, first one as moisture barrier and second one was used as a carrier for a number of additional ingredients, including vitamins, enzymes, phytochemicals and vegetable extracts.

Chocolate considered as a culprit for CHD, obesity and dental problems, also moves into healthy arena. The chocolate has neutral effect on cholesterol levels, low level of coffein, as compared to beverages, anti-oxidant activity of polyphenols, potential benefit of procyanidins and epicatechins on CVD. Replacement of a part of cocoa butter with PUFA; may further improve nutritional quality of chocolates.

Breakfast Cereals

Raspberries and blackberries, reservoirs of phytonutrients are considered as good option available for cereal product formulations. Initial studies conducted with DHA in bread are encouraging. Likewise, breads with garlic compounds and isothiocynates of cruciferous vegetables are prospective options. Soy flour is already important ingredient in bread, not only because of its nutritional significance, but excellent functional properties it impart.

Meat Products

Soy based meat analyogues, had clicked the market in early 70's. Texturizing properties of soy proteins, has offered the opportunity to develop a number of meat analogues. Addition of oat gum to meat patties resulted in decreased cooking loss. A comminuted beef product was developed that would have nutraceutical properties, exhibit resistance to rancidity and posses improved water holding capacity and cooking yield.

Safety Issues

The optimal levels of the majority of the biologically active components currently under investigation have yet to be determined. In addition, a number of animal studies show that some of the same phytochemicals (e.g.allyl isothiocynate) highlighted in this review for their cancer-preventing properties have been shown to be carcinogenic at high concentrations.

The benefits and risks to individuals and populations as a whole must be weighed carefully when considering the widespread use of physiologically-active functional foods. For example, what are the risks of recommending the increased intake of compounds (e.g. isoflavones) that may modulate estrogen metabolism? Soy phytoestrogens may represent a "double-edged sword" because of reports that genistein may actually promote certain types of tumors in animals. Knowledge of toxicity of functional food components is crucial to decrease the risk:benefit ratio.

Conclusion

Mounting evidence supports the observation that functional foods containing physiologically active components, either from plant or animal sources, may enhance health. It should be stressed, however, that functional foods are not a magic bullet or universal panacea for poor health habits. There are no "good" or "bad" foods, but there are good or bad diets. Moreover, diet is only one component of an overall life-style that can have an impact on health; other components include smoking, physical activity, and stress.

Health-conscious consumers are increasingly seeking functional foods in an effort to control their own health and well-being. The field of functional food, however, is in its infancy. Claims about health benefits of functional foods must be based on sound scientific criteria.

References

Finley,J.W. 2005. Proposed criteria for assessing the efficacy of cancer reduction by plant foods enriched in carotenoids, glucosinolates, polyphenols and selenocompounds. *Annals of Botany,* **95:** 1075-1096 pp.

Hasler, C.M. 1998. Functional Foods: Their role in disease prevention and health promotion. *Food Technology* **52**(11): 63-70.

Hasler, C.M. 2000. The changing face of functional foods. *Journal of American College of Nutrition* **19**(5): 499S-506S pp.

Milner, J.A. 2001. A historical perspective on garlic and cancer. *J.Nutr.* **131:** 10275 - 10315.

Setchell,K. 1998. Phytoestrogens. The biochemistry, physiology and implications for human health of soy isoflavones. *American J.of Clinical Nutrition,* **68:** (Suppl) 1333S-1465.

Song,K.and Milner,J.A.2001. The influence of heating on anticancer properties of garlic. *J.Nutr.* **131:** 10545 -10547.

□□□

Chapter 38

Production and Evaluation of Prebiotics

S.K. TOMAR*, VIKAS SANGWAN*, BABAR ALI*, PRASAD S PATIL* AND R.R.B. SINGH**

*Dairy Microbiology Division, National Dairy Research Institute, Karnal

**Dairy Technology Division, N.D.R.I., Karnal

Introduction

In 1995, Gibson and Roberfroid defined a prebiotic as a "nondigestible food ingredient that beneficially affects the host by selectively stimulating the growth and/or activity of one or a limited number of bacteria in the colon, and thus improves host health." This definition only considers microbial changes in the human colonic ecosystem. Later, it was considered timely to extrapolate this into other areas that may benefit from a selective targeting of particular microorganisms and to propose a refined definition of a prebiotic as (Gibson *et al.*, 2004) a selectively fermented ingredient that allows specific changes, both in the composition and/or activity in the gastrointestinal microflora that confers benefits. These definitions have attracted, and still continue to attract, a great deal of interest in the field of nutrition both in scientific research and industrial application in food products. Consequently and over the years, prebiotic activity has been attributed to many food components, particularly oligosaccharides and polysaccharides, but sometimes without due consideration to the criteria required. In particular it must be stressed that not all dietary nondigestible carbohydrates and certainly not all dietary fibers are prebiotics.

Therefore there is a need to establish clear criteria for classifying a food ingredient as a prebiotic.Indeed, such classification requires a scientific demonstration that the ingredient is :

- Resistant gastric acidity
- Not hydrolyzed by mammalian enzymes
- Not absorbed in the upper gastrointestinal tract
- Fermented by the intestinal microflora
- Selectively utilized by intestinal bacteria potentially associated with health and well-being (Gibson and Roberfroid 1995)

Prebiotic Production

Near about 20 different prebiotics are in world market today and these can be produced by either of the three main methods

- Extraction from Natural Sources
- Enzyme Processing
- Chemical Production

Extraction from Natural Sources

There are different sources from which we can extract these prebiotic oligosaccharides (van *et al.*, 2000) Fructans (e.g. inulin) are the most prominent example which are common to edible parts of a variety of plants such as Jerusalem artichoke, chicory root, dahlia tubers, garlic, asparagus root, and salsify.(Tungland 2003). Soybean OS raffinose and stachyose are commercially produced by direct extraction from soybean whey (Crittenden *et al.*, 1996). Prebiotics are also being isolated from Milk, Barley, Bran, Whole grains, Legumes such as beans, vegetables such as leeks, onions, spinach, burdock, fruits such as bananas, berries.

Enzyme Processing

In most of the research work enzymes are used to degrade polysaccharides for the preparation of prebiotics (Sarbu *et al.*, 2003). Polysaccharides such as xylan, inulin, and starch are treated with specific enzymes either in discrete batches (Noguchi *et al.*, 1997) or (in some cases) continuously using chromatography matrices carrying immobilized enzymes (Nakajima *et al.*, 2002) to yield the preferred NDOs.

Apart from this method of enzymatic degradation prebiotic can also be synthesized by enzymatic synthesis from simple sugars, like lactose and sucrose, using enzymes, such as glycosyl transferases and glycosidases (Perugino *et al.*, 2004). Glycosyl transferase catalyzes glycosyl transfer to an acceptor other than water.(Ly *et al.*, 1999) They are classified according to the sugar transferred from donor to acceptor and by the acceptor specifity. The regio- and stereospecifity for the acceptor molecule and the high yields that can be obtained are attractive features of these catalysts. However, their limited availability hampers their use (Perugino *et al.*, 2004) Glycosidases, used for the production of galactoololigosaccharides, iso-maltooligosaccharides, and gentiooligosaccharides,(Crittenden *et al.*, 1996) can be separated into two classes, which either do or do not invert or the stereochemistry of the hydrolyzed bond. Their use is limited because of low product yields (Perugino *et al.*, 2004). The use of glycosynthases tries to overcome the problems associated with the two above-mentioned enzyme types. Such enzymes are specifically mutated glycosidases that efficiently synthesize OS but do not hydrolyze them (Jakeman *et al.*, 2002).

Chemical Production

The major pathways to obtain OS from polysaccharides are acid hydrolysis, thermomechanical degradation, and oxidative degradation (Sarbu *et al.*, 2003) Acid hydrolysis of polysaccharides has mostly been used for analytical purposes. Thermomechanical degradation includes ultrasonication and twin screw extrusion (cote *et al.*, 1999). Both methods reduce the polysaccharide molecular weight and solution viscosity. Ultrasonication is most effective at depolymerizing polysaccharides,(Machov´a *et al.*, 1999) but the treatment times are long and the equipment is expensive and not amenable to scaleup. Extrusion is relatively inexpensive and scalable, but produces some undesirable dark-colored side products. For oxidative degradation, reagents such as ozone and atmospheric oxygen (Strlic *et al.*, 2003) can be used. Recently, more environmentally friendly technologies, such as autohydrolysis, also referred to as hydrothermolysis, have gained interest. Useful OS can be produced from polysaccharide enriched materials by reaction with water through hydronium-catalyzed degradation (Paraj´o *et al.*, 2004) In addition to the use of water as the only chemical, several advantages have been associated with this process. These include low by-product generation, limited problems derived from equipment corrosion owing to the mild pH of reaction media, and reduction of operational costs(Carvalheiro *et al.*, 2004).

Evaluation of Prebiotics

According to the European Consensus on Scientific Concepts of Functional Foods in Europe (Diplock *et al.*, 1999) the final demonstration of these prebiotic attributes should include *in vivo* nutritional feeding trials in the targeted species (i.e., humans, livestock, or companion animals) using validated methodologies that are supported by sound science. All aforementioned criteria are important but the demonstration of selectivity in the stimulation of growth and/ or activity of bacteria remains the most contentious and is difficult to fulfill. Indeed, it requires reliable and quantitative microbiological analysis of a wide variety of bacterial genera including total aerobes/ anaerobes, bacteroides, bifidobacteria, clostridia, enterobacteria, eubacteria and lactobacilli after anaerobic sampling of suitable biological materials, most usually feces, but sometimes also biopsies of colonic materials. Simply reporting *in vitro* fermentation in cultures of single microbial strains or even an increase *in vitro* in a limited number of bacterial genera in complex mixtures of bacteria (e.g., fecal slurries) is not proof of a prebiotic effect as it does not take bacterial interactions into account.

Regarding the stimulation of bacterial activity, patterns of production of organic acids, gases, and enzymes have been used as biomarkers of specific bacterial genera. However, these have not yet been validated and changes should be interpreted with caution. Moreover, it is also important that the rationale behind a claimed prebiotic effect is elucidated through mechanistic explanations of effect. In this context, several bacterial genes specific for the metabolism of oligosaccharides have recently been identified. In particular, this is the case for a gene, in bifidobacteria, that codes for an enzyme that specifically hydrolyzes inulin type fructans, thus explaining the selectivity in the action of these prebiotics (Schell *et al.*, 2002).

Testing Methodologies

A scheme for the evaluation of a candidate prebiotic has been proposed by referring to the criteria just described (Gibson *et al.*, 1999). Collection of biologically meaningful data on different prebiotics requires standardized testing methodologies that remain essential if we are to have confidence in any health claims on prebiotic functional foods.

Nondigestibility

In vitro methods are mainly used for the determination of resistance to acidic conditions (occurs in the stomach) and enzymatic (salivary, pancreatic, and small intestinal) hydrolysis (Ziesenitz *et al.*, 1987). Products are recovered by using such methods and these products are then quantified using chemical, physicochemical, or enzymatic methods (Dahlqvist *et al.*, 1984).

In *in-vivo* model experimental animals are used for the measurement of fecal recovery of an oral dose given in germ-free conditions . Other , more invasive methods involve intubation into the gastrointestinal system of living anaesthetized rats (Nilsson *et al.*, 1988). In human volunteers, direct or indirect methods are applied after oral administration of the candidate prebiotic. These methods involve either the direct recovery of undigested molecules in distal ileum and in feces or an indirect assessment that neither glycemia nor insulinemia is significantly increased following oral administration (Molis *et al.*, 1996). In some cases proctocolectomized individuals, the so-called ileostomy patients are also used as a model (Bach *et al.*, 1995). This model is widely accepted as a valuable alternative to study the small intestinal excretion of nutrients.

Fermentation in Large Intestine

The most commonly used in vitro models which are being used for studying anaerobic fermentation of carbohydrates by mixed bacterial populations are batch- and continuous-culture fermentation systems. In such methodologies, disappearance of the candidate prebiotic is quantified as a function of time. Batch-culture vessels are inoculated with pure cultures of different but selected species of bacteria and the carbohydrate to be studied. Use of fecal slurry is a better alternative in place of preparing the mixture of pure culture of different species of bacteria. Multichamber continuous culture systems have been developed to reproduce physical, anatomical, and nutritional characteristics of gastrointestinal regions (Macfarlane *et al.*, 1998, Gmeiner *et al.*, 2000) These models which are most exclusively used to study fermentation by mixed bacterial populations are useful for predicting both the extent and site of prebiotic fermentation.

Humans, companion animals and livestock can be used to study *In vivo* fermentation of nondigestible carbohydrates. In rats, the prebiotic which is to be investigated is added to food or drinking water but can also be administered by intubation. Animals are then anesthetized and killed at predetermined time intervals. Fecal samples and the contents of the gastrointestinal segments are collected for analysis analysis of fermentation products like gases and short chain fatty acids such as acetate, propionate, butyrate, lactate. One interesting model by which we can study carbohydrate fermentation in experimental animals is the heteroxenic rat harboring a human fecal flora (Szilit *et al.*, 1993).

Selective Stimulation of Growth of Intestinal Bacteria

With the development in the field of prebiotics the methodology for investigating functionality in particular flora compositional changes as a response to selective fermentation is also developed. Much of the early (and some of the current) literature describes studies performed on pure cultures. However, such studies failed to establish that the test carbohydrate is selectively metabolized by the intestinal microflora. The use of a fecal sample, which ensures that a representative range of bacterial species is exposed to the test material is a more meaningful in vitro method for studying prebiotic oligosaccharides. Study of the changes in populations of selected genera or species can then establish whether or not the fermentation is selective. The use of feces provides an accurate representation of events occurring in the distal colon. However, more proximal areas will have a more saccharolytic nature, and both the composition and activities of the microbiota indigenous to the colon are variable, dependent on the region sampled. This has been confirmed through studies on sudden death victims, where the colon contents were sampled shortly following death (Macfarlane *et al.*, 1998).

Identification of the genera and species present is a major problem with the use of fecal samples. Traditionally, this identification has been done by culturing samples on selective agars followed by morphological and biochemical tests designed to confirm culture identities. This is sufficient to establish that a prebiotic selectively enriches defined "desirable" organisms and depletes "undesirable" organisms but does not give a true picture of the population changes occurring. This is unavoidable using selective culture because it is estimated that only 50% of the diversity present in the human colon has yet been characterized (Suau *et al.*, 1999).

A much more reliable approach involves the use of molecular methods of bacterial identification. These have advantages over culture-based technologies in that they have improved reliability and can encompass the full flora diversity. These technologies include:

- Fluorescence in Situ Hybridization
- Polymerase Chain Reaction
- Denaturing/Temperature Gradient Gel Electrophoresis (DGGE or TGGE)
- Direct Community Analysis

Fluorescence in Situ Hybridization (FISH)

For carrying out bacterial enumeration in a quick, culture-independent and reliable manner modern techniques are now available such as fluorescence *in situ* hybridisation (FISH). It is a cytogenetic technique used to detect and localize the presence or absence of specific DNA sequences on chromosomes. FISH uses fluorescent probes that bind to only those parts of the chromosome with which they show a high degree of sequence similarity. Fluorescence microscopy can be used to find out where the fluorescent probe bound to the chromosomes. FISH is often used for finding specific features in DNA for use in genetic counseling, medicine, and species identification. FISH can also be used to detect and localize specific mRNAs within tissue samples. In this context, it can help define the spatial-temporal patterns of gene expression within cells and tissues. By targeting highly conserved areas of the rRNA, specific groups of bacteria can be distinguished from others in a mixed culture. A host of phylogenetic probes are currently available for the enumeration of faecal bacteria, whilst more are being designed and validated. Groups targeted include *Bacteroides* spp. (Manz *et al.* 1996), *Bifidobacterium* spp. (Langendijk *et al.* 1995) and *Lactobacillus* and *Enterococcus* spp. (Harmsen *et al.* 1999), *Eubacterium* (Franks *et al.* 1998), and *Clostridium* (Tuohy *et al.* 2001).

Polymerase Chain Reaction (PCR)

The polymerase chain reaction (PCR) is a technique used to amplify a single or few copies of a piece of DNA across several orders of magnitude, generating thousands to millions of copies of a particular DNA sequence. The method relies on thermal cycling, consisting of cycles of repeated heating and cooling of the reaction for denaturation, annealing and amplification of the DNA. Primers (short DNA fragments) containing sequences complementary to the target region along with a DNA polymerase are key components to enable selective and repeated amplification.

Due to the ambiguity inherent in using purportedly selective agars, only tenuous identifications of bacteria can be made using this methodology. Identifications can be made at a molecular level by using bacterial ribosomes. Genes encoding 16S subunits of the bacterial ribosomes are comprised of both conserved and variable regions, and sequencing of the 16S rRNA gene enables bacterial identifications to be made. By using PCR, segments of this gene can be amplified to a level whereby their sequence can be determined (Steffan & Atlas, 1991).

Denaturing/Temperature Gradient Gel Electrophoresis (DGGE or TGGE)

Another method to evaluate the genetic diversity of the intestinal microflora is denaturing gradient gel electrophoresis or temperature-gradient gel electrophoresis it separate amplified DNA fragments of the same size based on the extent of the sequence divergence between different PCR products (Muyzer & Smalla, 1998). The polymerase chain reaction of environmental DNA can generate templates of differing DNA sequence that represent many of the dominant microbial organisms. However, since PCR products from a given reaction are of similar size (bp), conventional separation by agarose gel electrophoresis results only in a single DNA band that is largely non-descriptive. DGGE can overcome this limitation by separating PCR products based on sequence differences that results in differential denaturing characteristics of the DNA.

A whole community PCR is carried out and partial 16S rDNA sequences are amplified from the different bacterial species present. Separation occurs due to the decreased electrophoretic mobility of the partially melted, doublestranded DNA molecule in polyacrylamide gels containing either a temperature or chemical denaturant gradient

(Muyzer & Smalla, 1998). Identification can be carried out either by excising fragments from the gel and sequencing them, or by comparing their motility with that of known control sequences. As with FISH, both culturable and unculturable populations can be characterised and this relatively rapid technique also offers the potential of monitoring gut flora over time (Zoetendal *et al.* 1998).

Direct Community Analysis

Direct community analysis is used to achieve characterisation of both the culturable and non-culturable components of a microbial mixture. 16S rRNA diversity of the sample of interest characterized by this technique. The total bacterial DNA is extracted from the sample and partial 16S rDNA genes are amplified (using universal primers) via PCR (Suau *et al.* 1999). The purified amplification products are subsequently cloned into *E. coli*, and clones containing the 16S rDNA inserts are sequenced and identified by comparison with database 16S rDNA sequences.

References

Carvalheiro, F., Esteves, M.P., Paraj'o, J.C., Pereira, H., and G'ýrio, F.M. 2004. Production of oligosaccharides by autohydrolysis of brewery's spent grain. *Bioresource Technol.* **91**(1):93–100.

Cote, G.L., and Willet, J.L. 1999. Thermomechanical depolymerization of dextran. *Carbohydr. Polym.* **39**(2):119–126.

Crittenden, R.G.(1996) Production, properties and applications of food-grade oligosaccharides. *Trends Food Sci. Technol.* **7:** 353–361

Dahlqvist, A.and Nilsson, U.(1984) Cereal fructosans: Part 1. Isolation and characterization of fructosans from wheat flour. *Food Chem.* **14:** 103–112.).

Diplock, A.T ., Aggett, P.J. , Ashwell, M., Bornet, F., Fern, E.B.and Roberfroid, M.B. (1999) Scientific concepts of functional foods in Europe: Consensus document. *Br. J. Nutr.* **81** (Suppl.1): S1–S27.

Franks AH, Harmsen HJM, Raangs GC, Jansen GJ, Schut F & Welling GW (1998) Variations of bacterial populations in human feces measured by fluorescent in situ hybridization with group-specific 16S rRNA-targeted oligonucleotide probes. *Applied Environmental Microbiology* **64:** 3336–3345.

Gibson, G.R. Rastall, R.A.and Roberfroid, M.B.Pr ebiotics.(1999) In: *Colonic Microbiota, Nutrition and Health.* (Gibson, G.R.and M.B.Roberfr oid, M.B., eds.), pp.101–124.Kluwer Academic Publishers, Dordrecht, The Netherlands.

Gibson GR, Berry Ottaway P, Rastall RA. (2000) Prebiotics: new developments in functional foods. Oxford: Chandos Publishing Limited.

Gibson, G. R. and Roberfroid, M. B. (2004), Dietary modulation of the human colonic microbiota: Updating the concept of prebiotics, *Nutr. Res. Rev.*, **17:** 259.

Gmeiner, M., Kneifel, W., Kulbe, K. D., Wouters, R., De Boever, P., Nollet, L., and Verstraete, W.(2000) Influence of a symbiotic mixture consisting of *Lactobacillus acidophilus* 74–2 and a fructooligosaccharide preparation on the microbial ecology sustained in a simulation of the human intestinal microbial ecosystem (SHIME reactor). *Appl. Microbiol. Biotechnol.* **53:** 219–223

Harmsen HJM, Elfferich P, Schut F & Welling GW (1999) A 16S rRNA-targeted probe for detection of lactobacilli and enterococci in fecal samples by fluorescent in situ hybridization. *Microbial Ecology Health and Disease* **11:** 3–12.

Jakeman, D.L., and Withers, S.G. (2002). Glycosynthases: new tools for oligosaccharides synthesis. *Trends Glycosci. Glycotechnol.* **14**(75): 13–25.

Langendijk PS, Schut F, Jansen GJ, Raangs GC, Kamphuis GR, Wilkinson MHF & Welling GW (1995) Quantitative fluorescent *in situ* hybridisation of *Bifidobacterium* with genus-specific 16S rRNA-targeted probes and its application in fecal samples. *Applied Environmental Microbiology* **61:** 3069–3075.

Ly, H.D., and Withers, S.G. (1999). Mutagenesis of glycosidases. *Annu. Rev. Biochem.* **68:**487–522.

Macfarlane GT, Macfarlane S, Gibson GR.(1998) Validation of a three-stage compound continuous culture system for investigating the effect of retention time on the ecology and metabolism of bacteria in the human colonic microbiota. Microb Ecol. **35:** 180 187.

Machov´a, E., Kvapilov´a, K., Kogan, G., and ¡ Sandula, J. (1999). Effect of ultrasonic treatment on the molecular weight of carboxymethylated chitinglucan complex from *Aspergillus niger*. *Ultrasonics Sonochem.* **5:** 169–172.

Manz W, Amann R, Ludwig W, Vancanneyt M & Schleifer K-H (1996) Application of a suite of 16S rRNA-specific oligonucleotide probes designed to investigate bacteria of the phylum *Cytophaga-Flavobacter-Bacteroides* in the natural environment. *Microbiology* **142:** 1097–1106.

Muyzer G & Smalla K (1998) Application of denaturing gradient gel electrophoresis (DGGE) and temperature gradient gel electrophoresis (TGGE) in microbial ecology. *Antonie Van Leewenhoek* **73:** 127–141.

Molis, C., Flourié, B., Ouarne, F., Gailing, M. F., Lartigue, S., Guibert, A., Bornet, F.and Galmiche, J.P. (1996) Digestion, excretion, and energy value of fructooligosaccharides in healthy humans. *Am. J. Clin. Nutr.* **64:** 324–328.)

Nakajima, N., Ishihara, K., and Matsuura, Y. (2002). Dietary-fiberdegrading enzymes from a human intestinal *Clostridium* and their application to oligosaccharide production from nonstarchy polysaccharides using immobilized cells. *Appl. Microbiol. Biotechnol.* **59**(2–3):182–189.

Nilsson, U., Oste, R., Jagerstad, M. and Birkhed, D. (1988) Cereal fructans: *In vitro* and *in vivo* studies on availability in rats and humans. *J. Nutr.* **118:** 1325–1330.)

Noguchi, Y., Ikeda, K., Masatsuji, E., and Seko, M. (1997). Xylanase process for producing the same method for the treatment of pulp and production of xylo oligosaccharides. *United States Patent US 5658765.*

Paraj´o, J.C., Garrote, G., Cruz, J.M., andDom´ýnguez, H. (2004). Production of xylooligosaccharides by autohydrolysis of lignocellulosic materials. *Trends in Food Sci. Technol.* **15:**115–120.

Perugino, G., Trincone, A., Rossi, M., and Moracci, M. (2004). Oligosaccharide synthesis by glycosynthases. *Trends Biotechnol.* **22**(1):31–37.

Playne MJ, Crittenden R.(1996) Commercially available oligosaccharides. *Bull Int Dairy J.* **313:** 10–22.

Roberfroid M.B. *(2008) General Introduction: Prebiotics in Nutrition. Handbook of prebiotics.,* 1-12.

Sarbu, A., Gon¸calves, F., and Norberta de Pinho, M. (2003). Radical initiated oxidative degradation of oat-spelts xylan. *Carbohydr. Polym.* **53**(1):1–8.

Sarbu, A., Gon¸calves, F., and Norberta de Pinho, M. (2003). Oat spelts xylan molecular mass estimation by size exclusion chromatography. *Carbohydr. Polym.* **53**(3):297–304.

Steffan RJ & Atlas RM (1991) Polymerase chain reaction: applications in environmental microbiology. *Annual Review of Microbiology* **45:** 137–161.

Strlic, M.,Kocar, D.,Kolar, J., Rychl´y, J., and Pihlar,B. (2003). Degradation of pullulans of narrow molecular weight distribution – the role of aldehydes in the oxidation of polysaccharides. *Carbohydr. Polym.* **54**(2):221– 228.

Suau A, Bonnet R, Sutren M, Godon J-J, Gibson GR, Collins MD & Doré J (1999) Direct analysis of genes encoding 16S rDNA from communities reveals many novel molecular species within the human gut. *Applied Environmental Microbiology* **65:** 4799–4807.

Szilit O, Andrieux C.(1993). Physiological and pathological effects of carbohydrate fermentation. In: Simopoulos AP, editor. World Review of Nutrition and Dietetics. Vol 74: Intestinal Flora, Immunity, Nutrition and Health. Basel: Karger. p. 88–122.

Tungland, B.C. (2003). Fructooligosaccharides and other fructans: structures and occurrence, production, regulatory aspects, food applications And nutritional health significance. In: Eggleston, G. and C^ot´e, G.L. *Oligosaccharides in food and agriculture*. Washington DC: ACS Press, 135–152.

Tuohy KM, Kolida S, Lustenberger AM & Gibson GR (2001) The prebiotic effects of biscuits containing partially hydrolysed guar gum and fructo-oligosaccharides – a human volunteer study. *British Journal of Nutrition* **86:** 341–348.

Van Laere, K.M.J. (2000). Degradation of structurally different nondigestible oligosaccharides by intestinal bacteria: glycosylhydrolases of *Bifidobacterium adolescentis. PhD Dissertation, Wageningen University, Wageningen, The Netherlands.*

Ziesenitz, S.C.and Siebert, G.(1987) *In vitro* assessment of nystose as a sugar substitute. *J. Nutr.* **117:** 846–851.

Zoetendal EG, Akkermans ADL & De Vos WM (1998) Temperature gradient gel electrophoresis analysis of 16S rRNA from human fecal samples reveals stable and host-specific communities of active bacteria. *Applied Environmental Microbiology* **64:** 3854–3859.

□□□

Chapter 39

Genomics of Probiotic Action

POONAM, SARANG DILIP POPHLY, AND S.K .TOMAR
Dairy Microbiology Division, National Dairy Research Institute, Karnal

Introduction

While the sequencing of the human genome has increased our understanding of the role of genetic factors in health and disease, each human being harbors many more genes than those in their own genome. These belong to our commensal and symbiotic intestinal microorganisms - our intestinal 'microbiome' - which play an important role in maintaining human health and well-being. A more appropriate image of ourselves would be drawn if the genomes of our intestinal microbiota were taken into account. The microbiome may contain more than 100 times the number of genes in the human genome and provides many functions that humans have thus not needed to develop themselves (Backhed *et al.,* 2005). The indigenous intestinal microbiota provides a barrier against pathogenic bacteria and other harmful food components. It has also been shown to have a direct impact on the morphology of the gut and many intestinal diseases can be linked to disturbances in the intestinal microbial population (Gaurner *et al.,* 2003).

The indigenous microbiota of an infant's gastrointestinal tract is originally created through contact with the diverse microbiota of the parents and the immediate environment. During breast feeding, initial microbial colonization is enhanced by galacto-oligosaccharides in breast milk and contact with the skin microbiota of the mother. This early colonization process directs the microbial succession until weaning and forms the basis for a healthy microbiota. The viable microbes in the adult intestine outnumber the cells in the human body by tenfold, and the composition of this microbial population throughout life is unique to each human being. During adulthood and aging the composition and diversity of the microbiota can vary as a result of disease and the genetic background of the individual (Salminen *et al.*, 2005)

Current research into the intestinal microbiome is focused on obtaining genomic data from important intestinal commensals and from probiotic microorganims. This genomic information indicates that gut commensals not only derive food and other growth factors from the intestinal contents but also influence their human hosts by providing maturational signals for the developing infant and child, as well as providing signals that can lead to an alteration in the barrier mechanisms of the gut (Backhed *et al.*, 2004). The availability of the human genome sequence has enabled us to better understand the genetic basis of many aspects of human health and disease. However, to fully understand the human genotype and its relationship with susceptibility to disease we need better information on how environmental and developmental factors interact with the genome to influence health.

Genomics of *Lactobacillus* and *Bifidobacteria*

The genus *Lactobacillus* has more than 100 cultured species (and probably more that are poorly culturable or non-culturable) and is noteworthy for its extreme phylogenetic, phenotypic and ecological diversity. However, the real extent of *Lactobacillus* diversity is not fully known and culture-independent 16s rRnA gene surveys of complex ecosystems (for example, the human gut microbiota) are expected to uncover novel phylotypes that belong to the genus *Lactobacillus*. Of the 14 sequenced and published *Lactobacillus* genomes, 8 (*Lactobacillus acidophilus, L. casei, L. fermentum, L. gasseri, L. johnsonii, L. reuteri, L. salivarius* and *L. plantarum*) are from cultures or species that are considered to be probiotic (Ventura *et al.*, 2009). Interestingly, 11% of the overall coding capacity of the *L. salivarius* genome is present on pmP118, the first megaplasmid described in lactic acid bacteria. This megaplasmid encodes biologically important features such as a locus for bacteriocin production, a bile salt hydrolase and two genes that complete the phosphoketolase pathway, officially reclassifying this organism as a facultative heterofermenter (Claesson *et al.*, 2006).

The genus *Bifidobacterium* is small, with 30 characterized species and a low level of phylogenetic and genomic diversity (Ventura *et al.*, 2006). Bifidobacteria were originally isolated from a breast-fed infant and 30 species have since been isolated from the GIT contents of mammals, birds and insects. Those Bifidobacterial species that have been isolated from the human intestine have attracted the interest of genomic researchers owing to their probiotic properties. However, of the Bifidobacterial taxa described so far, genomes of only three species, which belong to the *Bifidobacterium longum* and *Bifidobacterium adolescentis* groups, have been sequenced to completion (Ventura *et al.*, 2009). The availability of six genome sequences provides genetic evidence that bifidobacteria are prototrophic and therefore well adapted to growth in an environment such as the human colon, which contains low concentrations of some growth substrates for example, vitamins, amino acids and nucleotides (Ventura *et al.*, 2007). These bifidobacterial genome sequences harbour genes for the synthesis of at least 19 amino acids and they encode all of the enzymes that are needed for the biosynthesis of pyrimidine and purine nucleotides, as well as those that are required for the synthesis of the B vitamins, folic acid, thiamine and nicotinate (Schell *et al.*, 2002). Annotation and pathway prediction revealed that bifidobacterial species possess the genetic information that is required to shunt many monosaccharides or disaccharides into the fructose-6-phosphate pathway.

Molecular mechanism of *Modus operandi* of Probiotic

Mechanism of Survival and Persistence

Probiotic bacteria encounter various environmental conditions upon ingestion by the host and during transit in the GIT. Firstly, they need to survive the harsh conditions of the stomach. Humans secrete approximately 2.5 liters of gastric juice each day, generating a fasting pH of 1.5, increasing to pH 3 to 5 during food intake (Cotter and Hill, 2003). Given that the liver excretes approximately 1 liter of bile each day into the small intestine, exposure to bile represents another challenge for bacteria entering the GIT (Begley *et al.*, 2005).

In addition to coping with acid and bile, the contribution of other stress responses to the survival capacity of lactobacilli in the GIT should not be overlooked, although they remain rather unexplored. These certainly include oxidative and osmotic stress. Moreover, it can be anticipated that interactions with other microbes and with cells of the immune system and the various antimicrobial products that they produce can also impose a serious threat for the probiotic microbes (Lebeer *et al.*, 2009). Interestingly, the phenomenon of cross-adaptation is often observed, i.e., that adaptation to one stress condition also protects against another stress factor, implying some common mechanisms (De Angelis and Gobbetti *et al.*, 2004). Multiple genome-wide studies and downstream functional analyses have now been performed to characterize various stress responses in lactobacilli which are summarized below.

Maintaining Integrity of the Cell Envelope

The different macromolecules constituting the cell membranes and cell walls of lactobacilli have been shown to contribute to maintaining cell integrity during stress to various degrees. For example, low pH causes a shift in the fatty acid composition of the cell membrane of an oral strain of *L. casei* (Fozo *et al.*, 2004). Similarly, bile salts have been shown to induce changes in the lipid cell membrane of *L. reuteri* CRL1098 (Taranto *et al.*, 2003). Additionally, D-Ala esters of LTA and WTA have been suggested to be necessary for maintaining cell integrity at low pH and in the presence of bile. In addition, genes involved in PG biosynthesis and organization have been identified for acid responses in lactobacilli (Wall *et al.*, 2007).

Repair and Protection of DNA and Proteins

A number of proteins that play a role in the protection or repair of macro- molecules such as DNA and proteins also seem to be essential for acid and bile resistance. Bile acids have been shown to induce DNA damage and the activation of enzymes involved in DNA repair (Begley *et al.*, 2005). In *L. reuteri*, it was observed by microarray analysis that the level of expression of *dps* (DNA protection during starvation) increased after bile exposure. Perhaps even more vital in the general stress response are chaperones that intervene in numerous stresses for important tasks such as protein folding, renaturation, protection of denatured proteins, and removal of damaged proteins. A 2D proteomic approach to study acid adaptation in *L. delbrueckii* subsp. *bulgaricus* identified three highly induced proteins, i.e., GroES, GroEL, and DnaK

(Lim *et al.*, 2000). Similar approaches were used to demonstrate DnaK, DnaJ, GrpE, GroES, and GroEL production in *L. acidophilus* as a response to acid adaptation (Lorca *et al.*, 2002).

Two-Component and Other Regulatory Systems

Mechanisms to specifically sense the presence of certain stress factors and regulate gene expression in response to these stimuli are also crucial for bacterial survival under adverse conditions. Although these mechanisms are not well characterized for lactobacilli, they often involve two-component regulatory systems (2CRSs). Various studies have shown a role for 2CRSs in stress responses of lactobacilli e.g in *L. sakei* 23K, the disruption of the *rrp-1* and *rrp-48* genes, encoding RRs, resulted in an increased susceptibility to low pH (Morel-Deville *et al.*, 1998). Klaenhammer and coworkers identified a 2CRS (LBA1524-LBA1525) in *L. acidophilus* NCFM that was similar to the acid-related system LisRK from *Listeria monocytogenes.* (Azcarate-Peril *et al.*, 2005).

Transport and Hydrolysis of Bile

Some bacteria use efflux systems belonging to the family of multidrug resistance (MDR) transporters to export bile. In their screening for bile-responsive genes in *L. plantarum* WCFS1, Bron and coworkers identified three possible exporter proteins, including a putative MDR transporter gene (lp_3160) (Bron *et al.*, 2004). Several bacteria are also known to enzymatically modify bile salts. Bile salt hydrolases (BSHs) are generally intracellular enzymes that catalyze the hydrolysis of the amide bond between the steroid moiety and the amino acid side chain of the bile acids. BSH activity is found primarily in organisms isolated from the GIT of mammals (*Bifidobacterium* species, *L. acidophilus, L. gasseri, L. johnsonii,* and some strains of *L. plantarum*) (Begley *et al.*, 2005).

Adaptation to Human Gut

Intestinal lactobacilli compensate for their auxotrophy by encoding multiple genes for transporters. Their genomes also contain genes that encode acid and bile resistance, capacity for uptake of macromolecules, metabolism of complex carbohydrates and cell-surface proteins that interact with the intestinal mucosa (Pfeiler and Klaenhammer, 2007). *L. acidophilus* has adapted to the gut ecological niche by retaining the functional gene sets that are absent from *L. helveticus,* emphasizing the importance of these gene sets for probiotic functionality and niche adaptation by autochthonous lactobacilli that naturally reside in the GIT. Bron and coworkers used a modified *in vivo* expression technology to identify 72 genes that are expressed by *L. plantarum* in the mouse GIT, most of which were associated with survival or adaptation (Bron *et al.*, 2004).

Bifidobacteria have been presumed to have an ecological advantage owing to their capacity to metabolize complex sugars that are derived from the diet as well as from the host. Genome annotation has confirmed that genes that are required for the breakdown of complex sugars are abundant in sequenced bifidobacterial genome (Hooper *et al.*, 1999). More than 8% of annotated bifidobacterial genes encode enzymes that are involved in carbohydrate metabolism (Ventura *et al.*, 2007). Bifidobacteria use a 'docking station' to sequester and capture high-molecular-weight carbohydrate molecules such as xylose- and arabinose-containing polysaccharides and bind these to their cell surface (Van den Broek *et al.*, 2008), presumably to avoid losing them to nearby competitors.

Adherence

In addition to tolerating the harsh physical-chemical environment of the GI tract, adherence to intestinal mucosal cells is also frequently considered to be a desirable property for a probiotic strain. Attachment would be necessary for colonization and any direct interactions between the probiotic and host cells leading to the competitive exclusion of pathogens and/or modulation of host cell responses (Marco *et al.*, 2006).

Genomics-based approaches, in particular, are starting to reveal numerous bacterial cell-surface-associated proteins with intestinal cell and mucus binding functions. For example, homology searches led to the identification of several adhesion factors in *Lactobacillus acidophilus* (Buck *et al.*, 2005). Moreover, *in silico* genotype-phenotype matching was used to identify a mannose-specific adhesin of *Lactobacillus plantarum* (Pretzer *et al.*, 2005). As enteropathogenic *Escherichia coli* are known to bind to epithelial cells via mannose receptors, probiotic strains with similar adherence capabilities could inhibit pathogen attachment and colonization at these binding sites and thereby protect the host against infection.

Cell surface properties of bacteria can be expected to play a key role in the primary interaction with host cells or tissues; hence, the cell wall of probiotics has received considerable attention in general. For example, the *in silico* secretome of *L. plantarum* was predicted to contain at least 12 proteins directly involved in adherence to host components such as collagen and mucin. Similarly, putative adhesin proteins were predicted for the *L. acidophilus* NCFM genome, and mutation analysis of fibronectin- and mucin-binding proteins confirmed their role in bacterial adherence to epithelial cells (Caco2) *in vitro* (Buck *et al.*, 2005).

Mechanism of Health Promoting Effects

The mechanism underlying the variety of health benefits of probiotic lactobacilli and bifidobacteria remains poorly understood (Marco *et al.*, 2006). These effects are likely to involve both microbe-microbe and microbe-host interactions (Figure 1). Studies related to microbe-microbe interactions are generally focused on the inhibition of detrimental activities of pathogenic bacteria (Servin, 2004), while studies related to microbe-host interactions address diverse aspects, including strengthening of intestinal barrier function and modulation of the innate and adaptive immune system. The understanding of probiotic effector molecules involved in these activities and their mode of action and specificity is still in its infancy. Nevertheless, several (postgenomic) studies illustrate the potential of this field and highlight specific probiotic functionalities.

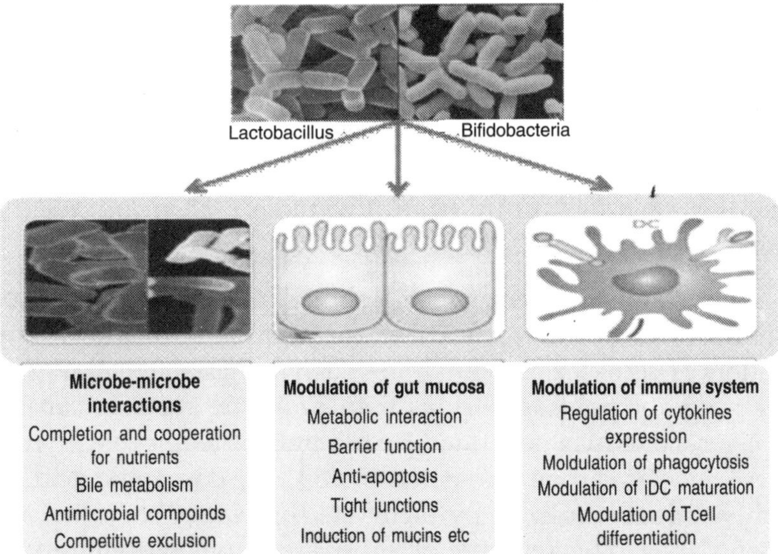

Figure 1: Schematic representation of the different modes of interaction that contribute to health promoting effects of probiotic bacteria.

Microbe-Microbe Interactions

An important mechanism of probiotic action is the capacity of probiotics to beneficially affect the host by direct effects on the microbiota.. Traditionally, most attention is given to the antipathogenic properties of probiotics. However, synergistic interactions could also occur between probiotics and endogenous beneficial members of the microbiota.

Probiotic inhibition of pathogenic microbes in the intestinal tract may involve a variety of mechanisms, including competition for the use of nutrients, production of antimicrobial compounds, or competition for specific adhesion sites (competitive exclusion). Nutrient-based interactions of probiotic bifidobacteria and lactobacilli with the endogenous microbiota have been elegantly illustrated in germ-free mouse models. This study highlighted the differential nutrient-polysaccharide targeting by the commensal *Bacteroides thetaiotaomicron* upon cocolonization with *Bifidobacterium longum, B. animalis,* or *L. casei* and also illustrated the cooperative modulation of host mucosa physiology (Sonnenburg *et al.,* 2006). A recent study exemplified the importance of antimicrobial peptide (bacteriocin) production by *L. salivarius* UCC118 for the protection against the invasive food-borne pathogen *Listeria monocytogenes.* In contrast to the wild-type strain, a bacteriocin-negative derivative of *L. salivarius* could not protect mice against infection by *Listeria* (Corr *et al.,* 2007). The competitive exclusion concept has drawn a lot of attention to the capacity of probiotic bacteria to adhere to the intestinal mucosa. A decade ago *L. plantarum* adherence to mannosyl moieties exposed on human mucosa had been associated with a competitive exclusion mechanism involved in prevention of infection by enterotoxigenic *Escherichia coli* (ETEC) (Adlerberth *et al.,* 1996).

Additionally, synergistic metabolic interactions could occur, particularly in the lower part of the GIT. For instance, primary degraders of complex carbohydrates such as *Bacteroides* may release a wider range of polysaccharides than probiotic bacteria can utilize. The oligosaccharides generated by the numerous glycosylhydrolases of these species can be fermented by, e.g., LAB to lactic acid, which can then be metabolized by lactic acid-utilizing bacteria (Xu and Gordon, 2003). The removal of substrates from the vicinity of the producer cells may also be advantageous for thermodynamic reasons, driving reactions that are otherwise energetically unfavorable.

Modulation of Gut Epithelium

The intestinal epithelial cells are presumed to be the first and most important target cells of probiotic action. Some factors that can beneficially influence the optimal functioning of the gut epithelium, related to its nutritive and its barrier function are summarized below

Metabolic Interactions

Metabolic interactions between lactobacilli and the host could modulate the primarily nutritive function of the epithelium, but these interactions are difficult to delineate. A holistic approach was recently used by Martin and coworkers which showed that probiotic exposure can modify the microbiota and alter hepatic lipid metabolism, coupled with lowered plasma lipoprotein, increased triglyceride levels, and apparent stimulated glycolysis (Martin *et al.*, 2007). They also observed changes in a diverse range of other pathways, including amino acid metabolism, methylamines, and short chain fatty acids such as butyrate. Butyrate is a particularly important source of energy for the colonic mucosal cells, and it was suggested to be necessary for the maintenance of the colonic epithelium.

Preservation of Epithelial Barrier Function

Alterations in epithelial barrier function are implicated in a variety of intestine- related disorders including enteric infections, IBD, food allergy, autism, and stress (Turner, 2006). Probiotic lactobacilli are suggested to strengthen the epithelial barrier by various mechanisms such as the induction of mucin secretion (Mack *et al.*, 2003), enhancement of tight-junction functioning (Johnson-Henry, 2008; Otte and Podolsky, 2004), upregulation of cytoprotective heat shock proteins (Petrof *et al.*, 2004) and prevention of apoptosis of epithelial cells (Yan *et al.*, 2007). In some cases, parts of the signaling pathways have been identified, but the *Lactobacillus* effector molecules and their corresponding receptors mediating these effects are largely unknown.

Immunomodulatory Effects

The capacity of probiotic bacteria to modulate the innate and adaptive immune system is generally considered a plausible mechanism of probiotic activity. However, this area is complicated by the complexity of the human immune system (Guigoz *et al.*, 2008). Consequently, immunomodulation by probiotics can be anticipated to depend strongly on the choice of the probiotic strain and the appropriate selection of potentially responsive subjects. Our understanding of the molecular interactions that drive immunomodulation by bacteria (probiotics) *in vivo* in humans is limited, and the available data is derived almost exclusively from *in vitro* assays using blood-derived immune cells.

Nevertheless, a few promising candidate probiotic effector molecules have been discovered and in some cases included in validation studies using animal models. As an example, the genome mining of the *B. longum* NCC2705 identified a gene encoding a potential serine protease inhibitor (serpin) (Schell *et al.*, 2002), which effectively inhibits human neutrophil elastase and might play a role in health-promoting regulation of innate immunity in the intestine by dampening excessive responses that are seen, for example, in inflammatory bowel disease.

An important conceptual health benefit of probiotics is based on their capacity to promote immunotolerance by priming dendritic cells to stimulate regulatory T-cell differentiation. This characteristic varies among probiotic *Lactobacillus* species, which might depend on their recognition by the C-type lectin receptor DC-SIGN (DC-specific intercellular adhesion molecule 3-grabbing nonintegrin), which is proposed to play a decisive role in regulatory T-cell stimulation (Smits *et al.*, 2005). A recent study using mutant strains of *L. acidophilus* NCFM identified surface layer protein A (SlpA) as the first probiotic DC-SIGN ligand functionally involved in the modulation of dendritic cells and T-cells (Konstantinov, 2008).

The importance of the bacterial cell wall in probiotic-immune interactions has also been illustrated using a *L. plantarum dlt* mutant that is incapable of incorporating d-alanyl into its teichoic acids. The mutant lipoteichoic acids (LTAs) were not only depleted of d-alanylation, but also appeared to be threefold longer and displayed an increased degree of glucosylation (Palumbo *et al.*, 2006), which was associated with a major impact on the immunomodulating capacity of the mutant strain compared with the wild type. The mutant displayed a strongly reduced induction of Toll-like receptor-2-dependent pro inflammatory responses and a concomitant increased stimulation of interleukin-10 (IL-10) in blood-derived immune cells. *Dlt* mutations in other lactobacilli generated alternative outcomes, including the unaltered immunomodulation in a *L. rhamnosus* GG *dlt* mutant that displayed decreased survival under intestinal conditions

(Perea Velez., 2007), a phenotypic trait that was shared by an *L. reuteri* 100-23 *dlt* mutant (Walter *et al.*, 2007). These studies establish that specific immunomodulatory capacities of a probiotic strain may depend on strain- or species-specific molecular characteristics.

Conclusion

Most probiotic strains marketed today were originally selected for their superiority in a variety of easily measurable phenotypes (e.g. ability to tolerate bile salts or survive GI tract passage) and not necessarily for their unique ability to confer defined health benefits. If consumer support for probiotics is to continue, it will be crucial to identify the precise mechanisms by which probiotics influence human health. Such studies should be accelerated by omics approaches, including genomics and functional analyses. Molecular interaction models are currently being developed, although more are required, to monitor the activation of cellular and systemic responses *in vivo* in animal models and in feeding trial participants through the measurement of previously validated biomarkers. The combination of validated molecular models with functional and comparative genomics-based approaches should enable selection of the most appropriate probiotic strain for a particular health benefit or should enable improvement of strain processing and administration regimes that optimize established health effects. This might allow the selection of specific probiotics for a particular human genotype, by analogy with personalized genomic medicine efforts.

Several issues regarding the sequences of complete probiotic bacterial genomes remain unresolved so far, only a limited number of completed probiotic bacterial genome sequences are available, and these only partially represent the total biodiversity of probiotic bacteria residing in the human gut. In this context, understanding the human gut microbiome will be an important challenge for the future (Turnbaugh *et al.*, 2007). Furthermore, sequencing the genomes of environmental organisms and carrying out metagenomic surveys of diverse gut environments (human versus animal GITs, for example) will provide not only an improved understanding of microbial biodiversity but also insights into the evolution of bacterial factors that may be crucial for the establishment of commensals (probiotics) in these different gut niches (Ley *et al.*, 2008). Advances in molecular diagnostic using noninvasive biological samples (for example, urine, feces, and blood), and the development of predictive biomarkers for the risk of disease, may enable a more rational design of individual dietary needs. Such knowledge, in combination with an improved understanding of probiotic-microbiota-host interaction and communication, could provide molecular criteria that predict susceptibility of individual subjects to specific probiotic supplementation and may thus be employed as a priori criteria to rationally enhance the success rates of probiotic application.

References

Adlerberth, I., Ahrne, S., Johansson, M. L., Molin, G., Hanson, L. A., et al. (1996). A mannose-specific adherence mechanism in *Lactobacillus plantarum* conferring binding to the human colonic cell line HT- 29. *Appl. Environ. Microbiol.,* **62:** 2244–2251.

Azcarate-Peril, M. A., McAuliffe, O., Altermann, E., Lick, S., Russell, W. M., et al. (2005). Microarray analysis of a two-component regulatory system involved in acid resistance and proteolytic activity in *Lactobacillus acidophilus. Appl. Environ. Microbiol.,* **71:** 5794–5804.

Bäckhed, F., Ding, H., Wang, T., Hooper, L. V., Koh, G. Y., et al. (2004). The gut microbiota as an environmental factor that regulates fat storage. *Proc. Natl. Acad. Sci. USA.,* **101:** 15718-15723.

Bäckhed, F., Ley, R. E., Sonnenburg, J. L., Peterson, D. A., Gordon, J. I. (2005). Host bacterial mutualism in the human intestine. Science, **307:** 1915-1920.

Begley, M., Gahan, C. G. M., Hill, C. (2005). The interaction between bacteria and bile. *FEMS Microbiol. Rev.,* **29:** 625–651.

Bron, P. A., Grangette, C., Mercenier, A., de Vos, W. M., Kleerebezem, M. (2004). Identification of Lactobacillus plantarum genes that are induced in the gastrointestinal tract of mice. *J. Bacteriol.,* **186:** 5721–29.

Bron, P. A., Marco, M., Hoffer, S. M., Van Mullekom, E., de Vos, W. M., et al. (2004). Genetic characterization of the bile salt response in *Lactobacillus plantarum* and analysis of responsive promoters in vitro and in situ in the gastrointestinal tract. *J. Bacteriol.,* **186:** 7829 7835.

Buck, B. L., Altermann, E., Svingerud, T., Klaenhammer, T. R. (2005). Functional analysis of putative adhesion factors in *Lactobacillus acidophilus* NCFM. *Appl. Environ. Microbiol.,* **71:** 8344-8351.

Claesson, M. J., Li, Y., Leahy, S., Canchaya, C., van Pijkeren, J. P., et al. (2006). Multireplicon genome architecture of *Lactobacillus salivarius. Proc. Natl. Acad. Sci.* USA., **103:** 6718–6723.

Corr, S. C., Li, Y., Riedel, C. U., O'Toole, P. W., Hill, C., et al. (2007). Bacteriocin production as a mechanism for the antiinfective activity of *Lactobacillus salivarius* UCC118. *Proc. Natl. Acad. Sci.* USA, **104:** 7617–7621.

Cotter, P. D., Hill, C. (2003). Surviving the acid test: responses of gram-positive bacteria to low pH. *Microbiol. Mol. Biol. Rev.*, **67:** 429–453.

De Angelis, M., Gobbetti, M. (2004). Environmental stress responses in *Lactobacillus*: a review. *Proteomics*, **4:** 106–122.

Fozo, E. M., Kajfasz, J. K., Quivey, R. G. (2004). Low pH-induced membrane fatty acid alterations in oral bacteria. *FEMS Microbiol. Lett.*, **238:** 291–95.

Guarner, F., Malagelada, J. R. (2003). Gut flora in health and disease. *Lancet*, **361:** 512-519.

Guigoz Y, Dor´e, J., Schiffrin, E. J. (2008). The inflammatory status of old age can be nurtured from the intestinal environment. *Curr. Opin. Clin. Nutr. Metab. Care*, **11:** 13–20.

Hooper, L. V., Xu, J., Falk, P. G., Midtvedt, T., Gordon, J. I. (1999) A molecular sensor that allows a gut commensal to control its nutrient foundation in a competitive ecosystem. *Proc. Natl. Acad. Sci. USA*, **96:** 9833–9838.

Johnson-Henry, K. C., Donato, K. A., Shen-Tu, G., Gordanpour, A., Sherman, P. A. (2008). *Lactobacillus rhamnosus* strain GG prevents enterohemorrhagic *Escherichia coli* O157:H7 induced changes in epithelial barrier function. *Infect. Immun.*, **76:** 1340–1348.

Konstantinov, S. R., Smidt, H., de Vos, W. M., Bruijns, S. C., Singh, S. K., et al. (2008). S layer protein A of *Lactobacillus acidophilus* NCFM regulates immature dendritic cell and T cell functions. *Proc. Natl. Acad. Sci. USA*, **105:** 19474–19479.

Lebeer, S., Vanderleyden, J., De Keersmaecker, S. C. J. (2008). Genes and molecules of Lactobacilli supporting probiotic action. *Microbiol. Mol. Bio. Rev.*, **72**(4): 728–764.

Ley, R. E., Hamady, M., Lozupone, C., Turnbaugh, P. J., Ramey, R. R. (2008). Evolution of mammals and their gut microbes. *Science*, **320:** 1647–1651.

Lim, E. M., Ehrlich, S. D., Maguin, E. (2000). Identification of stress inducible proteins in *Lactobacillus delbrueckii* subsp *bulgaricus*. *Electrophoresis*, **21:** 2557–2561.

Lorca, G. L., de Valdez, G. F., Ljungh, A. (2002). Characterization of the protein synthesis dependent adaptive acid tolerance response in *Lactobacillus acidophilus*. *J. Mol. Microbiol. Biotechnol.*, **4:** 525–532.

Mack, D. R., Ahrne, S., Hyde, L., Wei, S., Hollingsworth, M. A. (2003). Extracellular MUC3 mucin secretion follows adherence of *Lactobacillus* strains to intestinal epithelial cells *in vitro*. Gut, **52:** 827–833.

Marco, M. L., Pavan, S., Kleerebezem, M. (2006). Towards understanding molecular modes of probiotic action. *Curr. Opin. Biotechnol.*, **17:** 204–210.

Martin, F. P., Dumas, M. E., Wang, Y., Legido-Quigley, C., Yap, I. K., et al. (2007). A top-down systems biology view of microbiome-mammalian metabolic interactions in a mouse model. *Mol. Syst. Biol.,* **3:** 112.

Morel-Deville, F., Fauvel, F., Morel, P. (1998). Two-component signal transducing systems involved in stress responses and vancomycin susceptibility in *Lactobacillus sakei. Microbiology,* **144:** 2873–83.

Otte, J. M., Podolsky, D. K. (2004). Functional modulation of enterocytes by gram-positive and gram-negative microorganisms. *Am. J. Physiol. Gastrointest. Liver Physiol.,* **286:** G613–G626.

Palumbo, E., Deghorain, M., Cocconcelli, P. S., Kleerebezem, M., Geyer, A., et al. (2006). D-alanyl ester depletion of teichoic acids in *Lactobacillus plantarum* results in a major modification of lipoteichoic acid composition and cell wall perforations at the septum mediated by the Ccm2 autolysin. *J. Bacteriol.,* **188**(3): 709–715.

Perea V´elez, M., Verhoeven, T. L., Draing, C., Von Aulock, S., Pfitzenmaier, M., et al. (2007). Functional analysis of D-alanylation of lipoteichoic acid in the probiotic strain *Lactobacillus rhamnosus GG. Appl. Environ. Microbiol.,* **73:** 3595–3604.

Petrof, E. O., Kojima, K., Ropeleski, M. J., Musch, M. W., Tao, Y., et al. (2004). Probiotics inhibit nuclear factor-kappa B and induce heat shock proteins in colonic epithelial cells through proteasome inhibition. *Gastroenterology,* **127:** 1474–1487.

Pfeiler, E. A., Klaenhammer, T. R. (2007). The genomics of lactic acid bacteria. *Trends Microbiol.,* **15:** 546–53.

Pretzer, G., Snel, J., Molenaar, D., Wiersma, A., Bron, P. A., et al. (2005). Biodiversity-based identification and functional characterization of the mannose-specific adhesin of *Lactobacillus plantarum. J. Bacteriol.,* **187:** 6128-6136.

Salminen, S., Nurmi, J., Gueimonde, M. (2005). The genomics of probiotic intestinal microorganisms. *Genome Biology,* **225**(6): 1-4.

Schell, M. A., Karmirantzou, M., Snel, B., Vilanova, D., Berger, B., et al. (2002). The genome sequence of *Bifidobacterium longum* reflects its adaptation to the human gastrointestinal tract. *Proc. Natl. Acad. Sci. USA,* **99:** 14422–14427.

Servin, A. L. (2004). Antagonistic activities of lactobacilli and bifidobacteria against microbial pathogens. FEMS Microbiol. Rev., **28:** 405–440.

Smits, H. H., Engering, A., van der Kleij, D., de Jong, E. C., Schipper, K., et al. (2005). Selective probiotic bacteria induce IL-10-producing regulatory Tcells in vitro by modulating dendritic cell function through dendritic cell-specific intercellular adhesion molecule 3-grabbing nonintegrin. *J. Allergy Clin. Immunol.,* **115:** 1260–1267.

Sonnenburg, J. L., Chen, C. T., Gordon, J. I. (2006). Genomic and metabolic studies of the impact of probiotics on a model gut symbiont and host. *PLoS Biol.*, **4:** 413.

Taranto, M. P., Murga, M. L. F., Lorca, G., de Valdez, G. F. (2003). Bile salts and cholesterol induce changes in the lipid cell membrane of Lactobacillus reuteri. *J. Appl. Microbiol.*, **95:** 86–91.

Turnbaugh, P.J., Ley, R. E., Hamady, M., Fraser-Liggett, C. M., Knight, R., et al. (2007). The human microbiome project. Nature, **449:** 804–810.

Turner, J. R. (2006). Molecular basis of epithelial barrier regulation—from basic mechanisms to clinical application. *Am. J. Pathol.*, **169:** 1901–1909.

Van den Broek, L. A., Hinz, S. W., Beldman, G., Vincken, J. P., Voragen, A. G. (2008). Bifidobacterium carbohydrases-their role in breakdown and synthesis of (potential) prebiotics. *Mol. Nutr. Food Res.*, **52:** 146–163.

Ventura, M., Canchaya, C., Fitzgerald, G. F., Gupta, R. S., Van Sinderen, D. (2007). Genomics as a means to understand bacterial phylogeny and ecological adaptation: the case of bifidobacteria. *Antonie Leeuwenhoek*, **91:** 351-372.

Ventura, M., O'Flaherty, S., Claesson, M. J., Turroni, F., Klaenhammer, T. D., et al. (2009). Genome-scale analyses of health promoting bacteria: probiogenomics. *Nature*, **7:** 61-73.

Ventura, M., Canchaya, C., Del Casale, A., Dellaglio, F., Neviani, E., et al. (2006). Analysis of bifidobacterial evolution using a multilocus approach. *Int. J. Syst. Evol. Microbiol.*, **56:** 2783–2792.

Wall, T., Bath, M., Britton, R. A., Jonsson, H., Versalovic, J., et al. (2007). The early response to acid shock in *Lactobacillus reuteri* involves the ClpL chaperone and a putative cell wall-altering esterase. *Appl. Environ. Microbiol.*, **73:** 3924–3935.

Walter, J., Loach, D. M., Alqumber, M., Rockel, C., Hermann, C., et al. (2007). D-alanyl ester depletion of teichoic acids in *Lactobacillus reuteri* 100–23 results in impaired colonization of the mouse gastrointestinal tract. *Environ. Microbiol.*, **9:** 1750–1760.

Xu, J., Gordon, J. I. (2003). Honor thy symbionts. *Proc. Natl. Acad. Sci. USA*, **100:** 10452–10459.

Yan, F., Cao, H. W., Cover, T. L., Whitehead, R., Washington, M. K., et al. (2007). Soluble proteins produced by probiotic bacteria regulate intestinal epithelial cell survival and growth. *Gastroenterology*, **132:** 562–575.

□□□

Chapter 40

Nutritional and Functional Properties of Oats

SEEMA AND RAMESHWAR SINGH
Dairy Microbiology Division, National Dairy Research Institute, Karnal

Introduction

With urgently renewed interest in "eating for health" in the developed world, alongside the endemic obesity problem, much activity is focused on delivering soluble fibres to the consumer via the food industry. The today's diet is overall fibre deficient, especially so for beneficial dietary fibres. The global importance of cereal crops to the human diet and moreover to the written history of man and agriculture cannot be over stated. Cereal grains are the fruit of plants belonging to the grass family (*Gramineae*). The sustenance provided by cereals is frequently mentioned in the Bible, and they are by many other criterea the most important group of food crops produced in the world. Cereals are grown over 73% of the total harvested area and contribute over 60% of the world food production. Cereal crops are energy dense, containing 10 000-15 000 kJ/Kg, about 10-20 times more energy than most succulent fruits and vegetables. Nutritionally, they are important sources of dietary protein, carbohydrates, the B complex of vitamins, vitamin E, iron, trace minerals and fiber. It has been estimated that global cereal consumption directly provides about 50 percent of protein and energy necessary for the human diet, with cereals providing an additional 25 percent of protein and energy via livestock intermediaries. That is why it is said that they have a huge potential to fill the "health gap". Although dried cereal grains constitute living cells that respire, when kept in an appropriate environment, whole grains can be stored for many years. Major cereal crops produced worldwide include wheat, rice, maize and barley, oat, sorghum, and millet. Asia, America, and Europe produce more than 80 percent of the world's cereal grains. Cereal foods have a very important place in the nation's diet and are obvious and major sources of healthy soluble and insoluble fibres.

Oats: Unique Among Cereals

Oat, like all other grain varieties, belongs to the Poaceae family and is known as "Jai" or "Javi" in Indian subcontinent (Butt *et al*, 2008). The history of oats can be traced back to about 2000 B.C. The first evidence of oats was found in Egypt and among the lake dwellers of ancient Switzerland (Coffman, 1961). Oats may have spread there as a weed in wheat and barley crops. Oat cultivation began much later than that of wheat and barley. Oats descended from a number of diploid (14 chromosomes) and tetraploid species. Like wheat, it is an annual grass that is believed to be Asiatic in origin. During early growth, the oat plant consists of leaves and a shortened stem, giving a rosette type of plant. The tillers grow into additional "branch plants" or tillers and under favorable conditions, the plant can form up to 30 tillers. The main stem and tillers can reach up to 2 or more ft, depending on variety and growing condition. The most commonly cultivated and most popular species in the world today is *Avina sativa* making up more than 75% of the world's cultivars (Coffman, 1961).

Scientific classification

Kingdom : Plantae
Division : Magnoliophyta
Class : Liliopsida
Order : Poales
Family : Poaceae
Genus : Avena
Species : A. sativa

Oats remain an important grain crop for people in marginal ecologies throughout the developing world, and in developed economies for specialist uses. In many parts of the world oats are grown for use as grain as well as for forage and fodder, straw for bedding, hay, silage and chaff. Livestock grain feed is still the primary use of oat crops.

Oats are better adapted to variable soil types and can perform better on acid soils than other small grain cereals crops. They are mostly grown in cool moist climates and they can be sensitive to hot, dry weather from head emergence through to maturity. For these reasons, world oat production is concentrated from the 40th latitude in the southern hemisphere to the 60th latitude in the northern hemisphere. Thus including countries such as Australia, China, North America, Scandinavian, Russia and related countries (Schrickel, 1986). Russia is the country that had by far the highest production of oats (32.8%) followed by Canada, U.S.A., Australia, Poland and Ukraine.

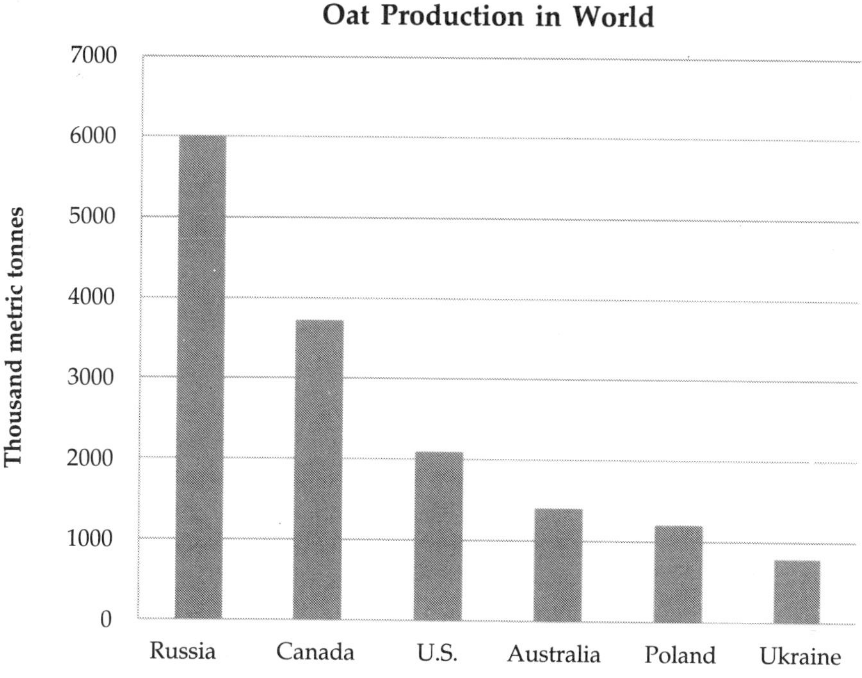

Figure 1.1 : "Top 6" of the world's greatest oat producing countries in thousand metric tonnes in the year 2010 (www.nationmaster.com/graph/agr_gra_oat_pro-agriculture-grains-oats-production)

Oat Grain

Oat grain has always been an important form of livestock feed. It has soft kernel and lipid distributed throughout the seed, which makes its milling process more difficult than wheat and corn. To prevent from atmospheric oxidation, the oat is given a hydrothermic treatment before processing. Hull (husk) of oat grain is about 25–30% of the seed. The grain is dehulled before use, whereas husk after processing may be used in food industry. Unprocessed hull contain silicate particles, which have barbed nature and can irritate the mouth, esophagus and gastrointestinal tract. Oat groat or whole grains (after removal of hull) contain all three parts of the grain - the germ, endosperm, and bran, rich in all valuable nutrients such as soluble fibers, proteins, unsaturated fatty acids, vitamins, minerals, and phytochemicals. These nutrients work together to provide the maximum health benefits. The dietary fiber complex with its antioxidants and other phytochemicals is effective against cardiovascular disease and some types of cancer (Jacobs *et al*, 1998a., Jacobs *et al*, 1998b., Jacobs *et al*, 1998., Slavin *et al*, 2000., Thompson, 1994).

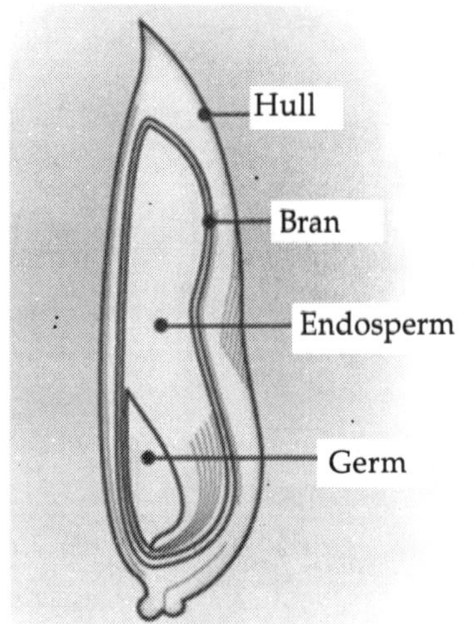

Bran: The bran consists of the coarse outer layers of the grain. It contains the greatest amount of fiber and is a rich source of vitamins and nutrients.

Endosperm: The endosperm is the largest section and the middle layer of the grain. It is the main energy source and dietary fiber constitute 75% of the endosperm cell wall (Miller *et al*, 1995).

Germ: The germ (the embryo in its early stages) is the heart of the grain and the smallest part of the grain, but it is loaded with nutrients.

Nutritional Value of Oats

Oat grain is characterised by a good taste, dietetic properties and an activity stimulating metabolic changes in the body. All this makes its nutritive value high for both people and animals (Brand and Merwe, 1996; Lia *et al.*, 1997; Peltonen-Sainio *et al.*, 2004; Peterson, 2004). In comparison with other cereals, oat grain is characterised by a larger amount of total protein and crude fat and a smaller one of crude fibre. The characteristic feature of protein is its good amino acid composition with a high nutritive value. Due to the difference in structure of oat proteins, they are officially recognised as safe in contrast to other cereals (Lockhart and Hurt, 1986). A high level of fat is a good source of essential unsaturated fatty acids. Studies by Brand and Merwe (1996) and Petkov *et al.* (2001) proved that oat have a higher nutritive protein value than that of other cereals, although lysine is still the limiting amino-acid.

Nutrients	Whole grain oat	Oat bran
Protein	15 - 17 %	15 - 18 %
Starch and sugars	59 - 70 %	10 - 50 %
Fat	4 - 9 %	5 - 10 %
Total dietary fibre	5 - 13 %	10 - 40 %
ß-Glucan	2 - 6 %	5 - 20 %

Beta Glucan from Oats and its Physiological Effects

What β-glucan is?

Oat β-glucan a mixed-linkage polysaccharide of D-glucose units, and an important part of oat dietary fibre. β-Glucan is a major component of the water extractable fibre fraction (Manthey *et al.* 1999). The bonds between D-glucopyranosyl units in β-glucan are either β-1,3 or β-1,4 linkages. The 1,3-linkages break up the uniform structure of the β-D-glucan molecule and make it soluble and flexible. Structural analysis shows that β-glucan is composed primarily of β-1,3 linked cellotriosyl and cellotetraosyl units, but there are also regions that are more cellulose-like in character, with 4 or more consecutive β-1,4 linked glucose units (Wood *et al.* 1993). The molar ratio of cellotriose to cellotetraose units (DP3:DP4) is typically 1.5 - 2.3 as far as oat β-glucan is considered (Miller and Fulch,1995).

Beta-glucan with B1→4 and B1 →3 linkages

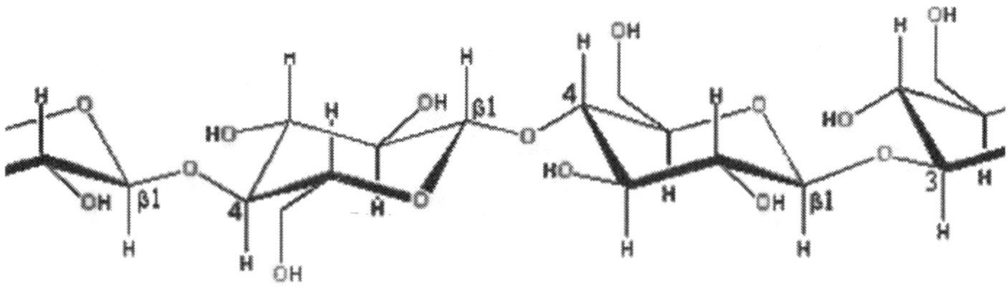

The soluble β-glucans make viscous, shear thinning solutions even at low concentrations. The viscosity is related to the molecular weight and is strongly dependent on the concentration (Åman *et al.* 2004, Anttila *et al.* 2004). Lyly *et al.* (2004) observed that with an oat β-glucan preparation of high-molecular weight (average MW = 2,000,000), the maximum practicable concentration of β-glucan in a soup was 0.5 w-%. With lower molecular weight preparations (MW ? 200,000) of oat, it was possible to add up to 2.0 w-% β-glucan to the soup.

Where β-glucan is located in oat grain?

β-glucan usually comprises 3.6 – 5.1 % (dw) of the oat whole grain (Hampshire 2004). The several factors that affect the concentration of β-glucan include cultivar variation and the growing conditions (e.g., nitrogen level, temperature and rainfall) (Asp *et al.* 1992, Brunner and Freed 1994). β-glucans are located throughout the starchy endosperm. They are concentrated in the bran, more precisely in the aleurone and sub-aleurone layer.

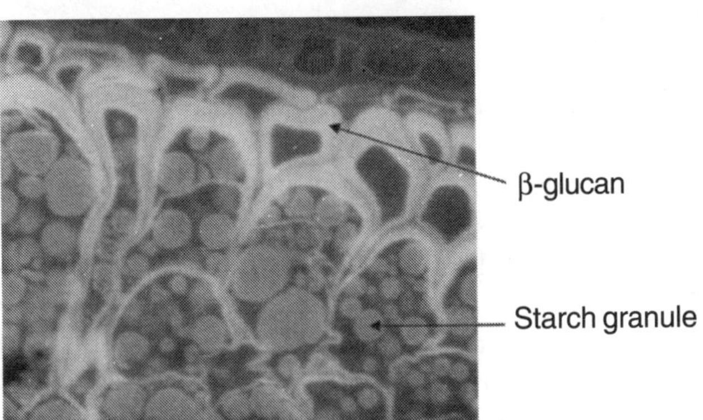

The content of dietary fibre and β-glucan varies between different oat products. Typical concentrations (% of dry weight) are presented in the table below.

	Oat endosperm flour	Whole grain products	Conventional oat bran products	Oat bran concentrates	Beta-glucan isolates
Dietary fibre (%)	5 - 10	10 - 12	15 - 20	20 - 35	80 - 100
Beta-glucan (%)	1 - 3	4 - 5	8 - 12	15 - 22	Up to 80

Therapeutic Attributes of Oats

Oat Against Type 2 Diabetes

Insulin resistance, the main component and causative factor of metabolic syndrome is an essential target in the therapeutic approach of oat β-glucans. The dietary intake of total as well as soluble and insoluble fibre is inversely associated with insulin resistance, supporting evidence that a high intake of dietary fibre is associated with enhanced insulin sensitivity (Galisteo, 2008; Ylonen *et al,* 2003). Oat β-glucans have been used in several clinical trials to reduce glucose. Studies showed that oat β-glucan lowered postprandial glycemia (Tappy *et al* 1996; Jenkins *et al* 2002). It has also shown that oat bran flour was more effective than oat bran crisp explained by the three times higher of β-glucan content in oat bran flour (Saris 2003; Tapola *et al* 2005) against type 2 diabetes. After ingestion of the oat (bran flour or crisp), the blood glucose levels were lower at 15, 30, and 45 min but higher at 90 min after 12.5 g glucose loading (Tapola *et al* 2005). These changes reduce the feeling of hunger caused by rapid decrease in blood glucose (Ludwig 2003; Saris 2003). Thus, β-glucans may decrease appetite and reduce food intake.

Mechanism of action : Oat β-glucans favor increased glucose uptake into skeletal muscle and improves insulin sensitivity by delaying stomach emptying because of high viscosity and stability with pH (Butt *et al,* 2008) so that dietary glucose is absorbed more gradually (Kiho *et al,* 1995).

The comparative reduction of glucagon response after the oat extracts may be partially responsible for the reduction in glucose concentrations (Judith *et al,* 1995). Another possible mechanism for oat β-glucans to reduce blood glucose level is mediated by signal pathway through PI3K/Akt activation. Decreased PI3K/Akt activity has been shown to play a key role in the pathogenesis of type 2 diabetes. β-glucans have been demonstrated to increase PI3K/Akt through several receptors (Hsu *et al* 2002; Chen and Seviour 2007). These receptors stimulated by β-glucans include Dectin-1, complement receptor 3 lactosylceramide, scavenger and toll like receptors; each induces specific signal pathways. The interaction between β-glucans and their receptors has been studied by Brown (2006) and Chen and Seviour (2007). The most relevant fact is that administration of β-glucans could restore decreased PI3K in diabetes. For example, mushroom extract have been shown to activate PI3K pathway (Li *et al* 2006; Lee *et al* 2008). These β-glucans can bind to dectin-1 receptor to stimulate the signal pathway (Underhill *et al* 2005; Trinidad *et al* 2006; Brown 2006; Olsson and Sundler 2007). Another β-glucan lentinan binds to scavenger receptors which can also activate PI3K pathway (Rice *et al* 2002; Mineo *et al* 2003). A study in stroke-prone spontaneously hypertensive rats suggests that dietary fibre (mainly soluble) supplementation effectively prevents insulin resistance by increasing skeletal muscle plasma membrane content of the insulin-responsive glucose transporter type 4 (GLUT-4) by a mechanism different from the activation of the phosphatidylinositol-3 kinase pathway (Songs *et al,* 2000). A hypothesis highlighting this concern suggests that a series of fatty acids stimulate peroxisome proliferator activated receptor (PPAR) γ, whose activation has been reported to increase GLUT-4 content in adipocytes (Park *et al,* 1998). Therefore, these authors speculate that it is likely that short chain fatty acids (SCFAs) such as propionic and butyric acids, which result from anaerobic bacterial fermentation of soluble fermentable dietary fibre in the colon (Cummings *et al,* 1987), increase muscle GLUT-4 via PPARã (Songs *et al,* 2000).

Oat Against Dyslipidemia

β-glucans has been shown to decrease LDL cholesterol and increase HDL to alleviate possibly dyslipidemia and reduce CVD (Anderson 1995; Reyna-Villasmil *et al* 2007; Kapur *et al* 2008). The required dose of β-glucan for a single food is 0.75 g/serving. Oats were first found to have a cholesterol-lowering effect and the active component was identified as β-glucans (Kerckhoffs *et al* 2002). On the basis of numerous (Butt *et al.,* 2008) clinical studies, the US Food and Drug Administration (FDA) permitted the use of a claim that oat soluble fibre has the ability to reduce the risk of heart disease. Oats reduced both serum total choleterol and LDL cholesterol compared with control (Davidson *et al* 1991; Van Horn *et al* 1991; Gerhardt and Gallo 1998; Kerckhoffs *et al* 2003; Karmally *et al* 2005; Naumann *et al* 2006; Reyna-Villasmil *et al* 2007; Theuwissen and Mensink 2007). In 20 hypercholesterolemic male patients, oat bran was shown to be better than wheat bran in lowing cholesterol (Anderson *et al* 1991).

Mechanism of action: The mechanism for oat bran soluble fibres to lower LDL is considered to be mediated by bile acids binding property of β-glucans. Therefore β-glucans increase exclusion of bile acids (Marlett 1997; Ellegard and Anderson 2007) and this in turn activates cholesterol 7á-hydroxylase and up-regulates low-density lipoprotein receptor (LDLR) and thus increase the transport of LDL into hepatocytes and the conversion of cholesterol into bile acids (Nilsson *et al* 2007). It has been reported that every 1 % reduction in LDL is associated with a decreased risk for CHD of 1% to 3% (Reynolds, 2000). In addition, oat β-glucans have also been shown to be fermented by human fecal microbiota to produce short-chain fatty acids such as acetate, propionate, and butyrate and after absorption into the portal vein both acetate and propionate inhibit hepatic cholesterol synthesis (Wright *et al*, 1990). This in turn also produce the hypocholesteolemic effect (Han *et al* 2004; Drzikova *et al* 2005; Alminger and Eklund-Jonsson 2008; Hughes *et al* 2008).

Oat Against Hypertension

A clinical trial with food containing oat β-glucans showed reduced blood pressure in the subjects with body mass index above medians (31.5 kg/m2) (Maki *et al* 2007). This is further demonstrated by another trial that also showed effective role of oat (He *et al* 2004). However, no pure β-glucan has been used for these studies and thus it is difficult to identify which components are effective (Chen and Raymond, 2008).

In a randomized, double-blind, placebo-controlled trial in 110 participants aged 30 to 65 years who had untreated but higher than optimal blood pressure or Stage 1 hypertension (He *et al*, 2004), the effects of supplemented soluble fibre intake from oat bran (mean dietary intake of total, soluble and insoluble fibre increased by 10.6, 5.8 and 4.9 g/day in the high-fibre group and -0.1, 0.2 and -0.5 g/day in the low-fibre group, respectively) on blood pressure for 12 weeks were analyzed; both systolic and diastolic blood pressure were significantly reduced in the study participants who consumed a high-fibre diet (mean reduction by 3.4 and 2.2 mmHg, respectively).

Mechanism of action: Several mechanisms for a potential effect of dietary fibre intake on blood pressure have been hypothesized. Dietary fibre has numerous effects on the digestion and absorption of foods. Insulin resistance and its concomitant compensatory hyperinsulinemia have been suggested as major underlying pathogenetic mechanisms for the development of hypertension (Ferrannini *et al*, 1987). In this way, effectiveness of both soluble and insoluble fibres in reducing insulin resistance and insulin levels in both diabetic and healthy persons (McKeown *et al*, 2002., King *et al*, 2005., Qi *et al*, 2005) would contribute to treating or preventing hypertension. Intake of dietary fibre has also been shown to efficiently reduce body weight (Rigaud *et al,* 1990) and weight gain (Liu *et al,* 2003). Increased body weight is a strong risk factor for hypertension (Neter *et al,* 2003).

Oat Against Proinflammatory State

There are interesting reports about the effects of dietary fibre intake on the acute-phase response marker CRP and the inflammatory cytokines IL-6, IL-18 and TNF-á whose levels are found abnormally elevated in the metabolic syndrome and some studies suggest an association between consumption of dietary fibre and the anti-inflammatory adiponectin. Recently, a study in diabetic women (Qi *et al,* 2006) indicated that after adjustment for age, body mass index (BMI), lifestyle and dietary covariates, intake of cereal fibres was associated with significantly decreasing trends of CRP (C reactive protein) and TNF-á receptor 2 among women with type 2 diabetes.

Mechanism of action: Oat intake effect favoring weight loss could be involved as most weight loss studies have shown that decreasing body weight was related to decreases in CRP (Dietrich *et al,* 2005), TNF-α of obese patients (Dandona *et al,* 1998) and to increases in adiponectin circulating levels of obese women (Esposito *et al,* 2003). In a study, King (2005) suggested that dietary fibre decreases lipid oxidation, which in turn is associated with decreased inflammation. These anti-inflammatory effects of dietary fibre could also be attributed, in the case of fermentable fibres, to the production of butyrate. Clinical trials suggest that butyrate has anti-inflammatory properties in pathologies such as inflammatory bowel disease (Patz *et al,* 1996), and it has been described for anti-inflammatory activity in different human cellular types, such as macrophages and monocytes (Segain *et al,* 2000).

Oat Against Vascular Injury

Vascular injury may contribute to the pathogenesis of CVD (Liuba and Pesonen 2005; Kibos *et al* 2007) associated with type 2 diabetes. Dietary fibres may increase collagen synthesis (Wei *et al* 2002). The study has shown that β-glucan can help in healing of wound in db/db mice (Berdal *et al* 2007).

Oat Based Products

Non-fermented Oat Based Products

1. *Oat Bran* : Oat bran is the food produced by grinding clean oat groats or rolled oats and separating the resulting oat flour by

sieving, bolting, and/or other suitable means into fractions such that the oat bran fraction is not more than 50% of the starting material and has a total b-glucan content of at least 5.5% (dry weight basis) and a total dietary fiber content of at least 16.0% (dry weight basis), such that at least one-third of the total dietary fiber is soluble fiber. Oat bran is consumed as a hot cereal and is a main ingredient in many R TE cereals. Additionally, oat bran is used in a wide variety of bread, granola, and cookies products.

2. *Oat flour or oat meal*: Oat flour or oat meal is produced by grinding flakes or groats into flour for use as an ingredient in a wide variety of food products. Oat flour is a major component of infant foods. In many instances, this is a baby's first introduction to solid foods. Oat flour is a major constituent in granola bars, South Arnerican beverage products, and pancake mixes. Additionally, oatmeal is used as a thickening agent in soups, sauces, and gravies and as a meat extender.

3. *Rolled oat flakes*: Rolled oats, which are produced by flaking whole groats, are the thickest of the standard oat-flake products; flake thickness varies from 0.51 to 0.91 mm depending upon the desired end use. The thicker flakes require longer cooking periods and maintain flake integrity for extended periods.

4. *Instant oat flakes* : Instant oat flakes are produced from "instantized" steel-cut groats. Before cutting, the groats have been subjected to a proprietary commercial process that results in rapid cooking of the flakes. The flakes are typically 0.28 to 0.46 mm thick.

5. *Steel cut oat groats*: Steel-cut groats are produced by sectioning groats into several pieces; they are used in the preparation of flakes and flour and as a specialty ingredient.

6. *Quick oats*: Quick oats are flakes produced from steel-cut groats. In this process, oat groats are typically cut into three to four pieces before the fmal steaming and flaking process. Quick oats, which are usually 0.36 to 0.61 mm thick, require less cooking time than the whole oat flake.

7. Baby oats are also produced from steel-cut groats, but the flakes are thinner and have a finer granulation than quick oats. These smaller, thinner oat flakes cook more rapidly than quick oats and have a smoother texture.

8. *Cold cereals or ready-to-eat (RTE)* products represent the second major use for oat products. The high fat and fiber content of oats limits the utilization of oats in this application, but a number of economically important products have been successfully introduced into the marketplace.

9. *Bread containing oat*: Oat ingredients are used as ingredients in a wide variety of bread and baked process. These ingredients provide unique flavourand moisture retention characteristics as well as enhancing the nutritional benefits of these products. It has been demonstrated that oat flour stabilized the fat component in breads. For the productionof bread the baker can use different types of oat fractions, rolled oat, oat flour and oat bran.

Fermented oat Based Products

1. *Oatly* (CeBa Food AB, Lund, www.oatly.com) is a healthy alternative to milk and soya, with all the benefits of oats. Oatly is free from added sugar, low in saturated fat and contains oat fibre which, when eaten as part of a healthy diet, can help to lower your cholesterol.The product is milk- and soya-free, and most of them are entirely vegetable-based. Oat kernels and water are mixed and ground and a patented mix of natural enzymes is added. This gives a milk-like consistency, and some of the insoluble fibres are withdrawn while leaving the water-soluble fibre in the product. Finally, the mixture is homogenised to make the product smooth and even. Oatly products can be used in the same way as traditional dairy products such as milk and cream.

2. *Oatrim* (ConAgra, www.conagrafoods.com, Quaker Oats, www.quakeroats.com, Rôhne-Poulenc) is a fat replacer with 5 % b-glucan and partly hydrolysed starch. Preparation of Oatrim involves the conversion of starch in oat flour or bran into amylodextrins by enzymatic hydrolysis. Oatrim consists primarily of amylodextrins and soluble fibre (b-glucan usually at 5 - 10 % by weight) with low amounts of lipids, proteins and minerals. Oatrim can be used in most foods as a powder or gel.

3. In *OatWell®* oat bran, b-glucans are present in high contents due to a careful stabilization process of oat bran. CreaNutrition / Swedish oat fibre (www.oatingredients.com) develops, produces and sells OatWell® oat bran products rich in ß-glucan, up to 22 %. Cereals with added OatWell® oat bran can be used as fortification in breakfast mueslis or designed as RTE cereals.

4. *Primaliv™* (Skåne mejerier, Malmö, Sweden, www.skanemejerier.com); Primaliv™ consists of a cup of 200 ml is delivering 4 grams of OatWell® b-glucans, equivalent to the amount in three plates of oat-porridge. In 2002, Skånemejerier's PrimaLiv became the first product in Sweden to be approved and marked in compliance of the new rules of the food industry for health statements.

5. *YOSA®* (Bioferme, Finland, www.bioferme.fi). YOSA® is a, non-dairy oat product made from oat and oat bran. The oats are fermented with a combination of probiotic bacteria (inc. Bifidobacterium lactis BB-12™ and Lactobacillus acidophilus LA-5™). These probiotics have scientifically proven benefits that can help regulate your digestive system. YOSA® has been granted the Heart Symbol the Finnish Heart Association and the Finnish Diabetes Association.

6. *Velle* (by Velle Oats, St. Petersburg, Russia) is a drinkable yogurt. It is the first drinkable probiotic cereal-based 'smoothie' which was introduced in 2004 at Hi Europe fair, Amsterdam

7. *Oat based probiotic drink* (Angelov *et al*, 2006). In this a whole-grain oat substrate is fermented with lactic acid bacteria *Lactobacillus plantarum* to obtain a drink, combining the health benefits of a probiotic culture with the oat prebiotic beta-glucan.

8. *Oat bio lacto* is an oats and fat-free milk based functional food product. This is yogurt like product which supports oat beta-glucan, oat lipid unsaturated fatty acids and living probiotic lactic acid bacteria (Lactobacillus acidophilus T_{20} and Bifidobacteria and commercial starter culture ABT-1) and is considered as polyfunctional nutrition (Bekers *et al*, 2001).

References

Alminger, M., Eklund-Jonsson, C. (2008). Whole-grain cereal products based on a high-fibre barley or oat genotype lower post-prandial glucose and insulin responses in healthy humans. *Eur. J. Nutr.;* **47**: 294–300.

Aman, P., Rimsten, L., Andersson, R. (2004). Molecular weight distribution of b-glucan in oat-based foods. *Cereal Chem*. **81:** 356-360.

Anderson, J. W., Gilinsky, N. H., Deakins, D. A. *et al*. (1991). Lipid responses of hypercholesterolemic men to oat-bran and wheat-bran intake. *Am. J. Clin. Nutr.;* **54**: 678–683.

Anderson, J. W. (1995). Dietary fibre, complex carbohydrate and coronary artery disease.

Anttila, H., Sontag-Strohm, T., Salovaara, H. (2004). Viscosity of beta-glucan in oat products. *Agric Food Sci.,* **13:** 80-87.

Asp, N. G., Mattsson, B., Onning, G. (1992). Variation in dietary fibre, b-glucan, starch, protein, fat and hull content of oats grown in Sweden 1987-1989. *Eur. J. Clinical Nutr.,* **46:** 31-37.

Berdal, M., Appelbom, H. I., Eikrem, J. H. *et al.* (2007). Aminated b-1,3-Dglucan improves wound healing in diabetic db/db mice. *Wound Repair Regen;* **15:** 825–32.

Brand, T. S., Merwe, J.P., (1996). Naked oats *(Avena nuda)* as a substitute for maize in diets for weanling and grower-finisher pigs. *Animal Feed Science and Technology.,* **57:** 139–147.

Brown, G. D. (2006). Dectin-1: a signalling non-TLR pattern-recognition receptor. *Nat. Rev. Immunol.;* **6**: 33–43.

Brunner, B. R., Freed, R. D. (1994). Oat grain b-glucan content as affected by nitrogen level, location, and year. *Crop Sci.,* 34: 473-476.

Butt, M.S., Nadeem, M.T., Khan, M.K.I., Shabir, R. (2008). Oat : unique among the cereals. *Eur. J. Nutr.,* **47**: 68-79.

Chen, J., Raymond, K. (2008). B glucans in the treatment of diabetes and associated cardiovascular risks. *Vascular Health and Risk Management;* **4**(6): 1265-1272.

Chen, J., Seviour, R. (2007). Medicinal importance of fungal b-(1–3) and (1–6)-glucans. *Mycol. Res.;* **111**: 635–52.

Coffman, F. A. (1961). Oats and oats improvement. *American society of Agronomy* Madison, Wisconsin.

Cummings, J. H., Englyst, H. N. (1987). Fermentation in the human large intestine and the available substrates. *Am. J. Clin. Nutr.;* **451**: 1243–1255.

Dandona, P., Weinstock, R., Thusu, K., Abdel-Rahman, E., Aljada, A., Wadden, T. (1998). Tumor necrosis factor-á in sera of obese patients: fall with weight loss. J Clin Endocrinol Metab; **83:** 2907–10.

Davidson, M. H., Dugan, L. D., Burns, J. H. *et al.* (1991). The hypocholesterolemic effects of b-glucan in oatmeal and oat bran. A dose-controlled study. *JAMA;* **265**: 1833–1839.

Dietrich, M., Jialal, I. (2005). The effect of weight loss on a stable biomarker of inflammation, C-reactive protein. *Nutr Rev.,* **63**: 22–28.

Drzikova, B., Dongowski, G., Gebhardt, E., Habel, A. (2005). The composition of dietary fibre rich extrudates from oat affects bile acid binding and fermentation in vitro. *Food Chem.;* **90** : 181–192.

Ellegard, L., Andersson, H. (2007). Oat bran rapidly increases bile acid excretion and bile acid synthesis: an ileostomy study. *Eur. J. Clin. Nutr.;* **61**: 938–945.

Esposito, K., Pontillo, A., Di Palo, C., Giugliano, F., Masella, M., Marfella, R. *et al.* (2003). Effect of weight loss and lifestyle changes on vascular inflammatory markers in obese women. *JAMA;* **289:** 1799–1804.

Ferrannini, E., Buzzigoli, G., Bonadonna, R., Giorico, M. A., Oleggini, M., Graziadei, L. *et al.* (1987). Insulin resistance in essential hypertension. *N. Engl. J. Med.;* **317**: 350–357.

Galisteo, M., Duarte, J., Zarzuelo, A. (2008). Effects of dietary fibres on disturbances clustered in the metabolic syndrome. *J. of Nutr. Biochem.;* **19** : 71-84.

Gerhardt, A. L., Gallo, N. B. (1998). Full-fat rice bran and oat bran similarly reduce hypercholesterolemia in humans. *J. Nutr.;* **128**: 865–869

Hampshire, J. (2004). Variation in the content of nutrients in oats and its relevance for the production of cereal products. In: Proceedings of the 7th International Oat Conference. Helsinki, Finland. Peltonen-Sainio, P., Topi-Hulmi, M. (eds.) Jokioinen: MTT Agrifood Research Finland, Agrifood Research Reports. **51**: 126.

Han, K. H., Sekikawa, M., Shimada, K. *et al.* (2004). Resistant starch fraction prepared from kintoki bean affects gene expression of genes associated with cholesterol metabolism in rats. *Exp. Biol. Med.* (*Maywood*); **229**: 787–792

He, J., Streiffer, R. H., Muntner, P. *et al.* (2004). Effect of dietary fibre intake on blood pressure: a randomized, double-blind, placebo-controlled trial. *J. Hypertens.;* **22**: 73–80.

Hsu, M. J., Lee, S. S., Lin, W. W. (2002). Polysaccharide purified from Ganoderma lucidum inhibits spontaneous and Fas-mediated apoptosis in human neutrophils through activation of the phosphatidylinositol 3 kinase/Akt signaling pathway. *J. Leukoc. Biol.;* **72**: 207–16.

Hughes, S. A., Shewry, P. R., Gibson, G. R. *et al.* (2008). In vitro fermentation of oat and barley derived b-glucans by human faecal microbiota. *FEMS Microbiol. Ecol.;* **64**: 482–93.

Jacobs, D. R., Marquart, L., Slavin, J., Kushi, L. H. (1998a). Whole-grain intake and cancer: an expanded review and metaanalysis. *Nutr. Cancer.;* **30:** 85–96.

Jacobs, D. R., Marquart, L., Slavin, J., Kushi, L. H. (1998b). Whole grain intake may reduce the risk of ischemic heart disease death in postmenopausal women: The Iowa Women's Health Study. *Am. J. Clin. Nutr.;* **68**: 248–257.

Jacobs, D. R., Meyer, K. A., Kushi, L. H., Folsom, A. R. (1998). Whole-grain intake may reduce the risk of ischemic heart disease death in postmenopausal women: the Iowa Women's Health Study. *Am. J. Clin. Nutr.;* **68:** 248–257.

Jenkins, A.L., Jenkins, D.J., Zdravkovic, U. (2002). Depression of the glycemic index by high levels of b-glucan fi ber in two functional foods tested in type 2 diabetes. *Eur. J. Clin. Nutr.;* **56**: 622–8.

Judith, H., Daniel, J. S., Behall, K.M. (1995). Diets containing soluble oat extracts improve glucose and insulin responses of moderately hypercholesterolemic men and women. *Am. J. Clin. Nutr.;* **61**: 379–384.

Kapur, N. K., Ashen, D., Blumenthal, R. S. (2008). High density lipoprotein cholesterol: an evolving target of therapy in the management of cardiovascular disease. *Vasc. Health Risk Manag.;* **4**: 39–57.

Karmally, W., Montez, M. G., Palmas, W. *et al.* (2005). Cholesterol-lowering benefi ts of oat-containing cereal in Hispanic americans. *J. Am. Diet Assoc.;* **105**: 967–970.

Kerckhoffs, D. A., Brouns, F., Hornstra, G. *et al.* (2002). Effects on the human serum lipoprotein profile of b-glucan, soy protein and isoflavones, plant sterols and stanols, garlic and tocotrienols. *J. Nutr.;* **132**: 2494–2505.

Kerckhoffs, D. A., Hornstra, G., Mensink, R. P. (2003). Cholesterol-lowering effect of b-glucan from oat bran in mildly hypercholesterolemic subjects may decrease when b-glucan is incorporated into bread and cookies. *Am. J. Clin. Nutr.;* **78**: 221–217.

Kibos, A., Campeanu, A., Tintoiu, I. (2007). Pathophysiology of coronary artery in-stent restenosis. *Acute Card Care.;* **9:** 111–119.

Kiho, T., Sobue, S., Ukai, S. (1994). Polysaccharides in fungi: antidiabetic activity of an acidic polysaccharide from the fruiting bodies of *Tremella aurantia. Biol. Pharmaceut. Bull.;* **18**: 1627-1629.

King, D. (2005). Dietary fibre, inflammation, and cardiovascular disease. Mol Nutr Food Res; **49:** 594–600.

Lee, J. Y., Kim. J. Y., Lee, Y. G. *et al.* (2008). Molecular mechanism of macrophage activation by exopolysaccharides from liquid culture of Lentinus edodes. *J. Microbiol. Biotechnol.;* **18**: 355–64.

Li, B., Allendorf, D. J., Hansen, R. *et al.* (2006). Yeast b-glucan amplifi es phagocyte killing of iC3b-opsonized tumor cells via complement receptor 3-Syk-phosphatidylinositol 3-kinase pathway. *J. Immunol.;* **177**: 1661–1669.

Lia, Å., Andersson, H., Mekki, N., Juhel, C., Senft, M., Lairon, D., (1997). Postprandial lipemia in relation to sterol and fat excretion in ileostomy subjects given oatbran and wheat test meals. *American Journal of Clinical Nutrition.;* **66:** 357–365.

Liu, S., Willett, W. C., Manson, J. E., Hu, F. B., Rosner, B., Colditz, G. (2003). Relation between changes in intakes of dietary fibre and grain products and changes in weight and development of obesity among middle-aged women. *Am. J. Clin. Nutr.;* **78**: 920–927.

Liuba, P., Pesonen, E. (2005). Infection and early atherosclerosis: does the evidence support causation? *Acta Paediatr.*; **94:** 643–51.

Lockhart, H. B. and Hurt, D. H. (1986). Nutrition in oats. In: Oats : Chemistry and technology. (F.H. Webster ed.) pp. 297-308. *American Association of Cereal Chemists.,*; U.S.A.. ISBN : 0-913250-30-9.

Ludwig, D. S. (2003). Dietary glycemic index and the regulation of body weight. *Lipids;* **38:** 117–21.

Lyly, M., Salmenkallio-Marttila, M., Suortti, T., Autio, K., Poutanen, K., Lähteenmäki, L. (2004). The sensory characteristics and rheological properties of soups containing oat and barley beta-glucan before and after freezing. *Lebensm-Wiss Technol.*; **37:** 749-761.

Maki, K. C., Galant, R., Samuel, P. *et al.* (2007). Effects of consuming foods containing oat b-glucan on blood pressure, carbohydrate metabolism and biomarkers of oxidative stress in men and women with elevated blood pressure. *Eur. J. Clin. Nutr.*; **61:**786–95.

Manthey, F. A., Hareland, G. A., Huseby, D. J. (1999). Soluble and insoluble dietary fiber content and composition in oat. *Cereal Chem.*; **76**: 417-420.

Marlett, J. A. (1997). Sites and mechanisms for the hypocholesterolemic actions of soluble dietary fi ber sources. *Adv. Exp. Med. Biol.*; **427**: 109–21.

McKeown, N. M., Meigs, J. B., Liu, S., Wilson, P. W., Jacques, P. F. (2002). Wholegrain intake is favourably associated with metabolic risk factors for type 2 diabetes and cardiovascular disease in the Framingham Offspring Study. *Am. J. Clin. Nutr.*; **76**: 390–398

Miller, S. S., Fulcher, R. G. (1995). Oat endosperm cell walls: II. Hot-water solubilization and enzymatic digestion of the wall. *Cereal Chem.*; **72:** 428-432.

Miller, S. S., Fulcher, R. G., Sen, A., Arnason, J. T. (1995). Oat endosperm cell walls: Isolation, composition, and comparison with other tissues. *Cereal Chem.*; **72**: 421–427.

Mineo, C., Yuhanna, I. S., Quon, M. J. *et al.* (2003). High density lipoproteininduced endothelial nitric-oxide synthase activation is mediated by Akt and MAP kinases. *J. Biol. Chem.*; **278**: 9142–9149.

Naumann, E., Van Rees, A. B., Onning, G. *et al.* (2006). B-glucan incorporated into a fruit drink effectively lowers serum LDL cholesterol concentrations. *Am. J. Clin. Nutr.*; **83**: 601–615.

Neter, J. E., Stam, B. E., Kok, F. J., Grobbee, D. E., Geleijnse, J. M. (2003). Influence of weight reduction on blood pressure: a meta-analysis of randomized controlled trials. *Hypertension.*; **42:** 878–884.

Nilsson, L. M., Abrahamsson, A., Sahlin, S. *et al.* (2007). Bile acids and lipoprotein metabolism: effects of cholestyramine and chenodeoxycholic acid on human hepatic mRNA expression. *Biochem. Biophys. Res. Commun.;* **357**: 707–711.

Olsson, S., Sundler, R. (2007). The macrophage b-glucan receptor mediates arachidonate release induced by zymosan: essential role for Src family kinases. *Mol. Immunol.;* **44**: 1509–1515.

Park, K. S., Ciaraldi, T. P., Lindgren, K., Abrams-Carter, L., Mudaliar, S., Nikoulina, S. E. (1998). Troglitazone effects on gene expression in human skeletal muscle of type II diabetes involve up-regulation of peroxisome proliferator-activated receptor-gamma. *J. Clin. Endocrinol. Metab.;* **83**: 2830–2835

Patz, J., Jacobsohn, W. Z., Gottschalk-Sabag, S., Zeides, S., Braverman, D. Z. (1996). Treatment of refractory distal ulcerative colitis with short chain fatty acid enemas. *Am J Gastroenterol.;* **91**: 731–734

Peltonen-Sainio, P., Kontturi, M., Rajala, A. (2004). Impact dehulling oat grain to improve quality of on-farm produced feed. I. Hullability and associated changes in nutritive value and energy content. *Agricultural and Food Science.;* **13**: 18–28.

Peterson, D. M. (2004). Oat – a multifunctional grain. In: Peltonen-Sainio, P., Topi- Hulmi, M. (Eds.), Proceedings, 7th International Oat Conference. Agrifood Research Reports 51. MTT Agrifood Research, Jokioinen, Finland.; 21–26.

Petkov, K., Biel, W., Kowieska, A., Jaskowska, I. (2001). The composition and nutritive value of naked oat grain (*Avena sativa var. nuda*). *Journal of Animal and Feed Sciences.;* **10** (2): 303–307.

Qi, L., Rimm, E., Liu, S., Rifai, N., Hu, F. B. (2005). Dietary glycemic index, glycemic load, cereal fibre, and plasma adiponectin concentration in diabetic men. *Diabetes Care.;* **28**: 1022–1028.

Qi, L., Van Dam, R. M., Liu, S., Franz, M., Mantzoros, C., Hu, F. B. (2006). Wholegrain, bran, and cereal fibre intakes and markers of systemic inflammation in diabetic women. *Diabetes Care.;* **29**: 207--211.

Reyna-Villasmil, N., Bermudez-Pirela, V., Mengual-Moreno, E. *et al.* (2007). Oat-derived b-glucan signifi cantly improves HDLC and diminishes LDLC and non-HDL cholesterol in overweight individuals with mild hypercholesterolemia. *Am. J. Ther.;* **14**: 203–212.

Reynolds, H. R., Quite, E., Hunninghake, D. B. (2000). Whole grain oat cereal lowers serum lipids. *Topics Clin. Nutr.;* **15**: 74-83.

Rice, P. J., Kelley, J. L., Kogan, G. *et al.* (2002). Human monocyte scavenger receptors are pattern recognition receptors for (1–3)-b-D-glucans. *J. Leukoc. Biol.;* **72**: 140–146.

Rigaud, D., Ryttig, K. R., Angel, L. A., Apfelbaum, M. (1990). Overweight treated with energy restriction and a dietary fibre supplement: a 6-month randomized, double-blind, placebo-controlled trial. *Int. J. Obes.;* **14**: 763–769.

Saris, W. H. (2003). Glycemic carbohydrate and body weight regulation. *Nutr. Rev.;* **61**: S10–16.

Schrickel, D. J. (1986). Oats production, value and use. In: *Oats : Chemistry and technology*. (F.H. Webster ed.). *American association of Cereal Chemists, USA.,* ISBN: 0- 913250-30-9.

Segain, J. P., Raingeard, de la., Bletiere, D., Bourreille, A., Leray, V., Gervois, N., Rosales, C. *et al.* (2000). Butyrate inhibits inflammatory responses through NF kappa B inhibition: implications for Crohn's disease. *Gu.;* **47:** 397–403.

Slavin, J., Marquart, L., Jacobs, D. Jr. (2000). Consumption of whole grain foods and decreased risk of cancer: Proposed mechanisms. *Cereal Food World.;* **45**: 54–58.

Song, Y. J., Sawamura, M., Ikeda, K., Igawa, S., Yamori, Y. (2000). Soluble dietary fibre improves insulin sensitivity by increasing muscle GLUT-4 content in stroke-prone spontaneously hypertensive rats. *Clin. Exp. Pharmacol. Physiol.;* **27**: 41–45.

Tapola, N., Karvonen, H., Niskanen, L., Mikola, M., Sarkkinen, E. (2005). Glycemic Response of Oat-bran products in type 2 diabetic patients. *Nut. Metabolism Cardiovas.;* **15**(4): 255-261.

Tappy, L., Gugolz, E., Wursch, P. (1996). Effects of breakfast cereals containing various amounts of b-glucan fi bers on plasma glucose and insulin responses in NIDDM subjects. *Diabetes Care.;* **19**: 831–834.

Theuwissen, E., Mensink, R. P. (2007). Simultaneous intake of b-glucan and plant stanol esters affects lipid metabolism in slightly hypercholesterolemic subjects. *J. Nutr.;* **137**: 583–588.

Thompson, L. U. (1994). Antioxidants and hormone-mediated health benefits of whole grains. *Crit. Rev. Food Sci. Nutr.,* **34:** 473–497.

Trinidad, A. G., De-la Puerta, M. L., Fernandez, N. *et al.* (2006). Coupling of C3bi to IgG inhibits the tyrosine phosphorylation signaling cascade downstream Syk and reduces cytokine induction in monocytes. *J. Leukoc. Biol.;* **79**: 1073–1082.

Underhill, D. M., Rossnagle, E., Lowell, C. A. *et al.* (2005). Dectin-1 activates Syk tyrosine kinase in a dynamic subset of macrophages for reactive oxygen production. *Blood.;* **106:** 2543–2550.

Van Horn, L., Moag-Stahlberg, A., Liu, K. A. *et al.* (1991). Effects on serum lipids of adding instant oats to usual American diets. *Am. J. Public Health.;* **81**: 183–188.

Wei, D., Zhang, L., Williams, D.L. *et al.* (2002). Glucan stimulates human dermal fibroblast collagen biosynthesis through a nuclear factor-1 dependent mechanism. *Wound Repair Regen.;* **10:** 161–168.

Wood, P. J. (1993). Physicochemical characteristics and physiological properties of oat (1>3),(1>4)-b-D-glucan. In: Wood, P.J. (ed.) Oat Bran. St. Paul, MN, USA: *Amer Assoc Cereal Chem.;* 83-112.

Wright, R. W., Anderson, J. W., Bridges, S. R. (1990). Propionate inhibits hepatic lipid synthesis. *Proc. Soc. Exp. Biol. Med.;* **195**: 26–29.

Ylonen. K., Saloranta, C., Kronberg-Kippila, C., Groop, L., Aro, A., Virtanen, M. (2003). Asoociations of dietary fibre with glucose metabolism in non-diabetic relatives of subjects with type 2 diabetes. *Diabetes care.;* **26:** 1979-1985.

□□□

Chapter 41

Diabetes, Postprandial Hyperglycemia and α-Glucosidase Inhibitors

PRITI DEVI*, SUMIT SINGH DAGAR, SANJAY KUMAR AND A.K. PUNIYA
Dairy Microbiology Division, National Dairy Research Institute, Karnal (India)

Introduction

Diabetes mellitus is a metabolic disease characterized by high blood sugar (glucose) level that results either from defects in insulin secretion (Type 1), or in its action (Type 2), and/ or both. Insulin, a pancreatic hormone triggers muscle and liver cells to take up glucose and store it for energy generation thus prevents hyperglycemia. World over 285 million people have diabetes whereas, in India it is amounting to 50.8 million, this number is estimated to rise to 435 million in world and 87 million in India by 2030 (www.idf.org., 2009). India has emerged as an unchallenged diabetic capital of world having largest subject of diabetics to the tune of 17.8% of total diabetic population. Urban population has a higher incidence of diabetes with 113 million people as compared to 78 million in rural areas. Out of the two types, Type 2 diabetes found in about 90% of total diabetic patients has become global epidemic in last two decades because of changing lifestyle and eating habits. Diabetes cost world economy very dearly i.e. $376 billion (11.6% of total world healthcare expenditure). Though India have the largest population of diabetics it spends only 1% of total diabetes spending worldwide i.e. $2.8 billion. Type 2 diabetes also called as lifestyle related diabetes is a complex metabolic disorder mainly due to insulin resistance starts with postprandial hyperglycemia and its persistence frequently results into diabetes. The other complications associated with postprandial hyperglycemia are oxidative dysfunction, organ failure, hypertension, hyperlipemia and cardiovascular disease (CVD) (Ceriello., 2004). Moreover the most common cause of mortality i.e. about 50% in diabetic patients is CVD (Vasudevan and Ballantyne., 2005). Therefore, the main challenge for the medical practitioners and scientists worldwide is to manage and normalize post-prandial glucose level in blood because it is the earliest abnormality associated with the development of diabetes and also responsible for elevated glycosylated haemoglobin level which contributes to type 2 diabetes associated complications (Baron., 1998). Therapies for type 2 diabetes consist of non pharmacological (diet, exercise & weight loss) & pharmacological approaches (Insulin secretagogues, Insulin, Insulin sensitizers & Alpha-glucosidase inhibitors). Moreover it has been proved that diabetic patients have an increased level of alpha glucosidase enzyme which leads to very high level of blood glucose after meal (Tandon *et. al.*, 1975). Thus alpha glucosidase inhibitors constitute an important class of antihyperglycemic drugs. But due to various side effects such as hypoglycemia, weight gain, and pulmonary edema and contraindications like severe liver and kidney dysfunction, cardiac dysfunction etc. associated with these synthetic antihyperglycemic agents (Cheng and Fantus., 2005) and due to economical constraints there is an immediate need for natural antihyperglycemic agents. This chapter covers types of diabetes, introduction to postprandial hyperglycemia with emphasis on its treatment and at last various antidiabetic or antihyperglycemic drugs and their mechanism of action.

Diabetes

Diminishing traditional lifestyle and dietary patterns that have sustained people over generation resulted in rapid increase in global prevalence of diabetes which is a serious concern for individuals and society all over the world. According to recent diabetic atlas approximately 6.6% of world's population suffers from diabetes (www.diabetesatlas.org). Diabetes is referred as a metabolic disorder where the body either produces little insulin or becomes progressively resistant to its action thus results in high blood sugar level. Generally it with a stage called as postprandial diabetes in which blood sugar levels just after meals traverse the limits while fasting blood glucose remains normal. Persistence of postprandial hyperglycemia for a long time progress to diabetes. Diabetes also called as lifestyle related disease is primarily concentrated in urban areas (113 million) as compared to rural areas (78 million) and this characteristic is destined to increase in future as a result of rapid industrialization, increased urbanization, westernization, economic development and public unawareness.

Types of Diabetes

Diabetes is categorized mainly into three types i.e. Type 1, Type 2 & gestational diabetes:

Type 1 diabetes is caused due to low insulin secretion by pancreatic β-cells. This is considered as an autoimmune disease which leads to the destruction of pancreatic β-cells by T-cells of immune system. This type of diabetes accounts for 10% of total diabetic population of world. This is also called insulin dependent diabetes as insulin is the only available therapy & juvenile diabetes because of its predominance among children.

Type 2 diabetes is the most common type accounting for 85-95% of total diabetic population in developed countries and accounts for even higher percentage in developing countries. In type 2 diabetes the normal relationship between blood glucose level and insulin secretion gets imbalance. This is because of insulin resistance or low insulin sensitivity, so after meal when blood glucose level increases body produces more insulin to combat this increase which further progress into more insulin resistance and ultimately beta cell destruction. The earliest abnormality associated with type 2 diabetes is postprandial hyperglycemia which remains unnoticed because regular detection for diabetes involves measurement of fasting blood glucose. This stage is called as postprandial diabetes or pre-diabetes; a condition that occurs when a person's blood glucose level is higher than normal but not high enough for a diagnosis of type 2 diabetes (Baron, 1998).

Gestational diabetes : This type of diabetes account for 4% of all pregnant women. In gestational diabetes women during pregnancy have higher blood glucose level than normal which may further progress into type 2 diabetes.

Some more types of diabetes that have been classified based on their cause are congenital diabetes (due to genetic defects of insulin secretion), cystic fibrosis-related diabetes and steroid diabetes (induced by high doses of glucocorticoids).

Postprandial Hyperglycemia (PPG)

In normal individual glucose levels are maintained within a very narrow range (i.e. less than 110 mg/dl) throughout the day by continuous secretion of insulin in very low amount, which regulate the hepatic glucose metabolism. In addition, postprandial rise in glucose level is closely regulated by rapid augment in insulin secretion so that it rarely rises above 140 mg/dl (7.8 mmol/ml). Glucose excursion is therefore maintained within safe limit through the action of insulin on its target organs. Pancreatic β-cells respond to increase in plasma glucose concentrations in two phases. In first phase insulin releases for 0-10 minutes and is very rapid while in second phase release is steady but long lasting (Cherrington., 1999). First phase insulin secretion is the key determinant of PPG, People with impaired glucose tolerance have faulty early phase insulin response which resulting into a progressive rise in postprandial glucose level or postprandial hyperglycemia (Del., 2003 and Rendell and Jovanovic., 2006). Control of postprandial hyperglycemia is important because it is the first abnormality of glucose homeostasis associated with type 2 diabetes. It is also associated with elevated glycosylated haemoglobin HbA1c and contributes to the complications (microvascular & macrovascular) of diabetes (Baron., 1998). Despite the increasing recognition of detrimental effects of PPG current medical practices focuses on fasting blood glucose rather than reduction of postprandial values. In current medical practices fasting blood glucose level is usually used to diagnose diabetes, which is not a good predictor of postprandial hyperglycaemia; a more accurate measure of the metabolic defect underlying type 2 diabetes (Bastyr *et. al.*, 2000). Postprandial blood glucose levels may be elevated while fasting levels are normal, constituting an early stage in type 2 diabetes that can be termed 'postprandial diabetes'. Reduction of PPG levels has shown benefit in reducing cardiovascular events. For this reason, it is imperative to focus on suppression of PPG elevation in treatment not only of diabetic patients, but also individuals with impaired glucose tolerance. Continuous glucose monitoring will increase the emphasis on PPG as our primary treatment target. The widespread adoption of PPG as our target will lead to improvement in HbA1c levels and reduction in the deleterious clinical effects of hyperglycemia (LeRoith and Smith *et. al.*, 2005; Davidson., 2005).

Postprandial Hyperglycemia & Diabetes

There is a high correlation between postprandial glucose level and HbA1c (Goldstein *et. al.*, 2002). In fact postprandial glucose levels contribute more to the elevation of HbA1c value than the fasting blood glucose (Avignon *et. al.*, 1997; Monnier *et. al.*, 2003; Rendell and Jovanovic., 2006). High HbA1c level have a well established relationship with micro and macrovascular diseases in diabetes. According to Diabetes Control and Complication Trial, United Kingdom Prospective Diabetes Study (UKPDS) intensive treatment to lower HbA1c results in the reduction of retinopathy, nephropathy and neuropathy. Moreover Shiraiwa *et. al.*, (2005) reported that post-prandial hyperglycemia, rather than fasting glycemia levels, is a stronger predictor of the incidence of each diabetic microangiopathy. However the relationship between HbA1c and cardiovascular disease in diabetics is still unestablished. Three different studies conducted by the Diabetes Control and Complications Trial in type 1 diabetes mellitus, the UKPDS, and the Veterans Affairs Cooperative Study on Glycemic Control and Complication in type 2 diabetes mellitus shows unimpressive reduction in myocardial infarction and strokes with improved HbA1c than microvascular diseases. While high level of PPG are much stronger predictor of CVD than high fasting glucose (Ceriello., 2003; Gerich., 2003). Other risk factors for CVD are also related to postprandial hyperglycemia and play an important role like insulin resistance is associated with increased cardiovascular diseases and insulin resistance almost always coexist with high PPG (Katsumori *et. al.*, 1995). Various research trials like Kumamoto study and STOP-NIDDM conducted for lowering of PPG level has reported reduction in event of CVD (Shichiri et. al 2000; Chiasson *et. al.*, 2002).

Mechanism of Vascular Damage by Postprandial Hyperglycemia

Potential mechanisms by which postprandial hypeglycemia causes CVD are abnormal vascular activity like increased secretion of endothelin 1 & abnormal vasodilation (Ceriello *et. al.*, 2004), Aldolase reductase activity (Yamamoto *et. al.*, 2000), Accumulation of advanced glycation end products, hypercoagulability Sakamoto *et. al.*, 2000, Excess oxidative activity (Ceriello *et. al.*, 1998), Cytokine and intracellular adhesion molecular activity and Activation of protein kinase C, vascular endothelial growth factor and nuclear transcription factor may all be promoted by Post-hyperglycemia.

Abnormal Vasodilation

There are abnormalities in vascular reactivity observed in diabetes. Endothelial function is altered early in diabetes. Elevated glucose concentrations are also associated with an increase in the secretion of endothelin 1 (Park *et. al.*, 2000), in diabetes mellitus and obesity, both endothelium-dependent and endothelium-independent vasodilation are diminished (Dandona P & Aljada A., 2004; Prior *et. al.*, 2005). Acetylcholine-induced endothelium-mediated vasodilation is also diminished in diabetes mellitus (Goodfellow *et. al.*, 1996; McVeigh *et. al.*, 1992). Similarly, insulin-induced vasodilation is reduced in both type 2 diabetes mellitus and obesity. There is a reduction in nitric oxide synthesis in diabetes. Insulin stimulates nitric oxide release (Calver *et. al.*, 1992) and also increases the expression of nitric oxide synthetase. Postprandial hyperglycemia suppresses endothelium-dependent vasodilation. Reduction of postprandial hyperglycemia with insulin aspart improved flow-mediated vasodilation. Hyperglycemia inhibits nitric oxide synthetase expression. The effect of hyperglycemia on endothelial dysfunction is counteracted by arginine, acetylcholine, or nitroprusside administration (Ceriello *et. al.*, 2004).

Aldose Reductase Abnormalities

Aldose reductase uses NADPH to reduce glucose to sorbitol. The accumulation of sorbitol leads to a decrease in nicotinamide adenine dinucleotide (NADPH), myo-inositol, and Na+,K+-dependent adenosine triphosphatase and glutathione. These biochemical changes cause endothelial injury (Guerci *et. al.*, 2001).

Glycation

Hyperglycemia gives rise to a direct chemical reaction of glucose with amino groups, which proceeds to an irreversible conversion to advanced glycation end products. These molecules accumulate in poorly controlled diabetic individuals. One of the key sites of deposition is in the vascular basement membrane. A potential relationship exists between advanced glycation end product concentrations and vascular damage (Yamamoto *et. al.*, 2000).

Hypercoagulability

Diabetes is a hypercoagulable state. Hypercoagulability accelerates the development of atherosclerosis. Hyperglycemia contributes to the acceleration of the clotting process in diabetes. Platelet aggregability is markedly increased by hyperglycemia. Hyperglycemia appears to stimulate the cascade of clotting factors. Hyperglycemia has been shown to shorten fibrinogen half-life and stimulate thrombin formation Carr ME (2001).

Endothelial Inflammatory Activity

Atherosclerosis involves an inflammatory response of the endothelium. There are a number of adhesion molecules that activate the interaction between leukocytes and the endothelium. In addition to the stimulation of the clotting process, postprandial hyperglycemia appears to stimulate the adhesion molecules. The concentration of the intracellular adhesion molecules intracellular adhesion molecule 1 (ICAM-1), vascular cell adhesion molecule 1 (VCAM-1), and E-selectin increase in response to an oral glucose tolerance test. Acute hyperglycemia increases the production of factors that accelerate inflammation such as interleukin 6, interleukin 18, and tumor necrosis factor. Repaglinide treatment, which reduced PPG levels more than glyburide treatment, led to a greater decrease in interleukin 6 (P = .04) and C-reactive protein (P = .02) levels in the same study that showed greater carotid intima medial regression with repaglinide (Esposito *et. al.,* 2004; Plutzky J., 2004).

Excess Oxidative Activity

Diabetes is a state of increased oxidative activity. The reduced antioxidant activity in diabetes is directly related to postprandial hyperglycemia Ceriello A., (1998). The oxidation of low-density lipoprotein is a key step in lipid-induced atherosclerosis. Low-density lipoprotein oxidation is increased by hyperglycemia, presumably accelerating damage to the vascular wall (Diwadkar *et. al.,* 1999).

Hyperglycemia and Protein Kinase C Activation

It has been suggested that the common pathway for endothelial abnormalities in diabetes is activation of protein kinase C. Activation of protein kinase C may lead to induction of several growth factors including transforming growth factor b and vascular endothelial growth factor as well as nuclear transcription factor. Although these changes are thought to impact most on microangiopathy, the effects on the endothelial cell may result in damage to large blood vessels (Booth *et. al.,* 2002).

Postprandial Glucose as a Priority

Despite the increasing recognition that elevation of PPG is detrimental, current-day practice patterns do not focus on reduction of postmeal values. The goal of treatment is to lower HbA1c to nondiabetic levels. Self-blood glucose monitoring is used as a tool to guide therapy to lower HbA1c and has proven to be beneficial, even in patients who are not insulin treated (LeRoith D and Smith DO., 2005; Martin *et. al.*, 2006). Typically, patients are advised to measure their fasting glucose values and then test before lunch, supper, and bedtime (Monnier *et. al.*, 2004). They are asked to chart these measurements, either on paper or using computer programs to download values stored on glucose oxidase meters. These numerical data are used to choose the treatment approach, essentially ignoring control of postprandial glucose. Yet, several studies show that a focus on PPG leads to improvement in HbA1c values. The use of PPG values to adjust repaglinide dosage led to improved HbA1c levels in metformin-treated patients (Gerstein *et. al.*, 2004). In a study of type 1 diabetic patients, an algorithm based upon postprandial glucose values demonstrated improved glycemic control over that achievable by use of preprandial glucose values (Zisser *et. al.*, 2005). In type 2 diabetes mellitus, an insulin treatment regimen directed at lowering postprandial hyperglycemia had a greater effect on reduction of HbA1c than an approach focused on fasting glucose values.

Antihyperglycemic Threrapies

On viewing the complications of diabetes different therapies are being used since ages for its treatment. The treatment is categorized according to the type of diabetes.

Type 1 Diabetes

In type 1 diabetes (T1D) major problem is less insulin secretion due to degeneration of pancreatic β-cells therefore insulin therapy is given either by injection or pump. As it is also well documented that type 1 diabetes is an autoimmune disease which leads to loss of insulin producing β-cells via autoaggressive T-cells and pancreatic macrophage infiltration so regenerative approaches like induction of tolerance by deletion, destruction & anergy of autoimmune T-cells can be used to induce tolerance to self antigens. Among these Regulatory T-cells (Treg) act as bystander suppressor which dampens the function of autoaggressive T-cells (Battaglia *et. al.*, 2006). Thus there role in induction of tolerance and their antigen specific induction are of special interest to the researchers (Jaeckel *et. al.*, 2008). Similarly as umbilical cord provides haematopoetic stem cells which may initiate β-cells regeneration (Hess *et. al.*, 2003) therefore its transfusions in children with type 1 diabetes have shown to decelerate the loss of insulin production (Haller et. al. 2008). Moreover stem cells transplantation from bone marrow was investigated by scientist and showed increase in β-cells function, maintenance of HbA1c (Voltarelli et al., 2007). But the major problem with the above said approaches is the limited availaibility of donor organs. So to circumvent these problems novel approaches targeting beta-cell regeneration or neogenesis are currently underway. Pleuripotent embryonic stem cells from pancreatic and non-pancreatic tissues have been proposed as an alternative to provide an unlimited source of insulin-secreting cells (Bonner- Weir et al., 2008; Seissler & Schott, 2008). Despite major advances in stem-cell therapy and regenerative medicine, clinical trials examining the efficacy in humans with autoimmune diabetes are critical.

Type 2 Diabetes

Lifestyle Modification

Type 2 diabetes (T2D) covers 90-95% of all diabetes cases. It is possible to slow the progression of diabetes to overt diabetes by promotion of dietary control and exercise activity which directly address risk factors for diabetes because in T2D minimizing the risk factors associated with T2D are prime targets for its prevention. The Da Qing IGT and diabetes study by Pan *et. al.*, (1997) reported a significant reduction in incidence rate of T2D (44% diet group, 41% exercise group, and 46% diet-plus-exercise group compared to 68% controls) and Indian diabetes prevention programme (IDPP) reported a relative risk reduction of 28.5 % with lifestyle modification (Ramachandran et al., 2006). Dietary management is more than a mere reduction of calorie intake. Decreasing carbohydrate intake is key for management of T2D and this can be maintained by ingesting carbohydrates with low potency to raise glucose level for example according to Tappy *et. al.*, (1996) Consumption of a food with low glycemic index leads to lower postprandial hyperglycemia. Similarly Jenkins *et. al.*, (2002) reported that Supplementation of β-glucan fibers to meals help in reducing postprandial hyperglycemia.

Pharmacological Strategies

Five different types of oral antihyperglycemic agents are used for the treatment of T2D. These drugs have different mechanism of action and can be used as singly or in combination with one another.

- Insulin secretogogues (Sulphonylureas)
- Insulin sensitizers
- Biguanide
- Intestinal Lipase Inhibitors (for obese patients only)
- Alpha-glucosidase inhibitors

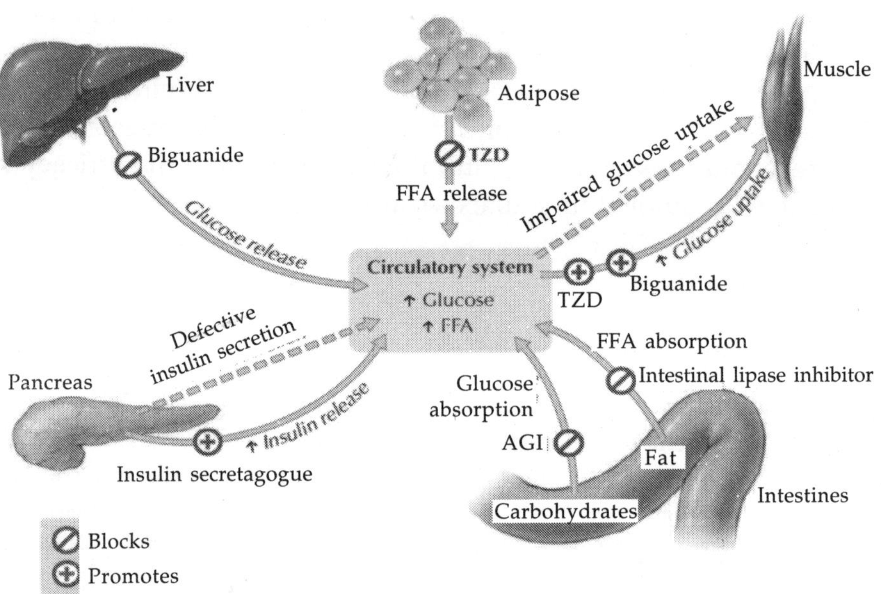

Figure 1: Major target organs and actions of orally administered antihyperglycemic agents in type 2 diabetes mellitus. TZD = thiazolidinedione; FFA = free fatty acid; AGI = a-glucosidase inhibitor.

Insulin secretagogues can be divided into 2 subclasses: sulfonylureas and non-sulfonylureas. Sulphonylureas (insulin secretors) work by stimulating the pancreas to release more insulin. For example: Tolbutamide (Orinase®), Tolazamide (Tolinase®), Glimepiride (Amaryl®), Glipizide (Glucotrol®, Glucotrol XL®), Glburide (DiaBeat®, Glynase Pre-Tab®, Macronase®) are the main agents of sulphonylureas available in market. Non-obese type 2 diabetic are usually started on sulphonylureas. Sulphonylureas act on receptors located on the plasma membrane of β-cells. The receptors are K_{ATP} channels and binding of sulphonylureas blocks the channels, leading to influx of calcium through calcium channels, membrane depolarization and finally insulin exocytosis (Tyrberg and Levine., 2001). Sulfonylureas are predominantly metabolized by the liver and cleared by the kidneys. Several metabolites of glyburide are partially active, so that if clearance is impaired in the kidney, the accumulating metabolites can have a significant hypoglycemic effect. In contrast, gliclazide and glimepiride are metabolized by the liver to inactive metabolites. Sulphonylureas are very effective in decreasing blood glucose, HbA1c level and produces an average reduction in HbA1c concentrations of about 1.0%–1.5%. Non sulphonylurea have the similar mechanism of action like sulphonylurea but have fewer side effects. This difference is due in part to the shorter duration of action and in part to the glucose- dependent insulinotropic effects. But they may have severe liver and kidney dysfunction.

Another very popular group of oral antidiabetic component is insulin sensitizers (thiazolidinediones TZDs;) that works to overcome insulin resistance by making body cells more sensitive to insulin for example: Pioglitazone (Actos®) and Rosiglitazone (Avandia®). The extensive studies of TZDs at the molecular level indicate that they act on Peroxisome proliferator-activated receptor gamma (PPAR-γ). PPAR-γ is a nuclear receptor that mediates adipocytes differentiation and insulin sensitivity. TZDs activate PPAR-γ, increase glucose transporter (GLUT-4) synthesis and differentiate stem cells into adipose tissue. PPAR-γ is abundantly expressed in adipose tissue and is also present in skeletal muscles, which is the major organ of glucose utilization (Kao *et al.*, 2000; Hauner., 2002). TZDs improve insulin sensitivity, particularly in the peripheral tissues. Although unproven, this appears to be mainly mediated through an effect on adipocytes, since there are fewer PPARγ receptors in muscle tissue. In the adipocyte, differentiation is enhanced, lipolysis is reduced, and levels of circulating adipo-cytokines or "adipokines" are altered, namely a decrease in tumour necrosis factor-α and leptin and an increase in adiponectin. The recruitment of a greater number of smaller adipocytes, which is associated with improved lipogenesis and storage, results in a reduction in circulating free fatty acids. All these effects i.e. Decreased TNF-a and free fatty acid levels and increased adiponectin levels are expected to enhance insulin sensitivity (Aldhahi and Hamdy., 2003). Among the side effects of TZDs weight gain, edema and anaemia are common while contraindications involve severe liver dysfuntions and cardiac heart failure.

Biguanides (inhibitors of hepatic glucose production) work by reducing the amount of glucose produced by liver. The mechanism of action of Biguanides (e.g. metformin, phenformin, buformin) is still not clear. They act mainly by decreasing hepatic glucose output via decreasing gluconeogenesis and may also increase uptake of glucose by skeletal muscles (Kirpichnikov *et. al.,* 2002). Recently metformin was found to activate adenosine monophosphate activated protein kinase (AMPK). Activation of AMPK results in the phosphorylation and inhibition of acetyl-coenzyme A carboxylase, which catalyzes the rate limiting step of lipogeneis. Metformin also cause a decrease in expression of sterol-regulatory-element-binding-protein-1 (SREBP-1), a transcription factor implicated in the pathogenesis of insulin resistance, dyslipidemia and diabetes. Decreased SREBP-1 expression results in decreased gene expression of lipogenic enzymes, which further contributes to decreased triglyceride synthesis and hepatic steatosis (Zhou *et. al.,* 2001). Obese patients with type 2 diabetes are usually started on biguanides namely Metformin (Glucophase® and Glucophase XR®). Major side effects of biguanides are gastrointestinal and in rare cases lactic acidosis while contraindication involves moderate to severe liver, cardiac and renal dysfunction.

Orlistat is an intestinal lipase inhibitor that acts as a selective inhibitor for gastric and pancreatic lipases and thereby inhibits the hydrolysis of dietary fat into absorbable free fatty acids and monoglycerides. The marked reduction in the absorption of fat results in decreased energy intake and weight loss. There is negligible systemic absorption of the drug. Metabolism occurs within the gastrointestinal wall, and fecal excretion is the main route of elimination (Guerciolini., 1997). Orlistat's role in the treatment of diabetes is limited to obese patients and improves HbA1c concentrations by 0.3%-0.9%. Since blood glucose lowering effect of orlistat is relatively small, it should be used in combination with other OHAs (Hollander *et. al.,* 1998; Hanefeld and Sachse., 2002; Kelley *et. al.,* 2002). The side effects of orlistat include flatulence with discharge, oily spotting, faecal urgency, steatorrhea, increased frequency of defecation and fecal incontinence (Guerciolini., 1997).

α-Glucosidase Inhibitors

Therapeutic approaches for the treatment of type 2 diabetes, such as sulphonylureas, metformin and insulin therapy are effective in decreasing fasting glucose levels but except for insulin therapy they have little effect on postprandial hyperglycemia. Increasing importance of postprandial hyperglycemia and little effectiveness of these oral antihyperglycemic agents have created a challenge to the researchers for development of new drugs for postprandial hyperglycemia (Baron, 1998). Among these new agents inhibitors of enzymes involved in digestion process has gained much importance in past few years. α-glucosidase inhibitors are one of these inhibitors and due to their unique mechanism of action there has been an increased interest in identifying α-glucosidase inhibitors that can be used as an important tool for understanding biochemical processes and as prospective therapeutic agents for postprandial hyperglycemia (Faridmoayer and Scaman, 2005; Markad *et. al.*, 2006; Liu *et. al.*, 2007). a-glucosidase inhibitors do not target a specific pathophysiologic aspect of diabetes but competitively inhibit enzymes in the small intestinal brush border that are responsible for the breakdown of oligosaccharides and disaccharides into monosaccharides (Lebovitz., 1997). It works primarily on α-glucosidase (EC 3.2.1.20, 3.2.1.10, 3.2.1.48 and 3.2.1.106), which are predominant in the proximal half of the small intestine and catalyzes the release of α-D-glucopyranose from the non-reducing ends of various carbohydrate substrates (Frandsen & Svensson, 1998). These enzymes also play an important role in the biochemical processes of glycoproteins and glycolipids (Bertozzi & Kiessling, 2001). Presence of α-glucosidase inhibitor for example: Acarbose (Precose®) and Miglitol (Glyset®) in diets inhibit the activity of α-glucosidase thus delays intestinal absorption of carbohydrates shifted to more distal parts of the small intestine and colon. This retards glucose entry into the systemic circulation and lowers postprandial glucose levels. α-Glucosidase inhibitors act locally at the intestinal brush border and are not absorbed in intestine but get excreted in faeces. Efficiency of α-glucosidase inhibitor to improve HbA1c concentrations is by 0.5%-1.0%. They also have beneficial effects on insulin resistance. α-glucosidase inhibitors are generally synthesized chemically. However they can be isolated naturally from plants, food products, or by microorganisms. Several chemical synthetic compounds, such as sulfonamide, xanthone derivatives, and deoxy salacious, have been reported to exhibit inhibitory effects against α-glucosidase activity (Seo et al., 2005; Muraoka et al., 2006; Liu et al., 2007). In addition, the salacinol from *Salacia reticulata* (Yoshikawa *et. al.*, 2002), *Punica granatum* flower extract (Li et al., 2005) and the water extract of douchi (Chinese traditional food) (Chen *et. al.*, 2007) also exhibits α-glucosidase inhibitory activity. Nevertheless, these natural α-glucosidase inhibitors are not easily produced at large scale (Fujita *et. al.*, 2003; Li et al., 2005; Chen et al., 2007) while chemically synthesized α-glucosidase inhibitors normally cause hepatic disorders (Diaz-Gutierrez., 1998), other negative gastrointestinal symptoms (bloating, diarrhoea and flatulence) Bayraktar *et. al.*, (1996) Inzucchi., (2002) Chiasson *et. al.*, (2002). In addition there are reports for an increased incidence of renal tumors and serious hepatic injury and acute hepatitis by Acarbose (Cheng and Fantus., 2005).

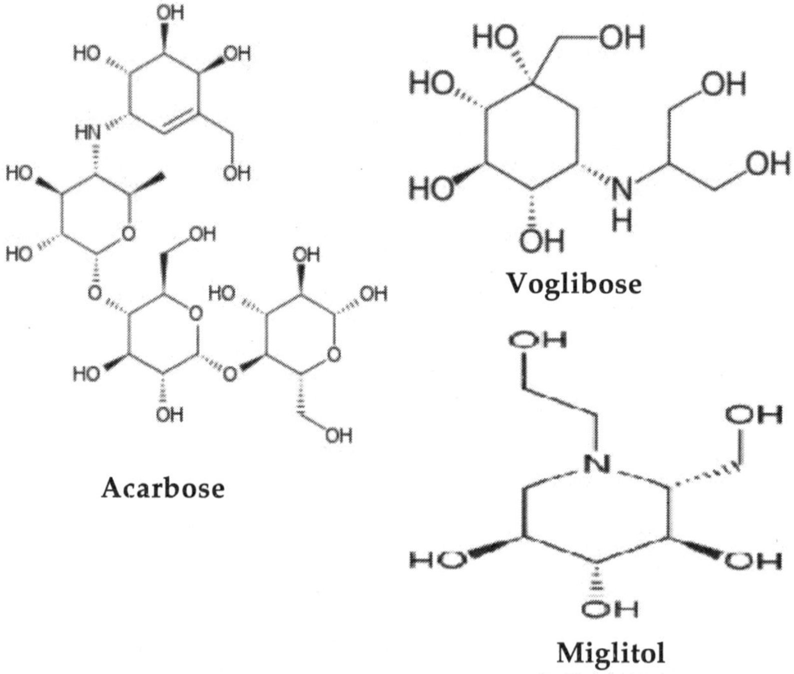

Acarbose

Voglibose

Miglitol

Figure 2: Chemical structures of three commercially available a-glucosidase inhibitors.

On contrary α-glucosidase inhibitors synthesis by microorganisms could be an effective strategy to produce cost-effective and productive α -glucosidase inhibitors. It has been reported that some microorganisms, including species of *Streptomyces* (Iwasa *et. al.*, 1970), *Actinoplanes* (Schmidt *et al.*, 1977) and *Flavobacterium saccharophilium*

(Kameda *et. al.*, 1980), *Bacillus subtilis* (Zhu *et al.*, 2008) were able to synthesize α-glucosidase inhibitors. Due to fast-growing characteristic of these microbes, there has been increased interest in identifying α -glucosidase inhibitors producing microorganisms.

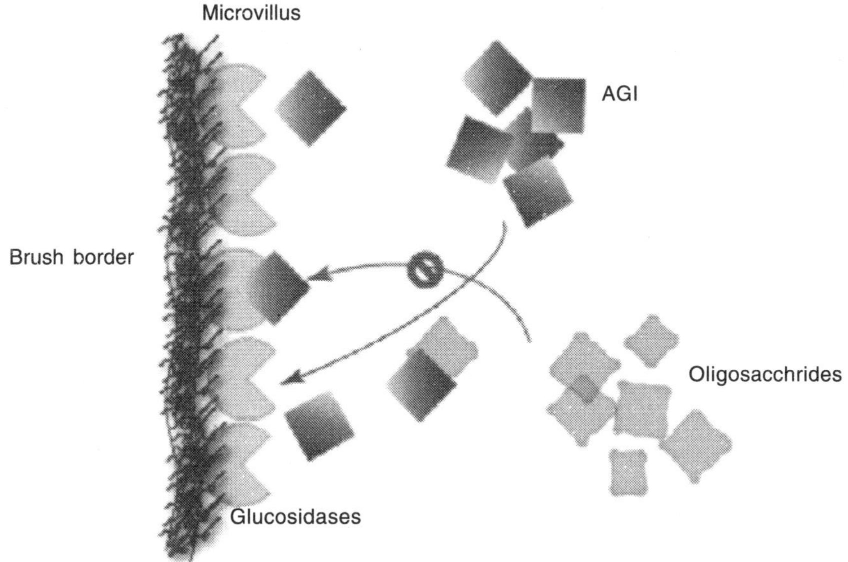

Figure 3: Mechanism of action of α-glucosidase inhibitors.

Conclusion

It is recognized that elevation of plasma glucose is detrimental. Numerous studies have demonstrated that lowering HbA1c reduces the incidence of microvascular disease and diabetic neuropathy. However, HbA1c reflects the mean plasma glucose over a prolonged period. It has now been recognized that the transient rise in blood glucose, which occurs after a meal, may have very specific deleterious effects on blood vessels. There is a marked increase in cardiovascular risk in individuals with postprandial hyperglycemia. Reduction of PPG levels has shown benefit in reducing cardiovascular events. For this reason, it is imperative to focus on suppression of PPG elevation in treatment not only of diabetic patients, but also individuals with IGT. Promotion of changes in dietary and exercise activity has proven to be the most effective means of preventing the progression of IGT to diabetes. Pharmacologic agents including sulfonylureas, metformin, disccharidase inhibitors, and thiazolidinediones are all effective in reducing PPG elevation. Exogenous insulin is the required management for patients with type 1 diabetes mellitus and for individuals with advanced type 2 diabetes mellitus. New synthetic insulin analogues, which have rapid absorption and disappearance times, permit more aggressive management of mealtime hyperglycemia. Inhaled insulin that also manifests rapid absorption and disappearance also promises to improve the insulin management of postprandial hyperglycemia. Continuous glucose monitoring will increase the emphasis on PPG as our primary treatment target. The widespread adoption of PPG as our target will lead to improvement in HbA1c levels and reduction in the deleterious clinical effects of hyperglycemia.

Need for Further Study

'Finally, economic strategies need to be developed that could minimize the treatment cost as well mortality rate occurring due to diabetes. Although many strategies and alpha glucose inhibitors are available in market, but due to high cost and side effects, further research on natural inhibitors is desirable.

References

Aldhahi W and Hamdy O 2003 Adipokines, inflammation and the endothelium in diabetes. *Current Diabetic Reports* **3:** 293-298.

Avignon A, Radauceanu A and Monnier L 1997 Nonfasting plasma glucose is a better marker of diabetic control than fasting plasma glucose in type 2 diabetes. *Diabetes Care* **20:** 1822-1826.

Baron A D 1998 Postprandial hyperglycaemia and a-glucosidase inhibitors. *Diabetes Research and Clinical Practice.* **40:** 51–55.

Bastyr EJ, Stuart CA, Brodows RG, Schwartz S, Graf CJ and Zagar A 2000 Therapy focused on lowering postprandial glucose, not fasting glucose, may be superior for lowering HbA1c. IOEZ Study Group. *Diabetes Care* **23:** 1236-1241.

Battaglia M, Stabilini A, Draghici E, Gregori S, Mocchetti C and Bonifacio E 2006 Rapamycin and interleukin-10 treatment induces T regulatory type 1 cells that mediate antigen-specific transplantation tolerance. *Diabetes* **55:** 40-49.

Bayraktar M, Van TDH and Adalar N 1996 A comparison of acarbose versus metformin as an adjuvant therapy in sulfonylurea-treated NIDDM patients. *Diabetes Care* **19:** 252-254.

Bertozzi CR and Kiessling LL 2001 Chemical glycobiology. *Science* **291:** 2357–2364.

Bonner-Weir S, Inada A, Yatoh S, Li WC, Aye T and Toschi E 2008 Transdifferentiation of pancreatic ductal cells to endocrine beta-cells. *Biochemical Society Transaction* **36:** 353-356.

Booth G, Stalker TJ, Lefer AM, Scalia R. Mechanisms of amelioration of glucose-induced endothelial dysfunction following inhibition of protein kinase C in vivo. *Diabetes* 2002; **51:** 1556- 64.

Calver A, Collier J, Vallance P. Inhibition and stimulation of nitric oxide synthesis in the human forearm arterial bed of patients with insulin-dependent diabetes. *J Clin Invest* 1992; **90:**2548- 54.

Carr ME. Diabetes mellitus: a hypercoagulable state. *J Diabetes Complications* 2001; **15:** 44-54.

Ceriello A 2003 The possible role of postprandial hyperglycaemia in the pathogenesis of diabetic complications. *Diabetologia* **46:** M9-M16.

Ceriello A, Bortolotti N, Motz E, Crescentini A, Lizzio S and Russo A 1998 Meal generated oxidative stress in type 2 diabetic patients. *Diabetes Care* **21:** 1529-1533.

Ceriello A, Cavarape A, Martinelli L, Da Ros R, Marra G and Quagliaro L 2004 The post prandial state in type 2 diabetes and endothelial dysfunction: effects of insulin as part. *Diabetic Medicine* **21:** 171- 175.

Ceriello A, Cavarape A, Martinelli L, Da Ros R, Marra G, Quagliaro L, et al. The post prandial state in type 2 diabetes and endothelial dysfunction: effects of insulin aspart. *Diabet Med* 2004; **21:**171- 5.

Chen J, Cheng YQ, Yamaki K, and Li LT 2007 Anti a-glucosidase activity of Chinese traditionally fermented soybean (douchi). *Food Chemistry* **103:** 1091–1096.

Cheng YY and Fantus IG 2005 Oral antihyperglycemic therapy for type 2 diabetes Mellitus. *Canadian Medical Association Journal* **172:** 213-226.

Cherrington AD 1999 Control of glucose uptake and release by the liver in vivo. *Diabetes* **48:** 1198- 1214.

Chiasson JL, Josse RG, Gomis R, Hanefeld M, Karasik A and Laakso M 2002 STOP-NIDDM Trail Research Group. Acarbose for prevention of type 2 diabetes mellitus: the STOP-NIDDM randomized trial. *Lancet* **359:** 2072-2077.

Dandona P, Aljada A. Advances in diabetes for the millennium: diabetes and the endothelium. *Medscape Gen Med* 2004; **6:** 6.

Davidson J 2005Strategies for improving glycemic control: effective use of glucose monitoring. *American Journal of Medicine* **118:** 27S-32S.

Del PS 2003 Loss of early insulin secretion leads to postprandial hyperglycaemia. *Diabetologia* **46:** M2-M8.

Diaz-Gutierrez FL, Ladero JM and Diaz-Rubio M 1998 Acarbose-induced acute hepatitis. *American Journal of Gastroenterology* **93:** 481.

Diwadkar VA, Anderson JW, Bridges SR, Gowri MS, Oelgten PR. Postprandial low density lipoproteins in type 2 diabetes are oxidized more extensively than fasting diabetes and control samples. *Proc Soc Exp Biol Med* 1999; **222:** 178- 84.

Esposito K, Giugliano D, Nappo F, Marfella R. Regression of carotid atherosclerosis by control of postprandial hyperglycemia in type 2 diabetes mellitus. *Circulation* 2004; **29:**2978- 84.

Faridmoayer A and Scaman CH 2005 Binding residues and catalytic domain of Saccharomyces cerevisiae processing a-Glucosidase I. *Glycobiology* **15:** 1341–1348.

Frandsen TP and Svensson B 1998 Plant a-glucosidases of the glycoside hydrolase family 31. Molecular properties, substrate specificity, reaction mechanism, and comparison with family members of different origin. *Plant Molecular Biology* **37:** 1–13.

Gerich JE 2003 Clinical significance, pathogenesis, and management of postprandial hyperglycemia. *Archive of Internal Medicine* **163:** 1306-1316.

Gerstein HC, Garon J, Joyce C, Rolfe A,Walter CM. Pre-prandial vs. post-prandial capillary glucose measurements as targets for repaglinide dose titration in people with diet-treated or metformin-treated type 2 diabetes: a randomized controlled clinical trial. *Diabet Med* 2004; **21:** 1200-3.

Goldstein DE, Rohlfing CL, Wiedmeyer HM, Little RR, England JD and Tennill A 2002 Defining the relationship between plasma glucose and HbA1c: analysis of glucose profiles and HbA1c in the Diabetes Control and Complications Trial. *Diabetes Care* **25:** 275-278.

Goodfellow J, Ramsey MW, Luddington LA, Jones CJ, Coates PA, Dunstan F, et al. Endothelium and inelastic arteries: an early marker of vascular dysfunction in non–insulin dependent diabetes. *BMJ* 1996; **312:**744-745.

Guerci B, Bohme P, Kearney-Schwartz A, Zannad F, Drouin P. Endothelial dysfunction and type 2 diabetes. Part 2: altered endothelial function and the effects of treatments in type 2 diabetes mellitus. *Diabetes Metab* 2001; **27:** 436-447.

Guerciolini R 1997 Mode of action of orlistat. *International Journal of Obesity* **21:** S12-S23.

Haller MJ, Viener HL, Wasserfall C, Brusko T, Atkinson MA, and Schatz DA 2008 Autologous umbilical cord blood infusion for type 1 diabetes. *Experimental Hematology* **36:** 710-715.

Hanefeld M and Sachse G 2002 The effects of orlistat on body weight and glycaemic control in overweight patients with type 2 diabetes. *Diabetes Obesity and Metabolism* **4:** 415-423.

Hauner H 2002 The mode of action of thiazolidinediones. Diabetes/Metabolism *Research and Review* **18:** S10-S15.

Hess D, Li L, Martin M, Sakano S, Hill D and Strutt B 2003 Bone marrow derived stem cells initiate pancreatic regeneration. *Nature Biotechnology* **21:** 763-770.

Hollander PA, Elbein SC, Hirsch IB, Kelley D, McGill J and Taylor T 1998 Role of orlistat in the treatment of obese patients with type 2 diabetes. *Diabetes Care* **21:** 1288-1294.

Inzucchi SE 2002 Oral antihyperglycemic therapy for type 2 diabetes. Journal of American *Medical Association* **287:** 360-372.

Iwasa T, Yamamoto H, and Shibata M 1970 Studies on validamycins, new antibiotic I. *Streptomyces hygroscopicus var. limoneus.* Validamycin producing organism. *Journal of Antibiotechnology* **23:** 595–602.

Jaeckel E, Mpofu N, Saal N and Manns MP 2008 Role of regulatory T cells for the treatment of type 1 diabetes mellitus. *Hormone and Metabolic Research* **40:** 126-136.

Jenkins AL, Jenkins DJ, Zdravkovic U, Wursch P and Vuksan V 2002 Depression of the glycemic index by high levels of beta-glucan fiber in two functional foods tested in type 2 diabetes. *European Journal of Clinical Nutrition* **56:** 622-628.

Kameda Y, Asano N, Teranishi M, and Matsui K 1980 New cyclitols, degradation of validamycin A by *Flavobacterium saccharophilum. Journal of Antibiotechnology,* **33:** 1573–1574.

Kao PC, Wu TJ, Ho LLT and Li XJ 2000 Review: Current and new approaches in the management of diabetes mellitus. *Annals of Clinical and Laboratory Science* **30:** 339-345.

Katsumori K, Wasada T, Kuroki H, Arii H, Saeki A and Aoki K 1995 Prevalence of macro- and microvascular diseases in non-insulin dependent diabetic and borderline glucose-intolerant subjects with insulin resistance syndrome. *Diabetes Research and Clinical Practices* **29:** 195- 201.

Kelley DE, Bray GA, Pi-Sunyer FX, Klein S, Hill J and Miles J 2002 Clinical efficacy of orlistat therapy in overweight and obese patients with insulin-treated type 2 diabetes: a 1-year randomized controlled trial. *Diabetes Care* **25:** 1033-1041.

Kirpichnikov D, McFarlane SI and Sowers JR 2002 Metformin: an update. *Annals Internal Medicine* **137:** 25-33.

Lebovitz HE 1997 Alpha-glucosidase inhibitors. *Endocrinology Metabolism Clinics of North America* **26:** 539-551.

LeRoith D and Smith DO 2005 Monitoring glycemic control: the cornerstone of diabetes care. *Clinical Therapeutics* **27:** 1489-1499.

LeRoith D, Smith DO. Monitoring glycemic control: the cornerstone of diabetes care. *Clin Ther* 2005; **27:**1489- 1499.

Li Y, Wen S, Kota BP, Peng G, Li GQ and Yamahara J 2005 Punica granatum flower extract, a potent a-glucosidase inhibitor, improves postprandial hyperglycemia in Zucker diabetic fatty rats. *Journal of Ethnopharmacology* **99:** 239–244.

Liu Y, Ma L, Chen WH, Wang B and Xu ZL 2007 Synthesis of xanthone derivatives with extended p-systems as a-glucosidase inhibitors: Insight into the probable binding mode. *Bioorganic & Medicinal Chemistry* **15:** 2810–2814.

Markad SD, Karanjule NS, Sharma T, Sabharwal SG, and Khavale DD 2006 Synthesis and evaluation of glycosidase inhibitory activity of N-butyl 1-deoxy-D-gluco-homonojirimycin and N-butyl 1-deoxy-L-ido-homonojirimycin. *Bioorganic and Medicinal Chemistry* **14:** 5535–5539.

Martin S, Schneider B, Heinemann L, Lodwig V, Kurth HJ, Kolb H, et al. Self-monitoring of blood glucose in type 2 diabetes and longterm outcome: an epidemiological cohort study. *Diabetologia* 2006; **49:**271 - 278.

McVeigh GE, Brennan GM, Johnston GD, McDermott BJ, McGrath LT, Henry WR, et al. Impaired endothelium-dependent and independent vasodilation in patients with type 2 (non–insulindependent) diabetes mellitus. *Diabetologia* 1992; **35:**771- 776.

Monnier L, Colette C, Lapinski H, Boniface H. Self-monitoring of blood glucose in diabetic patients: from the least common denominator to the greatest common multiple. *Diabetes Metab* 2004; **30:**113-119.

Monnier L, Lapinski H and Colette C 2003 Contributions of fasting and postprandial plasma glucose increments to the overall diurnal hyperglycemia of type 2 diabetic patients: variations with increasing levels of HbA1c. *Diabetes Care* **26:** 881-885.

Muraoka O, Yoshikai K, Takahashi H, Minematsu T, Lu G and Tanabe G 2006 Synthesis and biological evaluation of deoxy salacinols, the role of polar substituents in the side chain on the a-glucosidase inhibitory activity. *Bioorganic and Medicinal Chemistry* **14:** 500–509.

Pan XR, Li GW, Hu YH, Wang JX, Yang WY and An ZX 1997 Effects of diet and exercise in preventing NIDDM in people with impaired glucose tolerance. The Da Qing IGT and Diabetes Study. *Diabetes Care* **20:** 537-544.

Park JY, Takahara N, Gabriele A, Chou E, Naruse K, Suzuma K, et al. Induction of endothelin-1 expression by glucose: an effect of protein kinase C activation. *Diabetes* 2000; **49**:1239-1248.

Plutzky J. Inflammation in atherosclerosis and diabetes mellitus. *Rev Endocr Metab Disord* 2004; **5**: 255-259.

Prior JO, Quinones MJ, Hernandez-Pampaloni M, Facta AD, Schindler TH, Sayre JW, et al. Coronary circulatory dysfunction in insulin resistance, impaired glucose tolerance, and type 2 diabetes mellitus. *Circulation* 2005; **111**: 2291-2298.

Ramachandran A, Snehalatha C, Mary S, Mukesh B, Bhaskar AD and Vijay V 2006 The Indian Diabetes Prevention Programme shows that lifestyle modification and metformin prevent type 2 diabetes in Asian Indian subjects with impaired glucose tolerance (IDPP-1). *Diabetologia* **49**: 289-297.

Rendell MS and Jovanovic L 2006 Targeting postprandial hyperglycemia. *Metabolism Clinical and Experimental* **55**: 1263–1281.

Sakamoto T, Ogawa H, Kawano H, Hirai N, Miyamoto S and Takazoe K 2000 Rapid change of platelet aggregability in acute hyperglycemia: detection by a novel laser light scattering method. *Journal of Thrombosis and Haemostasis* **83**: 475-479.

Schmidt DD, Frommer W, Junge B, Muller L, Wingender W, Truscheit, E 1977 a-Glucosidase inhibitors, new complex oligosaccharides of microbial origin. *Naturwissenschaften* **64**: 535–536.

Seissler J and Schott M 2008 Generation of insulin-producing beta cells from stem cells— perspectives for cell therapy in type 1 diabetes. *Hormone and metabolic research* **40**: 155-161.

Seo WD, Kim JH, Kang JE, Ryu HW, Curtis-Long MJ and Lee HS 2005 Sulfonamide chalcone as a new class of a-glucosidase inhibitors. *Bioorganic and Medicinal Chemistry Letters* **15**: 5514–5516.

Shichiri M, Kishikawa H, Ohkubo Y and Wake N 2000 Long-term results of the Kumamoto Study on optimal diabetes control in type 2 diabetic patients. *Diabetes Care* **23**: B21 -B29.

Shiraiwa T, Kaneto H, Miyatsuka T, Katz K, Yamamoto K, Kawashima A, Kanda T, Suzuki M, Imano M, Matsuhisa M, Hori M and Yamasaki M 2005 Post-prandial hyperglycemia is an important predictor of the incidence of diabetic microangiopathy in Japanese type 2 diabetic patients. *Biochemical and Biophysical Research Communications* **336**: 339-345.

Tandon RK, Srivastva LM and Pandey SC 1975 Increased disccharidase activity in human diabetics. *American Journal of Clinical Nutrition* **28**: 621-625.

Tyrberg B and Levine F 2001 Current and future treatment strategies for type 2 diabetes: the b-cell as a therapeutic target. *Current Opinion in Investigating Drugs* **2:** 1568-1574.

Vasudevan AR and Ballantyne CM 2005 Cardiometabolic risk assessment: an approach to the prevention of cardiovascular disease and diabetes mellitus. *Clinical Cornerstone* **7:** 7-16.

Voltarelli JC, Couri CE, Stracieri AB, Oliveira MC, Moraes DA and Pieroni F 2007 Autologous nonmyeloablative hematopoietic stem cell transplantation in newly diagnosed type 1 diabetes mellitus. *Journal of the American Medical Association* **297:** 1568-1576.

Yamamoto Y, Yamagishi S and Yonekura H 2000 Roles of the AGE-RAGE system in vascular injury in diabetes. *Annals of New York Academy of Sciences* **902:** 163-170.

Yamamoto Y, Yamagishi S, Yonekura H, et al. Roles of the AGERAGE system in vascular injury in diabetes. *Ann NY Acad Sci* 2000; **902:** 163-170.

Zhou G, Myers R, Li Y, Chen Y, Shen X and Fenyk-Melody J 2001 Role of AMP-activated protein kinase in mechanism of metformin action. *The Journal of Clinical Investigation* **108:** 1167-1174.

Zhu YP, Yin LJ, Cheng YQ, Yamaki K, Mori Y, Cheng YS and Li-Te Li 2008 Effects of sources of carbon and nitrogen on production of a-glucosidase inhibitor by a newly isolated strain of *Bacillus subtilis* B2. *Food Chemistry* **109:** 737–742.

Zisser H, Jovanovic L, Doyle III F, Ospina P, Owens C. Run-to-run control of meal-related insulin dosing. *Diabetes Technol Ther* 2005; **1:** 48-57.

www.diabetesatlas.org

www.idf.org (International Diabetes Federation)

□□□

Chapter 42

Potential Herbs for Incorporation into Dairy Foods

SHAIK ABDUL HUSSAIN, REKHA RANI AND R. R. B. SINGH
Dairy Technology Division, National Dairy Research Institute, Karnal

Introduction

Enormous changes in life style, eating habits and shifting rural habitations have a significant effect on diets in the developing countries leading to assorted increase in health problems. With millions of rupees being spent on health care each year, consumer's desire for food products with desired health benefits continues to grow. Furthermore, people want to take food with desired health benefits rather than taking medicine separately.

Nutraceuticals often referred to as phytochemicals are natural bioactive, chemical compounds that have health promoting, disease preventing or medicinal properties. Nutraceuticals may range from isolated nutrients, herbal products, dietary supplements and diets to genetically engineered "designer" foods and processed products such as cereals, soups and beverages. Nutraceuticals are found in a mosaic of products emerging from (a) the food industry, (b) the herbal and dietary supplement market, (c) pharmaceutical industry, and (d) the newly merged pharmaceutical/ agribusiness/ nutrition conglomerates

(Dureja, *et al.*, 2003). The ascribed health benefits of nutraceuticals are countless. Various products are claimed not only to reduce the risk of cancer and heart disease but also to prevent or treat hypertension, high cholesterol, excessive weight, osteoporosis, diabetes, arthritis, macular degeneration (leading to irreversible blindness), cataracts, menopausal symptoms, insomnia, diminished memory and concentration, digestive upsets and constipation and not to mention headaches (Stauffer, 1999). Nutraceuticals are marketed in various forms as a single substance or as combination preparations (Stephen, 1998).

Categories of Nutraceuticals

Nutraceuticals can be grouped into the following three broad categories (Dureja *et al.*, 2003):

1. Substances with established nutritional functions, such as vitamins, minerals, amino acids and fatty acids - *Nutrients*
2. Herbs or botanical products as concentrates and extracts - *Herbals*
3. Reagents derived from other sources (*e.g.*pyruvate, chondroitin sulphate, steroid hormone precursors) serving specific functions, such as sports nutrition, weight-loss supplements and meal replacements – *Dietary supplements.*

Herbs

Herbs have been used as food and as medicine for centuries. Since the beginning of human civilization, herbs have been an integral part of the society, valued for both their culinary and medicinal properties. Herbs have played a significant role in maintaining human health, improving the quality of human life and served us with valuable components of seasoning, beverages, cosmetics, dyes and medicines. From 1960, interest in "natural health" and the use of herbal products has increased (Trevelyan, 1993). Herbal nutraceuticals, an important category of nutraceuticals, are commonly used by people who seek conventional health care. Herbs harbor a wide variety of active phytochemicals including the flavonoids, terpenoids, lignans, sulfides, polyphenols, carotenoids, coumarins, saponins, plant sterols, uramins and phthalates have been identified (Craig, 1999).

Categories of Herbs

Herbs can be broadly categorized into three categories based on their use and toxicity. The first category is the "food herbs" which are gentle in action, have very low toxicity and are unlikely to cause an adverse response. Food herbs can be utilized in substantial quantities over long periods of time without any acute or chronic toxicity (allergic responses with foods are possible, as are unique idiosyncratic reactions, and even common foods such as grapefruit juice, broccoli, and okra can interact with medications). The second category is the "medicine herbs" which are stronger acting and need to be used as prescription drug for specific medical conditions (with a medical diagnosis) and usually for a limited period of time. These herbs may often cause adverse reactions and in some cases, drug interactions. The last category is the "poison herbs" which have strong potential for either acute or chronic toxicity and need to be utilized strictly under supervisions of trained clinicians (Winston, 1992). Traditional Chinese Medicine (TCM) also categorizes herbs into three classes basing on safety or potential toxicity. The upper class (superior) drugs are non-toxic and are tonic remedies. The middle class (ministerial) drugs may have some mild toxicity and they support the superior medicines. The last category is the lower class (inferior) remedies that are toxic and used only for specific ailments for limited periods of time. A clear understanding of herb's benefits and possible risks as well as a clearly defined patient diagnosis are essential for the practitioner to safely and effectively counsel patients about the safe and effective choices in herb use (Winston and Dattner, 1999; Tierra and Tierra, 1999; Dan, 2003).

Active Ingredients in Herbs

Herbs have many constituents including vitamins, minerals and active ingredients that have a variety of medicinal benefits (Table 1). These active components include: alkaloids, anthraquinones, bitters, flavonoids, saponins, tannins and essential oils.

Alkaloids

An alkaloid is any biologically active and heterocyclic chemical compound which contains nitrogen may possess some pharmacological activity and, in many cases have medicinal or ecological use (Aniszewski, 1994). According to the database of the National Library of Medicine, alkaloids are defined as nitrogenous bases and occur in animal and vegetable kingdoms, while some of them may have been synthesized (NCBI, 2005). Alkaloids come under a special group of chemicals which take part in the biological processes of plants, animals and microorganisms. Alkaloids can create intense physiological action by acting at different cellular levels of organisms, hence are widely used in the medical fields as curative drugs. Alkaloids are only slightly soluble in water but soluble in ethanol, benzene, ether and chloroform and some alkaloids can also be highly toxic, even in very small doses (Becker *et al.*, 1986). Physiological action of these alkaloids varies according to the species of the plants from which they are extracted. Some alkaloids act as a catalyst to other healing agents without being involved themselves. Natural plant alkaloids are used by pharmaceutical industry for the development of antimalarian agents (quinine and chloroquinine), anticancer agents (taxol, vinblastine and vincristine) and agents promoting blood circulation in the brain (vincamine) (Pelletier, 1983).

The alkaloid piperine containing spices namely the black, white, green (*Piper nigrum* L.) and long pepper (*Piper longum* L.) are widely used in food. Other alkaloid plants used in food are capsicum peppers such as chilli or red pepper (*Capsicum annuum* L.), peruvian pepper (*Capsicum baccatum* L.), ají pepper (*Capsicum chinese* Jacq.), bird pepper or tabasco (*Capsicum frutescens* L) and rocoto pepper (*Capsicum pubescens* Ruizet Pav.). According to recent studies, piperine is non-toxic and has a great deal of physiological activity. It has been recently documented that piperine interacts with a mammalian protein. This alkaloid is efficiently taken by calyx of bovine beta-lactoglobulin, which is the major whey protein in milk (Zsila *et al.*, 2005). Dietary supplements containing ephedrine alkaloids are on the market (FDA, 1996; Brevoort, 1996). Ephedrine alkaloids have sympathominetic effects and cause weight loss and enhanced athletic performance. However, there are risks connected with the use of these alkaloids as supplements and determining a risk-free dosage is very essential (FDA, 2002). However, some alkaloids are used as additional components of food. The most well known is the use of the quinine as a bitter in tonic water according to an established procedure. Theophylline is an important component of black tea and caffeine is a well-known component of coffee and both are known to posses pharmacological activity and have become a part of our day to day life (Aniszewski, 2007).

Anthraquinones

Anthraquinones represent a large family of compounds having diverse biological properties (Srinivas *et al.*, 2007). They act to stimulate muscular contraction of the large intestine and so have a laxative effect.

Herbs such as dock, cascara, senna, rhubarb and aloe contain anthraquinones. Anthraquinones like emodin (Zhang *et al.*, 1995; Kamie *et al.*, 1998; Huang *et al.*, 2007), rhein (Kuo *et al.*, 2004; Huang *et al.*, 2007) from rhubarb and aloe-emodin from *Aloe Vera* (Kuo *et al.*, 2002; Chen *et al.*, 2004; Acevedo-Duncan *et al.*, 2004) have been found to have anti-cancer properties. Aloe-emodin has been found to inhibit cell growth in several tumor cells including human lung carcinoma (Lee *et al.*, 2001), hepatoma (Yeh *et al.*, 2003; Kuo *et al.*, 2002) and leukemia cell lines (Chen *et al.*, 2004). Aloe-emodin also has a specific in vitro and in vivo antineuroectodermal tumor activity (Elsohly, *et al.*, 2007). Aloe latex contains anti-inflammatory anthraquinones, which are used in healing and arresting pain.

Bitters

Bitters are a mixed group of bitter-tasting compounds found in many naturally occurring herbs. Bitter ingredients mainly affect the digestive tract, stimulating the secretion of digestive juices and enzymes in the stomach and the flow of bile from the liver. They enhance appetite and improve digestion as well as absorption of nutrients from food (Harrison and Bartels, 2006). They are prescribed for people with poor appetite, a sluggish bowel, gall bladder and liver problems, gastritis, and to aid convalescence after the flu and other illnesses and also acts as anti-oxidants (Liao *et al.*, 2008). Other beneficial effects of bitter herbs include antimicrobial (Tan and Vanitha, 2004), antineoplastic (anti-tumor), relaxing effect on the nervous system, anti-inflammatory action, antidiabetic action (Hui *et al.*, 2009) and they also act on the immune system (Tan and Vanitha, 2004). The beneficial action of the bitters starts in the mouth, hence for best effect they need to be tasted (despite our dislike of their effect on our tongues). Some common herbs containing bitter compounds are angelica *(Angelica archangelica)*, chamomile *(Matricaria chamomilla)*, dandelion *(Taraxacum)*, goldenseal *(Hydrastis Canadensis)*, horehound *(Marrubium vulgare)*, milk thistle *(Silybum marianum)*, peppermint *(Mentha piperita)*, rue *(Ruta)*, wormwood *(Artemisia absinthium)* and yarrow *(Achilles millefolium)*.

It is mentioned in Indain Ayurveda that bitter herbs such as *neem* (Azadirachta indica), *manjista or Indian madder (Rubia cordifolia Linn.), haridra* (*Curcuma longa- zingiberaceae)*, and *guduchi or* Heart-leaved moonseed (*Tinospora cordifolia*) have an affinity to the liver and spleen, and are commonly used for such ailments.

Flavonoids

Flavonoids are primarily present as glycosides in nature and represent a major group of natural antioxidants. Herbs (e.g. *Sophora japonica, Citrus grandis, Hypericum perforatum*), fruits (e.g. orange, grapefruit, apple, grape), vegetables (e.g. onion, kale, broccoli, green pepper, spinach, tomatoe) and soybeans are the major sources of flavonoids (Chao *et al.*, 2002). Flavonoids posses beneficial pharmacological effects including protective role against coronary heart diseases (Hertog *et al.*, 1993; Hertog *et al.*, 1995; Knekt *et al.*, 1996; de Groot and Rauen, 1998; Hertog and Hollman, 1996) and also shown anti-allergic (Middleton, 1998), antiviral (Wang *et al.*, 1998; Kaul *et al.*, 1985), anticancer (Knekt *et al.*, 1997; Fotsis *et al.*, 1997; Stefani *et al.*, 1999) as well as antioxidant properties (Pietta, 2000; Ghasemzadeh *et al.*, 2010). A variety of flavonoid products are either being actively developed or currently sold as dietary supplements and/ or herbal remedies. But, literature regarding absorption, distribution, metabolism, and excretion of flavonoids glycosides in animals is limited and hence, the biological outcomes of flavonoids in herbs have not been fully understood (Hollman and Katan, 1997a; Hollman and Katan, 1999).

Saponins

The saponins are naturally occurring surface-active glycosides produced by plants (Francis *et al.*, 2002). Herbs contain steroid saponins which are responsible for their health-promoting properties (Fenwick *et al.* 1991). Capsicum peppers, aubergine, fenugreek, yucca and ginseng are examples of common herbs that contain steroid saponins.

Saponins perform many useful pharmacological functions which include hypoglycaemic activity (Petit *et al.* 1993; Kim *et al.* 1998c; Lee *et al.* 2000a; Yoshikawa *et al.* 2001), lowering of serum cholesterol levels (Harwood *et al.* 1993; Potter *et al.* 1993; Matsuura, 2001; Wang *et al.*, 2008) and lowering of LDL-cholesterol levels (Potter *et al.* 1993; Harris *et al.* 1997; Morehouse *et al.* 1999; Matsuura, 2001). They stimulate the cell-mediated immune system (Kensil, 1996; Plohmann *et al.* 1997; Barr *et al.* 1998; Sjolander *et al.* 1998; Lacaille-Dubois *et al.* 1999; Oda *et al.* 2000), inhibit the growth of cancer cells (Rao and Sung, 1995; Konoshima *et al.* 1998; Marino *et al.* 1998; Mimaki *et al.* 1998a; Podolak *et al.* 1998). They possess antifungal activity (Delmas *et al.* 2000; Wang *et al.* 2000a; Miyakoshi *et al.* 2000), antioxidant effects (Yoshiki and Okubo, 1995; Yoshiki *et al.* 1998; Hu *et al.* 2002), virucidal activity

(Sindambiwe *et al.* 1998; Apers *et al.* 2000; Mengoni *et al.* 2002) and neurotrophic and neuroprotective effects (improves learning ability and cognitive functions) (Zhao and McDaniel, 1998; Kim *et al.* 1998a; Rudakewich *et al.* 2001; Yao and Li, 2001; Yao and Li, 2002). In some cases saponins consumption may lead to abortification, anti-implantation and they are antizygotic also (Tewary *et al.* 1973; Stolzenberg and Parkhurst, 1976; Quin and Xu, 1998). These harmful effects should be taken care of while dealing with saponins.

Tannins

Tannins are a group of complex phenol, polyphenol and flavonoid compounds that act as astringents, shrinking tissues and contracting structural proteins in the skin and mucosa (Nobre-Junior *et al.*, 2007). Tannins have the ability to prevent lipid peroxidation, superoxide formation and perform free radical scavenging activity (Bhattacharya *et al.*, 2002; Bors and Michel, 2002; Dufour *et al.*, 2002; Giovannelli *et al.*, 2000; Gyamfi and Aniya, 2002; Kaplan *et al.*, 2001; Lin *et al.*, 2001a; Lin *et al.*, 2001b; Mullen *et al.*, 2002; Nakagawa and Yokozawa, 2002; Wei *et al.*, 2002; Ying *et al.*, 2001). Tannins exhibit several useful health properties such as anti-inflammatory (Viana *et al.*, 1997), analgesic, anti-lymphocytic (Chen *et al.*, 2000; Giovannelli *et al.*, 2000; Kapu *et al.*, 2001; Okoli and Akah, 2000), antimicrobial (Adeleye *et al.*, 2003; Akiyama *et al.*, 2001, Cheng *et al.*, 2002, Elegami *et al.*, 2002; Ho *et al.*, 2001; Silva *et al.*, 2002), antileishmanial, immunomodulatory (Kiderlen *et al.*, 2001; Kolodziej *et al.*, 2001), neuroprotective (Nobre-Junior *et al.*, 2007), antihypertensive (Liu *et al.*, 2003), antidiarrhoeal (Abdullahi *et al.*, 2001) activities and are also useful in treatment of ulcerative colitis (Catherine Clinton, 2009).

Essential Oils

Essential oils (EOs) (also called volatile or ethereal oils) are natural mixtures of aromatic compounds obtained from plant materials (flowers, buds, seeds,leaves, twigs, bark, herbs, wood, fruits and roots) that are extracted by steam or solvent distillation (Guenther, 1948). Generally, the essential oils consist of a mixture of esters, aldehydes, ketones, terpenes and phenolic compounds and harbour the characteristic flavour and aroma of the particular spice or herb.

These oils are slightly enriched in terpenoids (Guenther, 1999), posseses antiseptic and anti-microbial action (Abed, 2007; Bansod and Rai, 2008) and enhances the body's ability to fight off a range of infections. They have anti-inflammatory and antispasmodic effects (chamomile, yarrow), anticancer effects (Zu *et al.*, 2010), acts as expectorants (thyme, hyssop), used in tonics enhancing the appetite and the digestion and absorption of food (rosemary, fennel, marjoram) and also stimulate the heart and circulatory system (ginger, rosemary, thyme).

Table 1: Active phytochemicals found in herbs

Herbal source	Active phytochemicals
Allium spp. (garlic, onions, leeks and chives)	Diallysulfide, disulfide and trisulfide
Labiatae family (basil, dill, fennel, marjoram, mint, rosemary, oregano, sage, thyme)	Monoterpenes, sesquiterpenes, and flavonoids. Rosemary and sage contain diterpenoids and ursolic acid (triterpenoid)
Umbelliferae family (anise, caraway, celery seed, cilantro, coriander, cumin, dill, fennel and parsley)	Coumarins, phthalides, polyacetylenes and terpenoids
Zingiberaceae family (turmeric and ginger)	Curcumin, gingerols and diarylheptanoids
Green tea	Epigallocatechin gallate and other catechins
Licorice	Glycyrrhizin and chalcones
Flaxseed	Lignans
Tarragon	Terpenoids

(Craig, 1999)

Some Common Herbs with Potential for Incorporation into Foods

1. Satavari (Asparagus Racemosus)

Satavari (*Asparagus racemosus*) belongs to the family *Liliaceae.* It is commonly called as *Shatavari, Satawar* or *Satmuli* in India. The plant grows throughout the tropical and subtropical parts of India and finds use in various medicinal preparations (Saxena and Bopana, 2009).

Biologically Active Constituents

The major active constituents of *Asparagus racemosus* are steroidal saponins (Shatavarins I–IV) that are present in the roots. Shatavarin IV is a biologically active glycoside (Asmari *et al.,* 2004) and consists of two molecules of rhamnose and one molecule of glucose. Other active compounds such as quercetin, rutin and hyperoside are found in the flowers and fruits; while diosgenin and quercetin-3 glucuronide are present in the leaves (Thomsen, 2002). Other active constituents of *Asparagus racemosus* include racemofuran an antioxidant compound (Wiboonpun *et al.,* 2004), isoflavone (8-methoxy-5, 6, 4'-trihydroxyisoflavone7-*o*-β-d-glucopyranoside) present in the roots of the plant (Saxena and Chourasia, 2001), asparagamine (polycyclic alkaloid) having remarkable anti-oxytocic activity (Sekine *et al.,* 1994), racemosol (Sekine *et al.,* 1997), sarsasapogenin and kaempferol (Ahmad and Jain, 1991).

Medicinal Uses

Ancient classical ayurvedic literature claims several therapeutic attributes for the roots of *Asparagus racemosus* (*Shatavari*) and has been specially recommended in cases of threatened abortion and as a galctogogue (Nadkarni, 1954). Beneficial effects of root of *Asparagus racemosus* are suggested in nervous disorders, dyspepsia, diarrhoea, dysentery, tumors, inflammations, hyperdipsia, neuropathy, hepatopathy, cough, bronchitis, hyperacidity and certain infectious diseases (Sharma *et al.,* 2000).

Asparagus racemosus have been reported to be an immunomodulatory and immunostimulant (Rege *et al.,* 1989). After challenging with *Bordetella pertussis,* animals treated daily with *Asparagus racemosus* aqueous root extract (100 mg/kg body weight) showed a significant increase in antibody titres to *Bordetella pertussis* as against the untreated animals. There was reduced mortality coupled with overall improved health status in treated animals which indicates the development of a protective immune response, confirming immunoadjuvant effects of *Asparagus racemosus.* Extracts and formulations prepared from *Asparagus racemosus* exhibited various immunopharmacological actions such as increase in white cell counts, haemagglutinating and haemolytic antibody titres in cyclophosphamide (CP)-treated mouse ascitic sarcoma (Diwanay *et al.,* 2004). Shive *et al.* (2000) and Gautam *et al.* (2009) also have observed the immuostimulant effects of *Asparagus racemosus.*

The hexane, aqueous and alcoholic extracts of the root at concentration of 200 mg /ml showed antibacterial activity against *Bacillus subtilis, Escherichia coli, Proteus vulgaris, Salmonella typhimurium, Pseudomonas aeruginosa* and *Staphylococcus aureus* (Ahmed *et al.*, 1998). Methanolic extract of the roots of *Asparagus racemosus* has shown considerable antibacterial efficacy under *in vitro* conditions against *Escherichia coli, Shigella dysenteriae, Shigella sonnei, Shigella flexneri, Vibrio cholerae, Salmonella typhi, Salmonella typhimurium, Pseudomonas putida, Bacillus subtilis* and *Staphylococcus aureus* (Mandal *et al.*, 2000) and the fresh juice of the plant showed fungitoxicity against *Helminthosporium sativum, Colletotrichum falcatum, Fusarium oxysporum* (Singh and Sharma, 1978) and *Rhizoctonia solanil* (Renu, 1983).

The aqueous solution of the crude alcoholic extract of the root at a dose of 10-20 mg has been found cardio protective in frog, mice and in rats also. The extract also caused an increase in bleeding time in the rabbit and slight diuretic effect in the rat (Roy *et al.*, 1971). 'Abana', a herbo-mineral formulation containing 10 mg *Asparagus racemosus* extract per tablet, was found to have significant hypocholesterolaemic effect in rats (Khanna *et al.*, 1991) and *Asparagus racemosus* root powder supplements has also been investigated for the reduction of cholesterol levels in hypercholesteremic rats by Visavadiya and Narasimhacharya (2005). Shatavari has been found to stimulate milk production in buffaloes (Patel and Kanitkar, 1969). The galactogenic effect has been confirmed by a clinical trial by Sharma *et al.* (1996). Joglekar *et al.* (1967) observed an increase in milk secretion after administration of *Asparagus racemosus* in the form of Ricalex® tablets (40 mg concentrated root extract per tablet) to women suffering from deficient milk secretion. Both the crude extract as well as a polysaccharide-rich fraction shatavari significantly inhibited lipid peroxidation and protein oxidation and also show partial protection against radiation-induced loss of protein thiols and inactivation of superoxide dismutase (Kamat *et al.*, 2000).

The methanolic extract of *Asparagus racemosus* roots played the role of an antioxidant by attenuating free radical induced oxidative damage in kainic acid (KA)-induced hippocampal and striatal neuronal damage in mice (Parihar and Hemnani, 2004). Rao (1981) demonstrated the inhibitory action of *Asparagus racemosus* on DMBA (7, 12 dimethyl benza anthracene) induced mammary carcinogenesis in rats. *Asparagus racemosus* is well known for its phytoestrogenic properties and use as a hormone modulator (Mayo, 1998). Rege *et al.*,

(1999) advocated that aqueous extract of *Asparagus racemosus* had shown adaptogenic effects by reversing the effects of cisplatin on gastric emptying, and also normalized cisplatin induced intestinal hypermotility. Nanal *et al.*, (1974) found *Satavari* to be extremely effective in the treatment of diarrhoea, dysentery and gastritis as described in Ayurvedic texts such as *Sushruta Samhita* and *Sharangdhar Samhita*. Ethanol and aqueous extracts of *Asparagus racemosus* roots exhibited significant anti-diarrhoeal activity against castor oil induced diarrhoea in rats demonstrating an activity similar to loperamide (a commercial anti-diarrhoeal drug) (Venkatesan *et al.*, 2005). *Asparagus racemosus* also was found to have an effect comparable to a modern allopathic drug metoclopramide which is a dopamine antagonist used in dyspepsia to reduce gastric emptying time (Dalvi *et al.*, 1990).

Toxicity

No toxic effects or mortality were observed with doses of *Asparagus racemosus* ranging from 50 mg/kg to 1 g/kg for four weeks. Acute and sub-acute (15-30 days administration) toxicity studies did not detect any changes in vital organ function tests (Rege *et al.*, 1999).

2. Ashwagandha (Withania somnifera)

Ashwagandha (*Withania somnifera*) belongs to the genus *Withania* and family *Solanaceae*. It is also known as Indian ginseng or winter cherry and belongs to major medicinal plants of India (Thakur *et al.*, 1987). The species name *somnifera* means "sleep-bearing" in Latin, indicating it was considered a sedative, but it has been also used for sexual vitality and as an adaptogen. It is found wild in the Mediterranean region in North America. In India, it is cultivated in Madhya Pradesh, Rajastan and other drier parts of the country (Thakur *et al.*, 1987).

Biologically Active Components

Different parts of *Ashwagandha* plant harbors wide variety of active constituents. Steroidal lactone withaferin A (Lavie *et al.*, 1965) is thought to be the primary pharmacological agent present in the roots and leaves of *Ashawagandha* and another steroidal lactone withanolides are currently being explored for their brain regenerative properties. Some of the other important withanolides present in *Ashwagandha* are 27- deoxywithaferin A, 17-hydroxy-27-deoxy withaferin-A, withanolide A, 17-hydroxy withaferin A, 27-hydroxy withanolide B, withanone and withanolide D. Other biologically active chemical constituents reported in *Ashwagandha* are alkaloids (isopelletierine, anaferine), saponins containing an additional acyl group (sitoindoside VII and VIII), and withanolides with a glucose at carbon 27 (sitoindoside IX and X) (Elsakka *et al.*, 1990; Abou-Douh, 2002; Choudhary *et al.*, 1995).

Health benefits of *Ashwagandha*

In a double-blind clinical trial, *Ashwagandha* (oral supplementation at a dosage of 3 grams daily for one year) when tested for anti-ageing property in a group of 101 healthy males of 50-59 years old showed a significant improvement in hemoglobin and red blood cell count. Erythrocyte sedimentation rate decreased significantly and there is significant improvement in sexual performance (Bone, 1996). Two new glycowithanolides, sitoindoside IX (1) and sitoindoside X (2) isolated from *Withania somnifera* have shown immunomodulatory and CNS effects (anti-stress, memory and learning) in doses of 100-400mg/kg body weight and 50-200mg/kg body weight in mouse and rat, respectively (Ghosal, *et al.*, 1989). Root extract of *Ashwagandha* had also shown immunomodulatory effects in mice (Ziauddin, 1996) and significant increase in hemolytic antibody responses toward human erythrocytes indicates its immunostimulatory activity. Dhuley (1997) confirmed the immunomodulatory activity of *Ashwagandha* in mice.

Active glycowithanolides of *Ashwagandha* given at a dose of 10 or 20 mg/kg intraperitoneally to mice have shown an antioxidant effect in the brain which may be responsible for its diverse pharmacological properties (Bhattacharya *et al.*, 1997). Oral administration of powdered root of *Ashwagandha* led to considerable reduction in inflammation and significant reduction in arthritis linked paw swelling and degenerative changes (Anbalagan and Sadique, 1981; Begum *et al.*, 1988). The steroidal lactone withaferin-A displayed significant antitumor and radio sensitizing effects, inhibiting tumor growth and increasing survival rate in Swiss mice inoculated with Ehrlich ascites carcinoma (Devi *et al.*, 1995; Sharad *et al.*, 1996). The administration of methanolic extract of *Ashwagandha* was found to significantly increase the WBC count in mice and reduce leucopenia induced by a sub-lethal dose of gamma-radiation (Kuttan, 1996). Antitumor activities of *Ashwagandha* have been confirmed by Devi *et al.* (1996), Menon *et al.* (1997) and Davis and Kuttan (1998).

Hypoglycemic, diuretic, and hypocholesterolemic effects of *Ashwagandha* root have been assessed in human subjects treated with a powder extract for 30 days. A decrease in blood glucose comparable to that of an oral hypoglycemic drug was observed. Significant increases in urine sodium, urine volume, and decreases in serum cholesterol, triglycerides, and low-density lipoproteins were also have been observed (Andallu and Radhika, 2000). A commercial root extract of *Ashwagandha* used repeatedly over nine days attenuated the development of tolerance to the analgesic effect of morphine and suppressed morphine-withdrawal jumps (Kulkarni and Ninan, 1997). Isolated constituents of *Ashwagandha* (sitoindosides VII-X and withaferin-A) increased cortical muscarinic acetylcholine receptor capacity, partly explaining the cognition-enhancing and memory-improving effects traditionally attributed to *Ashwagandha* (Schliebs *et al.*, 1997).

Toxicity

Ashwagandha appears to be very safe, 50% alcohol extract at a dose of 1000 mg/kg in rats did not shown any toxic effects (Aphale *et al.*, 1998).

3. Brahmi (Bacopa Monniera)

Bacopa monniera (family *Scrophulariaceae*) is also called as *Brahmi* in Indian sub-continent. *Brahmi* has been used by Ayurvedic medical practitioners in India for almost 3000 years as a drug to improve memory and intellect. It is currently recognized as effective herb in the treatment of mental illness and epilepsy (Russo and Borrelli, 2005). In Sri Lanka, *Brahmi* is prescribed for fevers and in the Philippines, it is used as a diuretic (Uphof, 1968). Dhawan and Singh (1996) of Central Drug Research Institute (India) conducted clinical trials in human volunteers and prepared a chemically standardized extract of *Brahmi* for clinical use.

Biologically Active Constituents

The active constituents responsible for the pharmacological effects of *Brahmi* include alkaloids, saponins, and sterols. Many active constituents viz. the alkaloids namely, brahmine (Bose and Bose, 1931) and herpestine (Chopra *et al.*, 1956), saponins namely, d-mannitol and hersaponin (Sastri *et al.*, 1959), and monnierin were isolated in India over 40 years ago. Other active constituents have since been identified, including betulic acid, stigmastarol, beta-sitosterol, as well as numerous bacosides and bacopasaponins. The bacosides A and B are the constituents responsible for *Brahmi's* cognitive effects (Kapoor, 1990; Chakravarty *et al.*, 2003; Hou *et al.*, 2002; Mahato *et al.*, 2000; Chakravarty *et al.*, 2001).

Medicinal uses of *Brahmi*

Two singe-blind open clinical studies have reported memory and learning enhancing effects of chronic *Brahmi* treatment in patients with anxiety neurosis (Singh and Singh, 1980; Singh and Dhawan, 1997; Sharma *et al.*, 1987). *Brahmi* plant (Malhotra and Das, 1959) and its alcoholic extract (Singh and Dhawan, 1982) have shown to enhance learning ability in rats. The alcoholic extract of *Brahmi* has also been found to inhibit the amnesia effects in rats (Dhawan and Singh, 1996). Subsequent studies indicated that the memory enhancing or beneficial effect of this plant (also known as the cognition-facilitating effect) is due to two active saponins; Bacosides A and B, present in the ethanol extract (Singh and Dhawan, 1992; Tripathy *et al*, 1996). It has been suggested that the bacosides induce membrane dephosphorylation, with a concomitant increase in protein and RNA turnover in specific brain areas (Singh *et al.*, 1990). Dave *et al.* (1993) reported that the commercial preparation of *Brahmi* has shown a remarkable nootropic activity. The mechanism of these pharmacological actions of Brahmi in enhancing memory power is not clearly known yet. *Brahmi* has been found to have anxiolytic properties with a distinct advantage over a commonly administered drug lorazepam (LZP) since it does not induce amnesia and has, instead, a memory-promoting action in animals and man (Singh and Dhawan, 1992; Dhawan and Singh, 1996). Shanker and Singh (2000) also reported that *Brahmi* extract possessed an anxiolytic effect.

Standardized methnolic extract of *Brahmi* has potential antidepressant activity (Sairam *et al.*, 2002). Another important use of *Brahmi* in traditional medicine is anticonvulsive action. Shanmugasundaram *et al.* (1991) have reported that the crude water extract of *Brahmi* controls epilepsy in experimental animals. Brahmi counteracts the effects of mental stress, neurosis and revitalizes sensory organs (Sivarajan and Balachandran, 1994), it acts as cardiotonic (Mathur *et al.*, 2002), broncho-vasodilatory (Channa *et al.*, 2003), hepatoprotective (Sumathi and Nongbri, 2008), muscle relaxant (Dar and Channa, 1997), cell stabilizing (Samiulla *et al.*, 2001), antiulcer

(Sairam *et al.*, 2001), antioxidant (Bhattacharya *et al.*, 2000), cognition-promoting (Nathan *et al.*, 2001; Roodernrys *et al.*, 2002) and effective against cigarette smoking induced brain damage (Anbarasi *et al.*, 2005 a; 2005b; 2006). *Brahmi* has been introduced into the market in India and other countries, alone or in association with other phytocomplexes, and utilized in the treatment of memory and attention disorders (Shukla *et al.*, 1987).

Toxicity

Aqueous crude extract or crude alcoholic extract of *Bacopa Monniera* has been found to be well tolerated and without any problematic reaction or side effects in regulatory pharmacological and toxicological studies (Martis *et al.*, 1992; Asthana *et al.* 1996; Russo and Borrelli, 2005). It was found that single oral doses of 20-200 mg or 100 and 200 mg once daily for 4 weeks were safe and did not produce any reaction or side effect. Even repeated doses of bacosides A and B were tolerated well with no side effects experienced by any subject (Singh and Dhawan, 1997).

4. Vidharikand (*Pueraria tuberosa*)

Pueraria tuberosa is one among more than 100 species of the genus *Pueraria* which is native to northern India, Pakistan and Nepal. *Pueraria* is one of the most versatile and promising of all the plants found in Asia. In Ayurveda, 'Vidari' is botanically equated to *Pueraria tuberosa* DC of the *Fabaceae* family and it is also known as *Vidharikand* and *Indian kudzu* (Rathore and Shekhawat, 2009).

Biologically active components

The major active components of *Pueraria tuberosa* include coumarin, alkaloids, volatile oils, anthocyanin, lupinoside (Palpu and Dhan, 2003) and isoflavonoids namely puerarin (Prasad *et al.*, 1984), genistin, genistein, daidzein and dadzin (Debra *et al.*, 2006; Kim *et al.*, 2003; Pandey *et al.*, 2007) and tuberosin (Joshi and Kamat, 1973).

Medicinal uses of *Pueraria tuberosa*

Pueraria tuberosa is used in traditional medicine as a fertility control agent, and as cardiotonic, diuretic and galactogogue. In Ayurveda, the flowers are used as cooling agent and as aphrodisiac, while roots act as a demulcent and refrigerant in fevers. It is also used to cure leprosy, diseases of blood and urinary discharges. It is employed as an emetic and also believed to be a lactagogue (Kritikar and Basu, 1933). In folk medicine the root tuber is applied for blood purification and to improve sperm production. The shade dried root powder controls overgrowth in stomach. The consumption of raw root for one month leads to sterilization in women (Venkata, 2006).

Vidharikand has been reported to contain lupinoside which acts as anti diabetic agent by improving insulin activity and decreasing palmitate thus inhibiting the glucose uptake (Palpu and Dhan, 2003). Recently, it has been demonstrated that puerarin reduced the atherogenic properties of dietary cholesterol in rats. It was observed that the decrease in total cholesterol was 25% when puerarin was given along with high cholesterol diet compared to animals fed with only high cholesterol diet (Yan *et al.*, 2006). *Indian kudzu* has been reported to show potent antioxidant activity against lipid peroxidation in egg yolk by chelating with iron and trapping of free radical but less than BHT because of its glycosidic nature and because it contains less free hydroxyl group (Pandey *et al.*, 2007). Tanwar *et al.* (2008) reported that *Pueraria tuberosa* possesed potent hypolipidemic activity as elucidated by concurrent feeding of *Pueraria tuberosa* extract to Wister rats. The cholesterol lowering effects of *Vidharikand* are mainly due to isoflavonoids present in it. *Micrococcus luteus, Bacillus cereus, Staphylococcus aureus, E. coli, Pseudomonas aeruginosa, Solmonella typhimurium* and *Candida albicans* have been inhibited by organic solvent extracts of *Vidharikand* (Ratnam and Raju, 2009). The flavonoids (puerarin, daidzein and genistein) induced lymphocyte proliferation in mouse spleen and also showed immunomodulatory effect through phagocytic activity (Lacaille-Dubois, 2005). Immunomodulatory activity of *Pueraria* has also been reported by Oh *et al.* (2000) and Venkat (2009). Antimutagenic activity of *Pueraria* has been roprted by Miyazawa *et al.* (2001). Recently, Rao *et al.* (2008) reported that aquous and alcoholic extracts of *Pueraria tuberosa* have shown Nootropic effects and they also reported that flavonoids are responsible for this activity.

Toxicological study

Toxicity studies with the butanol extract of *Pueraria tuberosa* in rats at a dose of 150 mg/ kg for 24 days have shown that it is generally safe (Shukla, 1995).

5. Triphala

Triphala is a combination of three active herb preparations viz. *Amalaki* (*Emblica officinalis*), *Haritaki* (*Terminalia chebula*) and *Bibhitaki*

(*Terminalia belerica*) (Frawley, 2003). These herbs have a well esteemed place in Ayurveda and are very widely used in treating various diseases. All these herbs are mixed in equal proportion by weight to obtain *Triphala*. It nourishes the nervous system, blood and muscles and increases digestion. *Triphala* is also widely taken for all eye diseases including the treatment of conjunctivitis, progressive myopia and cataracts.

a) Amalaki (Emblica officinalis)

Amalika (*Emblica officinalis*) also called as *Amla* is a fruit which is counted as among the highest known land sources of vitamin C, having 20 times that of an orange and 160-fold that those of apple.

Biologically Active Constituents

The fruit *Amalika* contains cytokinine-like substances identified as zeatin, zeatin riboside, zeatin nucleotide and phyllembin. The leaves contain gallic acid, besides ascorbic and music acid. The bark contains tannin identified as mixed type of proanthocyanidin. *Amla* also contains superoxide dismutase which exhibits antisenescent (anti-aging) activity (Anonymous, 2006).

Medicinal uses of *Amalika*

Emblica officinalis is highly valued in traditional Indian medicine (Scartezzini *et al.*, 2006). In Unani medicine the dried fruits of *Amla* are used to treat haemorrhage, diarrhoea and dysentery (Parrotta, 2001). Several other medicinal effects of *Amla* i.e. antioxidant activity (Ghosal *et al.*, 1996), antimicrobial properties (Ahmed *et al.*, 1998), adaptogenic (Rege *et al.*, 1999), antioxidant (Bhattacharya *et al.*, 1999), hepatoprotective (Jeena *et al.*, 1999), antitumor (Jose *et al.*, 2001), antiulcerogenic (Sairam *et al.*, 2002), anti-inflammatory (Sharma *et al.*, 2003), analgestic and antipyretic (Perianayagam *et al.*, 2005), hypocholestrolemic (Kim *et al.*, 2005) and diuretic (Anon., 2006),) have also been reported.

b) Haritaki (Terminalia chebula)

Haritaki or *Harada* is widely distributed throughout India, Burma and Sri Lanka. It is commonly known as Black Myrobalans and has traditionally been used in the treatment of asthma, sore throat, vomiting, hiccough, diarrhoea, bleeding piles, gout, and heart and bladder diseases. *Harada* contains B-complex vitamin that protects the lung from any damage. The dried ripe fruit of *Terminalia chebula* Retz. (*Combretaceae*) is used extensively in Ayurveda (Khan and Jain, 2009).

Biologically Active Constituents

The fruit is a rich source of tannins, amino acids, fructose, succinic acid and β-sitosterol. It contains shikimic, gallic, triacontanoic and palmitic acids, b-sitosterol, daucosterol, triethyl ester of chebulic acid and ethyl ester of gallic acid, terchebulin, punicalagin and teaflavin A, chebupentol, arjungenin, terminoic acid and arjunolic acid. Antioxidant constituents of the plant include phloroglucinol and pyrogallol and ferulic, vanillic, *p*-coumaric and caffeic acids (Kumar and Bhatnagar, 2007; Kumar *et al.*, 2009; Kumar *et al.*, 2010; Naik *et al.*, 2010).

Medicinal uses of *Haritaki*

Terminalia chebula along with *Asparagus racemosus* has been reported to protect gastric mucosa against pentagastrin and carbachol induced ulcers, by significantly reducing both severity of ulceration and ulcer index (Dahanukar *et. al*, 1983). A herbal combination of *Terminalia chebula* and *Glycyrrhiza glabra* (part of Abana extract formulation) have been reported to reduce cysteamine-induced duodenal ulcers in rats by increasing the b-glucuronidase activity in the Brunner's glands (Nadar and Pillai, 1989). *Terminalia chebula* has also been reported to protect epithelial cells against the cytopathic effects of Influenza-A virus and indomethacin-induced gastric ulcers in rats (Badmaev and Nowakowski, 2000). Khayyal *et al.* (2001) has reported the antiulcerogenic effect of *Terminalia chebula.* Combination of *Terminalia chebula* and chebulinic acid has been found to exhibit anticancer activity (Saleem, 2002).

c) Bibhitaki (Terminalia bellerica)

Bibhitaki (*Terminalia bellerica*), also called as *Behada* is a perennial herb mainly distributed in the tropical regions and commonly found in South-East Asia, including Thailand. This formulation, rich in antioxidants, is frequently used in Ayurvedic medicine to treat many diseases such as anemia, jaundice, constipation, asthma, fever and chronic ulcers. It possesses anti-mucal, laxative, astringent, digestive and anti-spasmodic properties. It is also used as a tonic and an expectorant and helps deal with allergies (Gilani *et al.*, 2008; Khan and Gilani, 2008).

Biologically Active Constituents

Terminalia bellerica is the richest source of vitamin A and b-sitosterol, gallic and ellagic acids, ethyl gallate, galloyl glucose, chebulagic acid and a cardiac glycoside, bellaricanin (Yadava and Rathore, 2001; Gilani *et al.*, 2008; Khan and Gilani, 2008; Khan and Gilani, 2010).

Medicinal uses of *Triphala*

The three herbs that make up *Triphala* have laxative actions. *Triphala* is not habit forming and is not known to have any side effects over long term use (Tierra, 1992). It works by stimulating the mucosa of the gastric-intestinal tract, improving and balancing digestion and further aids in digestion by eliminating toxins from the intestinal tract. People suffering from constipation will benefit from regular use of *Triphala* and it functions as metabolic regulator (Frawley, 2003). Alcoholic and water extracts of *Triphala* have successfully cured castor-oil induced diarrhea (Biradar *et al.*, 2007). *Triphala* has also showed hypolipidemic (Saravanan *et al.*, 2007) and cytotoxic and apoptotic activities (Kaur *et al.*, 2005).

Triphala has shown to be effective against cancer cells and *Helicobacter pylori* (Malckzadeh et. al, 2001; Rani and Khullar, 2004). It also acts as immunomodulator (Srikumar *et al.*, 2005). Effect of *Triphala* on experimental gouty arthritis in mice was evaluated by Sabina and Rasool (2008). The study compared *Triphala* to Indomethacin, an anti-inflammatory. Compared with the control group, a clear anti-inflammatory effect against Gouty Arthritis was observed in mice orally administered by *Triphala*.

Gamma-Radiation induced strand break formation in plasmid DNA (pBR322) was effectively inhibited by aqueous extract of *Triphala* and its constituents *Emblica officinalis, Terminalia chebula* and *Terminalia belerica*. These herbs also inhibited radiation induced lipid peroxidation in rat liver microsomes effectively. *Emblica officinalis* showed greater efficiency in lipid peroxidation and plasmid DNA assay, while *Terminalia chebula* had shown greater radical scavenging activity (Naik *et al.*, 2005). The cytotoxic effects of *Triphala* on a cell line of human breast cancer cells and a thymic lyphoma transplanted from a mouse were studied by Sandhya *et al.* (2006b). The study found that the viability of the cell lines decreased when the concentrations of *Triphala* was increased. Further, it was discovered that normal breast cells, along with other normal cells weren't significantly affected which strengthens the role of *Triphala* as a potential cancer inhibiter.

6. Arjuna (Arjuna terminalia)

The herb, *Arjuna* (*Arjuna terminalia*) which belongs to the family *Combertaceae* is native to India and has been known from Vedic period for its various therapeutic values (Nadkarni, 2000). It is also called by the names *Arjuna myrobalan* and *Arjun* (Kiritkar and Basu, 1935).

Biologically Active Components

Biologically active constituents of *Arjuna* include tannins containing catechin, gallocatechin, epicatechin, epigallocatechin (Rastogi and Mehrotra, 1993c; Kandil and Nassar, 1998), triterpinoids (Tripathi *et al.*, 1992; Patnaik *et al.*, 2007), phenolics (Anjaneyulu and Prasad, 1982a), saponins namely arjunic acid (Row *et al.*, 1970b), arjunolic acid (Rastogi and Mehrotra, 1993a), arjungenin, arjunglycosides I and II (Rastogi and Mehrotra, 1993b), arjunglucoside III (Rastogi and Mehrotra, 1993c), arjunglucoside IV, V (Wang *et al.*, 2010a), *Arjuna*sides A-E (Wang *et al.*, 2010b), termonic acid (Ahmad *et al.*, 1983), flavonoids (arjunon, arjunolone, luteolin) (Ghosal *et al.*, 1971; Sharma *et al.*, 1982), gallic acid, ellagic acid, oligomeric proanthocyanidins (OPCs), phytosterols (Row *et al.*, 1970b; Miller, 1998), β-sitosterol (Anjaneyulu and Prasad, 1982b), calcium, magnesium, zinc and copper (Kiritkar and Basu, 1935).

Medicinal Uses

In the Indian System of Medicine, the bark of *Arjuna* is used as astringent, cooling, aphrodisiac, cardiotonic, tonic, in fractures, ulcers, spermatorrhoea, leucorrhoea, diabetes, cough, tumour, excessive perspiration, asthma, inflammation and skin disorders etc (Warrier *et al.*, 1994; Dwivedi and Udupa, 1989).

Arjunolic acid from the bark of *Arjuna*, has been shown to provide significant cardiac protection in isoproterenol induced myocardial necrosis in rats (Sumitra *et al.*, 2001). *Arjuna* showed significant DPPH radical scavenging activity (similar to L-ascorbic acid) in wistar rats which can be used in the therapy of cardiovascular disease by exerting its beneficial effects via antioxidant activity (Munasinghe *et al.*, 2001). Karthikeyan *et al.* (2003) reported that chronic administration (6.75 mg/kg) of the alcoholic extract of *Arjuna* augments endogenous antioxidant compounds of the rat heart and also prevents the myocardium from isoproterenol induced myocardial ischemic reperfusion injury. Intravenous administration of 70% alcoholic extract of *Arjuna* (5 to15 mg/kg) produced dose-dependent hypotension in anaesthetized dogs which lends support for the claims of its traditional usage in cardiovascular disorders (Nammi *et al.*, 2003). The cardio protective effects of *Arjuna* have been also confirmed in animal models by Ramesh *et al.* (2004), Pawar and Bhutani (2005), Manna *et al.* (2006), Sinha *et al.* (2008), Singh *et al.* (2008) and Kumar *et al.* (2009).

The anti-inflammatory activity of *Arjuna* has been demonstrated by Tripathi *et al.* (2004), Halder *et al.* (2009). The anti-tumor effects of *Arjuna* have been reported by Sivalokasnathan *et al.* (2005), Kuo *et al.* (2005a), Kuo *et al.* (2005b), Saxena *et al.* (2007), Reddy *et al.* (2008), Verma and Vinayak (2009). The anti-ulcer effect of methanol extract of *T.arjuna* (TA) has been reported by Devi *et al.* (2007a), Devi *et al.* (2007b) and Devi *et al.* (2008). The wound healing activity of *Arjuna* has been documented by Mukherjee *et al.* (2003), Rane and Mengi (2003) and Chaudhari and Mengi (2006).

The cytoprotective role of Arjunolic acid against arsenic [Sodium arsenate; 1 mM]-induced cytotoxicity in isolated murine hepatocytes was found to be comparable to that of a known antioxidant, vitamin C suggesting that Arjunolic acid protects arsenic-induced cytotoxicity in murine hepatocytes (Manna *et al.*, 2007b). Earlier the hepatoprotective effects of *Arjuna* were reported by Manna *et al.* (2006), Sivalokanathan *et al.* (2006a) and Sivalokanathan *et al.* (2006b). The Antimicrobial activities of *Arjuna* extract on *E. coli*, *Pseudomonas aeruginosa*, *Bacillus subtilis*, *Staphylococcus aureus* and *Staphylococcus epidermidis* (Shinde *et al.*, 2009), *Klebsiella pneumoniae*, *Candida albicans*, *Streptococcus mutans*, *Enterococcus faecalis*, *Streptococcus bovis*, *Salmonella typhimurium*, (Khan *et al.*, 2009), *Salmonella typhi* (Rani and Khullar, 2004), *Klebsiella aerogenes*, *Proteus vulgaris* and *Pseudomonas aerogenes* (Perumal Samy *et al.*, 1998) and *Neisseria gonorrhoeae* (Petit *et al.*, 1996) have also been reported. The antioxidant and free radical scavenging capacities of *Arjuna* were reported by Chen *et al.* (2004) and Sun *et al.* (2008). Anti-diabetic activity (Raghavan and Kumari, 2006), antiviral activity (Cheng *et al.*, 2002), antiathersclerotic activity (Shailaa *et al.*, 1998), immunomodulatory activity (Halder *et al.*, 2009) and reproductive activity (Manna *et al.*, 2008) of *Arjuna* have also been documented.

Clinical trials have shown that alcoholic decoction of bark was beneficial in stable cases of ischemic heart disease (Anand, 1994).

Beneficial effects of *Arjuna* in modifying various coronary risk factors have been reported by Gupta *et al.* (2001) and Malik *et al.* (2009) during their clinical studies. Some clinical studies have also supported the antianginal activity of *Arjuna* (Dwivedi *et al.*, 2005 and Bharani *et al.*, 2002).

Aloe Vera

Aloe is a member of the tree, lily family (*Aloaceae*), and known as *Aloe barbadensis.* It is related to other members of the Lily family such as the onion, garlic and turnip. Throughout recorded history, *Aloe Vera* has been given a high ranking as an all-purpose herbal plant (Lee, 2006). More than 360 species of aloe are known worldwide and are classified into two groups according to their use: for production of extract as crude drugs and for production of gel as health functioning foods. Among the different varieties of *Aloe Vera* only a few species have been used commercially, others are merely of decorative value. The species for the production of crude drugs are *Aloe Vera* L. (*A. Barbadensis* Mill.), *A. ferox* Mill., *A. perryi* Baker, etc., and species for use in health functional foods are *Aloe Vera* L., *A. arborescens* Mill., *A. saponaria,* etc. Only two species are grown today commercially, with *Aloe barbadensis* Miller and *Aloe arborescens* being the most popular (Chauhan, *et al.,* 2007). *Aloe Vera* gel consists of more than 99 percent water. The remaining less than 1 percent of *Aloe Vera* gel consists of 200 different biologically active constituents which include a diverse mixture of antibiotics, pain inhibitors, cell growth stimulators, inflammation fighters, burns healers, capillary dilators, vasoconstrictor inhibitors and moisturizers. Thus Aloe possesses almost a complete drug store of medically useful ingredients (Waller, *et al.,* 1978).

Health Effects

The first somewhat detailed description of the pharmacological effects of Aloe is recorded in "The Greek Herbal", written by Dioscorides in the first century AD. According to such literature, Aloe has multiple pharmacological effects on healing wounds, burns and frostbite, constipation, insomnia, stomach disease, pain, hemorrhoids, itching, headache, hair loss, gum disease, kidney disease, blisters, sunburn and more (Lee, 2006). Reports credit *Aloe Vera* with antitumor (Pecere, *et al.,* 2000), antidiabetic (Rajasekaran *et al.,* 2006; Tanaka *et al.,* 2006; Kim *et al.,* 2009), healing wounds and burns (Davis, *et al.,* 1987; Schmidt and Greenspoon, 1991; Heggers and Winters, 1993), treatment of gastric ulcers (Teradaira, *et al.,* 1993; Suvitayavat *et al.,*

2004; Yusuf *et al.*, 2004; Eamlamnam *et al.*, 2006; Metowogo *et al.*, 2008) and peptic ulcers (Blitz *et al.*, 1963), possessing hypolipidemic (Kim *et al.*, 2009) and immunomodulatory properties (Womble and Helderman, 1988; Hart *et al.*, 1988; Winters, 1993; Tizard *et al.*, 1991; Im *et al.*, 2005; Madan *et al.*, 2008). Recently a research conducted by Sinnott *et al.* (2007) proved that Aloe polysaccharides (approx. MW 50,000 to 1,000,000) can be utilized better by human colonic bacteria.

Opportunities for Incorporation of Herbs into Dairy Foods

The increasing consumer demand for health foods/functional foods has prompted International health organizations and Governmental agencies to develop guidelines for use of herbal medicine. Also there is a considerable amount of research being carried out on various aspects of herbal plants to understand more about their properties and characterize their active components. *Ayurveda,* the Indian traditional medical science has found several ways in which the medicinal benefits of herbs could be conveyed via certain foods as carriers. It has also suggested several forms viz. paste, tray dried herbs, freeze dried herbs and solvent extracted herbal ingredients in which these herbs can be incorporated into foods. Milk and ghee (butter oil) are the most important among such carriers of herbal nutraceuticals. Milk fat, particularly Ghee has the characteristics to absorb all the medicinal properties of the herbs with which it is fortified, without losing its own qualities. Exploiting this concept several medicated ghee preparations have been developed and about 55-60 medicated ghee types are reported in Ayurvedic literature and they are used in the treatment of various diseases (Pandya and Kanwajia, 2002). Herbal ghee incorporating functional attributes of *Arjuna* has been developed for providing beneficial effects against cardiovascular diseases and the product is also nearly four times more stable to oxidative deterioration as compared to normal ghee (Rajanikant, 2005). Indian herbs such as turmeric (*Curcuma longa L.*), coriander (*Coriandrum sativum L.*), curry leaf (*Murraya koenigii L.*), spinach (*Spinacia oleracea*) and amla (*Emblica officinalis*), have been incorporated separately at 10% level into *sandesh* (a heat-desiccated product of coagulated milk protein mass or chhana, which is analogous to cottage cheese) (Bandyopadhyay *et al.*, 2007) to get the beneficial effects of these herbs upon *sandesh* consumption. Polyphenol-rich extracts of green tea (catechins) are being incorporated into fruit-flavoured milk drinks, as well as other products like chewing gum and sweet paste biscuits (Vasala, 2001; Pandey, 2001; Ravindran and Kalluparackal, 2001;

O'Connella and Fox, 2001; Bender and Bender, 2005). The addition of catechins, at 100-2000 mg/kg, to bifidobacteria-containing yoghurt has been reported to improve the survival of the bifidobacteria during storage (Akahoshi and Takahashi, 1996).

Herbal ingredients can also be incorporated into high fat products like cream, butter, table spreads besides other products like cheese and ice cream. However, the incorporation of these herbs into dairy products may have certain undesirable effects on their sensory properties like colour and appearance, flavour, body and texture which in turn could affect their overall acceptability. Any food product whether it is nutritional or functional, will be accepted by the consumers only when its aesthetic quality is good. Presently, the herbal ghee being marketed in the global market are mostly sold as medicine (medicinal ghee), and generally possess typical flavour, bitter or pungent taste and dark colour. Such therapeutic preparations are therefore not acceptable for routine use. Incorporation of these nutraceuticals into dairy food systems may therefore require use of technological modifications/alterations so that the sensory quality of the final product remains unaltered. Furthermore, very limited information is available for ascertaining the residual levels of these functional components in herbal ghee preparations. There also exists an ample scope to establish mechanisms through which these components impart health enhancing properties like anticarcinogenicity, antiatherogenicity etc. in human subjects. In many medicinal Ghee formulations available in the market, it is stated that herbs that are used have more beneficial effect when formulated along with ghee. The reasons are yet to be explored and established. Isolation of herbal components in concentrated form may lead us to explore ways to formulate several other fat based functional foods. Interactions between herbal and food constituents leading to changes in sensory, textural and other quality parameters of the foods as also their effect on human health have to be studied thoroughly. Accordingly, the scientific community must apply modern technologies to ensure acceptability, efficacy and safety of such herbal nutraceuticals based food preparations and their healthful aspects.

References

Abdullahi, A. L., Agho, M. O., Amos, S., Gamaniel, K. S. and Wambebe, C. (2001). Antidiarrhoeal activity of the aqueous extract of *Terminalia avicennoides* roots. *Phytotherapy Research*, **15**: 431-434.

Abed, K. F. (2007). Antimicrobial activity of essential oils of some medicinal plants from Saudi Arabia. *Saudi Journal of Biological Sciences*, **14(1)**: 53-60.

Abou-Douh, A. M. (2002). New withanolides and other constituents from the fruit of *Withania somnifera*. *Archiv der Pharmazie (Weinheim)*, **335**(6): 267-276.

Acevedo-Duncan, M., Russell, C., Patel, S. and Patel, R. (2004). Aloe-emodin modulates PKC isozymes, inhibits proliferation, and induces apoptosis in U-373MG glioma cells. *International Immunopharmacology*, **4**: 1775-1784.

Adeleye, I. A., Ogunniyi, A. A. and Omonigbehin, E. A. (2003). Antimicrobial activity of some local herbs on common skin pathogens. *Bioscience Research Communications*, **15**(3): 231-236.

Ahmad, M. U., Mullah, K. B., Norin, T. and Ulla, J. K. (1983). Terminoic acid, a new trihydroxytriterpene carboxylic acid from bark of *Terminalia arjuna*. *Indian Journal of Chemistry*, **22**: 738-740.

Ahmad, S., Jain, P. C. (1991). Chemical examination of Shatavari (*Asparagus racemosus*). *Bulletin of Medical and Ethnobotanical Research*, **12**: 157-160.

Ahmed, N., Mehmood, Z. and Mohammad, F. (1998). Screening of some Indian medicinal plants for their antimicrobial properties. *Journal of Etanopharmacology*, **62**: 183-190.

Akahoshi, R. and Takahashi, Y. (1996). Yoghurt containing bifidobacterium and process for producing the same. Patent: WO/1996/037113.

Akiyama, H., Fujii, K., Yamasaki, O., Oono, T. and Iwatsuki, K. (2001). Antibacterial action of several tannins agaisns *Staphylococcus aureus*. *Journal of Antimicrobial Chemotherapy*, **48**: 487-491.

Anand, V. (1994). Antianginal and cardio protective effects of *Terminalia arjuna*. *Journal of Association of Physicians of India*, **42**: 757-758.

Anbalagan, K., Sadique, J. (1981). Influence of an Indian medicine (Ashwagandha) on acute phase reactants in inflammation. *Indian Journal of Experimental Biology*, **19**: 245-249.

Anbarasi, K., Vani, G. and Devi, C. S. (2005b). Protective effect of bacoside A on cigarette smoking-induced brain mitochondrial dysfunction in rats. *Journal of Environmental Pathology, Toxicology and Oncology*, **24**(3): 225-34.

Anbarasi, K., Vani, G., Balakrishna, K. and Devi, C. S. (2006). Effect of bacoside A on brain antioxidant status in cigarette smoke exposed rats. *Life Sciences*, **78**(12): 1378-84.

Anbarasi, K., Vani, G., Balakrishna, K. and Shyamala Devi, C. S. (2005a). Effect of bacoside A on membrane-bound ATPases in the brain of rats exposed to cigarette smoke. *Journal of Biochemical and Molecular Toxicology*, **19**(1): 59-65.

Aniszewski, T. (1994). The biological basis of quinolizidine alkaloids. *Science of Legumes*, **1**: 1-24.

Aniszewski, T. (2007). Applications in "Alkaloids-Secrets of Life: Alkaloid Chemistry, Biological Significance, Applications and Ecological Role". pp: 192-194. Elsevier, Amsterdam, The Netherlands.

Anjaneyulu, A. S. R. and Prasad, A. V. R. (1982a). Chemical examination of the roots of Terminalia *Arjuna* (Roxb) Wight and Arnot. *Phytochemistry,* **21**: 2057-2060.

Anjaneyulu, A. S. R. and Rama-Prasad, A. V. (1982b). Chemical examination of the roots of *Terminalia arjuna*-characterization of two new triterpenoid glycoside. *Indian Journal of Chemistry,* **21**: 530-533.

Anonymous (2006). *Emblica officinalis.* http://www.ayurvediccure.com/amla.htm.

Apers, S., Baronikova, S., Sindambiwe, J. B., Witvrouw, M., De Clercq, E., Vanden Berghe, D., Van Marck, E., Vlietinck, A. and Pieters, L. (2000). Antiviral, haemolytic and molluscicidal activities of triterpenoid saponins from Maesa lanceolata: establishment of structure-activity relationships. *Planta Medica,* **67:** 528-532.

Aphale, A. A., Chhibba, A. D., Kumbhakarna, N. R., Mateenuddin, M., Dahat, S. H. (1998). Subacute toxicity study of the combination of ginseng (Panax ginseng) and ashwagandha (Withania somnifera) in rats: a safety assessment. *Indian Journal of Physiology and Pharmacology,* **42**(2): 299-302.

Asmari, S., Zafar, R., Ahmad, S. (2004). Production of sarsasapogenin from tissue culture of *Asparagus racemosus* and its quantification by HPTLC. *Iranian Journal of Pharmaceutical Research Supplement,* **2**: 66-67.

Asthana, O. P., Srivastava, J. S., Ghatak, A., Gaur, S. P. S. and Dhawan, B. N. (1996). Safety and tolerability of bacosides A and B in healthy human volunteers. *Indian Journal of Pharmacology,* **28**: 37.

Badmaev, V. and Nowakowski, M. (2000). Protection of epithelial cells against Influenza-A virus by a plant derived biological response modifier Ledretan-96. *Phytothererapy Research,* **4**: 245-9.

Bandyopadhyay, M., Chakraborty , R. And Raychaudhuri, U., (2007). Incorporation of herbs into sandesh, an Indian sweet dairy product, as a source of natural antioxidants. *International Journal of Dairy Technology,* **60**(3): 228-233.

Bansod, S. and Rai, M. (2008). Antifungal Activity of Essential Oils from Indian Medicinal Plants Against Human Pathogenic *Aspergillus fumigatus* and *A. niger. World Journal of Medical Sciences,* **3**(2): 81-88.

Barr, I. G., Sjolander, A. and Cox, J. C. (1998). ISCOMs and other saponins based adjuvants (Review). *Advanced Drug Delivery Reviews,* **32:** 247-271.

Becker, L. E., Butterfield, W. J. H., McGehee Harvey, A., Heptinstall, R. H. and Lewis, T. (1986). *International Dictionary of Medicine and Biology*. New York: John Wiley & Sons.

Begum, V. H. and Sadique, J. (1988). Long term effect of herbal drug Withania somnifera on adjuvant induced arthritis in rats. *Indian Journal of Experimental Biology*, **26**: 877-882.

Bender, A. E. and Bender, D. A. (2005). A Dictionary of Food and Nutrition. Oxford University Press, Oxford New York.

Bharani, A., Ganguli, A., Mathur, L. K., Jamra, Y. and Raman, P. G. (2002). Efficacy of *Terminalia arjuna* in chronic stable angina: a double-blind, placebo-controlled, crossover study comparing *Terminalia arjuna* with isosorbide mononitrate. *Indian Heart Journal*, **54**: 170-175.

Bhattacharya, A., Chatterjee, A., Ghosal, S. and Bhattacharya, S. K. (1999). Antioxidant activity of active tannoid principles of *Emblica officinalis* (amla). *Indian Journal of Experimental Biology*, **37**: 676-680.

Bhattacharya, A., Murugandam, A. V., Kumar, V., Bhattacharya, S. K. (2002). Effect of polyherbal formulation, EuMil, on neurochemical perturbations induced by chronic stress. *Indian Journal of Experimental Biology*, **40**: 161-163.

Bhattacharya, S. K., Satyan, K. S. and Ghosal, S. (1997). Antioxidant activity of glycowithanolides from *Withania somnifera*. *Indian Journal of Experimental Biology*, **35**: 236-239.

Biradar, Y. S., Singh, R., Sharma, K., Dhalwal, K., Bodhankar, S. L. and Khandelwal, K. R. (2007). Evaluation of anti-diarrhoeal property and acute toxicity of *Triphala* Mashi, an Ayurvedic formulation. *Journal of Herbal Pharmacotherapy*, 7(3-4): 203-12.

Blitz, J. J., Smith, J. W. and Gerard, J. R. (1963). *Aloe Vera* Gel in Peptic Ulcer Therapy: Preliminary Report. *Journal of the American Osteopathic Association*, **62**: 731-735.

Bone, K. (1996). Clinical Applications of Ayurvedic and Chinese Herbs. Monographs for the Western Herbal Practitioner. Australia: Phytotherapy Press, pp: 137-141.

Bors, W. and Michel, C. (2002). Chemistry of the antioxidant effect of polyphenols. *Annals of the New York Academy of Sciences*, **957:** 57-69.

Bose, K. C. and Bose, N. K. (1931). Observations on the actions and uses of Herpestis monniera. *Journal of The Indian Medical Association*, **1**: 60.

Brevoort, P. (1996). The U.S. Botanical Market. An Overview. *Herbal Gram*, 36: 49-57.

Catherine Clinton, N. D. (2009). Plant tannins: A novel approach to the treatment of ulcerative colitis. *Natural Medicine Journal*, **1**(3): 1-4.

Chakravarty, A. K., Garai, S., Masuda, K., Nakane, T. and Kawahara, N. (2003). Bacopasides III-V: three new triterpenoid glycosides from *Bacopa monniera*. *Chemical and Pharmaceutical Bulletin (Tokyo)*, **51**: 215-217.

Chakravarty, A. K., Sarkar, T., Masuda, K., Shiojima, K., Nakane, T. and Kawahara, N. (2001). Bacopaside I and II: two pseudojujubogenin glycosides from *Bacopa monniera*. *Phytochemistry*, **58**: 553-556.

Channa, S., Dar, A., Yaqoob, M., Anjum, S., Sultani, Z. and Atta-ur-Rahman (2003). Broncho-vasodilatory activity of fractions and pure constituents isolated from *Bacopa monniera*. *Journal of Ethnopharmacology*, **86**(1): 27-35.

Chao, P. D. L., Hsiu, S. L. and Hou, Y. C. (2002). Flavonoids in Herbs: Biological Fates and Potential Interactions with Xenobiotics. *Journal of Food and Drug Analysis*, **10**(4): 219-228.

Chaudhari, M. and Mengi, S. (2006). Evaluation of phytoconstituents of *Terminalia arjuna* for wound healing activity in rats. *Phytotherapy Research*, **20**: 799-805.

Chauhan, O. P., Raju, P. S., Khanum, F. and Bawa, A. S. (2007). *Aloe Vera* - therapeutic and food applications. *Indian Food Industry*, **26**(3): 43-51.

Chen, C. H., Liu, T. Z., Kuo, T. C., Lu, F. J., Chen, Y. C., Chang-Chien, Y. W. and Lin, C. C. (2004). Casuarinin protects cultured MDCK cells from hydrogen peroxide-induced oxidative stress and DNA oxidative damage. *Planta Medica*, **70**: 1022-1026.

Chen, H. C., Hsieh, W. T., Chang, W. C. and Chung, J. G. (2004). Aloe-emodin induced in vitro G2/M arrest of cell cycle in human promyelocytic leukemia HL-60 cells. *Food and Chemical Toxicology*, **42**: 1251-1257.

Chen, Y., Yang, L. and Lee, T. J. (2000). Roxyllin A inhibition of lipopolysaccharide-induced iNOS and COX-2 gene expression via suppression of nuclear factor-kappa B activation. *Biochemical Pharmacology*, **59**: 1445-1457.

Cheng, H. Y., Lin, C. C. and Lin, T. C. (2002). Antiherpes simplex virus type 2 activity of casuarinin from the bark of *Terminalia arjuna* Linn. *Antiviral Research*, **55**: 447-455.

Chopra, R. N., Nayar, L., Chopra, I. C. (1956). Glossary of Indian Medicinal Plants, vol. 32. Council of Scientific and Industrial Research, New Delhi.

Choudhary, M. I., Dur-e-Shahwar, Parveen, Z., *et al*. (1995). Antifungal steroidal lactones from *Withania coagulance*. *Phytochemistry*, **40**: 1243-1246.

Craig, W. J. (1999). Health-promoting properties of common herbs. *American Journal of Nutrition*, **70**(suppl): 491S-9S.

Dahanukar, S. A., Date, S. G. and Karandikar, S. M. (1983). Cryoprotective effect of *Terminalia chebula* and *Asparagus racemosus* on gastric mucosa. *Indian drugs,* **21**: 442-445.

Dalvi, S. S., Nadkarni, P. M. and Gupta, K. C. (1990). Effect of *Asparagus racemosus* (Shatavari) on gastric emptying time in normal healthy volunteers. *Journal of Postgraduate Medicine,* **36**: 91-94.

Dan, B. (2003). Chinese Herbal Medicine: Materia Medica (3rd edn). Eastland Press, Vista, CA, USA.

Dar, A. and Channa, S. (1997). Relaxant effect of ethanolic extract of *Bacopa monniera* on trachea, pulmonary artery and arota from rabbit and guinea pig. *Phytotherapy Research,* **11**: 323-325.

Dave, U. P., Chauvan, V. and Dalvi, J. (1993). Evaluation of BR-16 A (Mentat) in cognitive and behavioural dysfunction of mentally retarded children: a placebo-controlled study. *Indian Journal of Pediatrics,* **60**: 423-428.

Davis, L. and Kuttan G. (1998). Suppressive effect of cyclophosphamide-induced toxicity by *Withania somnifera* extract in mice. *Journal of Ethnopharmacology,* **62:** 209-214.

Davis, R. H., Kabbani, J. M. and Maro, N. P. (1987). *Aloe Vera* and wound healing. *Journal of the American Podiatric Medical Association,* **77**(4): 165-9.

de Groot, H. and Rauen, H. (1998). Tissue injury by reactive oxygen species and the protective effects of flavonoids. *Fundamental and Clinical Pharmacology,* **12**: 249-255.

Debra, L., Bemis, J. L., Capodice, J. E., Costello, G. C., Vorys, A. and Dev, S. (2006). A selection of prime ayurvedic plant drugs: ancient-modern concordance. Anamaya Pub, New Delhi, pp: 363.

Delmas, F., Di Giorgio, C., Elias, R., Gasquet, M., Azas, N., Mshvildadze, V., Dekanosidze, G., Kemertelidze, E. and Timon-David, P. (2000). Antileishmanial activity of three saponins isolated from ivy, alpha-hederin, beta-hederin and hederacolchiside A(1), as compared with their action on mammalian cells cultured in vitro. *Planta Medica,* **66**: 343-347.

Devi, P. U., Akagi, K. and Ostapenko, V. (1996). Withaferin A: a new radiosensitizer from the Indian medicinal plant *Withania somnifera. International Journal of Radiation Biology,* **69:** 193-197.

Devi, P. U., Sharada, A. C., Solomon, F. E. (1995). *In vivo* growth inhibitory and radiosensitizing effects of withaferin A on mouse Ehrlich ascites carcinoma. *Cancer Letters,* **95:** 189-193.

Devi, R. S., Kist, M., Vani, G. and Devi, C. S. (2008). Effect of methanolic extract of *Terminalia arjuna* against *Helicobacter pylori* 26695 lipopolysaccharide-induced gastric ulcer in rats. *Journal of Pharmacy and Pharmacology,* **60**: 505-514.

Devi, R. S., Narayan, S., Vani, G. and Shyamala-Devi, C. S. (2007a). Gastroprotective effect of *Terminalia arjuna* bark on diclofenac sodium induced gastric ulcer. *Chemico-Biological Interactions*, **167**: 71-83.

Devi, R.S., Narayan, S., Vani, G., Srinivasan, P., Mohan, K. V., Sabitha, K. E. and Devi, C. S. (2007b). Ulcer protective effect of *Terminalia arjuna* on gastric mucosal defensive mechanism in experimental rats. *Phytotherapy Research*, **21**: 762-767.

Dhawan, B. N. and Singh, H. K. (1996). Pharmacology of Ayurvedic nootropic Bacopa monniera, Abstr. No. NR 59. Int. Conv. Biol. Psychiat., Bombay.

Dhuley, J. N. (1997). Effect of some Indian herbs on macrophage functions in ochratoxin A treated mice. *Journal of Ethnopharmacology*, **58:** 15-20.

Diwanay, S., Chitre, D. and Patwardhan, B. (2004). Immunoprotection by botanical drugs in cancer chemotherapy. *Journal of Ethnopharmacology*, **90**: 49-55.

Dufour, C., da Silva, E., Potier, P., Queneau, Y. and Dangles, O. (2002). Gallic esters of sucrose as efficient radical scavengers in lipid peroxidation. *Journal of Agricultural and Food Chemistry*, **50**: 3425-3430.

Dureja, H., Kaushik, D. and Kumar, V. (2003). Developments in nutraceuticals. *Indian Journal of Pharmacology*, **35**: 363-372.

Dwivedi, S. and Udupa, N. (1989). *Terminalia arjuna*: Pharmacognosy, phytochemistry, pharmacology and clinical use: A review. *Fitoterapia*, **60**: 413-420.

Dwivedi, S., Agarwal, A., Agarwal, M. P. and Rajpal, S. (2005). Role of *Terminalia arjuna* in ischemic mitral regurgitation. *International Journal of Cardiology*, **100**: 507-508.

Eamlamnam, K., Patumraj, S., Visedopas, N. and Thong-Ngam, D. (2006). Effects of *Aloe Vera* and sucralfate on gastric microcirculatory changes, cytokine levels and gastric ulcer healing in rats. *World Journal of Gastroenterology*, **12**(13): 2034-2039.

Elegami, A. A., El-Nima, E. I., Tohami, M. S. and Muddathir, A. K. (2002). Antimicrobial activity of some species of the family *Combretaceae*. *Phytotherapy Research*, **16:** 555-561.

Elsakka, M., Grigorescu, E., Stanescu, U., *et al.* (1990). New data referring to chemistry of *Withania somnifera* species. *Revista medico-chirurgicala a Societatii de Medici si Naturalisti din Iasi*, **94**: 385-387.

Elsohly, M. A., Gul, W., Avula, B., Khan, I. A. (2007). Determination of the anthraquinones aloe-emodin and aloin-A by liquid chromatography with mass spectrometric and diode array detection. *Journal of AOAC International*, **90**(1): 28-42.

FDA (1996). Additional Market Review Information. Briefing Materials for Food Advisory Committee on Dietary Supplements Containing Ephedrine Alkaloids. Food and Drug Administration. Center for Food Safety and Applied Nutrition, USA. pp: 35.

FDA (2002). Final Rule Declaring Dietary Supplements Containing Ephedrine Alkaloids Adulterated Because They Present an Unreasonable Risk. Food and Drug Administration, USA. pp: 154.

Fenwick, G. R., Price, K. R., Tsukamoto, C. and Okubo, K. (1991). Saponins. In Saponins in Toxic Substances in Crop Plants, [FJP D'Mello, CM Duffus and JH Duffus, editors]. Cambridge: The Royal Society of Chemistry.

Fotsis, T., Pepper, M. S., Aktas, E., Breit, S., Rasku, S., Adlercreutz, H., Wahala, K., Montesano, R. and Schweigerer, L. (1997). Flavonoids, dietary-derived inhibitors of cell proliferation and *in vitro* angiogenesis. *Cancer Research,* **57:** 2916 -2921.

Francis, G., Kerem, Z., Makkar, H. P. S. and Becker, K. (2002). The biological action of saponins in animal systems. *British Journal of Nutrition,* **88**: 587-605.

Frawley, D. (2003). *Ayurvedic Healing - A Comprehensive Guide.* 2nd ed. Delhi: Montilal Banarsidass. Pp: 124.

Gautam, M., Saha, S., Bani, S., Kaul, A., Mishra, S., Patil, D., Salti, N. K., Suri, K. A. and Patwardhan, B. (2009). Immunomodulatory activity of *Asparagus racemosus* on systemic Th1/Th2 immunity: Implications for immunoadjuvant potential. *Journal of Ethnopharmacology,* **121**: 241-247.

Ghasemzadeh, A., Jaafar, H. Z. E. and Rahmat, A. (2010). Antioxidant Activities, Total Phenolics and Flavonoids Content in Two Varieties of Malaysia Young Ginger (*Zingiber officinale* Roscoe). *Molecules,* **15:** 4324-4333.

Ghosal, S., Lal, J. and Srivastava, R. (1989). Immunomodulatory and CNS effects of sitoindosides IX and X, two new glycowithanolides from Withania somnifera. *Phytotherapy Research,* **3**: 201-206.

Ghosal, S., Singh, S. and Bhattacharya, S. K. (1971). Alkoloids of Macuan Pruriens, Chemistry and Pharmacology. *Planta Medica,* **19**: 279.

Ghosal, S., Tripathi, V. K. and Choun, S. (1996). Active constituents of *Emblica officinalis*: Part1. The chemistry and antioxidant effect of two new hydrolysable tannins. Emblicanin A and B. *Indian Journal of Chemistry,* **35B**: 941-948.

Gilani, A. H., Khan, A., Ali, T. and Ajmal, S. (2008). Mechanisms underlying the antispasmodic and bronchodilatory properties of *Terminalia bellerica. Journal of Ethnopharmacology,* **116**: 528-538.

Giovannelli, L., Testa, G., De Filippo, C., Cheynier, V., Clifford, M. N. and Dolara, P. (2000). Effect of complex polyphenols and tannins from red wine onDNAoxidative damage of rat colon mucosa in vivo. *Europian Journal of Nutrition*, **39**: 207-212.

Guenther, E. (1948). The Essential Oils, Vol. 1, D. Van Nostrand Co., Inc., New York.

Guenther, E. (1999). The Essential Oils, Vol. 4, 2nd Edn., D. Van Nostrand Co., Inc., New York.

Gupta, R., Singhal, S., Guyle, A. and Sharma, V. N. (2001). Antioxidant and hypocholesterolaemic effects of *Terminalia arjuna* tree-bark powder: A randomized placebo-controlled trial. *Journal of Association of Physicians of India*, **49**: 231-235.

Gyamfi, M. A. and Aniya, Y. (2002). Antioxidant properties of Thonningianin A, isolated from the african medicinal herb, *Thonningia sanguinea. Biochemical Pharmacology*, **63**: 1725-1737.

Halder, S., Bharal, N., Mediratta, P. K., Kaur, I. and Sharma, K. K. (2009). Anti-inflammatory, immunomodulatory and antinociceptive activity of *Terminalia arjuna* Roxb bark powder in mice and rats. *Indian Journal of Experimental Biology*, **47**: 577-583.

Harris, W. S., Dujovne, C. A., Windsor, S. L., Gerrond, L. L. C., Newton, F. A. and Gelfand, R. A. (1997). Inhibiting cholesterol absorption with CP-88,818 (beta-tigogenin cellobioside; tiqueside): Studies in normal and hyperlipidemic subjects. *Journal of Cardiovascular Pharmacology*, **30**: 55-60.

Harrison, A. P. and Bartels, E. M. (2006). A Modern Appraisal of Ancient Etruscan Herbal Practices. *American Journal of Pharmacology and Toxicology*, **1**(1): 21-24.

Hart, L. A., van Enckevort, P. H., van Dijk, H., Zaat, R., de Silva, K. T. and Labadie, R. P. (1988). Two functionally and chemically distinct immunomodulatory compounds in the gel of *Aloe Vera. Journal of Ethnopharmacology*, **23**(1): 61-71.

Harwood, H. J., Jr, Chandler, C. E., Pellarin, L. D., Bangerter, F. W., Wilkins, R. W., Long, C. A., Cosgrove, P. G., Malinow, M. R., Marcetta, C. A., Pettini, J. L., Savoy, Y. E. and Mayne, J. T. (1993). Pharmacologic consequences of cholesterol absorption inhibition: alteration in cholesterol metabolism and reduction in plasma cholesterol concentration induced by the synthetic saponin b-tigogenin cellobioside (CP-88818; tiqueside). *Journal of Lipid Research*, **34**: 377-395.

Heggers, J. P. and Winters, W. (1993). Aloe and other topical antibacterial agents in wound healing. *Aloe Today / Aloecorp.*, **A4**: 8-11.

Hertog, M. G. L., and Hollman, P. C. H. (1996). Potential health effects of the dietary flavonol quercetin. *Europian Journal of Clinical Nutrition*, **50**: 63-71.

Hertog, M. G., Feskens, E. J., Hollman, P. C., Katan, M. B. and Kromhout, D. (1993). Dietary antioxidant flavonoid and risk of coronary heart disease: the Zutphen Elderly Study. *Lancet*, **342**: 1007-1011.

Hertog, M. G., Kromhout, D., Aravanis, C., Blackburn, H., Buzina, R., Fidanza, F., Giampaoli, S., Jansen, A., Menotti, A. and Nedeljkovic, S. (1995). Flavonoid intake and long-term risk of coronary heart disease and cancer in the seven countries study. *Archieves of Internal Medicine*, **155**: 381-386.

Ho, K. Y., Tsai, C. C., Huang, J. S., Chen, C. P., Lin, T. C. and Lin, C. C. (2001). Antimicrobial activity of tannin components from *Vaccinium vitis-idaea* L. *Journal of Pharmacy and Pharmacology*, **53**: 187-191.

Hollman, P. C. and Katan M. B. (1997). Absorption, metabolism and health effects of dietary flavonoids in man. *Biomedicine and Pharmacotherapy*, **51**: 305- 310.

Hollman, P. C. and Katan, M. B. (1999). Dietary flavonoids: intake, health effects and bioavailability. *Food and Chemical Toxicology*, **37**: 937- 942.

Hou, C. C., Lin, S. J., Cheng, J. T. and Hsu, F. L. (2002). Bacopaside III, bacopasaponin G, and bacopasides A, B, and C from *Bacopa monniera*. *Journal of Natural Products*, **65**: 1759-1763.

Hu, J., Lee, S. O., Hendrich, S. and Murphy, P. A. (2002). Quantification of the group B soyasaponins by high-performance liquid chromatography. *Journal of Agricultural and Food Chemistry*, **50:** 2587-2594.

Huang, Q., Lu, G., Shen, H. M., Chung, M. C., Ong, C. N. (2007). Anti-cancer properties of anthraquinones from rhubarb. *Med. Res. Rev.*, **27**: 609-30.

Hui, H., Tang, G. and Go, V. L. W. (2009). Hypoglycemic herbs and their action mechanisms. *Chinese Medicine*, **4**: 1-11.

Im, S. A., Oh, S. T., Song, S., Kim, M. R., Kim, D. S., Woo, S. S., Jo, T. H., Park, Y. I., Lee, C. K. (2005). Identification of optimal molecular size of modified Aloe polysaccharides with maximum immunomodulatory activity. *International Immunopharmacology*, **5**: 271-279.

Jeena, K. L., Joy, K. L. and Kuttan, R. (1999). Effects of *Emblica officinalis*, *Phyllanthus amarus* and *Picrorhiza kurroa* on N-nitrosodiethyl amine induced hepatocarcinogenesis. *Cancer Letters*, **136**: 11-16.

Joglekar, G. V., Ahuja, R. H. A. and Balwani, J. H. (1967). Galactogogue effect of *Asparagus racemosus*. *Indian Medical Journal*, **61**(7): 165.

Jose, J. K., Kuttan, Y. and Kuttan, R. (2001). Antitumor activity of *Emblica officinalis*. *Journal of Ethnopharmacology*, **75**: 65-69.

Joshi, B. S. and Kamat, V. N. (1973). Tuberosin, a new pterocarpan from *Pueraria tuberosa* DC. *Journal of the Chemical Society Perkin Transactions*, **1:** 907.

Kamat, J. P., Boloor, K. K., Devasagayam, T. P. and Venkatachalam, S. R. (2000). Antioxidant properties of *Asparagus racemosus* against damage induced by gamma-radiation in rat liver mitochondria. *Journal of Ethanopharmacology*, **71**: 424-35.

Kamei, H., Koide, T., Kojima, T., Hashimoto, Y., Hasegawa, M. (1998). Inhibition of cell growth in culture by quinones. *Cancer Biotherapy and Radiopharmaceuticals*, **13**: 185.

Kandil, F. E. and Nassar, M. I. (1998). A tannin anti-cancer promotor from *Terminalia arjuna. Phytochemistry*, **47**(8): 1567-1568.

Kaplan, M., Hayek, T., Raz, A., Coleman, R., Dornfeld, L., Vaya, J. and Aviram, M. (2001). Pomegranate juice supplementation to atherosclerotic mice reduces macrophage lipid peroxidation, cellular cholesterol accumulation and development of atherosclerosis. *Journal of Nutrition*, **131:** 2082-2089.

Kapoor, L. D. (1990). *Handbook of Ayurvedic Medicinal Plants*. Boca Raton, FL: CRC Press Inc, pp: 61.

Kapu, S. D., Ngwai, Y. B., Kayode, O., Akah, P. A., Wambebe, C. and Gamaniel, K. (2001). Anti-inflammatory, analgesic and anti-lymphocytic activities of the aqueous extract of *Crinum giganteum*. *Journal of Ethnopharmacology*, **78**: 7-13.

Karthikeyan, K., Bai, B. R., Gauthaman, K., Sathish, K. S. and Devaraj, S. N. (2003). Cardioprotective effect of the alcoholic extract of *Terminalia arjuna* bark in an in vivo model of myocardial ischemic reperfusion injury. *Life Sciences*, **73:** 2727-2739.

Kaul, T. N., Middleton, E. Jr. and Ogra, P. L. (1985). Antiviral effect of flavonoid on human viruses. *Journal of Medical Virology*, **15**: 71-79.

Kaur, S., Michael, H., Arora, S., Härkönen, P. L. and Kumar, S. (2005). The in vitro cytotoxic and apoptotic activity of *Triphala* – an Indian herbal drug. *Journal of Ethnopharmacology*, **97**(1): 15-20.

Kensil, C. R. (1996). Saponins as vaccine adjuvants. *Critical Reviews in Therapeutic Drug Carrier Systems*, **13**: 1-55.

Khan, A. and Gilani, A. H. (2008). Pharmacodynamic evaluation of *Terminalia bellerica* for its antihypertensive effect. *Journal of Food and Drug Analysis*, **16**: 6-14.

Khan, A. and Gilani, A. H. (2010). Antisecretory and analgesic activities of *Terminalia bellerica. African Journal of Biotechnology*, **9**(18): 2717-2719.

Khan, K. H. and Jain, S. K. (2009). Regular intake of *Terminalia chebula* can reduce the risk of getting typhoid fever. *Advanced Biotechnology*, **8**: 10-15.

Khan, R., Islam, B., Akram, M., Shakil, S. and Ahmad, A. *et al.*, (2009). Antimicrobial activity of five herbal extracts against Multi Drug Resistant (MDR) strains of bacteria and fungus of clinical origin. *Molecules*, **14**: 586-597.

Khanna, A. K., Chander, R., Kapoor, N. K. (1991). Hypolipidaemic activity of Abana in rats. *Fitoterapia (Italy)*, **62:** 271-275.

Khayyal, M. T, El-Ghazaly, M. A., Kenawy, S. A., Seif-el-Nasr, M., Mahran, L. G., Kafafi, Y.A., Okpanyi, S. N. (2001). Antiulcerogenic effect of some gastrointestinally acting plant extracts and their combination. *Arzneimittelforschung*, **51**: 545-553.

Kiderlen, A. F., Kayser, O., Ferreira, D. and Kolodziej, H. (2001). Tannins and related compounds: killing of amastigotes of Leishmania donovani and release of nitric oxide and tumour necrosis factor alpha in macrophages *in vitro. Zeitschrift für Naturforschung*, **56**: 444-454.

Kim, C., Shin, S., Ha H. and Kim, J. M. (2003). Study of substance changes in flowers of Pueraria thunbergiana Benth during storage. *Archieves of Pharmacal Research*, **26**: 210-213.

Kim, D. H., Jung, J. S., Suh, H. W., Huh, S. O., Min, S. K., Son, B. K., Park, J. H., Kim, N. D., Kim, Y. H. and Song, D. K. (1998a). Inhibition of stressinduced plasma corticosterone levels by Ginsenosides in mice: involvement of nitric oxide. *Neuroreport*, **9**: 2261-2264.

Kim, H. J., Yokozawa, T., Kim, K. H., Tohda, C., Rao, T. P. and Juneja, L. R. (2005). Influence of amla (*Emblica officinalis* Gaertn.) on hypocholesterolemia and lipid peroxidation in cholesterol-fed rats. *Journal of Nutritional Science and Vitaminology*, **51**(6): 413-418.

Kim, K., Kim, H., Kwon, J., Lee, S., Kong, H., Im, S. A., Lee, Y. H., Lee, Y. R., Oh, S. T., Jo, T. H., Park, Y. I., Lee, C. K. and Kim, K. (2009). Hypoglycemic and hypolipidemic effects of processed *Aloe Vera* gel in a mouse model of non-insulin-dependent diabetes mellitus. *Phytomedicine*, **16**: 856-863.

Kim, S. J., Kim, Y. Y., Ko, K. H., Hong, E. K., Han, Y. B., Kang, B. H. and Kim, H. (1998c). Butanol extract of 1 : 1 mixture of Phellodendron cortex and Aralia cortex stimulates PI3-kinase and ERK2 with increase of glycogen levels in HepG2 cells. *Phytotherapy Research*, **12:** 255-260.

Kiritkar, K. R. and Basu, B. D. (1935). Indian Medicinal Plants 2nd Ed. Pub. Lalit Mohan Basu. MB.49.Leader Road, Allahabad, India.2: pp1023.

Kirtikar, K. R. and Basu, B. D. (1933). Indian medicinal plants. Vol. I - IV. International Book Distributors, Dehra Dun.

Knekt, P., Jarvainen, R., Reunanen, A. and Maatela, J. (1996). Flavonoid intake and coronary mortality in Finland: a cohort study. *British Medical Journal*, **312**: 478-481.

Knekt, P., Jarvinen, R., Seppanen, R., Hellovaara, M., Teppo, L., Pukkala, E. and Aromaa, A. (1997). Dietary flavonoids and the risk of lung cancer and other malignant neoplasms. *American Journal of Epidemiology*, **146**: 223-230.

Kolodziej, H., Kayser, O., Kiderlen, A. F., Ito, H., Hatano, T., Yoshida, T. and Foo, L. Y. (2001). Antileshmanial activity of hydrolysable tannins and their modulatory effects on nitric oxide and tumour necrosis factor-alpha release in macrophages *in vitro*. *Planta Medica*, **67**: 825-832.

Konoshima, T., Takasaki, M., Tokuda, H., Nishino, H., Duc, N. M., Kasai. R. and Yamasaki, K. (1998). Anti-tumor-promoting activity of majonoside-R2 from Vietnamese ginseng, Panax vietnamensis HAet GRUSHV. (I). *Biological and Pharmaceutical Bulletin*, **21**: 834-838.

Kulkarni, S. K. and Ninan. I. (1997). Inhibition of morphine tolerance and dependence by*Withania somnifera* in mice. *Journal of Ethnopharmacology*, **57(3)**: 213-217.

Kumar, A., Lakshman, K., Jayaveera, K. N., Satish, K. V. and Tripathi, S. (2009). Estimation of rutin and quercetin Terminalia chebula by HPLC. *The Internet Journal of Aesthetic and Antiaging Medicine*, **2**(1):

Kumar, A., Lakshman, K., Jayaveera, K. N., Tripathi, S. and Satish, K. V. (2010). Estimation of gallic acid, rutin and quercetin Terminalia chebula by HPTLC. *Jordan Journal of Pharmaceutical Sciences*, **3**(1): 63-68.

Kumar, K. J. and Bhatnagar, S. P. (2007). Fuzzy Rule Based Systems in Classification of Terminalia Chebula Fruits of Complementary and Alternative Medicine. *Journal of Complementary and Integrative Medicine*, **4**(1): article2.

Kumar, S., Enjamoori, R., Jaiswal, A., Ray, R., Seth, S. and Maulik, S. K. (2009). Catecholamine-induced myocardial fibrosis and oxidative stress is attenuated by *Terminalia arjuna* (Roxb.). *Journal of Pharmacy and Pharmacology*, **61**: 1529-1536.

Kuo, P. L., Hsu, Y. L., Lin, T. C., Lin, L. T., Chang, J. K. and Lin, C. C. (2005a). Induction of cell cycle arrest and apoptosis in human non-small cell lung cancer A549 cells by casuarinin from the bark of *Terminalia arjuna* Linn. *Anticancer Drugs*, **16**: 409-415.

Kuo, P. L., Hsu, Y. L., Lin, T. C., Lin, L. T., Chang, J. K. and Lin, C. C. (2005b). Casuarinin from the bark of *Terminalia arjuna* induces apoptosis and cell cycle arrest in human breast adenocarcinoma MCF-7 cells. *Planta Medica*, **71**: 237-243.

Kuo, P. L., Hsu, Y. L., Ng, L. T., Lin, C. C. (2004). Rhein inhibits the growth and induces the apoptosis of Hep G2 cells. *Planta Medica*, **70**: 12-16.

Kuo, P. L., Lin, T. C. and Lin, C. C. (2002). The antiproliferative activity of aloe-emodin is through p53-dependent and p21- dependent apoptotic pathway in human hepatoma cell lines. *Life Sciences*, **71**: 1879-1892.

Kuttan, G. (1996). Use of *Withania somnifera* Dunal as an adjuvant during radiation therapy. *Indian Journal of Experimental Biology*, **34**: 854-856.

Lacaille-Dubois, M. A. (2005). Bioactive saponins with cancer related and immunomodulatory activity: Recent developments. *Studies in Natural Products Chemistry*, **32**(12): 209-246.

Lacaille-Dubois, M. A., Hanquet, B., Cui, Z. H., Lou, Z. C. and Wagner, H. (1999). A new biologically active acylated triterpene saponins from Silene fortunei. *Journal of Natural Products*, **62**: 133-136.

Lavie, D., Glotter, E. and Shvo (1965). Constituents of *Withania somnifera*-III -The side chain of Withaferin A. *Journal of Organic Chemistry*. **30**: 1774-1778.

Lee, H. Z., Hsu, S. L., Liu, M. C. and Wu, C. H. (2001). Effects and mechanisms of aloe-emodin on cell death in human lung squamous cell carcinoma. *Europian Journal of Pharmacology*, **431**: 287-295.

Lee, K. T., Sohn, I. C., Kim, D. H., Choi, J. W. and Kwon, S. H. (2000a). Hypoglycaemic and hypolipidemic effects of tectorigenin and kaikasaponin III in the streptozotocin-induced diabetic rat and their antioxidant activity in vitro. *Archives of Pharmacal Research*, **23**: 461-466.

Lee, S. K. (2006). Overview of Aloe study in New Perspectives on Aloe by Park, Y.I. and Lee, S.K. Springer Science+Business Media, LLC, 233 Spring Street, New York, NY 10013, USA.

Liao, H., Banbury, L. K. and Leach, D. N. (2008). Antioxidant activity of 45 Chinese herbs and the relationship with their TCM characteristics. *Evidence-based Complementary and Alternative Medicine*, **5**(4): 429-434.

Lin, C. C., Hsu, Y. F. and Lin, T. C. (2001a). Antioxidant and free radical scavenging effects of tannins of *Terminalia catappa* L. *Anticancer Research*, **21**(1A): 237-243.

Lin, C. C., Hsu, Y. F., Lin, T. C. and Hsu. H. Y. (2001b). Antioxidant and hepatoprotective effects of punicalagin and punicalin on acetaminophen-induced liver damage in rats. *Phytotherapy Research*, **15**: 206-212.

Liu, J. C., Hsu, F. L., Tsai, J. C., Chan, P., Liu, J. Y. H., Thomas, G. N., Tomlinson, B., Loe, M. Y. and Lin, J. Y. (2003). Antihypertensive effects of tannins isolated from traditional Chinese herbs as non-specific inhibitors of angiontensin converting enzyme. *Life Sciences*, **73:** 1543-1555.

Madan, J., Sharma, A. K., Inamdar, N., Rao, H. S. and Singh, R. (2008). Immunomodulatory properties of *Aloe Vera* gel in mice. *International Journal of Green Pharmacy*, **2**(3): 152-154.

Mahato, S. B., Garai, S. and Chakravarty, A. K. (2000). Bacopasaponins E and F: two jujubogenin bisdesmosides from *Bacopa monniera*. *Phytochemistry*, **53**: 711-714.

Malekzadeh, F., Ehsanifar, H., Shahamat, N., Levin, M., Colwell, R. R. (2001) Antibacterial activity of black myrobalan (*Terminalia chebula* Retz.) against *Helicobacter pyroli*. *International Journal of Antimicrobial Agents*, **18**(1): 85-8.

Malhotra, C. L. and Das, P. K. (1959). Pharmacological studies of Herpestis monniera Linn. (Brahmi). *Indian Journal of Medical Research*, **47**: 294-305.

Malik, N., Dhawan, V., Bahl, A. and Kaul, D. (2009). Inhibitory effects of *Terminalia arjuna* on platelet activation *in vitro* in healthy subjects and patients with coronary artery disease. *Platelets*, **20**: 183-190.

Mandal, S. C., Nandy, A., Pal, M. and Saha, B. P. (2000). Evaluation of antibacterial activity of *Asparagus racemosus* Willd. root. *Phytotherapy Research*, **14**: 118-119.

Manna, P., Sinha, M. and Sil, P. C. (2006). Aqueous extract of *Terminalia arjuna* prevents carbon tetrachloride induced hepatic and renal disorders. *BMC Complementary and Alternative Medicine*, **6**: 33-44.

Manna, P., Sinha, M. and Sil, P. C. (2008). Protection of arsenic-induced testicular oxidative stress by arjunolic acid. *Redox Report*, **13**: 67-77.

Manna, P., Sinha, M., Pal, P. and Sil, P. C. (2006). Aqueous extract of *Terminalia arjuna* prevents carbon tetrachloride induced hepatic and renal disorders. *BMC Complementary and Alternative Medicine*, **6**: 33-33.

Manna, P., Sinha, M., Pal, P. and Sil, P. C. (2007b). Arjunolic acid, a triterpenoid saponin, ameliorates arsenic-induced cyto-toxicity in hepatocytes. *Chemico-Biological Interactions*, **170**: 187-200.

Marino, S. D., Iorizzi, M., Palagiano, E., Zollo, F. and Roussakis, C. (1998). Starfish saponins. 55. Isolation, structure elucidation, and biological activity of steroid oligoglycosides from an Antarctic starfish of the family Asteriidae. *Journal of Natural Products*, **61**: 1319-1327.

Martis, G., Rao, A. and Karanth, K. S. (1992). Neuropharmaological activity of Herpestis monniera. *Fitoterapia*, **63**: 399-404.

Mathur, S., Gupta, M. M., Ram, M., Sharma, S. and Kumar, S. (2002). Herb yield and bacoside-A content of field-grown *Bacopa monnieri* accessions. *Journal of Herbs, Spices and Medicinal Plants*, **9** (1): 11-18.

Matsuura, M. (2001). Saponins in garlic as modifiers of the risk of cardiovascular disease. *Journal of Nutrition*, **131**: 1000S-1005S.

Mayo, J. L. (1998). Black cohosh and chasteberry: herbs valued by women for centuries. *Clinical Nutrition Insights*, **6**: 1-4.

Mengoni, F., Lichtner, M., Battinelli, L., Marzi, M., Mastroianni, C. M., Vullo, V. and Mazzanti, G. (2002). In vitro anti-HIV activity of oleanolic acid on infected human mononuclear cells. *Planta Medica,* **68**: 111-114.

Menon, L. G., Kuttan, R. and Kuttan, G. (1997). Effect of rasayanas in the inhibition of lung metastasis induced by B16F-10 melanoma cells. *Journal of Experimental and Clinical Cancer Research,* **16(4)**: 365-368.

Metowogo, K., Agbonon, A., Eklu-Gadegbeku, K., Aklikokou, A. K. and Gbeassor, M. (2008). Anti-ulcer and anti-inflammatory effects of hydro-alcohol extract of aloe buettneri A. Berger (Lilliaceae). *Tropical Journal of Pharmaceutical Research,* **7**(1): 907-912.

Middleton, E. J. (1998). Effect of plant flavonoids on immune and inflammatory cell function. *Advances in Experimental Medicine and Biology,* **439:** 175-182.

Miller, A. L. (1998). Botanical Influences on Cardiovascular Diseases. *Alternative Medicine Review,* **30**: 422-431.

Mimaki, Y., Kuroda, M., Kameyama, A., Yokosuka, A. and Sashida, Y. G., Francis *et al.,* (1998a). Steroidal saponins from the underground parts of Ruscus aculeatus and their cytostatic activity on HL-60 cells. *Phytochemistry,* **48**: 485-493.

Miyakoshi, M., Tamura, Y., Masuda, H., Mizutani, K., Tanaka, O., Ikeda, T., Ohtani, K., Kasai, R. and Yamasaki, K. (2000). Antiyeast steroidal saponins from Yucca schidigera (Mohave yucca), a new anti-food-deteriorating agent. *Journal of Natural Products,* **63**: 332-338.

Miyazawa, M., Sakano, K., Nakamura, S. and Kosaka, H. (2001). Antimutagenic Activity of Isoflavone from *Pueraria lobata. Journal of Agricultural and Food Chemistry,* **49** (1): 336-341.

Morehouse, L. A., Bangerter, F. W., DeNinno, M. P., Inskeep, P. B., McCarthy, P. A., Pettini, J. L., Savoy, Y. E., Sugarman, E. D., Wilkins, R. W., Wilson, T. C., Woody, H. A., Zaccaro, L. M. and Chandler, C. E. (1999). Comparison of synthetic saponin cholesterol absorption inhibitors in rabbits: evidence for a non-stoichiometric, intestinal mechanism of action. *Journal of Lipid Research,* **40**: 464-474.

Mukherjee, P. K., Mukherjee, K., Rajesh Kumar, M., Pal, M. and Saha, B. P. (2003). Evaluation of wound healing activity of some herbal formulations. *Phytotherapy Research,* **17**: 265-268.

Mullen, W., McGinn, J., Lean, M. E., MacLean, M. R., Gardner, P., Duthie, G. G., Yokota, T. and Crozier, A. (2002). Ellagitannins, flavonoids, and other phenolics in red raspberries and their contribution to antioxidant capacity and vasorelaxation properties. *Journal of Agricultural and Food Chemistry,* **50**: 5191-5196.

Munasinghe, T. C., Seneviratne, C. K., Thabrew, M. I. and Abeysekera, A. M. (2001). Antiradical and antilipoperoxidative effects of some plant extracts used by Sri Lankan traditional medical practitioners for cardioprotection. *Journal of Phytotherapy Research,* **15**: 519-523.

Nadar, T. S. and Pillai, M. M. (1989). Effect of Ayurvedic medicines on b-glucuronidase activity of Brunner's glands during recovery from cysteamine induced duodenal ulcers in rats. *Indian Journal of Experimental Biology,* **11**: 959-62.

Nadkarni, A. K. (1954). Indian Materia Medica. Popular Prakashan, Bombay, pp: 151-155.

Nadkarni, K. M. (2000). The Indian Materia Medica. Popular Prakashan, India, Vol 1, pp: 1198-1202.

Naik, D. G., Puntambekar, H. and Anantpure, P. (2010). Essential Oil of Terminalia chebula Fruits as a Repellent for the Indian Honeybee Apis florae. *Chemistry and Biodiversity,* **7**(5): 1303-1310.

Naik, G. H., Priyadarsini, K. I., Bhagirathi, R. G., Mishra, B., Mishra, K. P., Banavalikar, M. M. and Mohan, H. (2005). In vitro antioxidant studies and free radical reactions of *Triphala,* an ayurvedic formulation and its constituents. *Phytother Research,* **19**(7):582-586.

Nakagawa, T. and Yokozawa, T. (2002). Direct scavenging of nitric oxide and superoxide by green tea. *Food and Chemical Toxicology,* **40**: 1745-1750.

Nammi, S., Gudavalli, R., Babu, B. S., Lodagala, D. S. and Boini, K. M. (2003). Possible mechanism of hypotension produced by 70% alcoholic extract of *Terminalia arjuna* (L.) in anaesthetised dogs. *BMC Complementary and Alternative Medicine,* **3**(1): 5.

Nanal, B. P., Sharma, B. N., Ranade, S. S. and Nande, C. V. (1974). Clinical study of Shatavari (*Asparagus racemosus*). *Journal of Research in Indian Medicine,* **9**: 23-29.

Nathan, P. J., Clarke, J., Lloyd, J., Hutchison, C. W., Downey, L. and Stough, C. (2001). The acute effects of an extract of Bacopa monniera (Brahmi) on cognitive function in health normal subjects. *Human Psychopharmacology,* **16**(4): 345-351.

NCBI (2005). Alkaloids. NCBI. All Databases. *WEB: National Library of Medicine.*

Nobre-Junior, H. V., Maia, F. D., de Oliveira, R. A., Bandeira, M. A. M., do Ó Pessoa, C., Moraes, M. O., Cunha, G. M. A. and Viana, G. S. B. (2007). Neuroprotective Actions of Tannins from *Myracrodruon urundeuva* on 6-Hydroxydopamine-Induced Neuronal Cell Death. *Journal of Herbs, Spices and Medicinal Plants,* **13**(2): 41-57.

O'Connella, J. E. and Fox, P. F. (2001). Significance and applications of phenolic compounds in the production and quality of milkand dairy products: a review. *International Dairy Journal,* **11:** 103-120.

Oda, K., Matsuda, H., Murakami, T., Katayama, S., Ohgitani, T. and Yoshikawa, M. (2000). Adjuvant and haemolytic activities of 47 saponins derived from medicinal and food plants. *Biological Chemistry,* **381**: 67-74.

Oh, S. R., Kinjo, J., Shii, Y., Ikeda, T., Nohara, T., Ahn, K. S., Kim, J. H. and Lee, H. K. (2000). Effects of triterpenoids from Pueraria lobata on immunohemolysis: d-glucuronic acid plays an active role in anticomplementary activity in vitro. *Planta Medica,* **66**: 506-510.

Okoli, C. O. and Akah, P. A. (2000). Apilot evaluation of the anti-inflammatory activity of *Culcasia scandens,* a traditional antirheumatic agent. *Journal of Alternative and Complementary Medicine,* **6**: 423-27.

Palpu, P. and Dhan P. (2003). Herbal extract and compound lupinoside and analogues an anti-diabetic type II drug from plant *Pueraria tuberosa* U. S. patent. Application no.60/535,332.

Pandey, N., Chaurasia, J. K., Tiwari O. P. and Tripathi Y. B. (2007). Antioxidant properties of different fractions of tubers from *Pueraria tuberosa* Linn. *Food Chemistry,* **105**: 219–222.

Pandey, U. B. (2001). Garlic in "handbook of herbs and spices". Peter, K.V. (ed), pp: 187, Woodhead Publishing Limited, Cambridge, England.

Pandya, N. C. and Kanawjia, S. K. (2002) Ghee: A Traditional Nutraceutical. *Indian Dairyman,* **54**(10): 67-75.

Parihar, M. S. and Hemnani, T. (2004). Experimental excitotoxicity provokes oxidative damage in mice brain and attenuation by extract of *Asparagus racemosus. Journal of Neural Transmission,* **111**: 1-12.

Parrotta, A. J. (2001). Healing plants of Peninsular India. New York: CABI publishing, pp. 308.

Patel, A. B. and Kanitkar, U. K. (1969). *Asparagus racemosus* Willd-from bordi, as a galactogouge, in buffaloes. *Indian Veterinary Journal,* **46**: 718-21.

Patnaik, T., Dey, R. K. and Gouda, P. (2007). Isolation of Triterpenoid Glycoside from Bark of *Terminalia arjuna* using Chromatographic Technique and Investigation of Pharmacological Behavior upon Muscle Tissues. *E-Journal of Chemistry,* **4**(4): 474-479.

Pawar, R. S. and Bhutani, K. K. (2005). Effect of oleanane triterpenoids from *Terminalia arjuna*-a cardioprotective drug on the process of respiratory oxyburst. *International Journal of Phytomedicine and Phytopharmacology,* **12:** 391-393.

Pecere, T., Gazzola, M. V., Mucignat, C., Parolin, C., Vecchia, F. D., Cavaggioni, A., Basso, G., Diaspro, A., Salvato, B., Carli, M. and Palu, G. (2000). Aloe-emodin is a new type of anticancer agent with selective activity against neuroectodermal tumors. *Cancer Research,* **60**(11): 2800-2804.

Pelletier, S. W. (1983). The nature and definition of an alkaloid. In: *Alkaloids. Chemical and Biological Perspectives. Vol. One* (Pelletier, S. W., ed.), pp: 1-31. New York: John Wiley & Sons.

Perianayagam, J. B., Narayan, S., Gnanasekar, G., Pandurangan, A., Raja, S., Rajagopal, K., Rajesh, R., Vijayarakumar, P. and Vijaykumar, S. G. (2005). Evaluation of antidiarrheal potential of *Emblica officinalis. Pharmaceutical Biology,* **43**(4): 373-377.

Perumal Samy, R., Ignacimuthu, S. and Sen, H. (1998). Screening of 34 Indian medicinal plants for antibacterial properties. *Journal of Ethnopharmacology,* **62**: 173-182.

Petit, P. R., Sauvaire, Y., Ponsin, G., Manteghetti, M., Fave, A. and Ribes, G. (1993). Effects of a fenugreek seed extract on feeding behaviour in the rat: metabolic endocrine correlates. *Pharmacology Biochemistry and Behaviour,* **45**: 369-374.

Pettit, G. R., Hoard, M. S., Doubek, D. L., Schmidt, J. M., Pettit, R. K., Tackett, L. P. and Chapuis, J. C. (1996). Antineoplastic agents 338. The cancer cell growth inhibitory. Constituents of *Terminalia arjuna* (Combretaceae). *Journal of Ethnopharmacology,* **53**: 57-63.

Pietta, P. G. (2000). Flavonoids as antioxidants. *Journal of Natural Products,* **63**: 1035-1042.

Plohmann, B., Bader, G., Hiller, K. and Franz, G. (1997). Immunomodulatory and antitumoral effects of triterpenoid saponins. *Pharmazie,* **52**: 953-957.

Podolak, I., Elas, M. and Cieszka, K. (1998). In vitro antifungal and cytotoxic activity of triterpene saponosides and quinoid pigments from Lysimachia vulgaris L. *Phytotherapy Research,* **12**: S70-S73.

Potter, S. M., Jimenez-Flores, R., Pollack, J., Lone, T. A. and Berber-Jimenez, M. D. (1993). Protein saponin interaction and its influence on blood lipids. *Journal of Agricultural and Food Chemistry,* **41**: 1287-1291.

Prasad, A. V. K., Singh, A., Kapil, R. S. and Popli, S. P. (1984). Puetuberosanol, an epoxychalcanol from Pueraria tuberosa *Indian Journal of Chemistry,* **23B**: 1165-1167.

Quin, G. W. and Xu, R. S. (1998). Recent advances in bioactive natural products from Chinese medicinal plants. *Medical Research Reviews,* **18**: 375-382.

Raghavan, B. and Kumari, S. K. (2006). Effect of *Terminalia arjuna* stem bark on antioxidant status in liver and kidney of alloxan diabetic rats. *Indian Journal of Physiology and Pharmacology,* **50**: 133-142.

Rajanikant (2005). Develoment of process for herbal ghee (*Arjuna* ghee). M.Tech. thesis submitted to NDRI (Deemed University).

Rajasekaran, S., Ravi, K., Sivagnanam, K. and Subramanian, S. (2006). Beneficial effects of *Aloe Vera* leaf gel extract on lipid profile status in rats with streptozotocin diabetes. *Clinical and Experimental Pharmacology and Physiology,* **33**: 232-237.

Ramesh, C., Kavita, S., Khanna, A. K., Kaul, S. M., Anju, P. and Rashmi, S. (2004). *Terminalia arjuna* (Roxb.) protects rabbit heart against ischemic-reperfusion injury: Role of antioxidant enzymes and heat shock protein. *Journal of Ethnopharmacology,* **96**: 403-409.

Rane, M.M. and Mengi, S. A. (2003). Comparative effect of oral administration and topical application of alcoholic extract of *Terminalia arjuna* bark on incision and excision wounds in rats. *Fitoterapia,* **74**: 553-558.

Rani P. and Khullar, N. (2004). Antimicrobial evaluation of some medicinal plants for they're antienteric potential against multidrug resistance *Salmonella typhi. Phytotherapia Research,* **18**(8): 670-673.

Rani, P. and Khullar, N. (2004). Antimicrobial evaluation of some medicinal plants for their anti-enteric potential against multi-drug resistant *Salmonella typhi. Phytotherapy Research,* **18**: 670-673.

Rao, 1981. Inhibitory action of *Asparagus racemosus* on DMBA-induced mammary carcinogenesis in rats. *International Journal of Cancer,* **28**: 607–610.

Rao, A. V. and Sung, M. K. (1995). Saponins as anticarcinogens. *Journal of Nutrition,* **125**: 717S-724S.

Rao, N. V., Pijar, B., Nimbal, S. K., Shantakumar, S. K. and satyanarayana, S. (2008). Nootropic activity of tuber extract of Pueraria tuberosa (rxob). *Indian Journal of Experimental Biology,* **46**: 591-598.

Rastogi, R. P. and Mehrotra, B. N. (1993a). Compendium of Indian Medicinal Plants. Vol. 1, CSIR, New Delhi.

Rastogi, R. P. and Mehrotra, B. N. (1993b). Compendium of Indian Medicinal Plants. Vol. 2, CSIR, New Delhi.

Rastogi, R. P. and Mehrotra, B. N. (1993c). Compendium of Indian Medicinal Plants. Vol. 3, CSIR, New Delhi.

Rathore, M. S. and Shekhawat, N. S. (2009). Micropropagation of *Pueraria tuberose* (Roxb. Ex Willd.) and determination of puerarin content in different tissues. *Plant Cell, Tissue and Organ Culture,* **99**(3): 327-334.

Ratnam, K. V. and raju, R. R. V. (2009). Preliminary phytochemical and antimicrobial properties of *Pueraria tuberosa* (willd.) DC: a potential medicinal plant. *Ethnobotanical Leaflets,* **13**: 1051-1059.

Ravindran, P. N. and Kalluparackal, J. A. (2001). Black pepper in "handbook of herbs and spices". Peter, K.V. (ed), pp: 86-91, Woodhead Publishing Limited, Cambridge, England.

Reddy, T. K., Seshadri, P., Reddy, K. K., Jagetia G. C. and Reddy, C. D. (2008). Effect of *Terminalia arjuna* extract on adriamycin-induced DNA damage. *Phytotherapy Research,* **22**: 1188-1194.

Rege, N. N., Nazareth, H. M., Isaac, A., Karandikar, S.M. and Dahanukar, S.A. (1989). Immunotherapeutic modulation of interperitoneal adhesion by *Asparagus racemosus. Journal of Postgraduate Medicine,* **35**: 199-203.

Rege, N. N., Thatte, U. M. and Dahanukar, S.A. (1999). Adaptogenic properties of six rasayana herbs used in ayurvedic medicine. *Phytotherapy Research,* **3**: 275.

Renu, 1983. Fungitoxcity of leaf extracts of some higher plants against Rhizoctonia solani kuehm. *National Academi of Sciences Letters,* **6**: 245-246.

Roodenrys, S., Booth, D., Bulzomi, S., Phipps, A., Micallef, C. and Smoker, J. (2002). Chronic Effects of Brahmi (Bacopa monnieri) on Human Memory. *Neuropsychopharmacology,* **27**(2): 279-281.

Row, L. R., Murty, P. S., Subbarao, G. S. R., Sastry, C. S. P. and Rao, K. V. J. (1970a). Chemical examination of Terminalia species: part XII - isolation and structure determination of arjunic acid, a new trihydroxytriterpene carboxylic acid from the *Terminalia arjuna* bark. *Indian Journal of Chemistry,* **8**: 716-721.

Roy, R. N., Bhagwagar, S., Chavan, R. and Dutta, N. K. (1971). Preliminary Pharmacological studies on extracts of roots of *Asparagus racemosus*/ Shatavari, Willd. N.O.Liliaceae. *Indian Journal of Medical Research,* **6**(2): 132-138.

Rudakewich, M., Ba, F. and Benishin, C. G. (2001). Neurotrophic and neuroprotective actions of Ginsenosides Rb(1) and Rg(1). *Planta Medica,* **67**: 533-537.

Russo, A. and Borrelli, F. (2005). *Bacopa monniera,* a reputed nootropic plant: an overview. *Phytomedicine,* **12**: 305-317.

Sabina, E. P. and Rasool, M. (2008). An in vivo and in vitro potential of Indian ayurvedic herbal formulation *Triphala* on experimental gouty arthritis in mice. *Vascular Pharmacology,* **48**(1): 14-20.

Sairam, K., Dorababu, M., Goel, R. K. and Bhattacharya, S. K. (2002). Antidepressant activity of standardized extract of Bacopa monniera in experimental models of depression in rats. *Phytomedicine,* **9**: 207-211.

Saleem, A., Husheem, M., Härkönen, P. and Pihlaja, K. (2002). Inhibition of cancer cell growth by crude extract and the phenolics of *Terminalia chebula* Retz. fruit. *Journal of Ethnopharmacology,* **81**(3): 327-336.

Samiulla, D. S., Prashanth, D. and Amit, A. (2001). Mast cell stabilizing activity of *Bacopa monnieri*. *Fitoterapia*, **72**: 284-285.

Sandhya, T., Lathika, K. M., Pandey, B. N., Bhilwade, H. N., Chaubey, R. C., Priyadarsini, K. I. and Mishra, K. P. (2006b). Protection against radiation oxidative damage in mice by *Triphala*. *Mutation Research*, **609**(1): 17-25.

Saravanan, S., Srikumar, R. and Manikandan, S. (2007). Hypolipidemic effect of *Triphala* in experimentally induced hypercholesteremic rats. *Yakugaku Zasshi*, **127**(2): 385-388.

Sastri, M. S., Dhalla, N. S., Malhotra, C. L. (1959). Chemical investigation of Herpestis monniera Linn (Brahmi). *Indian Journal of Pharmacology*, **21**: 303-304.

Saxena, M., Faridi, U., Mishra, R., Gupta, M. M. and Darokar, M. P. *et al.*, (2007). Cytotoxic agents from *Terminalia arjuna*. *Planta Medica*, **73**: 1486-1490.

Saxena, S. and Bopana, N. (2009). In Vitro Clonal Propagation of *Asparagus racemosus*, a High Value Medicinal Plant. *Methods in Molecular Biology*, **547**(1): 179-189.

Saxena, V. K. and Chourasia, S. (2001). A new isoflavone from the roots of *Asparagus racemosus*. *Fitoterapia*, **72**: 307-309.

Scartezzini, P., Antognoni, F., Raggi, M. A., Poli, F. and Sabbioni, C. (2006). Vitamin C content and antioxidant activity of the fruit and of the Ayurvedic preparation of *Emblica officinalis* Gaertn. *Journal of Ethnopharmacology*, **104**(1-2): 113-118.

Schliebs, R., Liebmann, A., Bhattacharya, S. K., Kumar, A., Ghosal, S. and Bigl, V. (1997). Systemic administration of defined extracts from *Withania somnifera* (Indian Ginseng) and Shilajitu differentially affects cholinergic but not glutamatergic and GABAergic markers in rat brain. *Neurochemistry International*, **30**(2): 181-190.

Schmidt, J. M. and Greenspoon, J. S. (1991). *Aloe Vera* dermal wound gel is associated with a delay in wound healing. *Obstetrics and Gynecology*, **78**(1): 115-117.

Sekine, T., Fukasawa, N., Kashiwagi, Y., Ruangrungsi, N. and Murakoshi, I. (1994). Structure of asparagamine A, a novel polycyclic alkaloid from *Asparagus racemosus*. *Chemical and Pharmaceutical Bulletin*, **42**: 1360.

Sekine, T., Fukasawa, N., Murakoshi, I. and Ruangrungsi, N. (1997). A 9,10-dihydrophenanthrene from *Asparagus racemosus*. *Phytochemistry*, **44**: 763-764.

Shailaa, H. P., Udupa, S. L. and Udupa, A. L. (1998). Hypolipidemic activity of three indigenous drugs in experimentally induced atherosclerosis. *International Journal of Cardiology*, **67**: 119-124.

Shanker, G. and Singh, H. K. (2000). Anxiolytic profile of standardized Brahmi extract. *Indian Journal of Pharmacology*, **32**: 152.

Shanmugasundaram, E. R., Akbar, G. K. and Shanmugasundaram, K. R. (1991). Brahmighritham, an Ayurvedic herbal formula for the control of epilepsy. *Journal of Ethnopharmacology*, **33:** 269-276.

Sharad, A. C., Solomon, F. E. and Devi, P. U. (1996). Antitumor and radiosensitizing effects of withaferin A on mouse Ehrlich ascites carcinoma in vivo. *Acta Oncologica*, **35**: 95-100.

Sharma, P. C., Yelne, M. B. and Dennis, T. J. (2000). Database on Medicinal Plants Used in Ayurveda, Volume I. *Central Council for Research in Ayurveda and Siddha.* Yugantar Prakashan (P.) Ltd., New Delhi, pp: 418-430.

Sharma, P. N., Shoeb, A., Kapil, R. S. and Popli, S. P. (1982). Arjumolone, a new flavone from stem bark of Terminalia *Arjuna. Indian Journal of Chemistry,* **21B**: 263-64.

Sharma, R., Chaturvedi, C. and Tewari, P.V. (1987). Efficacy of Bacopa monniera in revitalizing intellectual functions in children. *Journal of Research and Education in Indian Medicine,* **1**: 12.

Sharma, S. K., Perianayagam, B. J., Joseph, A. and Christina, A. J. M. (2003). Antiinflamatory activity of ethanol an aqueous extracts of *Emblica officinalis* Geartn fruits. *Hamdard Medicus*, **46**: 75-78.

Sharma, S., Ramji, S., Kumari, S. and Bapna, J. S. (1996). Randomized controlled trial of *Asparagus racemosus* (shatavari) as a lactogogue in lactational inadequacy. *Indian Pediatrics*, **33**: 675-677.

Shinde, S. L., Junne, S. B., Wadje, S. S. and Baig, M. M. V. (2009). The diversity of antibacterial compounds of *Terminalia* species (Combretaceae). *Pakistan Journal of Biological Sciences,* **12**: 1483-1486.

Shive, C. L., Hofstelter, H., Arredondo, L., Shaw, C. and Forsthuber, T. G. (2000). The enhanced antigen-specific production of cytokines induced by pertussis toxin is due to donal expantion of T cells and not to altered effector functions of long-term memory cells. *European Journal of Immunology,* **30**: 242-331.

Shukla, B., Khanna, N. K. and Godhwani, J. L. (1987). Effect of Brahmi Rasayan on the central nervous system. *Journal of Ethnopharmacology,* **21**: 65-74.

Shukla, S. (1995). Toxicological Studies of Pueraria tuberosa, a Potent Antifertility Plant. *Pharmaceutical Biology,* **33**(4): 324-329.

Silva, O., Ferreira, E., Vaz Pato, M., Canica, M. and Gomes. E. T. (2002). *In vitro* anti-*Neisseria gonorrhoeae* activity of *Terminalia macroptera* leaves. *FEMS Microbiological Letters,* **211**: 203-206.

Sindambiwe, J. B., Calomme, M., Geerts, S., Pieters, L., Vlietinck, A. J. and Vanden Berghe, D. A. (1998). Evaluation of biological activities of triterpenoid saponins from Maesa lanceolata. *Journal of Natural Products,* **61**: 585-590.

Singh, G., Singh, A. T., Abraham, A., Bhat, B. and Mukherjee, A. *et al.,* (2008). Protective effects of *Terminalia arjuna* against Doxorubicin-induced cardiotoxicity. *Journal of Ethnopharmacology,* **117**: 123-129.

Singh, H. K. and Dhawan, B. N. (1982). Effect of Bacopa monniera extract on avoidance responses in rat. *Journal of Ethnopharmacology,* **5**: 205-214.

Singh, H. K. and Dhawan, B. N. (1992). Drugs affecting learning and memory. In: Tandon, P.N., Bijiani, V., Wadhwa, S. (Eds.), Lectures in Neurobiology, vol. 1. Wiley Eastern, New Delhi, pp: 189-207.

Singh, H. K. and Dhawan, B. N. (1997). Neuropsychopharmacological effects of the Ayurvedic nootropic Bacopa monniera Linn (Brahmi). *Indian Journal of Pharmachology,* **29**: S359-S365.

Singh, H. K., Srimal, R. C., Srivastava, A. K., Garg, N. K. and Dhan, B. N. (1990). Neuropsychopharmacological effects of bacosides A and B. Proceedings of the Fourth Conference on Neurobiology Learning Memory, Abstract No. 79. Irvine California.

Singh, L. and Sharma, M. (1978). Antifungal properties of some plant extracts. *Geobios,* **5**: 49-53.

Singh, R. H. and Singh, L. (1980). Studies on the anti-anxiety effect of the medyha rasayana drug Brahmi (Bacopa monniera Wettst.). *Research in Ayurveda and Siddha,* **1**: 133-148.

Sinha, M., Manna, P. and Sil, P. C. (2008). *Terminalia arjuna* protects mouse hearts against sodium fluoride-induced oxidative stress. *Journal of Medicinal Food,* **11**: 733-740.

Sinnott, R. A., Kalns, J. E., Boyd, S., Ramberg, J., Duncan, C., Kirchner, J. M. and Oubre, C. (2007). Human Colonic Bacteria Can Utilize Stabilized *Aloe Vera* Gel Polysaccharides and a Mixed Saccharide-Based Glyconutritional Dietary Supplement, Advanced Ambrotose™. Poster presented at 4th Annual Natural Suppliments Conference sponsored by the Scripps Centre for Integrative Medicine, San Diego, California, January 19-21, 2007.

Sivalokanathan, S., Ilayaraja, M. and Balasubramanian, M. P. (2006a). Antioxidant activity of *Terminalia arjuna* bark extract on N-nitrosodiethylamine induced hepatocellular carcinoma in rats. *Molecular and Cellular Biochemistry,* **281**: 87-93.

Sivalokanathan, S., Ilayaraja, M. and Balasubramanian, M. P. (2006b). Effects of *Terminalia arjuna* bark extract on apoptosis of human hepatoma cell line HepG2. *World Journal of Gastroenterology,* **12**: 1018-1024.

Sivalokanathan, S., Ilayaraja, M. and Balasubramanian, M. P. (2005). Efficacy of *Terminalia arjuna* (Roxb.) on N-nitrosodiethylamine induced hepatocellular carcinoma in rats. *Indian Journal of Experimentaal Biology,* **43**: 264-267.

Sivarajan, V. V. and Balachandran, I. (1994). Ayurvedic drugs and their plant sources. Oxford and IBH publishing Co., New Delhi, pp:97-99.

Sjolander, A., Cox, J. C. and Barr, I. G. (1998). ISCOMs: an adjuvant with multiple functions (Review). *Journal of Leukocyte Biology,* **64**: 713-723.

Srikumar, R., Parthasarathy, N. J. and Devi, S.R. (2005). Immunomodulatory activity of *Triphala* on neutrophil functions. *Biological and Pharmaceutical Bulletin,* **28**(8): 1398-1403.

Srinivas, G., Babykutty, S., Sathiadevan, P. P. and Srinivas, P. (2007). Molecular mechanism of emodin action: transition from laxative ingredient to an antitumor agent. *Medicinal Research Reviews,* **27**: 591-608.

Stauffer, J. E. (1999). Nutraceuticals. *Cereal Foods World,* **44**: 115-7.

Stefani, E. D., Boffetta, P., Deneo-Pellegrini, H., Mendilaharsu, M., Carzoglio, J. C., Ronco, A. and Olivera, L. (1999). Dietary antioxidants and lung cancer risk: a case-control study in Uruguay. *Nutrition and Cancer,* **34**: 100-110.

Stephen, A. M. (1998). Regulatory aspects of functional products, In "Functional foods: biochemical and processing aspects" Mazza, G. (Ed), pp:460. Technomic Publishing Co., Inc., Lancaster.

Stolzenberg, S. J. and Parkhurst, R. M. (1976). Blastocidal and contraceptive actions by an extract and compounds from endod (Phytolacca dodecandra). *Contraception,* **14:** 39-51.

Sumathi, T. and Nongbri, A. (2008). Hepatoprotective effect of Bacoside-A, a major constituent of *Bacopa monniera Linn. Phytomedicine,* **15**(10): 901-905.

Sumitra, M., Manikandan, P., Kumar, D. A., Arutselvan, N., Balakrishna, K., Manohar, B. M. and Puvanakrishnan, R. (2001). Experimental myocardial necrosis in rats: role of arjunolic acid on platelet aggregation, coagulation and antioxidant status. *Molecular and Cellular Biochemistry,* **277**: 135-142.

Sun, F. Y., Chen, X. P., Wang, J. H., Qin, H. L., Yang, S. R. and Du, G. H. (2008). Arjunic acid, a strong free radical scavenger from *Terminalia arjuna. American Journal of Chinese Medicine,* **36**: 197-207.

Suvitayavat, W., Sumrongkit, C., Thirawarapan, S. S. and Bunyapraphatsara, N. (2004). Effects of Aloe preparation on the histamine-induced gastric secretion in rats. *Journal of Ethnopharmacology,* **90**: 239-247.

Tan, B. K. H. and Vanitha, J. (2004). Immunomodulatory and Antimicrobial Effects of Some Traditional Chinese Medicinal Herbs: A Review. *Current Medicinal Chemistry,* **11**: 1423-1430.

Tanaka, M., Misawa, E., Ito, Y., Habara, N., Nomaguchi, K., Yamada, M., Toida, T., Hayasawa, H., Takase, M., Inagaki, M. and Higuchi, R. (2006). Identification of Five Phytosterols from *Aloe Vera* Gel as Anti-diabetic Compounds. *Biological and Pharmaceutical Bulletin*, **29**(7): 1418-1422.

Tanwar, Y. S., Goyal, S. and Ramawat, K. G. (2008). Hypolipidemic effects of tubers of Indian kudzu (*Pueraria tuberosa*). *Journal of Herbal Medicine and Toxicology*, **2** (1): 21-25.

Teradaira, R., Shimzato, M., Beppu, H. and Fujita, K. (1993). Antigastric ulcer effects in rats of *Aloe arborescens Miller var. natalensis* Berger extract. *Phytotherapy Research*, **7**: 534-536.

Tewary, P. V., Chaturvedi, C. and Pandey, V. B. (1973). Antifertility activity of Costus speciosus Sm. *Indian Journal of Pharmacology*, **35**: 114-115.

Thakur, R. S., Puri, H. S. and Hussain, A. (1987). *Major Medicinal Plants of India*. CIMAP, Lucknow (India) (a monograph on *Withania somnifera*).

Thomsen, M. (2002). Shatavari- *Asparagus racemosus*. http:// www.phytomedicine.com. au/ files/ articles/ shatavri.pdf.

Tierra, M. (1992). *Planetary Herbology*. Twin Lakes, WI: Lotus Press, pp: 166.

Tierra, M. and Tierra, L. (1999). Chinese Traditional Herbal Medicine, Vol (1). Lotus Press, Twin Lakes, WI, USA.

Tizard, I., Carpenter, R.H. and Kemp, M. (1991). Immuno-regulatory effects of a cytokine release enhancer (acemannan). *International Congress of Phytotherapy*, Oct Seoul, Korea 68.

Trevelyan, J. (1993). Herbal medicine. *Nursing Times*, **89**: 36-38.

Tripathi, V. K., Pandey, V.B., Udupa, K. N. and Rucker, G. (1992). Arjunolitin, a triterpene glycoside from Terminalia *Arjuna*. *Phytochemistry*, **31**: 349-351.

Tripathi, Y. B., Chaurasia, S., Tripathi, E., Upadhyay, A. and Dubey, G. P. (1996). Bacopa monniera Linn. as an antioxidant: mechanism of action. *Indian Journal of Experimental Biology*, **34:** 523-526.

Tripathi, Y. B., Reddy, M. M., Pandey, R. S., Subhashini, J., Tiwari, O. P., Singh, B. K. and Reddanna, P. (2004). Anti-inflammatory properties of BHUx, a polyherbal formulation to prevent atherosclerosis. *Inflammopharmacology*, **12**: 131-152.

Uphof, J. C. (1968). The dictionary of economic plants, vol. 62. Verlag von J Cramer, NewYork.

Vasala, P. A. (2001). Ginger in "handbook of herbs and spices", Peter, K.V. (ed), pp: 198-200. Woodhead Publishing Limited, Cambridge, England.

Venkat, P. (2009). Reverse Pharmacognosy: A Novel Strategy to Standardise ISM Drugs. 8th Annual Oxford-International Conference on the Science of Botanicals, Apr 6-9, 2009.

Venkata, R. K. (2006). Medico-botanical, phytochemicals and antimicrobial properties of certain rare and endemic medicinal plants from Gundlabrahmeswaram wild life sanctuary, Andhra Pradesh, India. Ph.D. Thesis, S.K. University, Anantapur.

Venkatesan, N., Thiyagarajan, V., Narayanan, S., Arul, A., Raja, S., Kumar, S. G. V., Rajarajan, T., Perianayagam, J. B. (2005). Anti-diarrhoeal potential of *Asparagus racemosus* wild root extracts in laboratory animals. *Journal of Pharmacology and Pharmaceutical Sciences*, **8**: 39-45.

Verma, N. and Vinayak, M. (2009). Effect of *Terminalia arjuna* on antioxidant defense system in cancer. *Molecular Biology Reports*, **36**: 159-164.

Viana, G. S. B., Bandeira, M. A. M., Moura, L. C., Souza-Filho, M. V. P., Matos, F. J. A. and Ribeiro, R. A. (1997). Analgesic and antiinflammatory effects of the tannin fraction from *Myracrodruon urundeuva* Fr. All. *Phytotherapy Research*, **11**: 118-122.

Visavadiya, N. P. and Narasimhacharya, R. L. (2005). Hypolipidemic and antioxidant activities in *Asparagus racemosus* in hypercholesteremic rats. *Indian Journal of Pharmacology* , **37**: 376-380.

Waller, G. R., Mangiafico, S. and Ritchey, C. R. (1978). A chemical investigation of *aloe barbadensis* miller. *Proceedings of the Oklahoma Academy of Science*, **58**: 69-76.

Wang, H. K., Xia, Y., Yang, Z. Y., Natschke, S. L. and Lee, K. H. (1998). Recent advances in discovery and development of flavonoids and their analogues as antitumor and anti-HIV agents. *Advances in Experimental Medicine and Biology*, **439**: 191-225.

Wang, N., Wan, J. B., Li, M. Y. and Wang, Y. T. (2008). Advances in studies on Panax notoginseng against atherosclerosis. *Chinese Traditional Herbal Drugs*, **5**: 787-791.

Wang, W., Ali, Z., Li, X. C., Shen, Y. and Khan, I. A. (2010a). Triterpenoids from two *Terminalia* species. *Planta Medica*, Apr. 9 [Epub ahead of print].

Wang, W., Ali, Z., Li, X. C., Shen, Y. and Khan, I. A. (2010b). 18, 19-secooleanane type triterpene glycosyl esters from the bark of *Terminalia arjuna*. *Planta Medica*, Jan. 28 [Epub ahead of print].

Wang, Y., McAllister, T. A., Yanke, L. J. and Cheeke, P. R. (2000a). Effect of steroidal saponin from Yucca schidigera extract on ruminal microbes. *Journal of Applied Microbiology*, **88**: 887-896.

Warrier, P. K., Nambiar, V. P. K. and Ramankutty, C. (1994). Indian Medicinal Plants. A Compendium of 500 Species. Vol. 1, Orient Longman Pvt. Ltd., Madras, India.

Wei, L., Chuncheng, Y., Huafeng, Z. and Rugang, Y. (2004). Preparation of aloe herbs health beverage. *Food Science China*, **25**: 207-209.

Wiboonpun, N., Phuwapraisirisan, P. and Tip-pyang, S. (2004). Identification of antioxidant compound from *Asparagus racemosus*. *Phytotherapy Research*, **18**: 771-773.

Winston, D. (1992). Medicine and Ethnobotany, in Tierra M, [Ed.], American Herbalism, The Crossing Press, Freedom, CA.

Winston, D. and Dattner, A. (1999). The American System of Medicine. *Clinics in Dermatology*, **17**(1): 53-56.

Winters, W. D. (1993). Immunoreactive lectins in leaf gel from Aloe barbadensis Miller. *Phytotherary Research*, **7**: S23-S25.

Womble, D. and Helderman, J. H. (1988). Enhancement of allo-responsiveness of human lymphocytes by Acemannan (Carrisyn). *Journal of Immunopharmacology*, **10**(8): 967-974.

Yadava, R. N. and Rathore, K. (2001). A new cardenolide from the seeds of *Terminalia bellerica. Fitoterapia*, **72**(3): 310-312

Yan, L. P., Chan, S. W., Chan, A. S. C., Chen, S. L., Ma, X. J. and Xu, H. X. (2006). Puerarin decreases serum total cholesterol and enhances thoracic aorta endothelial nitric oxide synthase expression in diet-induced hypercholesterolemic rats. *Life Sciences*, **79**(4): 324-330.

Yan, L. P., Chan, S. W., Chan, A. S., Chen, S. L., Ma, X. J. and Xu, H. X. (2006). Antioxidative activities and the total phenolic contents of tonic Chinese Medicinal Herbs. *Life Sciences*, **79**: 324-330.

Yao, X. H. and Li, X. J. (2001). Panax notoginseng saporins injection in treatment of cerebral infarction with a multicenter study. *Chinese Journal of New Drugs and Clinical Remedies*, **20**: 257-260.

Yao, X. H. and Li, X. J. (2002). Protective effects and its mechanism of panaxatriol saponins isolated from Panax notoginseng on cerebral ischemia. *Zhongguo Zhong Yao Za Zhi*, **27**(5): 371-373.

Yeh, F. T., Wu, C. H. and Lee, H. Z. (2003). Signaling pathway for aloe-emodin-induced apoptosis in human H460 lung nonsmall carcinoma cell. *International Journal of Cancer*, **106** (1): 26-33.

Ying, W., Sevigny, M. B., Chen, Y. and Swanson, R. A. (2001). Poly (ADP-ribose) glycohydrolase mediates oxidative and excitotoxic neuronal death. *Proceedings of National Academi of Sciences*, **98**: 12227-12232.

Yoshikawa, M., Murakami, T., Kishi, A., Kageura, T. and Matsuda, H. (2001). Medicinal flowers. III. Marigold (1): hypoglycaemic, gastric emptying inhibitory, and gastroprotective principles and new oleanane-type triterpene oligoglycosides, calendasaponins A, B, C, and D, from Egyptian Calendula officinalis. *Chemical and Pharmaceutical Bulletin,* **49**: 863-870.

Yoshiki, Y. and Okubo, K. (1995). Active oxygen scavenging activity of DDMP (2,3-dihydro-2, 5-dihydroxy-6-methyl-4H-pyran-4-one) saponin in soybean seed. *Bioscience Biotechnology and Biochemistry,* **59**: 1556-1557.

Yoshiki, Y., Kudou, S. and Okubo, K. (1998). Relationship between chemical structures and biological activities of triterpenoid saponins from soybean. *Bioscience Biotechnology and Biochemistry,* **62**: 2291-2299.

Yusuf, S., Agunu, A. and Diana, M. (2004). The effect of *Aloe Vera* A. berger (*Liliaceae*) on gastric acid secretion and acute gastric mucosal injury in rats. *Journal of Ethnopharmacology,* **93**: 33-37.

Zhang, L., Chang, C. J., Bacus, S. S., Hung, M. C. (1995). Suppressed transformation and induced differentiation of HER- 2/neu-overexpressing breast cancer cells by emodin. *Cancer Research,* **55**: 3890.

Zhao, R. Z. and McDaniel, W. F. (1998). Ginseng improves strategic learning by normal and brain-damaged rats. *NeuroReport,* **9**: 1619-1624.

Ziauddin, M. (1996). Studies on the immunomodulatory effects of Ashwagandha. *Journal of Ethnopharmacology,* **50**(2): 69-76.

Zsila, F., Hazai, E. and Sawyer, L. (2005). Binding of the pepper alkaloid to bovine beta-lactoglobulin: Circular dichroism spectroscopy and molecular modeling study. *Journal of Agricultural Food Chemistry,* **53**(26): 10179-10185.

Zu, Y., Yu, H., Liang, L., Fu, Y., Efferth, T., Liu, X. and Wu, N. (2010). Activities of Ten Essential Oils towards *Propionibacterium acnes* and PC-3, A-549 and MCF-7 Cancer Cells. *Molecules,* **15**: 3200-3210.

□□□

About the Editors

Dr. Sudhir Kumar Tomar, (sudhirndri@gmail.com) is a PhD and M.Sc in Dairy Bacteriology from National Dairy Research Institute, Karnal and is Presently Working Senior Scientist, Dairy Microbiology Division & Incharge Electron Microscopy Centre, National Dairy Research Institute (NDRI), Karnal with an experience of 27 years of teaching and research in the related field. Dr. Tomar has been a recipient of UNDP Sr. Research Fellowship and New Zealand Commonwealth Scholarship for his PhD and two year Post Doctoral Research Programmes respectively. He has been actively engaged in the area of bioprospecting native strains of lactic acid bacteria from indigenous dairy products and exploring them for their ability to produce functional biomolecules. He has authored more than 60 peer-review publications. He has guided 2 PhD and 4 M. Tech scholars and is presently engaged in supervising 3 PhDs and One M. Tech student. He is a Life member of a number of professional bodies.

Dr. Rameshwar Singh, (rsndri@gmail.com) has obtained his Master's degree from the National Dairy Research Institute, Karnal and PhD from Indian Veterinary Research Institute (IVRI), Izatnagar. Consequent up on his selection as a Scientist in Agricultural Research Services of ICAR, New Delhi, he worked at IVRI, Izatnagar for 17 years. Dr. Singh came to NDRI, Karnal in 1995 and is currently working as Head, Dairy Microbiology Division, Incharge, National Collection of Dairy Cultures and Registrar (Academics), NDRI Deemed University, Karnal. Dr. Singh has led a number of research projects to successful completion. He has guided 10 PhD scholars and 9 Masters Degree students for their dissertation. He has published over 100 papers in International & Indian Journals. Dr. Singh is member of different professional societies holding various offices from time to time. He has organized many conferences including three International Conferences, many workshops, summer schools and training programmes.

Dr. Ashish Kumar Singh, (aksndri@gmail.com) Senior Scientist, Dairy Technology Division, NDRI, Karnal, studied at G.B. Pant University of Agriculture and Technology, Pant Nagar for his Master's and PhD degrees in Food Technology and has more then 14 years of professional experience of academics, research and extension in the field of Food and Dairy Technology. His area of specialization is novel formulated and composite dairy foods and Dairy by-products. Mr. Singh has published more than 100 research papers, technical articles and review papers. He has guided 3 M. Tech (Dairy Technology) students at NDRI, Karnal and more than a dozen M.Sc (Food Sci. & Technology) students from other Indian Universities. He has been a recipient of prestigious Erasmus Mundus Scholarship and had an opportunity to teach at Dublin Institute of Technology, Ireland, Anhalt University of Applied Sciences, Bernburg, Germany and KaHo Sint-Lieven, Ghent, Belgium as visiting faculty under this programme.

Dr. Sumit Arora, (sumitak123@gmail.com) is a Senior Scientist in the Dairy Chemistry Division at NDRI, Karnal. He has an experience of over thirteen years as a Dairy Chemist and has published more than 50 research papers in International and National journals of repute. He had received the Erasmus Mundus Grant (Visiting Professors) for teaching and research guidance to International students of food science and nutrition under European Union's programme. Dr Arora is currently working in the area of functional foods with special health benefits.

Dr. R. R. B. Singh (rrb_ndri@rediffmail.com) is a Principal Scientist with National Dairy Research Institute, Karnal, India. He obtained Bachelor degree from the then Gujarat Agricultural University, Anand, Gujarat, India and obtained his post graduate degree in Dairy Technology from National Dairy Research Institute, Karnal, India. He specialized in thermal processing with focus on UHT processing of milk. His other areas of research interests have been absorption phenomenon in foods, application of reaction kinetics and artificial neural network in food quality modeling, traditional indian dairy products, nutraceuticals and functional foods. Dr. Singh has visited many countries to pursue his professional interests. He has authored more than 100 publications and is recipient of many awards including Dr. J.R. Patel Memorial Award and ICAR Outstanding Team Research Award. He is actively involved in many professional activities as Secretary of Dairy Technology Society of India and National Academy of Dairy Science (India).